A Manual on
ASTRONOMIC AND GRID AZIMUTH

R. B. Buckner, Ph.D.

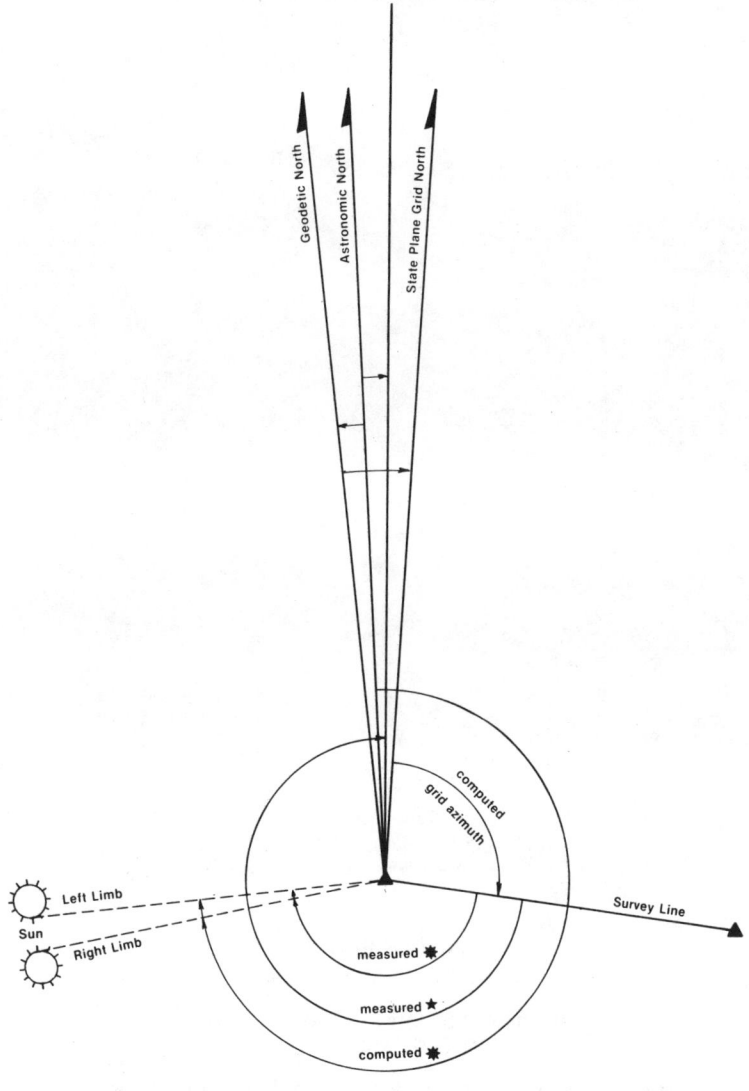

Landmark Enterprises
10324 Newton Way
Rancho Cordova, California 95670

First Printing August 1984
Second Printing June 1985

First Edition
© 1984 by R. B. Buckner

International Standard Book Number
ISBN O-910845-22-O

This book or any part thereof may not be reproduced
without the written consent of the Author.

PRINTED IN THE UNITED STATES OF AMERICA

PREFACE

This manual explains how to determine astronomic and state plane grid azimuth for the related purposes of preserving land survey evidence, avoiding confusion between surveys in adjacent areas, extending local survey control, resolving the meridian convergence problem, and integrating survey work within land data systems in the future.

The text presents a reliable and inexpensive method of determining astronomic azimuth to the equivalent of at least "third-order" accuracy. The technique requires less than $100 in accessory equipment (plus any ordinary theodolite and electronic calculator), usually less than one man-hour of total field and calculation time, and only one observer. The observer has a choice of either Polaris or the sun and uses the same equation for either choice. The development of the method is the culmination of over two decades of experience, teaching, and research by the author. Primary considerations in refining the techniques have been to minimize costs, time, investment in equipment, confusion in field techniques and calculations, and use of tables, but to not sacrifice accuracy. Toward this end, the hour angle equation is featured, the solar method has been refined, unnecessary tables have been avoided, and the solar eyepiece and stop watch are the primary accessories to be used.

Besides providing a clear explanation of the technique of achieving an inexpensive yet accurate azimuth, another primary objective in writing this work was to offer a *complete* manual on the subject so that the U.S. surveyor does not need to consult any other outside reference, even for conversion to grid azimuth. This goal has been accomplished, and at a cost only slightly greater than a government ephemeris for one year. Besides the ephemeris tables covering several years, this manual contains information on convergence angles and Laplace corrections for converting an astronomic azimuth to a state plane grid azimuth anywhere in the United States, and even has maps for identification of any SPC zone in the U.S. Many other inclusions, such as the calculator programs, error analyses, and presentation of related astronomic phenomena provide the desired completeness to this manual. The completeness is seen in another way -- in every case where an explanation of procedures is given, the precautions are also given, so that the surveyor understands what could go wrong, the magnitude of the effect if it does, and how to control the errors. This important focus provides a certain thoroughness not found in other texts covering the subject.

This manual serves more than one purpose. It is a "how to" reference for the practitioner who wants to learn how to do the field work and calculations, and to do so in such a way as to avoid mistakes and minimize errors. It is a text for college and technical school students. It is a research report summarizing several years of investigation of procedures, accuracies, and cost factors. It is an ephemeris of essential data for the azimuth problem. It is a handbook for anyone seeking an answer to a specific question on the subject. It is an appropriate handout manual for seminars and workshops. It is a useful reference for anyone preparing for or taking licensing exams. The arrangement and presentation of material, along with the tables, figures, graphs, index, solved examples, derivation of equations, definition of terms, explanation of the theory, analysis of errors, and many practical tips creates the all-purpose scope to the book.

The manual is suitable for surveyors and students anywhere, particularly in the Northern Hemisphere. It is not intended to be useful only in the U.S. The astronomic solutions can be achieved whether the state plane coordinates (SPC) aspects are applicable in a particular country or not.

Chapter 1 gives background explaining the history and reasons for accurate azimuth. The theory starts in Chapter 2. Procedures and theory are introduced and concisely explained in this chapter. Chapter 3 contains the initial aspects of the details of making some of the measurements. Having an understanding of these details leaves Chapter 4 for explaining field techniques. Each chapter builds on knowledge from previous ones and is arranged in a way intended to coax the reader into continuing, and to provide self-contained sections for easy reference. It may seem that aspects of the computational procedures are scattered among the chapters creating discontinuity, but it should be understood that the intent is to allow the reader time to absorb each concept without overwhelming him with too much at once. Separating the use of the ephemeris data and the solution of field observations into separate chapters (5 and 6) is an example of this philosophy. The subject matter of Chapter 7 is probably new to many. It explains how a surveyor can reference surveys to the SPC azimuth system without referencing to SPC control monuments. Chapter 8 is a concise but complete explanation of how to place state plane coordinates on survey corners when control monuments are close enough for such referencing. This chapter goes beyond the primary subject matter but was felt to be a desirable overlap in order to encourage use of the system.

The depth necessary for the manual to be called a complete text is developed in Appendixes A, B, and C. Appendixes A and B are suitable for anyone who wants to learn more than simply how to perform the field work and calculations using the author's specific technique. It supplements the main text, adding theory and related procedures for the student or the person desiring the information to pass registration exams or for general interest. The examples in Appendix C provide homework or practice problems for students and others.

Astronomic observations are highly useful for orientation of surveys. As presented in this manual, this means of orientation is the least time-consuming and most accurate of available methods, including the various gyro systems and direct positioning methods. It is by far the least expensive, considering instrumentation needed, personnel needs, and time costs. As the various satellite and inertial systems provide means for communities to densify control, the surveyor will still usually need to have a method of orienting local survey control to the new monumentation. Astronomic azimuth with subsequent conversion to SPC grid as explained in this manual offers the means to do so, and at a cost of less than the ordinary travel time to a local survey job site or a coffee break for the survey crew.

ACKNOWLEDGEMENTS AND DEDICATION

The author wishes to acknowledge the direct and indirect input and encouragement offered by several professional people, including Kenneth S. Curtis, Richard L. Elgin, Fareed Nader, David R. Knowles, G. Warren Marks, Ivan Mueller, Charles C. Campbell, Enoch J. French, Michael D. Abell, William E. Carter, LeRoy E. Doggett, Charles W. Elam, Stephen V. Estopinal, James M. Theriot, William G. Norton, Prentis E. Murphy, Rodney L. Crochet, and Roy Minnick. Thanks are also extended to James W. Arnold, Erricos Pavlis, and Mohamed Hanafy for their programming efforts, and to Eugene R. Hilligas, A.J. Meyers, and Brad Partridge for their assistance in checking some of the programs. Laura Brumfield is to be commended for her excellent job of typing.

The late Winfield Eldridge was influential in laying the foundation of thought and philosophy of the reasons and methods to preserve land survey evidence through accurate azimuth determination. My former employer, Charles S. Danner, through his consistent teaching and practice of these principles was most influential in causing these ideas to become practical rather than merely theoretical. It is to these two men that this book is dedicated.

R.B. Buckner, Ph.D.
Columbus, Ohio 43210
April 1984

CONTENTS

Chapters

1. INTRODUCTION
    - 1-1 History and Future of Azimuth Determination — 1
    - 1-2 Reference Meridians for Surveying — 4
    - 1-3 Consistency in Accuracy of Distance and Direction — 6
    - 1-4 Accuracy for Local Control Surveys — 8
    - 1-5 Preservation of Land Survey Evidence — 10
    - 1-6 Control Survey Advantages — 10
    - 1-7 Modern Astronomical Work — 11

2. THE THEORY OF ASTRONOMIC AZIMUTH
    - 2-1 The Celestial Sphere and Spherical Triangle — 14
    - 2-2 Solutions of the Spherical Triangle — 15
    - 2-3 Behavior of Polaris and the Sun — 15
    - 2-4 Basic Procedures to Determine Azimuth — 19

3. TIME, GEOGRAPHIC POSITION, ANGLE MEASUREMENT
    - 3-1 Time Measurement — 22
    - 3-2 Latitude and Longitude Determination — 25
    - 3-3 Angle Measurements — 31

4. FIELD OBSERVATIONS
    - 4-1 Planning and Preparation for Observations — 32
    - 4-2 Field Notes and Data — 39
    - 4-3 Making the Observations — 40
    - 4-4 Suggestions and Precautions — 48

5. THE EPHEMERIS AND ITS DATA
    - 5-1 General Comments — 55
    - 5-2 Declination of the Sun — 56
    - 5-3 Time Relationships and Equation of Time — 58
    - 5-4 Greenwich Hour Angle and Declination of Polaris — 60

CONTENTS

6. CALCULATIONS
   - 6-1 General Precautions in Calculations — 61
   - 6-2 Computation of Greenwich Hour Angle — 61
   - 6-3 Computation of Hour Angle — 64
   - 6-4 Procedures for Azimuth Calculation — 66
   - 6-5 Solar Azimuth Computations — 68
   - 6-6 Polaris Azimuth Calculations - Example P1 — 76
   - 6-7 Finding Polaris in Daytime - Example PP1 — 76
   - 6-8 Programmed Calculations — 78
   - 6-9 Summary and Review of Precautions — 79

7. GEODETIC AND GRID AZIMUTH
   - 7-1 The Laplace Correction — 81
   - 7-2 Mapping Angle in SPC Systems — 82
   - 7-3 Geodetic and Grid Azimuth Calculations — 89

8. USING STATE PLANE COORDINATES
   - 8-1 Introduction and Background — 107
   - 8-2 Fundamentals of SPC — 108
   - 8-3 Traversing Using SPC — 112
   - 8-4 SPC Control Data — 114
   - 8-5 Precautions in Using SPC — 116
   - 8-6 Uses and Advantages of SPC — 117

Appendixes

A. MISCELLANEOUS CONCEPTS AND DEFINITIONS
   - A-1 Terrestrial and Celestial Coordinate Systems — 118
   - A-2 Time Systems — 120
   - A-3 Derivations of Azimuth Solutions — 123
   - A-4 Refraction, Parallax, Sun's Semi-Diameter — 127
   - A-5 Miscellaneous Attachments for Celestial Observations — 133
   - A-6 Solar Altitude Method — 139

CONTENTS

- **B. ERROR ANALYSES AND SPECIFICATIONS**
  - B-1 Errors and their Nature 142
  - B-2 Error Sources in Astronomic Azimuth 142
  - B-3 Propagated Errors in Astronomic Azimuths 154
  - B-4 Specifications for Astronomic Azimuths 158
  - B-5 Conversion to Grid Azimuth 161
- **C. SAMPLE OBSERVATIONS** 164
- **D. ASTROGEODETIC CONVERSIONS AND LAPLACE CORRECTIONS** 183
- **E. CALCULATOR PROGRAMS FOR ASTRONOMIC AZIMUTH**
  - E-1 Comments on Programs 217
  - E-2 HP 67/97 Programs 217
  - E-3 HP 41C/41CV/41CX Programs 220
  - E-4 TI 58/58C/59 Programs 223
  - E-5 HP 85 Solar Program 226
- **F. EPHEMERIS TABLES** 229

REFERENCES 252

ADDRESSES 253

INDEX 254

LIST OF ILLUSTRATIONS

Figure

1.1	Accessories Needed for Observations	3
1.2	Consistency Between Distance and Direction	7
2.1	The Celestial Sphere	16
2.2	Movement of Polaris	18
2.3	Location of Polaris	18
2.4	Relationship Between Azimuth to Celestial and Terrestrial Targets	21
3.1	Map Scaling of Latitude and Longitude	28
3.2	Surveyor Making a Solar Observation	31
4.1	Stop Watches for Solar Observations	38
4.2	Solar Eyepiece Image of Sun's Limbs	41
4.3	Angular Field of View of a Theodolite Telescope	44
4.4	Ground Referencing for Polaris' Direction	45
5.1	Sun's Declination Changes During Orbit	57
5.2	Elliptical Orbit of the Earth and Equation of Time	58
6.1	Solar Field Notes, Afternoon Observation	69
6.2	Solar Computation, Afternoon Observation	70
6.3	Solar Field Notes, Morning Observation	73
6.4	Solar Computation, Morning Observation	74
6.5	Polaris Observation Field Notes	75
6.6	Polaris Observation Computation	77
6.7	Preliminary Calculation to Find Polaris	80
7.1	Deflection of the Vertical	81
7.2	Convergence and Mapping Angles	83
7.3	SPC Zone Boundaries	92-106
7.4	Conversions to Grid Azimuth, Example G4	91
8.1	State Plane Projections	109
8.2	Control Station Index Map	113
8.3	SPC Station Description	115
A.1	Equatorial System	120
A.2	Hour Angle Transformations	121
A.3	Solar and Sidereal Time	122
A.4	Spherical Triangle	123
A.5	Effect of Atmospheric Refraction	129
A.6	Sun's Parallax	129
A.7	Sun's Semi-Diameter	129

ILLUSTRATIONS

Figure

A.8	Roelofs Solar Prism and Solar Image	135
A.9	Diagonal Eyepiece Prism and Eyepiece Prism Set	135
A.10	Target for Night Observations	136
A.11	Striding Level	136
A.12	Image of the Sun Using Altitude Method	139
A.13	Solar Observation Computation, Altitude Method	140
A.14	Standard Time Zones and their Relationship to UTC	141
B.1	Effects of Bubble Centering Error	145
B.2	Vertical Angle to the Sun	147
B.3	Latitude Error Effects on Solar Azimuth	149
B.4	Longitude and Timing Error Effects on Solar Azimuth	150
B.5	Error in Solar Azimuth Due to Changing Diameter of the Sun	153
D.1a	Laplace Corrections for Illinois	184
D.1b	Laplace Corrections for Indiana	185
D.1c	Laplace Corrections for Louisiana	186
D.1d	Laplace Corrections for Ohio	187
D.1e	Laplace Corrections for Pennsylvania	188

LIST OF TABLES

Table

3.1	Astrogeodetic Deflection Data	30
4.1	Times of Sunrise and Sunset	34
7.1	Central Meridians of SPC Zones	86-87
7.2	Theta Angle Conversion Constants, ℓ	88
A.1	Sun's Semi-Diameter	132
B.1	Summary of Error Sources	163
D.2	Astrogeodetic Conversions and Laplace Corrections for the U.S. and Puerto Rico	189-216
F.1	Corrections for Sun's Declination	230
F.2	Ephemeris for January and December 1982	230
F.3	Ephemeris for July, September, October 1981	231
F.4	Ephemeris for 1984-1988	232-251

CHAPTER 1

INTRODUCTION

1-1 HISTORY AND FUTURE OF AZIMUTH DETERMINATION

  The ancient Egyptians and other cultures had knowledge of observational astronomy. The pyramids, dating back thousands of years, have errors in their bases of only a few minutes of arc with respect to true north. Historians have assumed that these ancient people knew how to determine true north from star observations. Evidence of other cultures having knowledge of such observation techniques is seen in the remains of Stonehenge, in England. Many feel that this site was an observatory of sorts. Among others who practiced astronomy were the Babylonians and the Greeks. The design of some ancient and medieval instruments reveals an interest in astronomy and the geometric shape of the earth during this period. In 240 B.C., Eratosthenes determined the circumference of the earth to be about 25,000 miles by analyzing sun angle data at two locations. The dioptra an ancient instrument described in a book of the same name by Heron of Alexandria was an instrument designed to measure both horizontal and vertical angles, the latter being used to determine angular heights of stars. The astrolabe, popular in medieval times but dating back to ancient times, was an instrument for determining various angular data related to celestial bodies. Eventually, similar instruments such as the geometric square, quadrant, cross-staff, sextant, and zenith sector were used in the Renaissance and later periods to determine earth positions and orientation by star observations.
  Interest in astronomy was spurred by the awakening interest in science and knowledge of all kinds at the end of the Dark Ages. Nicolaus Copernicus (1473-1543) revealed his theory that the sun was the center of the solar system with the earth revolving about the sun as well as rotating on its own axis. As interest in this theory and several other surveying-related developments emerged, determination of astronomic azimuth became part of the outgrowth of these interests. During the seventeenth century, in particular, several events happened to foster these developments. The telescope was invented in 1608. Then, cross-hairs were added to it in 1640 and it gradually began to be used in angle measuring instruments, replacing the peepsight. The science of geodesy or "trigonometric surveying" was growing during this and the next centuries. By the nineteenth century, scholars in both Europe and the United States had developed instruments and methods to determine many earth-related measurements, including determining accurate geographic position and astronomic azimuths. More complete star tables and almanacs began to appear as progress continued.

INTRODUCTION

The properties of lodestone are said to have been known since ancient times. The first known use of the compass in navigation was in the eleventh century. Through the Middle Ages, the compass gradually began to see more use in surveying, particularly in mine surveying. During the American Colonial period, surveyors on this continent adopted the compass (also called circumferentor) as their primary direction orientation instrument. This trend continued into the twentieth century in the United States. However, at the same time, interest in trigonometric or geodetic surveying represented another trend toward astronomical observations for azimuth determination. Occasionally, the knowledge of astronomical azimuth found its way into the practice of the "ordinary surveyor". In the layout of the public lands, surveyors were directed to orient with the compass, but to correct for declination in order to determine true north. Thus, the surveyors who laid out the township and other important lines needed to know some astronomy in order to determine this declination. The surveyors subdividing the townships usually did not use astronomy, merely calibrating their compass to the township lines previously established by others.

People of the more scientific school in the late and early nineteenth centuries were David and Benjamin Rittenhouse, Charles Mason and Jeremiah Dixon, Andrew Ellicott, Jared Mansfield, and William Austin Burt. The solar compass, invented in 1836 by Burt was one example of an attempt to overcome the shortcomings of the magnetic compass. It could solve the spherical triangle mechanically and would be used when the magnetic direction was erratic or needed checking. Later, attachments to the surveyor's transit accomplished a similar function. Gradually, the General Land Office began to abandon the compass, but the switch to astronomy has never been complete in land surveying practice.

The trend of the "ordinary surveyor" shunning the more precise or scientific methods and instruments actually began at the close of the Middle Ages, continued through the Renaissance, and took firm root in the United States in the nineteenth century. The result today is some lingering reluctance among surveyors to adopt methods that would upgrade the measurement quality of their work to benefit the public and fellow surveyors who must "follow in their footsteps." This attitude has endured in spite of the methods that have been made easily available. As affecting the subject matter of this book, the ease, reliability, reproducibility, low cost, and many advantages of astronomic or state plane grid azimuth have been largely ignored in favor of perpetuating the tradition of magnetic, (or magnetic corrected for declination), assumed, or old record bearings as adequate in modern surveys. Many surveyors today have been deluded by past habits and the values passed down through the generations, that astronomy falls under the heading of "too scientific" or unnecessarily precise or costly. This attitude has prompted at least one author to write that "very seldom in modern surveys does a surveyor find it necessary to make astronomical observations". (Reference 2 , p. 150). His reasoning is that old bearings are assumed to be correct and another surveyor in the future will always find those lines intact on the ground. This is a sad commentary to the state of the art of surveying in the United States and is a reinforcement of the theory that the practice in this country has evolved from Medieval England where the "ordinary surveyor" wasn't expected (nor did he desire) to delve into any more theory or mathematics than absolutely essential to complete the immediate task.

Now, as we are drawing close to the end of another century, there is much discussion of professionalism and modern practice. On the subject of this manual, the means to determine a highly accurate astronomic azimuth, at low cost, with subsequent correction to state plane grid is at the fingertips of all surveyors. This has been made possible through the ready availability of optical theodolites, accurate time signals, ephemeris tables, United States Geological Survey (USGS) maps, hand-held programmable and non-programmable calculators, short-wave radios (including an inexpensive crystal-controlled time "Kube"), such simple devices as a solar eyepiece filter and stop watch, and data furnished by NGS for certain aspects of the observation reductions. It will be demonstrated herein how **one** person can do the field work **and** calculations to arrive at an astronomic azimuth with a 90% certainty of 5 seconds or better, (with a choice of either Polaris or the sun), in less than one hour and less than $100 worth of equipment (1983 prices) beyond the cost of a theodolite, a calculator, and this manual. (See Figure 1.1). Then it will be shown how such an azimuth can be corrected to state plane grid azimuth so as to overcome the problem of meridian convergence and have all azimuths in an area oriented to a reproducible, common system.

Figure 1.1 Accessories Needed for Observations
(Solar Eyepiece Filter, Stop Watch, TimeKube)

We are now on the verge of still another technology that could completely revolutionize surveying practice, including even the horizontal traverse itself. First, we've seen the recent developments in gyro-instruments which determine "true" azimuth using forces related to the earth's rotation rather than astronomy. This method is still much more expensive and less accurate than the method to be described herein. The latest developments are in the area of Doppler, inertial, and Global Positioning Systems (GPS), which have potential to change practice dramatically, if they become feasible for use by the "ordinary" surveyor. Such systems are still too expensive and inaccurate to use on small, local projects and thus for the next few decades at least, it is felt that the method to be described in this manual is the best for most surveying practice of a local or even a small regional nature, due to its high accuracy, and almost insignificant cost. Even if the positioning systems described are used to densify regional control to first or second-order accuracy, local surveyors will still need to traverse from these high-order control monuments to survey local projects or to further densify control. Since many monuments established by automatic positioning systems will not be intervisible, the local surveyor will need an azimuth mark to commence local surveys from these monuments. Astronomic monuments with the method described herein will provide the accuracies needed for further densification of control and execution of local surveys.

1-2 REFERENCE MERIDIANS FOR SURVEYING

Magnetically Determined Azimuth

Magnetic north is the direction taken by a magnetic needle in the earth's magnetic field. An azimuth oriented to magnetic north is subject to several errors, particularly the basic reading error of the compass circle, being several minutes of arc uncertainty. Nearby metal objects, underground utilities, direct current electricity, and iron ore cause "local attractions", which can cause errors of several degrees of arc. Daily and annual changes in the magnetic declination (difference between true and magnetic north) cause further uncertainties.

"True" north, as determined by a compass is simply magnetic north corrected for magnetic declination. Because of the lack of precision in determining magnetic declination and in the compass itself, plus the other uncertainties cited above, true north as determined magnetically has uncertainties of probably 30 minutes or more of arc.

It is doubtful that there is any justification for using the compass for azimuth determination in modern surveys, due to the fundamental inaccuracy of this method in relationship to what is easily achievable by astronomy and because of the need for better, more reproducible azimuths in land surveys.

Assumed or Arbitrary Azimuth

Assumed or arbitrary azimuth is that as chosen as a matter of convenience based on the general orientation of land lines, street lines, or other base lines. By relating the direction of survey lines to such base lines, a plane azimuth is then assigned to one survey line and other azimuths calculated from angle measurements at the survey stations.

The value of direction during retracement of such surveys is low since it is dependent upon the existence of original physical ground monuments.

Such azimuths just as well not be given at all as they are meaningless and misleading and have value only as a means of computing interior angles. It would have been more appropriate to simply use the angles themselves on the plat and in the description if arbitrary systems are used.

When monument evidence is lost or meager, such azimuths are useless in retracement. At least two adjacent undisturbed monuments would need to exist, in order to commence a metes and bounds retracement survey.

Record Bearing

A special case of the convenience azimuth is seen when a survey has been referenced to a former surveyor's monuments and that surveyor's bearings between the monuments then used. Thus, instead of arbitrarily assigning azimuths, a line in the original survey is the base line and the numerical value used for the azimuth is the one used in the original description.

Record bearings or azimuths are no more reproducible than arbitrary ones if one of the above described reference meridians (magnetic, true determined from compass, arbitrary) was used by the original surveyor, or if he did not cite his reference system and it cannot be discovered. Using record bearings on retraced survey plats is a supposed convenience for title purposes, the assumption being that the original description thus remains unchanged. This is a fallacy since all bearings except the one used on the base line will probably be different on the retracement because of discrepancies between original and retraced angle measurements. Furthermore, attempts to resolve all discrepancies between adjoining parcel descriptions may be futile when different reference meridians have been used or when low precision methods were used on the original surveys.

Unless record bearings have been determined accurately on the original and referenced to a universally reproducible reference meridian, there may be no justification for propagating them in retracement surveys. It would seem better to correct the record for dimensions between found monuments so that future surveyors have more reliable evidence to use.

Grid Azimuth

Convergence of meridians. As lines of longitude proceed from the equator toward the poles of the earth, the distance between them becomes shorter. In the Northern Hemisphere, a parcel of land having lines of longitude as two of its boundaries and parallel north and south boundaries has a shorter dimension on the north than on the south. This difference increases with increase in latitude. In many cases it can be ignored, but not always, depending upon the extent of the survey in an east-west direction and how great the latitude is. Except for very limited areas, surveys cannot be referenced to true or geodetic north without compromising the rectangular characteristics of plane coordinates. To make calculations and to show dimensions on a rectangular system, grid north must be the reference.

State plane coordinate grid. Any arbitrary grid can be established for a survey provided that said survey is less than a few miles in extent where curvature of the earth does not affect accuracy. Such an arbitrary grid can suitably be employed to compute latitudes and departures as a check for blunders, but it provides little toward preservation of evidence. The State Plane Coordinate (SPC) grid lends most toward preservation of evidence since it is mathematically related to geodetic north which can be determined accurately through astronomic observations or by referencing surveys to established monuments on the system. Formulas and tables are published for each state by the National Geodetic Survey (hereafter referred to as NGS) in a series of special publications, which provide the required information to make the adjustment from geodetic to grid azimuth at any given location in any state. The necessary information is also included in this manual.

Besides the advantages of enhancing preservation of evidence, future computations could conceivably be decreased through use of an established grid direction. For instance regardless of whether SPC coordinates are used on the date of the survey, such coordinates can be added later with no need for a rotational correction if the grid direction of the system was used on the initial survey.

Grid azimuth from astronomic and as determined from SPC monuments and control data will be explained in Chapters 7 and 8.

True, Geodetic, and Astronomic Azimuth

The term "true azimuth" was used in the early years of surveying to distinguish between azimuths determined by astronomic observations and those determined by a magnetic compass. In relation to older surveys the term "true north" therefore has significance and meaning as applying to astronomic north. However, true azimuth could be interpreted as meaning geodetic azimuth which is not the same as astronomic azimuth. It has been recommended to discontinue the use of the term in favor of more definitive terms such as astronomic and geodetic in future surveys. (Reference 13).

The difference between geodetic and astronomic azimuth is related to the fact that different reference surfaces are used for each. If a survey instrument is leveled using gravity as a reference, then an astronomic observation made for azimuth from this instrument set-up, an astronomic azimuth has been determined. However, the direction of gravity at a particular point may not be the same as the direction of a normal to the theoretical spheroid used for geodetic calculations and determination of geodetic azimuth. This difference may be caused by mathematical differences between the shape of the spheroid and the actual shape of the earth and/or variations in the direction of gravity due to effects of changes in terrain, such as mountains, oceans, or features of lesser proportions. These latter effects are called local anomalies. The difference between the direction of the vertical as defined by gravity and a normal to the theoretical spheroid used in the United States, called Clarke's Spheroid, is called the deflection of the vertical. The NGS has determined the deflection of the vertical at various geodetic control stations throughout the country. These stations are called Laplace stations. The "Laplace correction", which is the difference between geodetic and astronomic azimuth can be calculated using data from the NGS observations at these stations. This will be explained in Section 7-1 and further discussed in other sections. Other aspects related to deflection of the vertical are discussed in Section 3-2.

1-3 CONSISTENCY IN ACCURACY OF DISTANCE AND DIRECTION

The Meaning of Consistency

Consistency between distance and direction means that they are determined with comparable accuracy. Distances within one part in 3000 would mean that

1-3 CONSISTENCY IN ACCURACY OF DISTANCE AND DIRECTION

directions should be to an accuracy of approximately ±1 min (or 1 divided by 3000). Or if distances are to an accuracy of then to be consistent, directions of all lines should be to an accu... ±20 seconds. (See Figure 1.2).

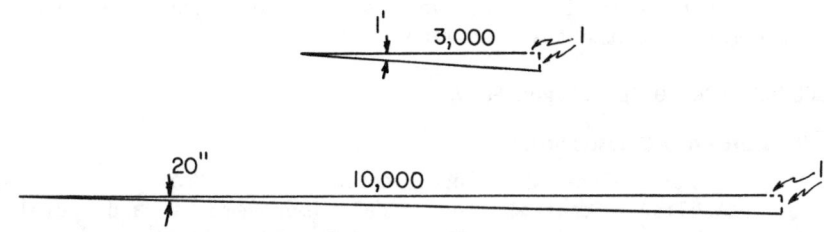

Figure 1.2 Consistency Between Distance and Direction

A Common Inconsistency

A commonly accepted standard of practice for measurements reads (Reference 1) "all tapes shall be calibrated to government standards..., and all measurements in the category of one part in 10,000 or greater shall be <u>made</u>..." (underscoring mine). On direction, the requirements shall read "bearings or angles on the map shall be <u>given</u> (underscoring mine) to the nearest 5 seconds...." Nowhere is it stated the accuracies needed for either angles or directions. One can "<u>give</u>" them to any number of figures but that is not the same as specifying an accuracy of <u>making</u> the measurements of either angles or directions, as is done with distances. This inconsistency is common in survey standards of practice.

Relationship between Angles and Directions

Extending this discussion one step further, it should be recognized that even if angular accuracy is specified, this does not assure good accuracy of directions. If the starting direction is inaccurate, all directions will contain the same initial inaccuracy in the same sense that distance measurements of the same line may check precisely but be inaccurate if the tape is too long or short or if the starting point is poorly identified. Angular accuracy of an angle is its degree of conformity with the actual value between two lines whereas the accuracy of a direction is its degree of conformity with an angle from a stated reference meridian.

In terms of preserving evidence, it can be seen that if the reference meridian can be accurately re-established, direction relocation is dependent upon only location of **one** starting point rather than **two** reliable starting points as must be done to re-establish a traverse angle.

Accuracy of the Starting & Other Directions

The accuracy of most lines in a survey depends upon both the accuracy of the starting direction and the accuracy of the angles measured from the

INTRODUCTION

starting line. Theoretically, the directional accuracy of any line will deteriorate as the number of angle points between it and the starting line increases. Therefore, the starting direction must be more accurate than the allowable directional accuracy of any line in the survey. If any distance is to contain an accuracy of one part in 10,000, consistency in directions would mean that the direction of *any* line in the survey must be the value given, as referenced to the meridian stated, ±20 seconds, including those lines most remote from the starting line.

1-4 ACCURACY FOR LOCAL CONTROL SURVEYS

Current Standards and Classification

The NGS defines third-order local horizontal control surveys as those "used to establish control for local improvements and developments, topographic and hydrographic surveys, or for such other projects for which they provide sufficient accuracy." The NGS further states that the work should be performed with sufficient accuracy to satisfy the standards for such surveys. The standard for Third-Order, Class I astronomic azimuth for traverse is a standard error of ±3". For Third-Order, Class II astronomic azimuth for traverse it is ±8" standard error. (Reference 23). These standard errors convert to 90% errors of approximately ±5" and ±13" for Class I and II surveys, respectively.[1]

According to NGS classification, Second-Order, Class II astronomic azimuth for traverse must be within a standard error of ±1.5". Such surveys are used to establish control along the coastline, inland waterways, and interstate highways. Such control contributes to the national network and is published as part of the network. It is recommended for controlling extensive land subdivision and construction. In other words, this class of survey, with its accuracy requirements for astronomic azimuth and other measurements is appropriate for the control network in a region or county where future remonumentation or cadastral improvements may be planned.

It is felt that control traverses of several miles to reference local surveys to the national network should be done with second-order accuracy. Surveys for shorter extensions, such as within a small city, should meet third-order, Class I standards. It seems appropriate that starting azimuths for local surveys could also be observed to Class I standards, even though the accuracy may deteriorate to no better than Class II equivalent standards, primarily because of shortness in sight distances, monumentation practices, and other characteristics of such surveys. The following sections discuss this further.

What is Appropriate for Local Surveys?

The NGS specifications cite starting azimuth accuracies for control surveys. Before arbitrarily selecting a standard for surveys related to property

---

[1] The 90% error (1.645σ) will be used throughout this manual, as it is felt to represent a level of certainty more appropriate than that contained in the standard error (68.3%). The term standard error (or standard deviation) is a statistical term and has no relationshp to the meaning of "standard" as used in "standards and specifications".

1-4 ACCURACY FOR LOCAL CONTROL SURVEYS

corners, let us relate azimuth requirements in terms of positional tolerances. An azimuth error of ±5" yields a positional error of ±0.24 feet in a line 10,000 feet long and ±0.03 feet in a line ¼ mile long. Positional error relates to the actual distance, or size of survey, as well as to the azimuth uncertainty. Thus, if positional errors (with respect to an origin such as state plane coordinates) in surveys are to be kept within a few hundredths of a foot, careful analysis of starting azimuth and traverse angle errors must be done, considering how the azimuth errors propagate in terms of all errors (including distance errors) and the line length. This is not easily done since each survey is different, geometrically, and it is impossible to define a "typical" survey. However, to illustrate how the azimuth errors might propagate, let us use a quarter-section of land on the public land system as an example. If the starting azimuth had been determined to a 90% error of ±5" (third-order, Class I) and the angles at each of the four sides were determined to be the same precision, the 90% error in any one of the azimuths of the four sides should be within ±10". This is derived from the fact that the azimuths of the other three sides are computed from angular measurements, which themselves contain errors. These errors propagate as random errors. There being four sides and errors of ±5", the expected maximum error in any azimuth thus computed would be $5\sqrt{4} = \pm 10"$. In ½ mile, the relative positional error of any point with respect to any adjacent point, due to azimuth uncertainty, would be approximately 0.13 foot, a relative error of 1:20,600.

The objective here is to illustrate something about how azimuth errors relate to positional errors, since these concepts translate directly into accuracy of state plane coordinates of a land survey corner if established from nearby control. Another objective is to illustrate how close the azimuths of adjacent surveys might match in deed descriptions if all surveys were done using astronomic or grid directions. Due to differences in configuration, survey size, and extent of local control, this cannot be stated absolutely. Let us propose, as stated at the conclusion of the last section, that third-order accuracy is desirable and appropriate for local land and subdivision surveys.

Practical Accuracy Achievable

In 1968, this author made several hundred field observations to research the relative precision of Polaris vs. solar observations for azimuth. Since that time, the author has conducted many seminars on this subject as well as on measurement theory, and has continued to research the precision and accuracy of azimuth determination and of theodolite angle measurements. The result of this has been a refinement in the methods to improve the accuracy, reduce the cost, and develop a procedure for practicing surveyors to achieve third-order astronomic azimuth with less than $100 equipment (beyond the theodolite) and less than one man-hour of time, usually with a choice between either the sun or Polaris. Figure 1.1 illustrates the accessory instruments.

Polaris observations will achieve the equivalent of third-order, Class I traverse accuracy[2] without very many refinements in techniques or cause for

---

[2]The word "equivalent" accuracy is used throughout this manual since the procedures described herein are not necessarily the same as used by NGS to achieve the stated accuracies.

concern from some error sources which affect solar observations. Solar observations using the hour angle method can as easily be used for this class of accuracy, but there are a few additional precautions that must be considered in some geographic areas. Solar observations, under a few ideal conditions, may even be suitable for second-order, Class II accuracy, but generally this accuracy is best achieved using Polaris observations. With a minimum of refinement in techniques and under some of the worst observation and local geographical conditions, either solar or Polaris observations will yield the equivalent of third-order, Class II accuracy. The main text of this manual describes the observational and calculation techniques for third-order equivalent accuracy. Usually, third-order, Class I can be assumed if the details of these chapters are followed. However, the error analyses leading to specifications in Appendix B are intended as a more complete precautionary reference to be used as a guide for perhaps deciding whether results are actually in this class, or what can or needs to be done in a particular situation to achieve this or higher accuracies using the choices of methods available.

1-5 PRESERVATION OF LAND SURVEY EVIDENCE

Every original survey will probably need to be retraced at some future time. As time passes, evidence of monuments may become lost or obliterated. During retracement, it is the surveyor's function to gather remaining evidence, analyze it, and replace the lost corners at points which have the highest probability of being the original positions of the monuments. Directions, as given in the original description of the land, are part of the evidence. The extent to which these directions are reproducible determines, in many cases, the certainty of corner positions established during retracement, as well as the time and cost for such retracements. Since astronomic azimuth is a common reference that is easily reproducible at low cost and high accuracy, it is ideal for use in all land surveys, whether original or retracements.

Astronomic azimuth is, in a sense, a reference tie, and is used in a manner very similar to bearing trees or other reference ties. One slight disadvantage of astronomic reference monuments is that they are constantly moving. Another possible disadvantage in some parts of the country is that cloud cover sometimes hampers observations. The field and computational procedures are not felt to be disadvantages, since they can be learned easily. The disadvantages of using astronomic monuments is more than overcome by the fact that, unlike ground reference ties, astronomic monuments are indestructible. They are as permanent as the universe itself, the very total of perpetuation in a real sense. They aren't subject to disturbance from nature and forces of man, as are terrestrial monuments. Since their movement and change is predictable and accurately recorded in ephemerides prior to the changes, they may be considered more predictable than anything we know in the physical world.

1-6 CONTROL SURVEY ADVANTAGES

Astronomic azimuths, converted to state plane grid azimuths, provide a control system of azimuths for land surveys, engineering design and constructions surveys, cadastral information systems, and a variety of interrelated systems for legal and land use purposes. The advantages, if all surveyors did their work using these common reference meridians, would be much better coordination and compatibility among surveys of all kinds. If all surveyors in a state or a region used the same highly accurate, reproducible meridian, surveys would check each other in retracement, and layout surveys could be done using old lines as base lines for orientation. When and if state plane coordinates were to be established on survey points, there would be no rotation needed since the correct azimuth system would already have been used. Furthermore, if all azimuths were of at least third-order accuracy, survey stations could probably be incorporated into a regional or local network of third-order control, depending on specifications for monumentation and other aspects of the survey.

The ideal orientation system is one that is accurate and reproducible at low cost, can be established anywhere without depending on other local survey lines, and provides the important advantage of compatibility with adjoining surveys. The azimuth system, established as described herein, approaches this ideal.

1-7 MODERN ASTRONOMICAL WORK

Optical Theodolites

Many surveyors, accustomed to using 30" or 1' vernier theodolites, probably would have trouble conceiving of how azimuths can be determined within a few seconds. Such accuracy is made possible, in part, by the precision provided by optical theodolites. Either the direction or repetition type of instrument can be used for third-order azimuth, if sufficient repetitions are made. It should be understood that the statements regarding low cost and accuracy of azimuth determination assume that the surveyor already possesses and uses an optical theodolite for angulation, that vernier instruments are not used, and that a solar eyepiece and a battery pack for internal illumination can be attached to the theodolite.

Time Determination

Time measurement is critical using the solar method. Sufficient precision, accuracy and convenience are possible now, because of refinements made in recent years in propagation of time signals. Timing in the field is also made more convenient because of the ready availability of small, inexpensive, light-weight transistor radios to receive the signals. At least one of the available radios is crystal controlled, so that tuning the transmitted signal on a radio dial is not necessary. Also, electronic and other precision mechanical stop-watches afford a variety of ways to measure time accurately, conveniently, and inexpensively in the field. These several improvements made in time signal propagation, radios, and stop watches in the last few decades, makes the determination of astronomic azimuth more convenient and accurate than ever before because the precision necessary in timing is easily achieved.

Calculators

Before the development of calculators which have pre-programmed trigonometric functions and storage capability, calculations for astronomic observations were just cumbersome enough to discourage their use by many surveyors. Now, especially since relatively inexpensive electronic calculators are available, the computations can be done quickly and free of many types of mistakes that formerly caused confusion for the person computing the observations. These developments remove another small obstacle which, at one time, gave surveyors reason to hesitate before making observations. If a programmable calculator is used, the computation time and chance for mistakes is reduced even more.

Simplified Ephemeris

Several ephemerides are available for astronomic calculations. Most of them contain an abundance of tables and other data not needed for the azimuth solution, resulting in high cost and/or unnecessary bulk in the book of tables furnished. None contain all of the tables with data listed to sufficient precision to solve the azimuth problem using the hour angle method. Some are missing one or two tables. Others have tabulations of data which are confusing to follow. At least one ephemeris uses zero hours as noon instead of midnight and this causes added confusion. Another emphasizes the altitude method rather then the hour angle method. All of these problems have been corrected in the print-outs of ephemeris data in this manual. These tables have been prepared to avoid any significant round-off errors and to minimize the need for the tables as much as practicable. This is made possible by listing all values to sufficient significant figures and programming several conversions and interpolations into the computations, rather than requiring the surveyor to use tables for such simple steps.

Quadrangle Maps

In the last few decades, USGS 7½-minute quadrangle maps have been prepared for most urban and rural areas of the country. These are frequently updated in urban areas by photogrammetry. The availability of up-to-date maps, at this scale rather than the older 15-minute series provides the surveyor with means to accurately scale latitude and longitude of the observation station.

NGS Data

For correction of astronomic azimuth to geodetic azimuth (which is a necessary intermediate step toward state plane grid azimuth) the Laplace correction must be applied. Furthermore, for solar observations, astronomic rather then geodetic latitude and longitude (which is scaled from the USGS maps) should be used in calculations to achieve maximum accuracy. The NGS has recently developed computer programs which predict these differences using input of approximate latitude and longitude. With these data, the practicing surveyor can achieve the equivalent of third-order Class I accuracy nearly anywhere in the U.S., using solar observations.

Hour Angle Method

Although not new, the solar hour angle method has not been emphasized or even mentioned in some surveying textbooks, the altitude method being taught instead. The reason is probably that surveyors have traditionally used vernier instruments where the precision provided by the hour angle method would have been lost in the reading system anyway, and also because time wasn't as easily and as precisely measureable as it is today. Other probable reasons are that the altitude method is somewhat easier to calculate and this would have been a consideration before the development of electronic calculators and computers. Other reasons why the altitude method is still more prevalent can simply be categorized under the heading of tradition, and the typical delay between the time when any system becomes less desirable and the time this fact is accepted to the extent that it is replaced in practice, in accepted standards, and books of instruction.

The solar altitude method is inferior due primarily to errors associated with the vertical angle measurement as related to mechanical errors in measuring the angle and the effect that the angle has on correction for atmospheric refraction, as well as uncertainties in the refraction of the atmosphere itself. The field observations using this method are awkward in that the horizontal and vertical angle must be observed simultaneously which requires centering the sun using special reticles or observing by positioning the sun in the four quadrants created by the cross-hair intersection of a theodolite. The main advantage of the altitude method is that time and longitude need not be determined accurately.

Even when a precise theodolite is used, research and experience have shown that agreement among repetitions of an azimuth solution using the altitude method seldom is much better than a standard deviation of ±15" and discrepancies often exceed 1' of arc. In comparison, the solar hour angle method consistently yields standard deviations of a few seconds of arc with differences between sets seldom exceeding 20" when the precautions and systems described herein are diligently employed.

The hour angle method for both Polaris and the sun is recommended since only horizontal angles and time must be observed in the field and no special tables equating altitude (h) of Polaris with time are required as is necessary when using a familiar simplified formula for Polaris. With programmable and other electronic calculators, the hour angle formula is easily used for both the sun and Polaris and this method is more pure - that is, it avoids certain round-off errors and assumptions made in the derivation of the simplified formula for Polaris.

The theory of the hour angle method is not new, but the specific field and computational techniques described herein are unique in many ways. They must be followed to assure results claimed by the author. The emphasis in development of the method has been in simplicity, improved precision, minimum time and expense, minimum need for expensive attachments and accessories, and minimum use of tables.

Need for Improved Standards

As movements are made toward integrating survey control work in regions and communities for multi-purpose cadastral maps, needs and desirability follow to improve the quality of surveys for land and engineering surveys. It soon may not be acceptable to perform surveys that cannot relate in position and azimuth to other nearby surveys. The expense of resurveying in order to integrate all surveys is unnecessary if original surveys are done using common systems, and, furthermore are performed to an accuracy standard high enough to be reliable. Following the methods described herein will assure that local systems of land and engineering surveys can, at any time, be coordinated with larger, regional control networks, and thus will be compatible with and usable in future cadastral mapping programs.

CHAPTER 2

THE THEORY OF ASTRONOMIC AZIMUTH

2-1 THE CELESTIAL SPHERE AND SPHERICAL TRIANGLE

General Theory and Definitions

The celestial sphere. The earth rotates from west to east. Thus, the apparent movement of the stars is from east to west. The stars are considered to be an infinite distance away and on the surface of the celestial sphere. It is this sphere that appears to rotate from east to west, the earth being the center of the sphere.

An extension of a line through the earth's North and South Poles intersects the celestial sphere at the north and south celestial poles. See Figure 2.1 for illustrations of properties of the celestial sphere. The following definitions are taken from Reference 16.

Observer's zenith and nadir. A plumb line passing through the observer's position and extending upward pierces the celestial sphere at the observer's zenith. The opposite point on the other side of the sphere is the nadir.

Observer's meridian. On the celestial sphere, a great circle passing through the celestial poles and the observer's zenith and nadir is known as the observer's meridian. The observer's meridian is also an imaginary line on the earth, being simply the observer's longitude.

Star's meridian. Another great circle on the celestial sphere passing through the north and south celestial poles and the star is called the star's meridian, or the "vertical circle through the body."

Celestial equator. The celestial equator is the intersection of the celestial sphere and the plane through the center of the earth which is perpendicular to the line joining the two poles.

Horizon. The horizon is located at the intersection of the celestial sphere and the plane tangent to the level surface at the observer's positon. In surveying, a line to the observer's horizon is defined when a transit or theodolite telescope has been made perpendicular to the plumb line as defined by the level on the instrument.

Azimuth, Z. The azimuth of a star is the clockwise angle on the horizon plane from the north point to the vertical circle through the star.

Altitude, h. The altitude of a star is the angular distance measured upward on the vertical circle through the star from the horizon to the star. (Surveyors recognize this as simply the vertical angle of inclination. Usually, however, the observed vertical angle must be corrected for refraction to arrive at the true altitude.)

2-2 SOLUTIONS OF THE SPHERICAL TRIANGLE

Local Hour Angle, LHA. The local hour angle is the angle measured westward along the celestial equator from the observer's meridian to the meridian of the star.

Meridian angle, t. The meridian angle is the same as local hour angle, except that it is measured either east or west. That is, if the local hour angle is less than 180°, then t=LHA and if the local hour angle is greater than 180°, then t=360°-LHA. Meridian angle is often called "hour angle".

Declination, δ, and polar distance, p. The declination of a star is the angular distance measured along the hour circle through the star from the celestial equator to the star. Polar distance is measured along the same hour circle, but from the celestial pole to the star. Declination=90°-polar distance.

The Spherical Triangle

The astronomic azimuth and other unknowns, if needed, can be determined through solution of a triangle on the celestial sphere. In the Northern Hemisphere, the three points of the triangle are the north celestial pole, the observer's zenith, and the star being observed. The three sides of the spherical triangle are (90°-δ) or p, (90°-h), and (90°-ϕ). The quantity, ϕ, is the observer's latitude. The two angles used in the solution of the spherical triangle are Z and t. See Figure 2.1.

2-2 SOLUTIONS OF THE SPHERICAL TRIANGLE

Using various combinations of the law of sines, the law of cosines, and other relationships used in spherical trigonometry, equations can be derived for the determination of azimuth. These equations are as follows:

$$\tan Z = \frac{\sin t}{\tan \delta \cos \phi - \sin \phi \cos t} \quad \text{(Hour angle method)} \quad (2.1)$$

$$\cos Z = \frac{\sin \delta - \sin \phi \sin h}{\cos \phi \cos h} \quad \text{(Altitude Method)} \quad (2.2)$$

$$\sin Z = \cos \delta \frac{\sin t}{\cos h} \quad (\sin p \text{ can be substituted for } \cos \delta) \quad (2.3)$$

Equation 2.3 is impractical since both t and h must be accurately determined. A simplified form of Equation 2.3 is sometimes used for Polaris, (See derivation in Appendix A for Equation A.12) but this equation will not be used herein since it necessitates additional ephemeris tables and is not as precise as the hour angle formula. The disadvantages of the altitude method (Equation 2.2) have been discussed. Only Equation 2.1, the hour angle method, will be applied in this manual for both Polaris and the sun. The above equations are derived in Appendix A where additional discussion is included.

2-3 BEHAVIOR OF POLARIS AND THE SUN

Polaris

This is the star closest to the north celestial pole. In the early 1980's, its polar distance was approximately 49 minutes. It is becoming closer to the celestial pole each year and will continue to do so for another century. This trend of decreasing polar distance can be observed by scanning that

Figure 2.1 The Celestial Sphere

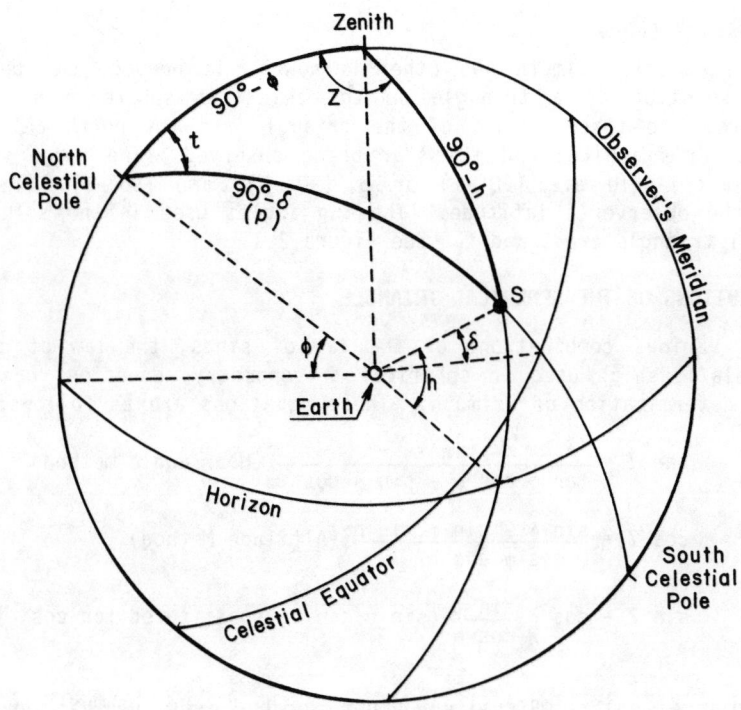

S = any celestial body
t = hour angle
φ = latitude
Z = azimuth
δ = declination
p = polar distance
h = altitude

column in the ephemeris tables of Appendix F. Due to its closeness to the celestial pole, its motion with respect to an observer on the earth is very slow. Its magnitude (brilliance and size) is ideal for observations from the standpoint of minimum pointing error, and its altitude throughout the United States is not excessive, enabling observations to be made without special attachments to the theodolite.

As the celestial sphere makes its apparent rotation, Polaris describes a small "diurnal" circle of radius equal to the polar distance, p, for the particular time of the year. It moves in a counterclockwise direction as observed from the earth as shown in Figure 2.2. In this figure, UC stands for upper culmination, LC, lower culmination, EE, east elongation, and WE, west elongation. The vertical line through UC and LC represents the observer's meridian. At culmination, Polaris is at the observer's meridian and its apparent movement in azimuth is largest. When exactly at elongation, it is farthest from the observer's meridian and it has no apparent movement in azimuth. It can be seen from Figure 2.2 that at culmination, the astronomic azimuth of Polaris is 0°. At elongation the altitude, h, of Polaris would theoretically be equal to the observer's latitude, ϕ.

Figure 2.3 shows the location of Polaris with respect to the stars in the constellation known as the Big Dipper. It should be noted that the orientation of this constellation tells the observer not only where to find Polaris, but also its approximate location and movement with respect to elongation and culmination. Chapter 6 will explain how to find Polaris during daytime with the aid of calculations.

It is noted that Polaris is the only bright star within several degrees of the north celestial pole. With a small amount of experience, and knowing approximate north and one's latitude, it can be identified at night with little chance for mistake. If one was to make such a mistake it could easily be resolved since any bright star other than Polaris would move away from the cross-hair in a matter of a few seconds of time whereas Polaris remains almost stationary, even at culmination, and several seconds time lapse will result in little movement.

The Sun

Unlike Polaris, the sun rises and sets. It circles the pole in much the same way as Polaris or any star, but has a much larger polar distance, resulting in the rising and setting. Since the sun is much closer to the earth, with the earth in orbit around it, its movement behavior differs in several ways to that of Polaris. For example, the daily declination change for Polaris is much smaller than that of the sun. Also, time relationships with respect to the stars differ from the relationship with respect to the sun, due to the great distance to the stars.

Some elementary facts about the sun's apparent rotation are that it rises in the east, sets in the west, and moves across the southern sky during the day in the Northern Hemisphere. At the time of the Equinoxes (March 21 and September 23), it rises and sets exactly on the "prime vertical" (the great circle at 90° to the observer's meridian). Stated differently, the sun rises and sets due east and west on these dates. The sun's declination changes from south to north at the Vernal (Spring) Equinox. This has the effect between March 21 and September 23 of the sun rising north of due east and

18

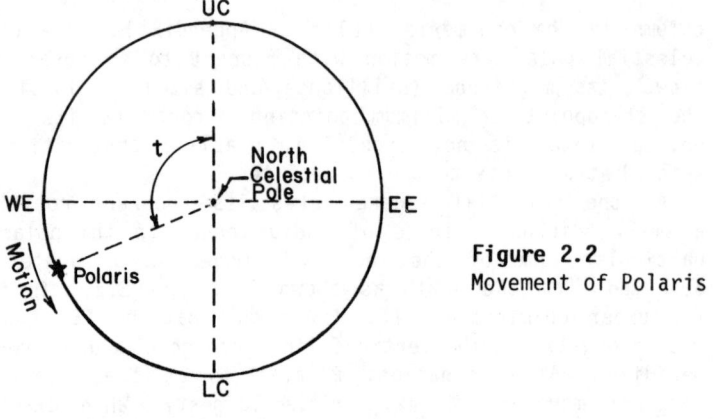

Figure 2.2
Movement of Polaris

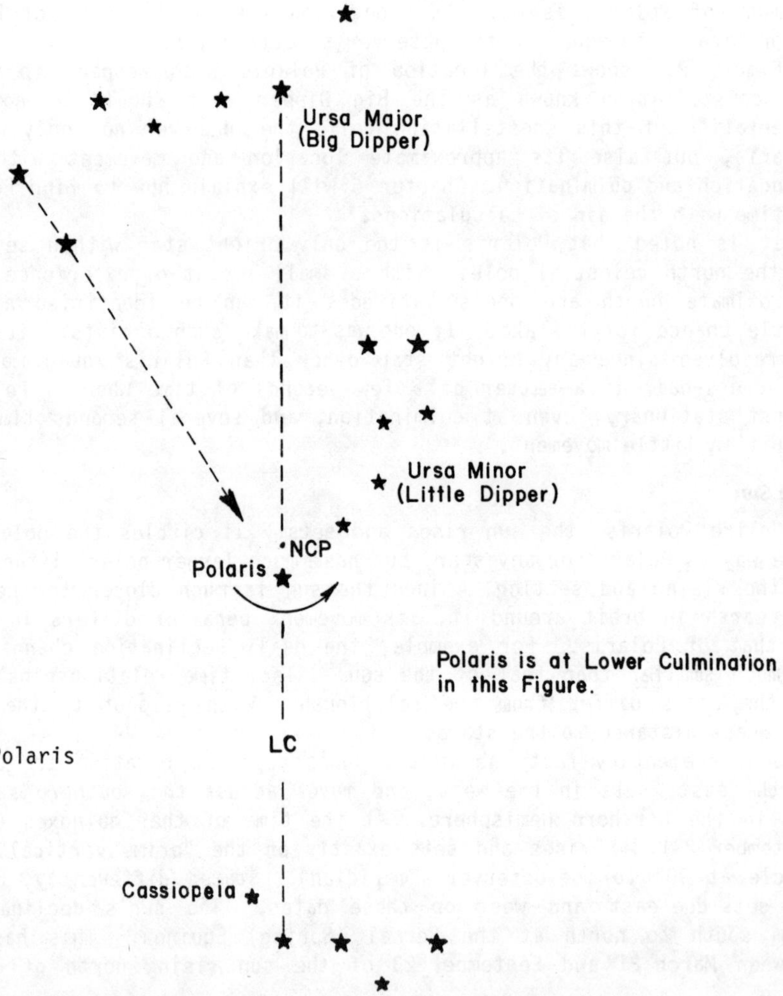

Figure 2.3
Location of Polaris

Polaris is at Lower Culmination in this Figure.

setting north of due west, its maximum northeast and northwest bearing occurring on June 22, at the Summer Solstice, when declination is maximum. After September 23, the declination is south or negative and this results in the maximum south component being on December 22, at the Winter Solstice, when declination is maximum. These facts aid the surveyor in planning field operations and in analyzing calculations since the quadrant of the sun can generally be decided based simply on common sense. For example, the only time the sun could be northeast would be if it was very early morning in the summer. Trite as it may seem, many mistakes can be avoided by understanding these elementary aspects of the sun's behavior.

Since the sun is so far from the celestial pole its movement is very rapid with respect to the observer. This is why time determination is so critical. Furthermore, its rate of change in azimuth per unit of time increases as it approaches noon (hour angle approaching zero). This is one reason why observations are undesirable near local noon (t=0°).

Time relationships are different with the sun as compared to the stars. Because the earth rotates around the sun once a year, for example, there is one less solar day than sidereal day in a year. This is further discussed in Chapter 5 and Appendix A. Also, because the earth revolves in an elliptical orbit rather than a circular one, the actual time it takes for any particular earthly point to rotate once on the earth's axis and face the sun again the next day varies. This variation, called "equation of time", is described more completely in Section 5.3. A complete understanding of these points is not essential to successfully make the observations, but is helpful to fully appreciate and enjoy the work.

2-4 BASIC PROCEDURES TO DETERMINE AZIMUTH

Field Observations

The determination of an astronomic azimuth involves a field observation and computations, which is no different than most problems in surveying--collection of data and reduction of data to usable measurements.

The field phase, using the system described in this manual involves sighting a ground station, then a celestial azimuth mark (either the sun or Polaris), thus measuring the horizontal angle from the ground station to the celestial body. Measurements are made in sets of direct and reverse telescope, as with any angle measurement. Several sets are observed, so as to achieve required precision, allow for rejections in case of mistakes, and in the case of the sun, an equal number of pointings are made to the right as to the left limbs (sides of the sun). The computations will yield an azimuth to the sun's center when limb pointings are properly averaged. Pointings are made directly to the star, since its size is ideal for pointing to its center.

Since celestial azimuth marks are constantly changing position as the earth rotates and revolves around the sun, time must also be observed--not time interval, but time according to elapsed time from midnight as zero hours for the observation date. This is done by tuning a time signal, broadcast on the short wave band of a radio. With the sun, a stop watch is required since time is needed within a fraction of a second for each pointing.

It is seen that the only field measurements needed for any astronomic observation for azimuth are horizontal angle and time. The only field equipment beyond the usual theodolite and targets is a stop watch for solar observations, a short-wave radio receiver, and an inexpensive solar eyepiece to block out the sun's brilliance when sighting its limbs. Additionally, a battery pack or flashlight may be desirable or necessary if one waits until dark for a Polaris observation.

Data Reduction

The solution of the equation for azimuth involves only simply algebra and plane trigonometry. The variables are latitude, ϕ, sun's declination, δ, and hour angle, t for the solution of azimuth Z.

Latitude is scaled from a USGS 7½ minute quadrangle, or similar scale map made to comparably high accuracy standards, all of which is explained in Chapter 3.

The declination of the sun and Polaris is listed in the ephemeris tables in Appendix F. For the sun, declination changes rapidly so that interpolation is necessary. The tables provide an easy means to accomplish this. The declination of Polaris changes very slowly so that interpolation in the tables is unnecessary.

The hour angle, t, is a function of the observed time, the longitude of the observation station, and "equation of time". Longitude is scaled from a map in the same manner that latitude is determined. Equation of time is listed in the ephemeris tables. The details of determination of t, explanation of equation of time, and solution of the equation for azimuth will be covered in Chapters 5 and 6.

A single computation is made for each observation. That is, for each timing-pointing made to the star's center or to a limb of the sun, an azimuth is computed. For Polaris, such an azimuth is not usable until averaged with one made with the telescope in the opposite position (direct and reversed averaging). Additionally, for solar observations, a result computed to the left limb must be averaged with a result to the right limb. For either the sun or Polaris, a backsight is made to the ground target first, then the foresight made to the celestial body. Thus, the computation of azimuth to the ground station requires subtracting the measured horizontal angle from the computed astronomic azimuth.

Discrepancies between direct and reverse pointings to Polaris should be very small, theoretically containing only the effects of instrumental and personal errors. However, it should be realized that an azimuth computation when a limb of the sun has been sighted is fictitious, theoretically containing a systematic error of half the sun's angular diameter, since values in the ephemeris assume that pointings were made to the sun's center. For a pointing to the left limb, a computed azimuth to a ground station will be too large by half the sun's diameter since the horizontal angle measured to the limb is too small by that amount. Likewise, the computed azimuth to the ground station will be too small when the right limb of the sun is used. These relationships are shown in Figure 2.4.

It should be understood that **the computed azimuth to the sun is unaffected by angular measurements to targets.**

2-4 BASIC PROCEDURES TO DETERMINE AZIMUTH

Summary

The purpose of this section has been to explain basic concepts so that the measurements and detailed procedures of Chapters 3 to 6 can be more easily understood. The principal point is that an azimuth determination by sighting a celestial azimuth mark is basically the same in principle as determining the azimuth to a ground station by sighting any azimuth mark--it involves merely measuring the angle between the mark of known value and the one whose value is desired. One main difference with celestial marks is that they are moving. Another is that one is very large and bright and pointing cannot be made directly to it. Because the targets are moving, computations are required so as to relate time and direction and to account for the need to measure to the two sides of the solar azimuth mark. Also, special field techniques are required to sight these rather unusual, moving azimuth marks and achieve accurate results. These field and computation techniques will next be detailed.

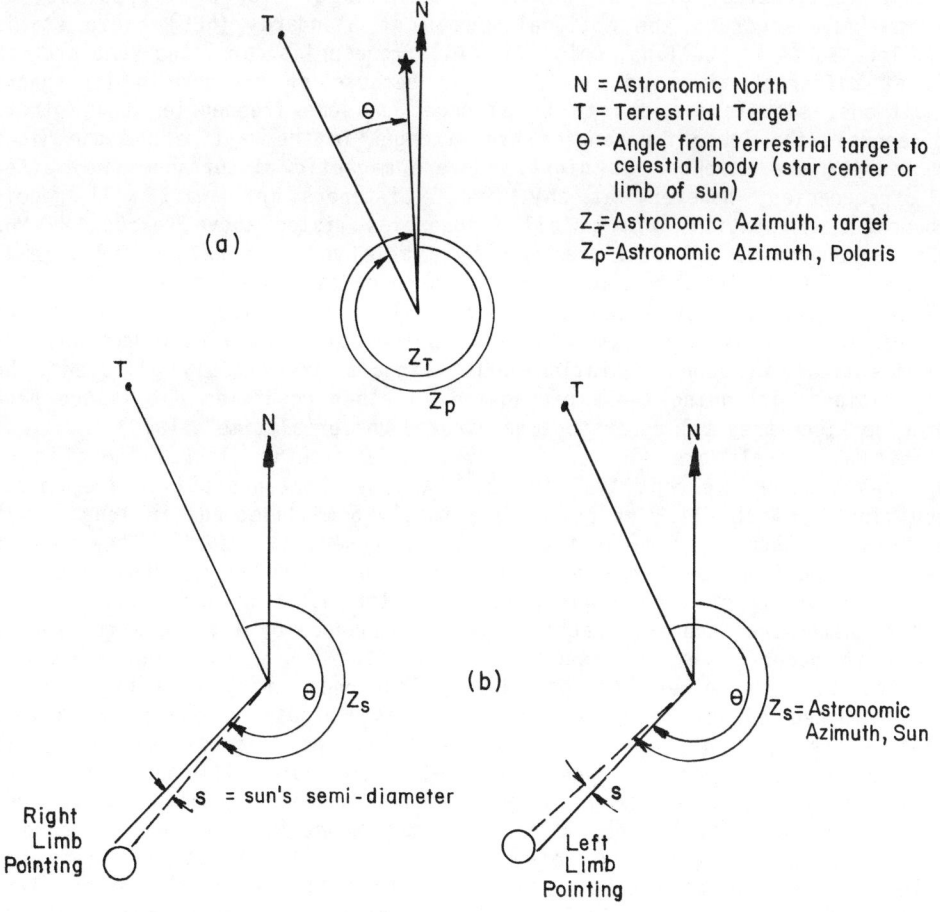

Figure 2.4 Relationship Between Azimuth to Celestial and Terrestrial Targets

CHAPTER 3

TIME, GEOGRAPHIC POSITION, ANGLE MEASUREMENT

3-1 TIME MEASUREMENT

NBS Time Signals

Time is broadcast over short wave frequencies of 2.5, 5, 10, 15, 20, and 25 MHz (megahertz) by the National Bureau of Standards (NBS) radio station WWV, located at Ft. Collins, Colorado. All frequencies carry the same program, but at different times of the day, and because of changes in ionospheric conditions, signals are apt to be stronger on some frequencies than others. In general, the lower frequencies are stronger in the daytime and the higher frequencies are stronger at night. Severe magnetic disturbances may affect all frequencies, however, at any time. Listeners in Hawaii will receive broadcasts on station WWVH on all frequencies listed above, except 25 MHz. Voice announcements of time are made by a male voice on WWV and by a female voice on WWVH. Similar time signals are broadcast in Canada over station CHU on frequencies of 7.335 and 14.670 MHz. The voice announcements are in French, as broadcast from CHU. The pulses propagated over WWV and WWVH are identical in sound. Station CHU's signals are audibly different, but all stations, including those propagated in other countries not listed here, coincide since they all refer to "Coordinated Universal time", (UTC).

The WWV signals are similar in sound to a ticking clock. There is one click each second, except that the 29th and 59th seconds pulses are omitted each minute. Each click or pulse is actually 5 milliseconds in length. The beginning of each pulse marks the zero of that second. Usually, the surveyor need not count or be concerned with the seconds, however. More important is the beginning of each minute, since a stop watch or correction will be used to determine seconds. Each minute is announced by a voice a few seconds before it occurs, the 59th second is skipped, then a tone much longer in duration than the other seconds (800 milliseconds) marks the beginning of the announced minute. The specific hour and minute announced at UTC is actually the time in the zone of Greenwich, England, and was formerly (prior to 1972) called Greenwich Mean Time (GMT). The UTC will be greater than local time in the continental United States by 5 hours in the Eastern zone, 6 hours in the Central zone, 7 hours in the Mountain zone, and 8 hours in the Pacific zone. When daylight savings time is in effect, one less hour would be added to watch time to arrive at UTC. These points are of little concern, however, since field data and computations following the procedures used in this manual will be performed by recording UTC directly from the radio broadcasts. This procedure avoids the necessity of considering zone corrections and avoids mistakes in this part of the computations. Ephemeris tables are prepared using zero hours UTC for each date for each variable. Time zones are given in Figure A.14 at the end of Appendix A.

The broadcasts may be heard by telephone. WWV can be heard by calling 303/499-7111. WWVH can be heard by calling 808/335-4363. The calls are automatically limited to three minutes. These are long distance toll calls for those outside the local dialing area.

Correction to Time Signals

Since 1972, refinements have been made in the broadcast signals to account for variations in the earth's rotation. The time scale broadcast is almost perfectly constant, being based on ultra-stable atomic clocks. For surveyors wishing the best possible accuracy for solar azimuth observations, the small correction to be described should be applied. It is broadcast over WWV and other stations, superimposed in code. These corrections, called DUT1 corrections (or ΔUT1 in some references), are encoded over WWV stations by using double clicks after the start of each minute. If one hears double clicks for the 1st, 2nd, and 3rd second, for example, the DUT1 correction is $+0.3^S$. Up through the 8th second, each double click represents $+0.1^S$ correction. To assign the negative sign to the correction, the code is to note what is heard starting with the 9th second. If no double clicks are heard during the first 8 seconds, one must count the double clicks, if any, from the 9th second onward to determine negative DUT1 and the amount. For example, if the 9th, 10th, 11th, and 12th clicks are doubled, the DUT1 correction is -0.4^S, each double click representing -0.1^S correction. On the Canadian broadcasts, the code is a much deeper single click than the remaining clicks, rather than a double click as used by WWV. The time used in the computations for an azimuth determined by solar observation is called UT1 where:

$$UT1 = UTC + DUT1 \tag{3.1}$$

UTC being as announced over the radio (and further determined with the aid of a stop watch), and DUT1 determined as explained, with correct sign.

The DUT1 corrections remain fairly constant from day to day but are likely to change by 0.1 seconds unexpectedly, whenever the International Time Bureau feels they should be corrected another 0.1 seconds to coordinate atomic time with the earth's rotation rate. It can be seen, that after a time, the corrections would exceed 1.0 second unless a second is added or skipped. The UT1 time scale is kept within the UTC scale $\pm0.9^S$ at all times by "leap seconds", added or subtracted as needed, at midnight on either or both June 30 or December 31, depending on how the earth's rotation rate is behaving each year. Thus, one may observe a DUT1 of $+0.7^S$ on June 30 and -0.3^S on July 1. But the UT1 time will be coordinated correctly since a full second would have been added at the end of June 30 in anticipation of DUT1 possibly exceeding $+0.9^S$ before December 31. Thus, the concern of the surveyor is only in how to correct for DUT1 at any time.

The DUT1 is only of concern for solar observations. Time need not be determined with this precision for Polaris observations, and it can be assumed that UT1=UTC.

In case a surveyor neglected to listen for the DUT1 correction or wished to check calculations made prior to the present data, the DUT1 is available for past dates from the National Bureau of Standards, Boulder, Colorado 80302 (Phone 303/499-1000).

Radio Receivers

Any short-wave radio can be used to receive the signals, the strength of the signal being somewhat proportional to the quality and price of the radio. A receiver is essential in the field for solar observations. It is also highly advantageous for Polaris observations to avoid errors and mistakes. Since price, bulk, and quality are of concern in a radio, one should be used that passes the test of being able to receive the time signals at any time and which is light and small enough to be portable. One need not spend several hundred dollars for a radio. For several years, a crystal-controlled "timeKube" has been sold by Radio Shack for $30 to $35 which has frequencies of 5, 10, and 15 MHz. Canadian sales include the 7.3 frequency. It weighs only 14 ounces and is less than 4.5 inches wide at its largest dimension. Using this timeKube, this author has never experienced a situation where the signal could not be audibly heard on at least one of the three frequencies when the battery was fresh and the radio held in such a configuration allowing maximum receipt of signal by the antenna. Experience ranges from California to Connecticut to Florida and in between. The advantages of the timeKube are its low cost with sufficiently high quality, light weight, small size and no tuning necessary. Other crystal controlled radios may be available, but if so, were unknown to the author at the time of writing of this manual.

Timing for Solar Observations

As is explained more fully in Appendix B, an error of 1^S in time for a solar observation can cause as much as 13-15" error in azimuth. It is therefore necessary to carefully control the time error. Good control is possible using a stop watch having a positive and reliable start/stop mechanism and which does not gain or lose time significantly for short runs.

The technique used is to listen to the radio announcement of time, get ready to start the watch when the voice announcement is heard, start it at the beginning of the 0th second signal, and stop it when the limb of the sun is tangent to the vertical cross-hair. The seconds read from the watch are then added to the broadcast hour and minute to yield UTC of the pointing.

Tests have shown that human reaction time in starting/stopping a stop watch is approximately $\pm 0.09^S$ standard deviation. (Reference 5). Watches reading to 0.01^S are not essential since the reading precision is lost in the human reaction time error. However, a clear scale or dial to resolve the reading to at least $\pm 0.1^S$ is necessary to control the effects of time error for third-order azimuth.

Modern digital wrist watches often have a start/stop feature, but unfortunately their button for operating this feature is too tiny to give confidence that the watch has actually been started or stopped properly. A mechanical watch with a larger button, more of a size than can be positively felt by a surveyor's finger is probably a better choice. Mechanical watches sometimes do not keep good time, however. They should be tested for gain or loss over a few minutes run by comparison with WWV time signal intervals. Usually, the error from this source is negligible if the time interval between starting and stopping is kept within a minute.

Whatever the style or brand of watch used, it is necessary to have one which runs accurately, has at least 0.1^S precision in reading, and has a button which is designed for precise human manipulation and which responds in a reliable way.

Solar observations must *never* be made by calling time to a time-keeper if precision in azimuth of 5" or better is expected. The observer (theodolite operator) *must* personally start and stop the watch. More will be said about timing techniques and precautions in Chapter 4.

Timing for Polaris Observations

An error in 1^S in time for a Polaris observation will cause a *maximum* error of 0.3" in azimuth (Polaris at culmination). There are many sources causing errors much larger than this. Therefore, split-second accuracy is unnecessary when making a Polaris pointing. A time $\pm 3^S$ is accurate enough if azimuths are to be of third-order precision. A stop watch is not essential. Nor is it necessary to have a short-wave receiver in the field. The second-hand or digital reading of an ordinary wrist watch can be read at a time signal, either in the field or before or after going to the field, and a watch correction applied to yield UTC. For example, if the watch read 4:12:15 at an announced time of 4:12:00, a constant amount of 15^S would be subtracted from each field observed time read from the watch.

An observer can read the watch himself by quickly glancing at it after making the pointing, or he can call "time" to a time-keeper at the instant of pointing to the star. Sufficient precision will result with either technique. If a watch is calibrated with the time signal in the office before going to and after returning from the field, an averaged or prorated watch correction should be used which considers the time lapses between calibrations and field observations and any discrepancy between watch corrections observed before and after the observations.

3-2 LATITUDE AND LONGITUDE DETERMINATION

USGS Map Sources

The exerpt (inset, next page) from a USGS information sheet (Reference 24) explains how to obtain maps. Besides the addresses listed in the exerpt, maps are also sometimes available in university and other libraries.

Care Needed in Scaling

For solar observations, geodetic latitude (ϕ) and longitude (λ) should be scaled from the map within ±1". An error analysis shows (Appendix B) that careful plotting of the instrument station on a 7½-minute USGS map with subsequent equally careful scaling of ϕ and λ for this plotted position, will yield this accuracy. The reader is encouraged to study the error analysis in order to appreciate the care necessary in plotting and scaling. Maps of the scale and accuracy of the 7½-minute USGS series are the minimum acceptable for ϕ and λ determination when solar observations are being employed. In some areas, the difference between geodetic position (scaled from map) and astronomic position (which should be used in the calculations) may be significant. This is further discussed subsequently.

For Polaris observations, ϕ and λ can be in error nearly a mile in position (approximately 50" in position) and the resulting error in azimuth will not exceed 1". Therefore, much less care and/or maps of scale smaller than the 1:24,000 of the 7½-minute USGS series can be used. However, as a matter of good practice, careful scaling should be done, just in case the plan is changed and solar observations prove to be more convenient, or in case ϕ and λ are later needed to ±1" for some other purpose.

INDEXES SHOW PUBLISHED TOPOGRAPHIC MAPS

Indexes for each State, Puerto Rico and the Virgin Islands of the United States, Guam, American Samoa, and Antarctica show available published maps. Index maps show quadrangle location, name, and survey date. Listed also are special maps and sheets, with prices, map dealers, Federal distribution centers, and map reference libraries, and instructions for ordering maps. Indexes and a booklet describing topographic maps are available free on request.

HOW MAPS CAN BE OBTAINED

Mail orders for maps of areas east of the Mississippi River, including Minnesota, Puerto Rico, the Virgin Islands of the United States, and Antarctica should be addressed to the Branch of Distribution, U. S. Geological Survey, 1200 South Eads Street, Arlington, Virginia 22202. Maps of areas west of the Mississippi River, including Alaska, Hawaii, Louisiana, American Samoa, and Guam should be ordered from the Branch of Distribution, U. S. Geological Survey, Box 25286, Federal Center, Denver, Colorado 80225. A single order combining both eastern and western maps may be placed with either office. Residents of Alaska may order Alaska maps or an index for Alaska from the Distribution Section, U. S. Geological Survey, Federal Building-Box 12, 101 Twelfth Avenue, Fairbanks, Alaska 99701. Order by map name, State, and series. On an order amounting to $300 or more at the list price, a 30-percent discount is allowed. No other discount is applicable. Prepayment is required and must accompany each order. Payment may be made by money order or check payable to the U. S. Geological Survey. Your ZIP code is required.

Sales counters are maintained in the following U. S. Geological Survey offices, where maps of the area may be purchased in person: 1200 South Eads Street, Arlington, Va.; Room 1028, General Services Administration Building, 19th & F Streets NW, Washington, D. C.; 1400 Independence Road, Rolla, Mo.; 345 Middlefield Road, Menlo Park, Calif.; Room 7638, Federal Building, 300 North Los Angeles Street, Los Angeles, Calif.; Room 504, Custom House, 555 Battery Street, San Francisco, Calif.; Building 41, Federal Center, Denver, Colo.; Room 1012, Federal Building, 1961 Stout Street, Denver Colo.; Room 1C45, Federal Building, 1100 Commerce Street, Dallas, Texas; Room 8105, Federal Building, 125 South State Street, Salt Lake City, Utah; Room 1C402, National Center, 12201 Sunrise Valley Drive, Reston, Va.; Room 678, U. S. Court House, West 920 Riverside Avenue, Spokane, Wash.; Room 108, Skyline Building, 508 Second Avenue, Anchorage, Alaska; and Federal Building, 101 Twelfth Avenue, Fairbanks, Alaska.

Commercial dealers sell U. S. Geological Survey maps at their own prices. Names and addresses of dealers are listed in each State index.

INTERIOR—GEOLOGICAL SURVEY, RESTON, VIRGINIA—1978

Plotting and Scaling Techniques

Only the position of the instrument station is necessary to know. The location of the ground azimuth mark need not be plotted except for the purpose of scaling the approximate azimuth from the instrument station to it. Since the determination of ϕ and λ for a solar observation requires a high degree of care, the following discussion will focus on the techniques required for that care. For Polaris observations, this care is not required, but recommended.

If the surveyor is fortunate enough to have an instrument station already plotted on the map (NGS horizontal control station, etc.), plotting of position is unnecessary. However, this is usually not the case and the surveyor must determine the position of a traverse or other station set as a part of the new survey. With care, a point can be plotted to perhaps ±0.01 to 0.02 inches by X and Y components from existing ground features such as roads, fences, and other well defined horizontal features. This means that the station must be known to ±20 to 40 feet on the ground, with respect to these same features. Such measurements can usually easily be made in the field by stadia or pacing along approximate "true" N-S and E-W directions.

The station is next plotted on the map using the 20-scale of an engineers scale (1"=2000' scale map) and cardinal directions for the N-S and E-W components determined by parallelism with the map margins. The next step is to determine ϕ and λ. This is done by carefully scaling from the point, perpendicularly, to the map margin lines, again estimating to ±0.01 inches. The 10-scale can be conveniently used for this. A component of ϕ and λ is thus determined in map inches from the lines of exact value (the margins) to the station. Next, the distance in inches is similarly scaled along the margin lines between 2½-minute marks for both ϕ and λ. After these four map measurements are made, ratios are used to determine the instrument station position. The example illustrated in Figure 3.1 demonstrates the calculation. The map measurement, in inches, from map margin to the station was 2.06 and 2.88, in ϕ and λ, respectively. The map measurement, in inches, for 2½ minutes was 7.62 and 5.86 for ϕ and λ, respectively. The calculations yield:

$$\frac{2.06}{7.62} (150") = 40", \qquad \frac{2.88}{5.86} (150") = 73.7"$$

to be added to the value of ϕ and λ represented by the east margin line and south margin line, respectively. Thus,

$$\phi_{sta} = 40°00'00" + 40" = 40°00'40"$$
$$\lambda_{sta} = 83°00'00" + 74" = 83°01'14"$$

Care must be taken that components of ϕ and λ are measured perpendicular to the margins. Mechanical parallel rules or large 90° triangles should be used with appropriate care. In adding the components of ϕ and λ as computed above, the fact that longitude increases to the west and latitude to the north must be observed. Often it is easier to scale from the north and west margins in which case the computed components would need to be subtracted from those margin line values.

It should be realized that any error in plotting or scaling ϕ or λ will result in a systematic error in the final average azimuth. No matter how precisely scaling is done, a mistake or poor location of the station will yield the same error in the azimuth each time it is calculated. Precision in scaling and in agreement of computed azimuths does not mean that the results are accurate. This possible systematic error can only be controlled and kept to a minimum acceptable by determining ϕ and λ within approximately ±1".

Astronomic vs. Geodetic Position

Unfortunately, geodetic and astronomic position are not the same. The difference is caused by the deflection of the vertical (local gravity anomalies--See Section 1-2). The ϕ and λ that should be used in the equations for astronomic azimuth are the astronomic ϕ and λ. But, a USGS map or NGS control data sheet gives geodetic ϕ and λ. For Polaris observations, the difference would cause an error in astronomic azimuth of much less than a second in nearly all areas, since the deflections rarely exceed more than a few seconds. Thus, even for observations in the highest mountains of the western U.S., where deflections of the vertical are maximum, geodetic positions can be used in Polaris azimuth calculations when only third-order azimuths are sought. However, unless geodetic positions are corrected to astronomic, solar azimuths cannot usually be considered of third-order equivalent accuracy. The corrections can be determined from data furnished by the NGS and are usually certain within an error of less than ±1". Thus, the accuracy achieved by careful map scaling is usually maintained in the astronomic position.

Figure 3.1 Map Scaling of Latitude and Longitude

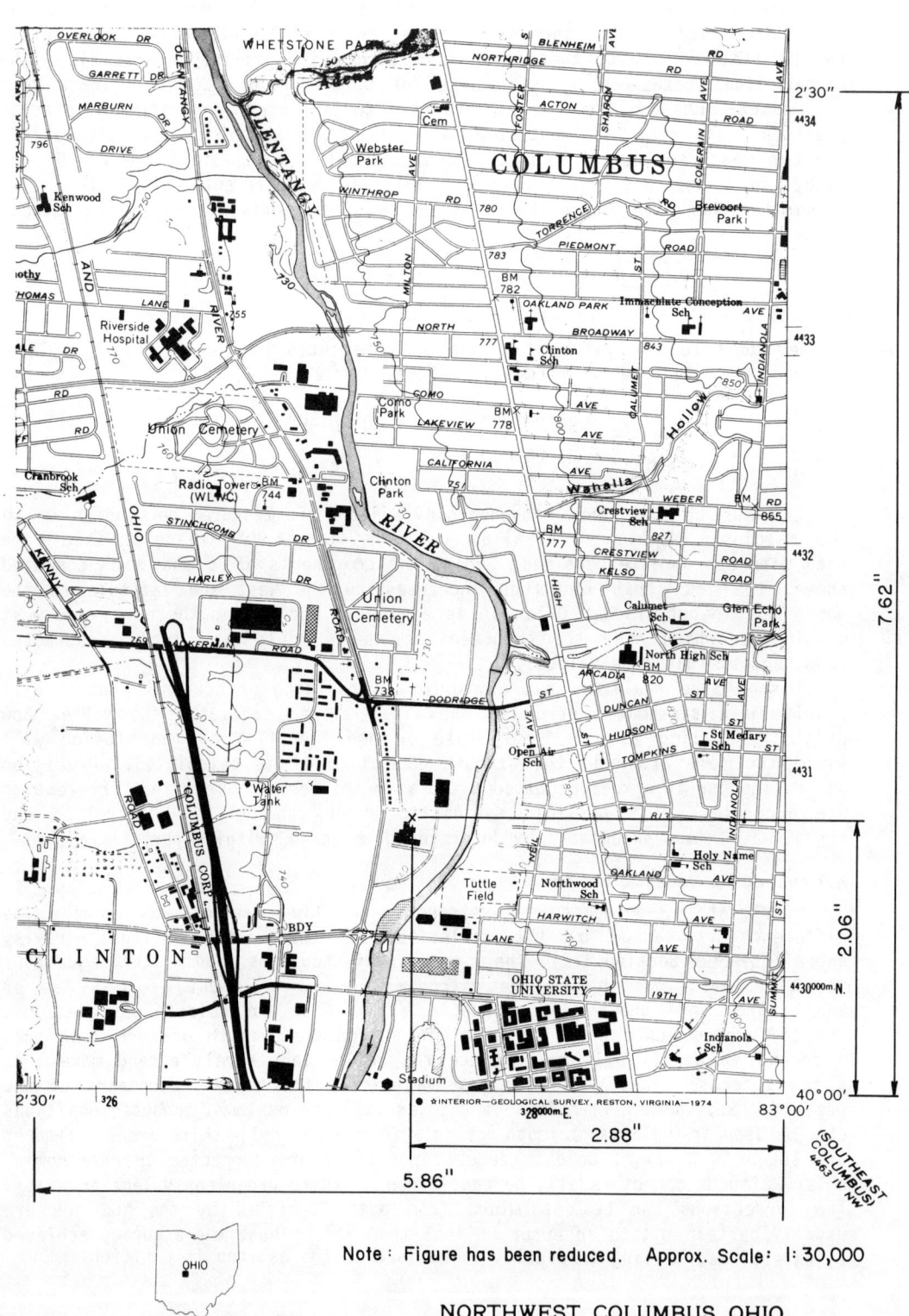

Note: Figure has been reduced. Approx. Scale: 1:30,000

NORTHWEST COLUMBUS, OHIO

An illustration of deflection of the vertical is shown in Figure 7.1. Section 7-1 includes additional explanation. The theory pertaining to ϕ and λ is as follows:

$$\xi = \phi_A - \phi_G = \text{deflection of the vertical in the meridian} \quad (3.2)$$
$$(\text{N-S, or latitude})$$

where

ϕ_A = astronomic latitude,

ϕ_G = geodetic latitude.

$$\eta = (\lambda_A - \lambda_G) \cos \phi_G = \text{deflection of the vertical in the} \quad (3.3)$$
$$\text{prime vertical (E-W, or longitude)}.$$

where

λ_A = astronomic longitude,

λ_G = geodetic longitude.

Thus,

$$\phi_A = \phi_G + \xi \quad (3.4)$$

and

$$\lambda_A = \lambda_G + \frac{\eta}{\cos \phi} \quad (3.5)$$

Also, in terms of the Laplace correction (See Section 7-1),

$$\lambda_A = \lambda_G + \frac{\text{Laplace Correction}}{\sin \phi} \quad (3.6)$$

In order to make the conversion using these equations, the surveyor needs to have values for ξ and η (or ξ and Laplace correction) at the observation station. At the time of writing of this manual, the NGS had prepared deflection prediction programs which output these values. The surveyor needs only to write to the NGS (Chief: National Geodetic Information Center, 6001 Executive Boulevard, Rockville, Md. 20852) and ask for the output. The surveyor must furnish the input (observation station latitude, longitude, elevation, as scaled from a map).

Table 3.1 shows output the NGS furnished for stations requested by the author. The LAT and LON columns are scaled geodetic ϕ and λ. Under the DEFLECTION heading are the ξ (or, $\phi_A - \phi_G$) values, the $(\lambda_A - \lambda_G) \cos \phi$ (or, η) values and the estimated error (standard deviation, SIG) in each.

TIME, GEOGRAPHIC POSITION, ANGLE MEASUREMENT

TABLE 3.1
Astrogeodetic Deflection Data

STATION NAME	LAT ϕ_G DD MM SS.SS	LON λ_G DDD MM SS.SS	$\phi_A - \phi_G$ DEFLECTION N-S	SIG	$(\lambda_A - \lambda_G)\cos\phi$ NAD27 E-W	(SEC.) SIG
FLORIDA 2	28 8 17.00	81 2 28.00	3.7	0.7	-2.6	0.7
FLORIDA 1	30 11 4.00	85 43 59.00	4.0	0.6	-1.8	0.6
LOUISIANA 2	29 57 19.00	89 58 47.00	5.7	0.7	-0.0	0.8
LOUISIANA 1	30 24 36.00	91 11 10.00	-0.7	0.6	1.2	0.7
PENNSYLVANIA	40 48 36.00	77 51 22.00	-0.1	3.3	-1.7	3.3
OHIO 4	39 49 28.00	82 31 54.00	0.6	0.6	-0.9	0.5
OHIO 3	40 0 40.00	83 1 14.00	0.6	0.4	-6.7	0.4
OHIO 1	40 0 16.00	83 1 17.00	0.5	0.4	-6.9	0.4
OHIO 2	40 6 20.00	83 1 10.00	0.7	0.4	-5.3	0.4
MICHIGAN	43 13 48.00	84 3 35.00	4.3	0.7	-5.8	0.7

Using Table 3.1 output as an example, the astronomic positions for "OHIO 2" would be:

$$\phi_A = (40°06'20") + 0.7" = 40°06'20.7" \tag{3.4}$$

$$\lambda_A = (83°01'10") + \left(\frac{-5.3"}{\cos\phi}\right) = 83°01'03.1" \tag{3.5}$$

An alternate method to estimate the corrections from geodetic to astronomic position is to interpolate for $\phi_A - \phi_G$ and $\lambda_A - \lambda_G$ between control stations listed in Appendix D. This can be done either by plotting isolines in the same manner used for Laplace corrections as explained in Section 7-1, or by simply investigating the values at the stations nearest to the observer's station. The uncertainties in this method are likely to be somewhat greater than allowing the NGS to make a specific estimate for each station using their deflection prediction program. However, for the equivalent of third-order, Class II azimuths, the data in Appendix D, if properly used, should suffice. Thus, the surveyor has complete data included in this manual for avoiding this systematic error. The reader is encouraged to determine these values using Appendix D, then write NGS for their independent estimates, in order to check the precision of using the interpolated values from the Appendix.

Appendix B will include a discussion of the possible error in solar azimuth caused by the error in ϕ and λ, including both the map scaling error and conversion to astronomic position using NGS data and the formulas introduced here. As a preliminary observation, it might be noted that the error values (SIG) are in the order of tenths of seconds in Table 3.1, except in the mountainous area of Pennsylvania. For most areas, the uncertainty in position should not be enough to lose the third-order accuracy in solar azimuths. More will be said on this point in Appendix B.

3-3 ANGLE MEASUREMENTS

Only an optical theodolite should be used for astronomic observations for azimuth if third-order accuracy is to be expected in results. Although a 1"-reading instrument is recommended, such results can generally be achieved using a repetition instrument with 10" or 20" direct reading (estimation perhaps to 2" to 5"), if sufficient repetitions are made. There are several errors more significant than the reading error and this is why a good repetition instrument is acceptable. The discussion of theodolite errors in Appendix B explains the above more completely.

Fundamental to the angle measuring process is that the operator must be thoroughly familiar with the instrument concerning proper set-up, reading, pointing, focussing, leveling, and other procedures. The instrument should be in good adjustment, have clean optics, not have "play" in the motions, and be mounted on a sturdy tripod of appropriate manufacture. This manual is not intended to provide training in theodolite operation. It is assumed that the reader understands the importance of knowing the instrument.

The procedures and precautions unique to astronomic azimuth data collection will be discussed in the next chapter.

Figure 3.2 Surveyor Making a Solar Observation

CHAPTER 4

FIELD OBSERVATIONS

4-1 PLANNING AND PREPARATION FOR OBSERVATIONS

Choice of Celestial Body

Polaris is usually observed at night or at dusk. It can be seen through the telescope in daylight only when the sky is very clear and when one knows where to point the telescope. A preliminary calculation, using planned observation time, can be made to know the approximate zenith angle to set on the theodolite circle. (This calculation is explained in Chapter 6). If this procedure is used, rather than waiting until dark, the surveyor must consider the added time necessary for this calculation. Also required is careful focussing of the objective lens and some means to determine where to scan the sky, horizontally, to bring the star into the telescope's field of view. Some may feel that it is easier and less time-consuming to wait until dark. But, then additional problems ensue with battery attachments and illumination of targets, not to mention the fundamental hazards, inefficiency, and inconvenience of working at night. For these reasons, the sun has advantages over Polaris as a choice, mainly in that it can be sighted during daytime hours and is easily visible, even during partly cloudy or somewhat hazy weather. The main disadvantage of using the sun is that latitude, longitude, and time must be determined more accurately. The computations are only slightly more complex. This is not a significant disadvantage, especially if computations are programmed.

In most of North America, the vertical angle to Polaris is large enough that small plate bubble centering errors cause significant azimuth errors, whereas the sun can be sighted when much lower on the horizon, thus controlling the effects of bubble centering errors. This is a consideration of more significance than may be realized. The error analysis in Appendix B explains these errors in detail. Suffice to say here that this disadvantage of Polaris observations is one of the factors that cause solar observations to be comparable in accuracy to Polaris observations when considering third-order equivalent accuracies.

All considered, if the techniques are learned properly, solar observations are recommended by the author for third-order astronomic azimuth in most areas not affected by large gravity anomalies and uncertainties in these deflections. The reader should try to gain equal skill and confidence in both Polaris and solar observations so as to have the flexibility to use either as circumstances dictate, and to possibly use one method to check the other in certain analysis and troubleshooting situations.

4-1 PLANNING AND PREPARATION FOR OBSERVATIONS

Planning Time of Observations

The best daytime hour for a Polaris sighting is late evening because the star appears clearer as night falls. But, a sighting can be planned just before sunrise if desired. Polaris can actually be sighted at any hour angle, day or night, if the sky is clear enough. This author found it at 3 P.M. in November, more than two hours before sundown, on a clear day. If a morning observation is planned, the surveyor must always take the risk of the sky becoming too bright to see Polaris clearly as the sun begins to rise. This is why evening observations are better.

Solar observations should not be made within approximately three hours of local noon (based on standard time) because of higher errors in azimuth due to a weaker solution of the spherical triangle and other factors. The best time for a solar observation is just after the sun rises and is high enough to be clear of morning fog and clouds. Refraction of the atmosphere has no effect on precision since vertical angles are not measured and used in the formula. However, the larger the vertical angle, the higher the error in azimuth caused by random errors in plate bubble centering. Therefore, the sun should be sighted as low on the horizon as possible.

Table 4.1 shows the times of sunrise and sunset. The sun changes vertical angle approximately 10° per hour of time just after rising and just before setting. This should help in planning when to be set-up for solar observations.

Whether the surveyor prefers morning, evening, or afternoon work, knowing both Polaris and solar techniques as explained in this manual affords much flexibility--the chance to use one method when the other fails, check one observation against another, mix Polaris and solar, etc. For example, if Polaris is planned in the morning and the sky becomes too bright to complete the observations as the sun rises, the surveyor needs only to place the solar eyepiece filter on the theodolite and point to the sun to complete the measurements. Likewise, as the sun sets and solar observations become difficult at the conclusion of a set of observations, the surveyor can do a preliminary calculation to find Polaris and point the telescope north to complete the measurements. The reader will notice that the January 14, 1982 solar and Polaris observations used in the upcoming examples were taken at the same instrument station only minutes apart. Some surveyors may wish to use one method to check the other in this way until confidence is gained in both. The point is, that knowing both solar and Polaris techniques gives much flexibility and weather then becomes the surveyor's only real problem.

Weather Conditions

Weather reports should be taken into consideration. Often, it is good judgment to make observations for azimuth while the sky is clear rather than risk the weather the next day. On the other hand, there might be a high probability of clear skies the next morning and waiting might be better than fighting clouds at quitting time. In the U.S., weather information, when available in a community is broadcast over frequencies of 162.40 or 162.55 MHz. Radio Shack sells at least two versions of a weather radio tuned to these frequencies. Weather bands are available on many portable radios. Such a radio can help minimize the frustrations caused by uncertainties of weather.

The view of Polaris is more easily blocked by clouds, pollution, and haze than is the view of the sun. One cannot expect to find Polaris in the daytime

Table 4.1

TIMES OF SUNRISE AND SUNSET

(Local Standard Time)

Date	20° Latitude Rises AM	Sets PM	30° Latitude Rises AM	Sets PM	40° Latitude Rises AM	Sets PM	50° Latitude Rises AM	Sets PM
Jan. 1	6:34	5:30	6:57	5:11	7:22	4:45	7:59	4:06
Feb. 1	6:36	5:51	6:50	5:36	7:09	5:19	7:35	4:53
March 1	6:20	6:04	6:27	5:59	6:35	5:51	6:46	5:41
March 21	6:02	6:12	6:03	6:12	6:03	6:12	5:56	6:18
April 1	5:53	6:14	5:51	6:18	5:45	6:24	5:37	6:32
May 1	5:31	6:24	5:18	6:36	5:01	6:54	4:36	7:19
June 1	5:21	6:36	5:00	6:56	4:33	7:22	3:55	8:00
June 22	5:21	6:43	4:59	7:05	4:31	7:33	3:51	8:13
July 1	5:24	6:43	5:02	7:05	4:34	7:33	3:55	8:12
Aug. 1	5:35	6:37	5:18	6:54	4:57	7:14	4:30	7:42
Sept. 1	5:44	6:15	5:36	6:23	5:27	6:32	5:16	6:41
Sept. 23	5:50	5:55	5:49	5:56	5:48	5:57	5:48	5:56
Oct. 1	5:50	5:48	5:53	5:46	5:55	5:43	6:00	5:38
Nov. 1	6:01	5:26	6:13	5:14	6:28	4:59	6:51	4:37
Dec. 1	6:19	5:19	6:38	5:00	7:02	4:36	7:38	4:01
Dec. 22	6:31	5:26	6:51	5:04	7:19	4:38	7:56	4:00
Jan. 1	6:34	5:30	6:57	5:11	7:22	4:45	7:59	4:06

(Source: Reference 25 and 26)

To use the table for a described rising or setting time:
1. Interpolate for the date of observation in the column closest to the latitude of the observation station.
2. Interpolate in the same manner, using the next higher or lower latitude column.
3. Use the above values and interpolate between them for observation latitude (nearest degree).
4. Adjust the time from step 3 by four minutes for each degree of longitude the observation station lies east or west of the central meridian of the time zone (See Figure A.14 for time zones) adding if west and subtracting if east of this meridian.
5. Add one hour if necessary for daylight savings time to have actual watch time.
6. Extrapolation can be used to determine times for latitudes a few degrees outside the range given in the table.

4-1 PLANNING AND PREPARATION FOR OBSERVATIONS

unless the sky is extremely clear and bright. Furthermore if it is hazy, it must be essentially dark (not dusk) before Polaris can be observed. However, the sun can often be sighted through thin clouds. Neither the sun nor Polaris can be observed when it is overcast. It is inadvisable to try to complete observations when skys are "mostly cloudy" unless it would be too expensive to return to the site for another observation.

Because of the constant threat of weather changing in most areas of the country, it is advisable to always have equipment on hand and to make astronomic observations on any job at the first convenient opportunity. Unnecessary added expense is realized when the surveyor has planned to make observations as the last measurement on the project, is caught in the rain, and must then return another day or wait the storm out.

Location of Instrument Station

The instrument station to be used for an astronomic observation should meet the following requirements. First, it should be located where there is a distant, permanent azimuth mark clearly visible. Second, it should be a station at the end of a long line in the traverse so as to introduce the best possible starting azimuth accuracy into the ground control system. Third, it should be located to maximize the possibilities for astronomic observations. This latter point must take into consideration the azimuth of the sun at the time of the year when the observation is to be made, both morning and afternoon, and the vertical angle to it and Polaris. A location would be undesirable if Polaris was obscured by a tall building north of the station or if trees were likely to block the sun morning and afternoon.

Other considerations in selection of the instrument station are safety from traffic, safety for curious passing motorists, other possible hazards, and convenience in transporting and using equipment such as radio and battery attachments.

One final consideration is the effect of the observations on curious residents and others. When surveyors are present, curiosity and questions are usually inevitable. When the surveyors begin to tune in time signals and gaze at the stars and sun, curiosity is sometimes aroused to the point of someone calling the police or stopping to ask a lot of questions, all of which delays observations and can cause complete failure of the mission if clouds are moving in or the sun is rapidly setting. Barking dogs can add to the confusion at night near farm houses. For these reasons, some isolation in location of the instrument station is also advisable.

The best instrument station would be located at one end of the longest traverse leg, in an open and accessible area away from traffic and potential disturbances, with a clear line of sight to a good azimuth mark or traverse station and to the celestial bodies to be sighted.

Selection of Ground Azimuth Mark

The azimuth mark can be either another survey monument or a prominent terrain feature on the horizon. Objects such as radio towers, church spires and water tank lights generally make good targets. One advantage of using such objects is that there is no need to erect a target, which saves time and eliminates the target centering error. Another advantage is that such a target is probably less subject to disturbance than a ground monument. A third advantage when a light on top of a water tank or radio tower is used, is that the target doesn't require illumination and can usually be observed day or night. However, there are disadvantages in using such targets. For instance, confusion and mistakes may result in future observations if there is more than one radio tower in the area or if there are two lights on top of a water tank, or changes are made in the lighting system. Or, perhaps the light isn't placed at the geometrical center of a spire or tower causing a surveyor to measure to a different point in the day than at night. Another disadvantage of towers is that they may sway in the wind if inadequately supported.

It should be kept in mind that the target has not served its purpose upon completion of the azimuth calculation. It will still need to be referenced to survey lines and possibly used in the future to orient other adjacent surveys. Therefore, a target must be selected that is not very likely to be mistaken in identification. Sometimes targets such as church spires are difficult to describe clearly as to the exact point sighted, especially if they are not vertical or symmetrical. Clear descriptions are important.

A target must be "well defined". This means that the image of the line or point being sighted must not be more than two or three times wider than the vertical cross-hair image, and preferably be straight and vertical if linear such as a church cross, lightning rod, or range pole. This characteristic of size and shape is important as it affects the random pointing error to the target.

Distance to the target is important in its effect on target definition, vertical angle, distortion due to heat waves, and effects of centering errors. For example, a radio tower only a few hundred feet away would probably be unsuitable because the light would be too big when viewed through the telescope, and the vertical angle to it could cause appreciable azimuth uncertainties due to bubble centering errors. The vertical angle to the target should not exceed about 5 to 10 degrees, so as not to introduce significant errors into the pointings due to bubble centering errors.. If a target is too distant, however, it may not be clear on a hazy day and heat waves could interfere with sightings. Whenever man-made targets must be erected behind or over monuments, distance is a consideration since centering errors result in azimuth errors and these errors are reduced by longer distance.

If sight distances are very short or if the target is not well defined or is poorly centered, precision of the entire observing and computing procedure could be lost due to target centering or pointing errors. For example, a one-inch diameter range pole subtends an angle of approximately 5" over a sight distance of 3400 feet. Error control is an important part of any measuring operation. These aspects are discussed more fully in Appendix B. Hopefully, the guidelines above will help in selecting good stations so as to minimize errors.

In summary, it can be stated that the best azimuth mark is one that is visible from the observation station, well defined, on the horizon at a long distance (but not so far as to be obscured by haze or heat), observable day or night, easily and clearly describable, as permanent in position as possible, and not highly subject to mistake in identification in the future.

Map Aids

USGS maps are used to scale latitude and longitude prior to computation of observations. Sometimes they are used before going to the field to scale

4-1 PLANNING AND PREPARATION FOR OBSERVATIONS

these values if a sighting is to be made to Polaris prior to dark. In such a case, the values of ϕ and λ are used to determine the zenith angle and approximate azimuth as explained in Section 6-1. The map can also be used for scaling the angle between a selected azimuth mark and the "true" north direction through the instrument station. This angle, scaled with a protractor, is very useful when searching for Polaris during the day.

Maps are also useful for identifying any NGS control that might be in the vicinity, scaling state plane coordinates, traverse planning, determining magnetic declination for possible use in locating Polaris with the aid of a compass, planning travel to control stations, and general planning of field work.

Equipment and Personnel

The following are needed in the field for astronomic observations for azimuth:

Solar Observations

1. Optical theodolite with tripod.
2. Target system (unless prominent terrain feature is to be used as a mark).
3. Short wave radio or timeKube with good batteries.
4. Solar eyepiece.
5. Stop watches (2 preferred) plus good batteries if electronic.

Polaris Observations, Daytime

1. Optical theodolite with tripod.
2. Target system (unless prominent terrain feature is to be used as a mark).
3. Wrist watch with sweep second hand or digital seconds calibrated to UTC.
4. Ephemeris to compute approximate vertical angle and azimuth.
5. USGS map to scale approximate azimuth from prominent map feature.

Polaris Observations, Night

1. Optical theodolite with tripod.
2. Target system (unless prominent terrain feature is to be used as a mark).
3. Wrist watch with sweep second hand or digital seconds calibrated to UTC.
4. Battery attachments with cords, lights, and batteries to illuminate theodolite circles, cross-hairs, and target system (or other illumination method).
5. Flashlight to aid in setting up, observing bubble centering, safety.

For solar observations, it is convenient to use two stop watches, starting them simultaneously and using one for the left limb and the other for the right limb sighting. Alternately, a watch with a "split-time" system can be used. Some acceptable timepieces are shown in Figure 4.1.

It is *very important* that an eyepiece filter specifically made for *solar* observations be used. Other ordinary darkening lenses **must not** be used since damage to the eye would result because the bright rays of the sun are insufficiently blocked out. A solar eyeiece is very dark and actually appears to be totally black when held toward any light except something as bright as the sun.

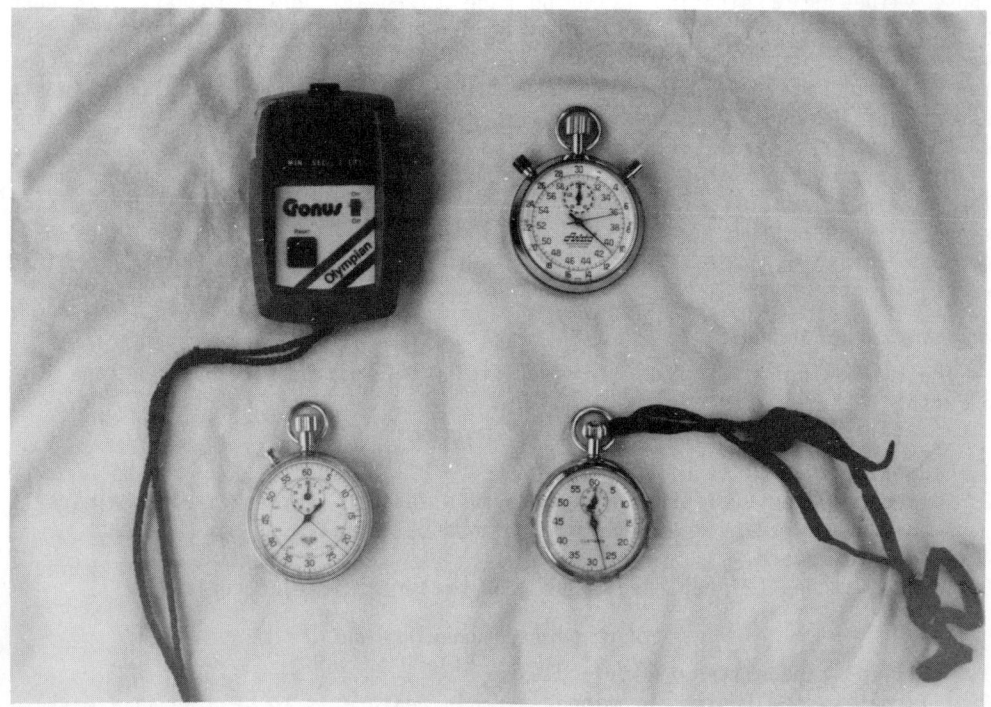

Figure 4.1 Stop Watches for Solar Observation

Although the short wave receiver is not essential to have in the field for Polaris observations, some may prefer to have it, especially if the watch is apt to gain or lose more than two or three seconds between the time of calibration and observation. However, the watch gain or loss can be corrected by calibrating before and after observations and some may feel it is handier to leave the radio in the office or motel room.

The ephemeris tables are not essential to have in the field if calculations for approximate vertical angle and azimuth of Polaris are made prior to leaving the office and *if* that planned observation time is kept. It is advisable, however, to have them in case changes are made in time or date of planned observations.

Illumination at night is often a problem, especially if cloud cover or other difficulties delay observations. It is always advisable to have plenty of spare batteries for all illumination systems. One should always check to be sure all lamps and cables are present in illumination systems. If specially made illumination systems are not available, an ordinary flashlight works satisfactorily to read circles and illuminate cross-hairs. Auto headlights or flashlights can be used to illuminate ground targets. However, this sometimes causes an apparent eccentricity error on a range pole if it is illuminated from the side because of a shadow on the opposite side.

Only **one** person is needed to make any astronomic azimuth observation since time and theodolite measurements are all done by the observer. However, an assistant to help set up and record usually makes operations work more efficiently. The observer should be a person fully knowledgeable in theodolite operation, error control, and astronomic details described in this manual, and have the reflexes and skill to point to the sun and use the stop watch with precision.

4-2 FIELD NOTES AND DATA

Field notes must contain the theodolite circle readings when sighting both the terrestrial and celestial targets and the time of each celestial pointing. Observations are made in sets. The notes should indicate whether the telescope was direct or inverted. For solar observations, the notes must indicate which limb of the sun was sighted at each pointing. If a second hand of a watch is calibrated with the time signal for Polaris observations, the watch and propagated signal readings should both be given in the notes, preferably both before and after the observations. For solar observations, the DUT1 correction must be included in the notes. Clear and complete instrument station and azimuth mark descriptions should appear. Instrument type, personnel, weather, date, job or project name, and location should be included. If the date is different locally than for UTC, this should be reflected in the notes. When local time is read from a watch, the fact that it is standard or daylight savings time should be indicated. Finally, for clarity, notes for solar observation should show whether the observations were in the morning or the afternoon.

Theodolite readings for one-second instruments need not be read to tenths of seconds since this precision is lost in pointing and other errors anyway. But, one-second readings are appropriate. Normally, repetition instruments should be read to one-tenth of the least count, depending upon what precision in reading is revealed by instrument tests.[3] The precision of instruments is sometimes a little better than stated in manuals and it is better to read to a tenth of the least count than to lose precision in results by reading too crudely. Stop watch readings should be to the nearest tenth of a second for solar observations. Times to the nearest second are good enough for Polaris observations. Always record readings to the correct number of figures so that round-off errors do not occur in calculations.

The sample set of notes shown in Figure 6.1 reveal all of the above and give a suitable format for the observations. It is noted that the angle between the ground mark and the celestial body is obtained by subtracting circle readings on each pointing, and that circle settings are changed on the direction theodolite between sets by approximately 180/n degrees where n is the planned number of sets. (One "set" is one pointing with telescope direct, one with telescope inverted or one face left, one face right pointing). Reduction of data will be explained in Chapter 6. The mechanics of making the observations is explained next.

---

[3]Reference 8, Chapter 6, gives a thorough discussion of errors in angles using theodolites. Chapter 4 of that reference text also shows how reading and pointing errors can be tested for most theodolites.

4-3 MAKING THE OBSERVATIONS

Solar Observations

The theodolite is centered over the station and leveled in the usual manner, being careful to set up in such a way, if possible, as to avoid climbing over a tripod leg between pointings to the sun and ground target. Close attention should be paid to centering the plate bubbles when pointing to either target.

The ground azimuth mark is first sighted, telescope in the direct position. The horizontal circle setting of a direction theodolite should be set first to a value close to 0°. The minutes and seconds will be arbitrary, whatever results when the reading is completed. Proper use of a direction theodolite does not include attempting to set the minutes and seconds to exactly zero. A repetition instrument would be set to 0°00'00" on the backsight. Section 4-4 includes additional comments regarding use of a repetition theodolite. The notes of Figure 6.1 show an observation taken January 14, 1982 on the Louisiana State University campus. Note that the initial backsight reading was 0°26'06" to the ground azimuth mark (WBRZ radio tower) with the telescope direct.

Next, the solar eyepiece is attached, the radio signal is tuned, the stop watches are readied, the circle motion of the theodolite is unclamped, and the theodolite is pointed toward the sun. To avoid excessive gazing toward the sun, this author has developed a technique of using the rough peepsight to quickly point approximately to the sun, then getting the eye behind the eyepiece (Remember the *solar eyepiece filter must be attached at this stage!*), and finding the sun's darkened image by scanning left and right and up and down with both motions unclamped. If the right eye is behind the eyepiece, closing the left eye avoids strain on it from the sun's brilliance and rays. When the sun is found through the telescope, it can be readied for observation by placing one limb close to the vertical cross-hair. A minor amount of objective lens focus may be necessary in order to see the sun's disk as clearly as possible.

If and whenever the voice announcement of time is heard during the above process of finding the sun's image, the observer should prepare to start the watches. Paused and ready with a watch in each hand, they are started the instant the distinctive tone of the 0th second of the announced minute is heard. After starting the watches, the observer returns to the process of finding the sun's image and aligning a limb of it close to the vertical cross-hair. The purpose of using two watches is that one will be used for the left limb pointing, the other for the right. The recorder must listen carefully for the announced UTC hour and minute and record these in the field notes. For the example of Figure 6.1 the first time at which the watches were started was 22:29 hours and minutes UTC. It is usually most convenient for the observer to have the stop watches attached to cords looped around his neck. This keeps them close-by for ready use and generally safe from damage.

Assuming that the watches have now been started and the observer has found the sun, he next aligns either the left or right limb of the sun so that it will cross the vertical cross-hair in a few seconds time. A final step is to check the plate bubble and quickly center it if necessary. The first "observation" is now ready to be made. As performed in the example, the vertical cross hair was aligned just to the right of the left limb of the sun. With an *erect* image telescope, morning or afternoon, the sun will always be moving to the right, so that this now puts the sun in a location to cross the left limb within a few seconds. Figure 4.2 shows what the observer sees. With motions unclamped and hands off of the theodolite, the observer readies the watch for stopping. Concentrating intently on the sun as it is about to cross the cross-hair, the observer stops one watch the instant the sun becomes tangent to the cross-hair. He then reads the minutes (if any) and seconds from the watch. The seconds should be recorded to tenths. In Figure 6.1, the time was 37.0 seconds. If the stop watch went past one minute, this should be carefully communicated to the recorder so that no mistake is made in minutes. Some surveyors may wish to have a separate column in the notes for UTC hour-minute and another for stop watch minute-seconds so as to minimize the possibility of mistakes in minutes of time. The last step in the observation is to read the horizontal circle of the theodolite and record this value in the field notes. For the example here this value was 236°10'58". The observer may wish to lower the telescope from its inclined position toward the sun before reading the horizontal circle. This avoids continuing to look toward the sun any more than is necessary, avoids having to make the reading in an uncomfortable position, and probably assures better precision in the reading as a result.

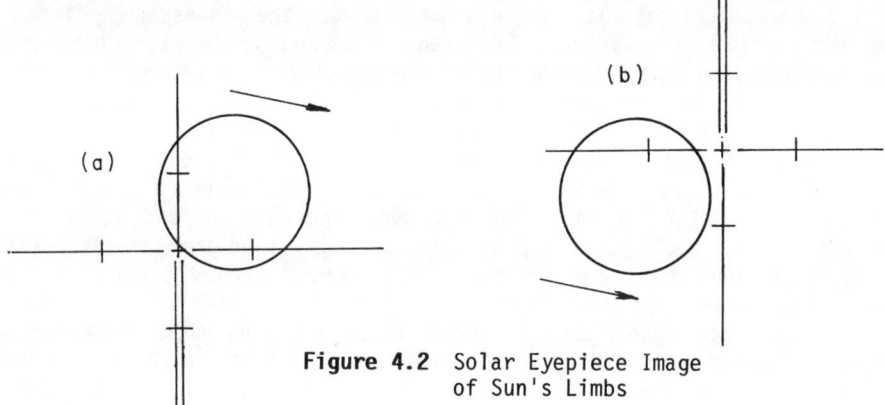

Figure 4.2 Solar Eyepiece Image of Sun's Limbs

This completes half of one "set" of observations. With the solar eyepiece still attached, the observer next plunges the telescope and sights the sun again, this time centering the image near the other limb of the sun. In the example, this would be the right limb. What the observer sees is shown in Figure 4.2(b). Actually, in this position, the vertical cross-hair cannot be seen until almost the very instant it becomes tangent because the area around the sun is so black. It takes a few trials to learn how to anticipate when the sun is about to become tangent. Beginners should allow themselves a couple of minutes to observe this movement before making their first timing. The observer next goes through the procedure of checking and centering the plate pubble, removing hands from the instrument and readying the stop watch. If more than a couple of minutes have passed on the watch and it is convenient to do so, the observer may wish to start the watch on a new time signal. This assures maximum accuracy in case a mechanical watch is losing or gaining time. The watch is stopped the instant tangency of the cross-hair and sun's limb is observed. The time on the watch is next read and recorded, the minutes especially noted so that no mistakes occur. In the example used here, we can assume that 2:12.0 minutes and seconds were read from the watch, which was added to the 22:29 hours and minutes when the watch was started to get 2:31:12.0 as the observation time. Or, perhaps the watch was restarted at 22:31 UTC and then ran only 12.0 seconds until stopped.

The horizontal circle reading is next observed and recorded as described previously. It was 56°57'30" as observed in this example. Note that it is 180° different than the first observation due to the fact that the telescope is plunged. This should be a routine check, so as to avoid mistakes. Observe also that, if 180° is added, the circle reading is larger than the first observation because the right limb was sighted the second time. It will always be larger on the second observation in a set if the left limb, then right limb sequence is used. If right, then left is used (or if telescope is inverted image) the second reading will be smaller. This check should be routinely made before unclamping the motion so as to avoid mistakes in reading. The exact difference won't be known because the sun is moving, so the check does not assure that no mistake occurred, but it will discover large reading mistakes and whether the correct limb of the sun was sighted the second time.

The horizontal circle is next unclamped, the telescope pointed toward the terrestrial target, the solar eyepiece removed, the plate bubbles checked and centered, and the theodolite circle read and recorded, telescope still inverted. In the example, this value was 180°25'42". This should compare favorably with the initial backsight reading by 180° plus or minus any random errors due to bubble centering and pointing and also due to inclination of the horizontal axis or error in line of sight adjustment. As work progresses through each set, the discrepancy between the direct and reverse readings on this target should be fairly consistent. Observing the consistency provides a check on operator care in minimizing random errors. More will be said about this later.

This completes one "set" of observations, called 1D and 1R (one-direct, one reverse telescope). Note that the limb L (left) or R (right) was given in the notes and that UTC was noted as the time reference. To commence Set 2, the instrument remains pointed to the ground target, telescope inverted, and the circle reading shifted approximately 180°/n where n=number of planned sets. In the example, it was advanced approximately 45° since 4 sets were planned. The collimation marks are next aligned, the circle read, and the value recorded. In the example, it was 225°31'13". From here on, the sequence is as for the first set. The watches are started, solar eyepiece attached, pointings to the left and right sun limbs made, watch stopped each time, times and circles read and recorded, and closure made back to the terrestrial target. For Set 2, the sequence of readings (Figure 6.1) was 225°31'13" (angle), 22:35:08.9 (time), 102°04'23" (angle), 22:36:13.0 (time), 282°46'52" (angle), 45°31'31" (angle), each time making appropriate checks on bubble centering, minutes of time, and angular values for solar and terrestrial target pointings.

The third set commences with the telescope direct, as it was when returning at the end of Set 2. The circle is advanced another 45° (approximate) and the sequence of pointings and timings repeated.

When completed, the surveyor should have data sets fairly evenly distributed around the theodolite circle, an equal number direct as inverted, an equal number to the left as to the right limb, with one each direct and inverted and one each left and right limb in each set.

Note that, in the example, the discrepancy between the direct and inverted readings on the terrestrial target is a fairly consistent 18" to 27" spread, the direct value, always being larger. This reveals good operator skill and that the discrepancy is caused primarily by instrument error which is compensated in the direct/reverse process. By observing these discrepancies a skilled recorder will be able to check for mistakes in minute/second readings when closing back to the ground target after the second set has been made or through prior knowledge of the instrumental error effects on the particular instrument.

One other important data item should be observed before leaving the site, or at least during the day sometime. This is the DUT1 correction to UTC. This correction was explained in Section 3.1. On January 14, 1982, DUT1=0.0 seconds. (There were no double clicks heard on the radio.) After describing the instrument station and ground target, and adding the date and other miscellaneous information in the notes, the observation is completed. Note that the observation of four sets in the example took approximately 15 minutes from initial to last solar pointing. This is typical when no problems exist and observers are experienced.

A few suggestions on variations of the program of operations described above, and additional precautions will be discussed in Section 4-4.

As will be further borne out throughout the discussions and results of computations and error analysis, four sets are a practical number to make. This will normally yield a mean value with a 90% error of ±5" or better and even if one set must be rejected due to mistakes discovered later. More specific and detailed specifications for the observations are summarized in Appendix B.

Polaris Observations-Daytime

Finding polaris. For a Polaris observation during daylight, the calculation explained in Section 6-1 yields an approximate value for zenith angle and azimuth at the planned time. The values change very slowly. Normally Polaris can be found even if the time is off an hour or more. For example, the zenith angle changes approximately 12 seconds of arc per minute of time (or 12 minutes of arc per hour) at elongation and the azimuth changes approximately 18 seconds of arc per minute of time (or 18 minutes of arc per hour) at culmination. The field of view of most instruments is at least 30 minutes of arc up-down and right-left as shown in Figure 4.3. This gives an appreciation of the tolerances required in calculating and using preliminary values for finding the star.

Remember that Polaris won't be found if the sky is hazy, even at dusk. If the sky is clear, however, the procedure can commence. The first step after setting up is to focus the eyepiece, then focus the objective lens to infinity. This infinite focus is **very important**. Polaris will not be found, even on the clearest day unless this is done. Objects that can be sighted to accomplish this are the moon, passing jet planes, and clear objects several miles away. Clouds and trees do not make good objects since they often aren't distinct enough, but they can be used if nothing else is available. The focus must not be touched, once set, until Polaris is found and its zenith angle and horizontal direction from some terrain feature noted. The place of infinite focus varies among individuals since eyesight and eyepiece focus varies. It is *not* at the extreme run of the focus knob as some may erroneously think.

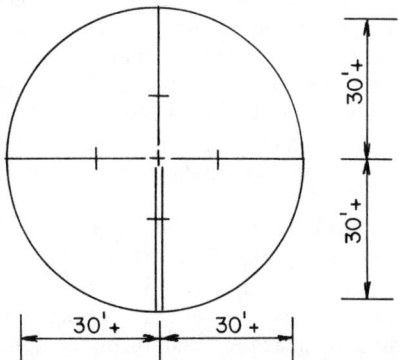

Figure 4.3 Angular Field of View of a Theodolite Telescope

After the focus to infinity is achieved, the calculated zenith angle is set on the instrument. If the observer knows the north direction within a few degrees from local ground features, it may be most efficient to very slowly scan the sky in that direction, using the horizontal fine motion screw (drive screw). If the zenith angle is correct within a few minutes of arc and the objective lens is focused precisely to infinity, Polaris can often be found in this manner. Alternatively, the preliminary azimuth can be added to or subtracted from the map angle scaled from a prominent terrain feature (possibly the chosen azimuth mark) and this angle then measured from this feature. If this procedure is used, the observer must remember *not* to change the infinite focus when sighting to the ground mark. After laying off this angle, the star should appear near the center of the telescope's field of view. A third method to use for approximate north is a magnetic compass with correction for magnetic declination. This may be the best method where there are no local land lines approximating north, nor any map features from which to measure an angle. When the compass is used, Polaris is sought as with the first method--by slowly scanning east-west at the appropriate zenith angle.

If Polaris is not found very soon using one of the above procedures, there may be a very slight focus adjustment needed, a mistake either in the preliminary calculations or use of the theodolite, thin clouds or haze blocking the view, or the theodolite optics are dirty or of poor quality. No rule of thumb can be suggested if Polaris is not found. If not found after checking calculations, focus, theodolite readings, and diligently looking right or left of the theoretical location, the best assumption is that it is too hazy and/or the sky too bright. This means a delay in operations. It is best to try in a few minutes, waiting until it is a little darker or clearer.

4-3 MAKING THE OBSERVATIONS

Once found, the observer should read and note the zenith angle, then lower the telescope to ground level and align a range pole or other object a slight distance away. This is for use as a reference so that the search in azimuth does not need to take place with each observation. In lieu of the range pole, the observer may be able to notice and memorize some feature on line at ground level to use as a reference in the same manner as the range pole. Figure 4.4 depicts this step.

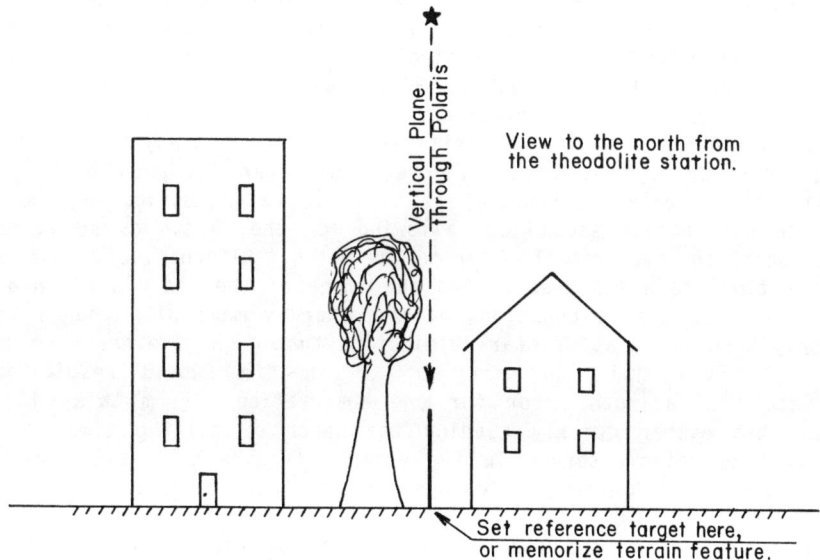

Figure 4.4 Ground Referencing for Polaris' Direction

Observations for azimuth. Now that the observer knows where to look for Polaris, the measurements can be made. Unless the watch correction has already been determined, it should now be made by calibrating a wrist watch with UTC, as explained in Section 3-1. In the example of Figure 6.5, it is seen that a watch reading of $40^m\ 40^s$ was observed at $43^m\ 00^s$ UTC, yielding a watch correction of $+2^m\ 20^s$. From the values noted, it appears as though these correction data were observed after the Polaris pointings were made, rather than before.

The observation procedure is similar to that used for the sun, except that pointings are made directly to the center of the star, there is no need for a special darkening eyepiece, and a stop watch is not necessary. The circle settings, starting with a setting near 0° and changing the circle by 180/n is done the same as explained in the last section on solar observations.

The backsight on the ground target is the first pointing in the program. For this example 0°45'19" was observed after focussing on the ground mark and leveling the plate bubble. After this reading is taken, the motion is unclamped and the theodolite pointed roughly to the ground reference mark set beneath Polaris. So as not to disturb the near infinite focus of the objective lens, the observer should point to this reference mark using the peepsight on the telescope. The telescope is next elevated to the zenith angle reading noted previously when Polaris was found. If Polaris is not seen immediately, a minor objective lens focus may be needed since the focus to the ground azimuth mark probably wasn't to infinity. When Polaris is in focus, it should be brought to a position, vertically, where its image is near the center of the field of view. This is to maximize clarity.

At this stage of the operations, the plate bubble should be checked and carefully centered. At an altitude of 45° an error in bubble centering of ½ division with a 20" sensitivity bubble would cause an azimuth error of 10". The observer must strive for near perfection (at least ±¼ division) in order to keep this error in control. Bubble centering is a little less critical at lower altitudes such as 30°. This error source is probably the most important in Polaris observations.

Next, a pointing is made to Polaris just as is done to any target, centering the vertical cross-hair directly over the star. When this has been accomplished, the observer immediately reads the watch, estimating the seconds reading to the nearest second and allowing for the second or so it may have taken to shift the eye from the star to the watch. Alternatively, the observer can call "time" to a timekeeper when the image of the star is aligned. When Polaris is at or near elongation, it will hardly move with change in time. It is only when it is at or near culmination that much care needs to be taken in timing. Even then, an error of 3^s in time would result in only approximately 1" azimuth error for one observation. From this, it can be realized that either quickly reading the watch or calling time to another person will be satisfactory. In the example (Figure 6.5) UTC time for the first pointing to Polaris, telescope direct, was 23:25:13 on January 14, 1982.

After the time is read and recorded, a final plate bubble check should be made. If the bubble appears to have moved, consideration should be given to taking a new pointing and a new time with the bubble centered.

After achieving a satisfactory pointing and timing with the bubble centered, the telescope can be lowered to a comfortable position and the circle read. In the example, the reading was 358°54'51".

Next, the telescope is plunged, pointed toward the ground reference beneath Polaris, elevated to a reading on the vertical circle of 360° minus the approximate zenith angle, and Polaris focussed and centered in the field of view. The pointing and timing procedure is just as described for the direct telescope position, the plate bubble being first carefully centered, the star sighted, the time read and recorded, bubble checked again, and the horizontal circle reading taken if the bubble has remained centered. In the example, for the first observation, telescope inverted, the time read was 23:27:20, and the horizontal circle reading was 178°54'00". At this point, agreement of the inverted and the direct reading should be examined. Since the star has moved very little, the discrepancy should be 180° plus or minus only a few seconds. But, because of instrumental errors a sure check on mistakes cannot be made. Large mistakes, such as 10' or degree reading mistakes might be found and corrected here. The discrepancy here was 51" which is probably all right, but the observer may want to check to see if the minutes were not 55 instead of 54 before unlocking the motion.

Next, the motion is unclamped and the telescope pointed toward the ground azimuth mark, bubble centered, precise pointing made, and the circle read. In the example, a direction reading of 180°44'43" was observed. It is noted that the degree value on the azimuth mark differs by 180° from the initial backsight reading as expected. The minutes and seconds should also agree favorably, except for the usual instrumental and random reading and pointing errors. It will be seen that the pattern of agreement upon closing back for all sets is much like it was in the solar observation. This is not surprising since this was the same instrument and, in fact, the same date, set-up, and observer. The value of checking the agreement of these closures is the same as discussed for solar observations and will not be further belabored here.

One "set" of Polaris observations are now completed in the discussion of this example. To commence Set 2, the instrument remains pointed to the ground target, telescope inverted, and the circle reading shifted 180°/n. From here, the sequence is as in the first set and resembles the program used for solar observations in reading and recording times and directions. For Set 2 (Figure 6.5) the reading sequence was 225°29'26" (angle), 23:29:47 (time), 223°38'06" (angle), 23:31:33 (time), 43°38'09" (angle), 45°29'56" (angle). The reader should recognize how the remainder of the data were read and recorded from having studied the same sequence in the solar observations. When completed, sets of one direct and one reverse should exist, fairly evenly distributed around the theodolite circle.

Before leaving the site, or upon returning to the office, the watch correction should be observed and recorded in the notes, even if it has already been done. There is no need to determine the DUT1 correction, since the accuracy in timing does not warrant doing so. In case the observation date differed locally as on UTC, this should be noted. Observe that in the example, in another 21 minutes the observations would have extended into January 15 but local time would still be January 14, approximately 6:00 P.M. central standard time. The ephemeris date used for calculating is the one on UTC time. Note that the observer and recorder made the conversion in the field and recorded UTC hours directly, even though a watch with local time was read. This was an easy conversion since the WWV announcer gave UTC hours and there was not really a need to manipulate times considering zone and other corrections.

As with solar observations, four sets are recommended for Polaris, although three would suffice for precision requirements if no mistakes were detected later.

Polaris Observations - Night

Section 2-3 explained how to locate Polaris at night using the Big Dipper and other references. There should be little chance for mistaken identity once one becomes familiar with the appearance of Polaris in the telescope. There are no other bright stars nearby.

Night observations differ little from daytime observations except that a preliminary calculation need not be done to find the star, thus saving time. Additional time is usually saved in the field initially since the star can be found quicker at night. Also, there is no need to set and use a ground reference beneath the star nor to note and set approximate zenith angles on the circle each time the star is to be sighted, since it can be seen directly. Battery attachments or other means of illumination of cross-hairs and theodolite circles will be needed. The observer should remain diligent in checking the plate bubble level with a flashlight at the critical times throughout the observations. It is easy to forget or neglect this when it is dark.

Most battery attachments and theodolite systems have a rheostat control so that just the right amount of light can be brought into the telescope to see the cross-hairs clearly, yet not fade the image of the star. Beginners should experiment with this adjustment so that definite and clear images of both the cross-hairs and the star are seen. If a battery system is not used, a flashlight pointed into the objective lens at an angle will serve the same purpose, although this method is more awkward. Likewise, a flashlight can be used to illuminate the circle of the theodolite if need be. Car headlights on targets serve to illuminate these if better systems are not available.

The program and sequence of observations are the same as described in the daytime procedure and will not be repeated. The total time for an observation is normally about 15 minutes for four sets, as in the example used in the last section. However, this is the time expended after operations are in process, and does not include "finding" time or target erection in the dark, etc. Usually, Polaris observations take a little longer than solar observations because of these added preliminaries or work in the dark.

What frequently happens is that Polaris is planned to be sighted at dusk when it is still daylight but the observation extends into the dark anyway and thus the full program is a mix of both daytime and night observations. This, of course, has no effect on results unless the surveyor is unprepared for night observations by not having flashlights and other illumination. Darkening hours especially affect the precision in reading the theodolite circles and one is often tempted to squeeze one more reading out of the fading twilight at the expense of lost precision in final results. Be prepared with flashlights for illumination, even when plans are to be completed before dark.

4-4 SUGGESTIONS AND PRECAUTIONS

General Comments

The previous three sections were intended to outline the details and give the important checks and precautions, all in the appropriate sequence, yet be free of too many distracting and unusual variations and suggestions. The following paragraphs are intended to round out the description of the observations, answer questions that may have arisen, give a few alternative procedures and suggestions, and complete the precautions recommended.

Variations in Sighting Program

As has been stated, a direction theodolite need not be used. There are several random errors in the observations which are large enough and propagate in such a way that computed results have nearly the same uncertainty regardless of the precision of the theodolite. Furthermore, the reading error effect is reduced when the mean of several sets is used, regardless of whether the instrument is direction or repeating. For these reasons, most repetition instruments are suitable for third-order astronomic azimuth. Suggestions as to reading precision requirements were made in Section 4-2.

Examples using the repetition theodolite are shown in Appendix C. Here, the orientation reading is set at 0°00'00", with the telescope in the direct position, at the beginning of each set. Equally suitable alternatives may be used. For instance, the telescope need not be redirected after pointing 1R, but may remain inverted, thus starting that set with 2R instead of 2D.

4-4 SUGGESTIONS AND PRECAUTIONS

This saves one field step and may be a small advantage in the opinion of some. In this case, the scale could not be set to exactly 180°00'00" or to 0°00'00" for azimuth mark pointing 2R. Others may wish to use the repetition instrument like a direction theodolite, changing the circle reading between succeeding sets, zeroing perhaps only on 1D. Since the repetition feature can't be used anyway because of the moving target on the foresight, the upper motion would then merely serve the same function as the circle drive knob of a direction theodolite.

All of the above methods yield an independent and different reading for all sightings to both the mark and the astronomic target. This randomizes the error in reading. Any program, whether using a repetition or a direction theodolite, should do this. Other methods of using one reading and pointing to the azimuth mark, then making multiple observations to the celestial body using the one pointing, does not randomize these errors and whatever error (or mistake) occurred in the initial reading and pointing will be the same for all observations. Thus, the precision advantage of the law of compensation is not being used. And, there is no check on a faulty reading or pointing to the mark when the series is computed. Similarly, particularly when using a repetition theodolite, some may be tempted to use the same reading and pointing at 1R for 2R, turning then to sight the celestial mark for 2D without changing the orientation on the azimuth mark. Again, results will be good in most respects, but the effects of reading and pointing are not entirely randomized and the sets are not entirely independent with checks for mistakes.

The objective of achieving third-order accuracy should not be jeopardized by using a method that does not randomize and minimize the effects of errors as much as possible. This is the underlying principle in the above discussions. It is felt that the methods recommended are as efficient as possible, keeping these error propagation principles in mind.

An acceptable and possibly preferable modification to the solar sighting program is to alternate between sighting the left and right sun limbs in each set. That is, if in Set 1 the sighting sequence is to the left limb first, then right, Set 2 would be right limb first, then left. This has one small advantage of having all direct telescope readings to one limb and all inverted readings to the other limb (1D→L, 1R→R; 2R→R, 2D→L, etc). The advantage is that when each observation is computed, the azimuths resulting from limb pointings do not contain different systematic errors due to instrument maladjustment and this helps to troubleshoot, investigate errors, and evaluate precision. Another advantage is that a small error related to the sun's semi-diameter and its changing altitude is compensated (See Appendix B) if limbs are alternated between sets. A small disadvantage in this modification is that the observer must keep alert as to which limb should come first, whereas the procedure demonstrated has the advantage of routine. It should be stated that if the instrument was in perfect or very good adjustment, the method demonstrated would be most field efficient and still have the same advantages later in investigating errors as would the method of alternating limb sightings between the sets. The choice is left to the reader.

It should be obvious that it would not matter whether the left or right limb of the sun is sighted first in any program with either the direction or repetition instrument, as long as an equal number of sightings are made to each.

It is not recommended to make all left limb pointings in sequence, then all right limb pointings; nor, to make all direct telescope observations in sequence, then all inverted telescope observations. The reason for not doing this is that clouds or other problems can force the surveyor to cease observations, perhaps leaving him with only half the data needed.

More on Timing

Many timepieces have a "split-time" feature--they can be started once and stopped to read more than once, the mechanism continuing to run until another hand or digital number is stopped for another "split". Some electronic timepieces have an infinite number of opportunities to read an intermediate time. This would seemingly be an advantage since calibration with the time signal would need to be done only once for the entire program of observations. However, for solar observations, such a procedure is not recommended because timing errors are not randomized. The theory is similar to that discussed concerning angle measurements. By starting a watch only once and then taking all other times from it, all observations have a built-in constant error according to the plus or the minus nature of that initial reaction. Stated differently, the error is random in nature but systematic in effect on results. This applies to any chronometer, however accurate and expensive. There is another potential error using the split-time method, which occurs if the timepiece gains or loses a few tenths of a second every few minutes (which is highly possible with hand-wound watches). In the interest of minimizing errors by taking advantage of laws of compensation, it is recommended that a watch be started and stopped for each solar pointing. The only exception the author might suggest is to use one starting of a split-time watch for one *set* of observations, thus having two times (direct and inverted pointing) from one start of the watch. This small sacrifice in precision might be made for the convenience of dangling only one watch around the neck instead of two.

The above precautions apply primarily to solar observations. Since high precision in time is not important for Polaris observations, a split-time watch can be used to advantage. If enough checks are made to assure that no mistake occurs in the initial starting of the timepiece on a time signal, a continuous running timepiece either with or without a split-time feature is advantageous because the watch correction can be avoided. Any chronometer set to UTC can be efficiently used for Polaris observations to avoid the watch correction. Naturally, a stop watch started on a time signal and stopped at the star sighting can be used for Polaris in the same manner as with the sun. The reason this isn't recommended is that the precision isn't needed and it probably takes more field time since watches must be started on time signals, and the signals do not usually occur at the most convenient time in the operations.

For solar observations the observer, theodolite, and time system are one unit. A helper should be used only for assistance such as note keeping and mechanical operations which do not affect measurements. The reaction time of two people should not be allowed to affect timing, for example. This could happen if times are called by the observer to a timekeeper who is either holding the stop watch or observing the movement of a chronometer. It also occurs when a well-meaning assistant starts the stop watch on the signal and then hands it to the observer. Only the observer must start *and* stop the watches and "time" should not be called to an assistant. The senses of one person, observing the sun limb tangency and directly communicating this through his own brain to the stop watch button is more precise, dependable, and consistent than when this visual occurrence must be audibly communicated and processed through the brain of another who must then react. Furthermore, when two people manipulate the stop watch, there is a difference in reaction time. When one person uses the stop watch (the observer) his own tendency to respond late or early will generally cancel in the process, thus compensating for "personal equation". It is less likely that such compensation would occur if more than one person uses the watches. Even if the chronometer is an expensive, highly accurate, quartz controlled instrument, calibrated to UTC within thousandths of seconds, the procedure of an observer calling "time" to a time-keeper is not as precise as the stop watch method advocated in this manual, in the opinion of the author, simply because precision is lost in the reaction time between two people, and the timekeeper must make a visual observation of moving hands or digits. The use of such a chronometer would be superior to the stop watch method only if the observer could, with a stop button, stop the instrument for a time as when using the stop watch. Any method will be subject to the reaction time of the observer, which is in the order of $\pm 0.1^s$ and a highly accurate time-keeping system will not overcome the human linkage with its personal error.

Care must be taken in reading minutes. In setting any mechanical wrist watch, the hands should be set so that the minute hand divides the space equally in fifths between the numbers when the second hand points up as it makes each revolution. Also, some stop watches have a small dial or other system for reading minutes (30-second sweep, etc.) which makes minute determination confusing. The surveyor must be especially careful not to make half-minute and other mistakes in reading such watches. (See Fig. 4.1). As has already been cautioned, counting minute revolutions on any stop watch and correctly adding these to the UTC watch start time must be done with care. A common mistake in reading any mechanical watch is to read the next higher minute when the seconds are approaching the "straight-up" position. For instance, a careless or tired observer might mistakenly read 34:52 as 35:52 simply because the minute hand is closer to 35 then 34. Experience has shown that one-minute timing mistakes are some of the most common mistakes made in azimuth observations, and that they usually can be traced to faults such as those described above.

Instrument Adjustment and Cleaning

If the line of sight and the horizontal axis are in adjustment, the direct and inverted reading to a ground target should differ by 180°, theoretically. Having these axes in adjustment makes it easier to check the control of random reading, pointing, and bubble centering errors. It also helps check readings to Polaris in that the star's movement can be observed in the series of reading values and if an interruption suddenly appears in an observed trend to increase or decrease, a mistake in reading would be a probable cause. If the instrument is not in good adjustment, the difference in direct and inverted readings could easily exceed the discrepancies caused by movement of the star. Thus, this check could not easily be made without considering the instrumental error, and this wouldn't be very efficient. Such is the case with the data of Figure 6.5 where the instrumental error was several seconds. It is recommended that instruments be checked for these adjustments according to instructions in instrument manuals and adjusted by qualified technicians if the maladjustment exceeds the normal spread in values due only to random errors. Lack of adjustment does not cause errors in results if the direct/inverted system is used, since errors compensate. These comments are made only as regards ease in making intermediate checks on data as it is gathered.

Since bubble centering is so important, bubble adjustment is important. Although a skilled instrument operator can theoretically keep a bubble "centered" even though it is out of adjustment, it is much less trouble to adjust the plate bubble so that no detectible movement is observed when the instrument is rotated about the vertical axis, using the bubble marks as reference. This adjustment is very simple and should be routinely performed by instrument operators, as needed.

Optical plummets should be checked and adjusted. Other theodolite adjustments are generally of small concern and the errors are compensated in the measuring process if recommended procedures are followed. Tripods must be in good adjustment and repair, so that there is no "play" in the legs. Targets must also be free of defects.

Cleanliness of the optics and motions is an important part of any angle measuring system. Precision can easily be lost if circles cannot be read clearly due to dirty optical systems or if the motions operate so stiffly that bubbles are forced off center whenever an instrument is repointed.

The results claimed in this manual are stated with the assumption that instruments are in good adjustment and well maintained.

Inverted Images

Most modern theodolite telescopes have erect images. Because many surveyors use those with an inverted image, it is appropriate to mention the peculiarities of the observations using such instruments. All statements in this manual concerning movement of the sun and Polaris, limb sighting, etc., assume that erect image instruments are being used. A regular user of an inverted image instrument will instinctively realize that left is right and vice versa, but a student may need to carefully think through these concepts. With an inverted image instrument the sun will appear to be setting in the morning and rising in the afternoon. It will appear to be moving to the left rather than to the right and when a sighting is made to what appears as the right limb, it is actually the left limb and vice versa. How this is recorded in the notes is left to the judgment of the surveyor, but the notes should reflect whether it was truly the left limb sighted or the *image* of the left limb (being actually the right limb). In analyzing these data, the field surveyor and the person doing the calculations should recognize the confusion that could result in troubleshooting and checking. For Polaris, one must remember that trends in the change of the direction will be reversed when an inverted image telescope is used. To avoid confusion, the field surveyor and person performing the calculations must understand the optical phenomena involved and the field surveyor must communicate everything clearly in the notes.

Appendix C contains one example where an inverted image telescope was used (Example C.9).

Marking Infinite Focus

A surveyor who uses a particular theodolite regularly and shares its use with few others may wish to mark the spot of infinite focus on the telescope. To do this, the procedure is as follows. First, focus the eyepiece. Then, focus the objective lens to infinity as described in Section 4-3. Make sure this is correct by eventually achieving a clear image of Polaris. With most theodolites, the objective lens focus knob is a knurled knob (sleeve) which rotates in alignment with the barrel of the telescope, behind the eyepiece. When the infinite focus has been achieved, take a very thin paint brush and paint a small, thin stripe extending across both the focus sleeve and the barrel of the telescope. Then, in the future, when the mark on the sleeve is aligned with the mark on the barrel, infinite focus will be achieved without needing to track jet planes or find the moon. Since eyesight varies among individuals, this mark probably won't appear in the same place for all instrument operators. One must be sure that the eyepiece is focussed the same each time, or the place of infinite focus on the barrel may be somewhat different.

Knowing the Instrument and Techniques

The importance of knowing the theodolite, error sources, possible magnitude of errors from various sources, and how to control errors, is sometimes overlooked in any surveying measurement operation, not just in astronomical observations. Most textbooks explain how to set-up and use instruments and what errors are, but seldom explain the refinements of application of the instrument to various projects and why the instrument operator must **know** the instrument and the applicable measurement theories. Hopefully, the reader of this manual will grow to be an expert in the astronomical azimuth problem and possibly also become a better all-around measurement professional. For more depth in this regard, readers are referred to Reference 8.

Results are very important in any surveying measurement operation. "Going through the motions" will often yield satisfactory results only because the instruments are so accurate and reliable that they often overcome some lack of operator skill. However, the best and most consistent results are gained by those who understand the desired relationships of the axes with each other, how to check or compensate for instrument errors, what the sources of errors are and their magnitude, and then how to minimize the errors. An operator who understands the instruments **and** the methods and can consistently apply them in the field will have consistent results. Technicians who have only been taught how to **use** instruments may not. Regarding astronomic observations, appreciation for such error sources as scaling ϕ and λ, timing, reading and pointing, sight distance effects, target size and centering, vertical angle effects, bubble centering, and instrument adjustment, care, and cleaning is important. An operator should have reflexes and alertness sufficient to make a solar pointing. Some observers may enjoy wondrous results when targets are as in conventional angulation but may have trouble working with a moving target like the sun. However, most experienced theodolite operators should quickly learn to achieve precise and accurate results with a small amount of practice and attention to the details in this chapter.

Commonly Misunderstood Aspects

1. A direction theodolite should not be set to exactly 0°00'00". It wasn't designed to do so. The knob which changes the circle reading (circle drive knob) is not a fine motion knob and doesn't have the same touch or precision as the larger collimation knob used to align the minutes and seconds or the fine motion knob used to set the degrees. Thus, a precise setting to exactly 0° cannot be made as when simply reading whatever results when the circle drive knob is set arbitrarily and the collimation knob used for precise setting.

2. There is no need to sight Polaris near elongation. In the past, when calculations weren't so easy and when ephemeris data and accurate time may not have been so readily available, surveyors would determine when Polaris was approaching elongation and then be there at those few minutes for sighting. The ephemeris gave an azimuth at elongation for the date and thus there was no need for calculations. With the 30" or 1' transits, and the rather imprecise value given in the ephemeris for azimuth, this method was accurate enough, even though the star was never exactly at elongation but actually moving toward or away from it and the azimuth thus varying slightly. Nowadays, with more precise instruments, better calculators and accurate time readily available, there is no advantage in planning observations at elongation and it is actually more of a disadvantage since elongation probably occurs at an inconvenient time.

3. The surveyor should not measure the vertical angle to the sun. There is no need for it. To do so means sacrificing precision in the timing and horizontal angle, both of which are important in the hour angle solution. The sacrifice in precision occurs because the observer must concentrate on keeping the horizontal cross-hair tangent to the upper or lower side of the sun, simultaneous with the left or right limb on the vertical cross-hair, using both vertical and horizontal fine motion screws, then somehow manage to stop the stop watch when this appears to have been achieved. Some may want to check hour angle results using the altitude formula by doing this, but to check a superior method using one that is inferior, and losing the precision in the superior method in the mechanical process of attempting to do so does not make sense. Vertical angle is not needed in the hour angle method. The observer should concentrate only on the sun moving across the vertical cross-hair. One approximate reading of the sun's altitude may be appropriate to read, for purposes of error analysis and control, but even this can easily be computed later (See Appendix B).

4. The surveyor should not "track" the sun with the fine motion screw. Some who feel they must do this have probably carried the practice over from the old altitude method, as discussed previously. The best precision in the hour angle method is to leave the hands completely off of the theodolite and wait with the stop watch until the sun crosses the vertical cross-hair. It isn't done the other way around. The only reason to adjust the vertical cross-hair position using the horizontal fine motion screw is if the image of the limb has already crossed or if the surveyor ascertains that it will take more than a few seconds to do so. Then, a small repointing is appropriate out of necessity or to save field time and run on the stop watch.

CHAPTER 5

THE EPHEMERIS AND ITS DATA

5-1 GENERAL COMMENTS

The ephemeris tables in Appendix F contain only the variables needed for solar and Polaris azimuth using the hour angle equation. They can also be used for solar altitude solutions for azimuth. Included are the declination, δ, for the sun, polar distance, p, for Polaris and a variable for each of the celestial bodies which is used to compute the hour angle, t, from observed UTC and other measurements. These tables have sufficient significant figures to prevent round-off errors in computations for either second or third-order azimuth, if used according to instructions.

Surveyors accustomed to using the solar altitude method or the several approximate methods for Polaris azimuth may at first feel that something is missing, but this isn't so. No other ephemeris is needed. Data such as the sun's parallax, atmospheric refraction, and temperature and pressure effects on refraction aren't needed since the altitude method isn't used. Furthermore, formulas can be used to solve for these if that method is used. The sun's semi-diameter isn't listed for each day since the observation program results in an average to the sun's center which generally compensates for the semi-diameter. Sun's semi-diameter is, however, given in Table A.1 with a discussion in Appendixes A and B as to the use of this table. There is no need for tables for right ascension and conversion to sidereal time in the method used in this manual, nor is there need for data for unfamiliar stars. To compute the preliminary altitude of Polaris so as to find the star, an equation is used, rather than a table. Likewise, many conversions and interpolations are done using the calculator, rather than with tables, following the trend to eliminate tables since the advent of electronic calculators. Hours-minutes-seconds to degrees-minutes-seconds is done quickly and with less likelihood for mistake using a calculator instead of a table. There is no need for tables for culmination and elongation of Polaris since the star is observed at any hour angle. The position of Polaris will be known as soon as the hour angle is computed and it need not be known any sooner to analyze results.

Hopefully, readers will understand that bulk, cost, and confusion have been minimized by providing only the data needed for the azimuth solution. As provided, the tables are consistent with the author's overall plan and intent of providing an efficient, accurate, precise, straightforward, easily understood azimuth solution, and eliminate the need to worry about ordering tables for several years.

After the year 1988, it is planned to either revise and update the manual, adding ephemeris tables for another decade or more, or to print the tables in a separate volume.

Readers may be aware that ephemeris data may be available in computerized form which could be programmed into computations for azimuth. Such programs were not included in this manual since it was felt that many surveyors would not use programmable methods or would not wish to have all variables programmed for other reasons related to their computation preference. The tables provided herein require so little time to extract the two values for each date and enter them into the computation steps that this would seem to be more efficient than adding program steps for an entire ephemeris. After extracting the values directly from the tables, it is intended that an electronic calculator be used to convert the data to azimuth. No hand calculations or interpolations are necessary. Another reason for not programming the ephemeris is that the printed tables provide an important method of learning some aspects of the astronomical theory when the daily and monthly changes in numerical data are scanned and studied. The next sections will reveal this.

The remainder of this chapter explains the variables in the ephemeris tables--what they mean and how to use, read and calculate with them in such a way as to avoid mistakes and computational errors. In the process, a better appreciation of some of the concepts introduced in Chapter 2 will be gained, as it will be necessary to return to some earth, sun, star, and time relationships.

5-2 DECLINATION OF THE SUN

Declination was defined in Section 2-1 and illustrated in Figure 2.1. The sun slowly changes position north and south becaue of the earth's tilt on its axis and its revolving around the sun. The following description of the earth's rotation and orbit will explain this behavior.

The earth revolves around the sun and rotates on an axis through its poles. One revolution takes a year or 365.2422 solar days. The earth's axis is inclined approximately 23°26.5' with respect to the orbital path around the sun. This inclination stays the same, the axis of the earth pointing the same direction with respect to the distant stars at all times. In Figure 5.1, the orbital plane of the earth is into the paper. To further explain the figure, the sun and earth would align on March 21 and September 22. From the definition of declination, it is realized that the maximum south declination, equivalent to the inclination of the earth's axis, would occur on December 22. The maximum north declination occurs on June 22. Imagining the sun and earth aligning in the figure, it can be realized that the sun's declination would be zero at those moments, since a line perpendicular to the paper, depicting the sun's rays, would pass through the earth on the equator.

A look at the ephemeris tables for any year reveals the above facts. By convention, north is positive declination; south is negative. The tabulated listings are for zero hours UTC (abbreviated UT in the tables). The slightly different value for the declination on any given date in different years relates to the fact that some years have 365 days, others 366, the effect of the accumulated approximate ¼ day each year occurring abruptly every four years in the ephemeris. This is not important regarding use or analysis of the data, however, but does reinforce the importance of using the correct year's values.

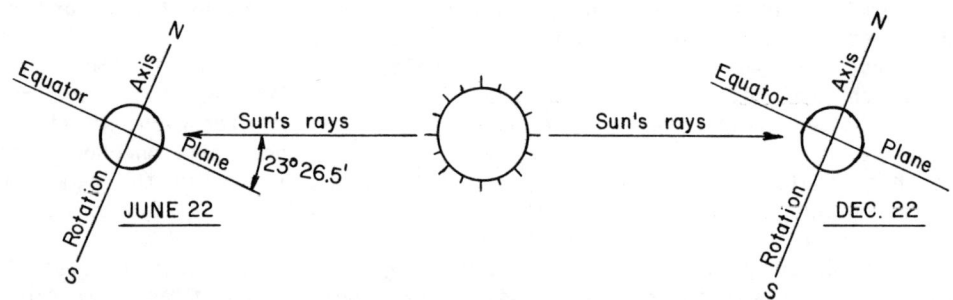

Figure 5.1 Sun's Declination Changes During Orbit

The rate of change in declination is not constant. It is most rapid at the equinoxes and slowest at the solstices. Therefore, a table for change per hour for each day is given. The value under this column (Δ, 1 hr.), in seconds of declination change, is the change per hour for the date on the line directly opposite the figure for Δ. This figure was computed by merely subtracting the declination at the end of the day (0 hours, next day) from the declination at the beginning of the day, dividing by 24 hours, and converting to seconds. For example, the declination on April 22, 1984 was +12°11'10" and +12°31'15" at the beginning and end of the day, respectively. This yields a declination *increase* of 20'05" or +1205" Dividing this by 24 hours yields +50.2" as shown in the table. To compute a declination for an observation time of UT=6:00:00 on this date, 50.2 would be multiplied by 6.00 and added to +12°11'10" yielding +12°16'11" for declination at observation time UT=6 hours exactly. Note that care must be taken to always add the declination at 0 hours to the change, **algebraically**, considering signs of **both**. Another example on March 13, 1984 would result in a declination at 6.00 hours of -2°50'13" The reader should check this solution to be sure that the algebraic sign is understood. As a check, one can always examine the declination at 0 hours for the next day to see if it is increasing or decreasing, numerically.

The change in declination should be computed to the nearest second. To avoid round-off errors, the UTC observation time should be entered into the calculation, in decimal hours, to at least three significant figures, and preferably four, when the Δ, 1 hr. column has three figures. A solar observation time of 17:18:39.7, reduced to decimal hours, is 17.31. On October 5, 1981, the change in declination was -57.7" per hour. The change to be added to the 0 hour value for that date is -57.7x17.31=-999"=-16'39". The calculation of declination will be illustrated further in Chapter 6.

When the date is close to the solstices and UT is close to 12 hours, better precision results if a small correction is first made to the declination before adding the change for elapsed time. This correction is explained at the beginning of Appendix F. Although this correction was ignored in the examples herein, it should be considered and applied when appropriate.

5-3 TIME RELATIONSHIPS AND EQUATION OF TIME

Thus far, the discussion of time has extended only to the aspects of determining UTC, the relationship of this time to local systems, and how to "time" an observation. But, further refinements are needed in order to arrive at the time of observation used for the calculations.

First, in review, recall that UTC relates to the time in the zone of Greenwich England, through which the 0° longitude line passes. Time zones in the United States are 5, 6, 7, and 8 hours west of Greenwich, in standard time increments. Daylight savings time changes the local time one hour ahead, so this results in only 4, 5, 6, and 7 hours difference in the same zones corresponding to standard time differences, Greenwich time always being later on the same date. Figure A.14 gives world time zones.

A circular globe measures 360° in longitude. The earth rotates 360° and faces the sun in 24 hours, using any longitude point as a reference. Thus, 15° of longitude rotates every hour. Time could be measured in degrees as easily as in hours, since it is a function of how much rotation of the earth has occurred in a day.

The earth revolves around the sun in an elliptical orbit. This causes its speed to increase and decrease. Since it is rotating on its axis at a constant rate and its speed along its orbit is constantly changing, it requires a different amount of time each day for any point on its surface to rotate once and face the sun. That is, a "day" actually varies in length when the true sun is used as a reference. This day of variable length is called **apparent solar day**. It would not be practical or possible to regulate all timepieces in the world to run at a variable rate, so we average the time over a year. This is called **civil time** or **mean time** and the 24 hour day of uniform time length is a **mean solar day**. A fictitious or "mean sun" is thus created that is always either ahead of or behind the true sun. The difference between the **apparent time** and the **mean time** is **equation of time**. (Section A-2 includes additional comments and definitions). The elliptical orbit of the earth and the approximate dates of maximum and minimum values of equation of time are shown in Figure 5.2.

Figure 5.2 Elliptical Orbit of the Earth and Equation of Time

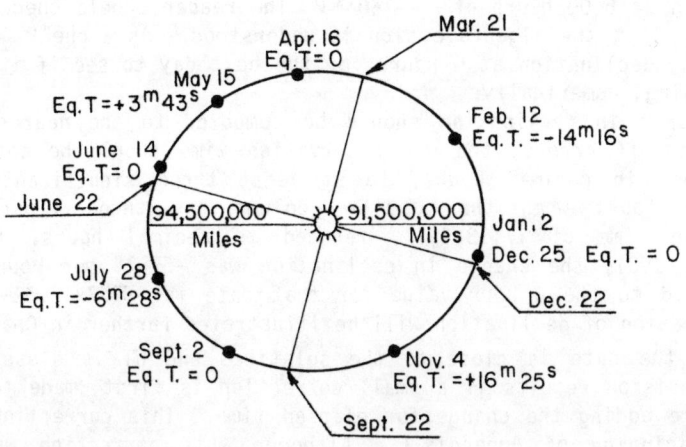

5-3 TIME RELATIONSHIPS AND EQUATION OF TIME

$$\text{Equation of Time} = \text{Apparent Time} - \text{Mean Time} \tag{5.1}$$

Since an observation is timed in the field using UTC, which is a uniform signal based on mean time, and the pointing has been made to the *true* sun, the observed time must be converted to what it should be on the true sun. Thus,

$$\text{Apparent Time} = \text{Mean Time} + \text{Equation of Time} \tag{5.2}$$

Clocks throughout the world are regulated to UTC. But, as has been stated in Section 3-1, even UTC varies slightly because of very small variations in orbital rate. The UTC signals are regulated to an ultra-accurate atomic clock that doesn't depend upon the earth or sun movements. To coordinate them with the true orbital variations, the DUT1 correction is provided. Recall that

$$\text{UT1} = \text{UTC} + \text{DUT1} \tag{3.1}$$

For solar observations, the values in the ephemeris assume that UTC has been corrected to UT1. Thus, the abbreviation UT in the tables is actually UT1 as corrected.

Examining the equation of time fluctuations in the ephemeris, it is seen that it reaches various maximum and minimum values throughout the year. The equation of time is dependent on the rate of speed of the earth in its orbit around the sun rather than location of the earth in the orbit, so the maximum and minimum values occur at entirely different dates than those for the declination. The changes in equation of time occur as they do because of the nature of the change in the speed of the earth in its orbit, the geometry of the ellipse, and the off-center location of the sun in this elliptical orbit. The mean sun and apparent sun coincide four times each year. It is seen that these dates of zero difference occur on April 15, June 13, September 1, and December 25, or very close to these dates.

It should be noted that, just as with the declination, the values of equation of time vary on a given date each year because of leap year shifts and other more minor reasons. Because of this, one must be cautious to use the ephemeris for the correct year.

Equation of time is computed in the same manner as declination, first entering the ephemeris table with the date and extracting the value for zero hours, then adding to it the value obtained by multiplying the change per hour (Δ, 1 hr.) by the decimal hours of UT (actually UT1), always performing *algebraic* addition. Generally, four significant figures should be used in the decimal hours to multiply by the number opposite the date in the column Δ, 1 hr. In Equation 3.1, UT1 is Greenwich "mean time". Furthermore, "apparent time" can be either Local Apparent Time, LAT, or Greenwich Apparent Time, GAT, depending on whether the local or the Greenwich time zone is the reference. Using a previous equation and these facts,

$$\text{GAT} = \text{UT} + \text{Eq.T.} \tag{5.3}$$

On April 22, 1984, at UT=6:00:00,

$$\text{Eq.T.} = +1^m\ 26.8^s + (+0.49)\ 6.000 = +1^m\ 29.7^s$$

and

$$\text{GAT} = 6:00:00 + 0:01:29.7 = 6:01:29.7$$

On March 13, 1984, at UT=6.000 hours

$$\text{Eq.T.} = -9^m\ 36.3^s + (+0.69)\ 6.000 = -9^m\ 32.2^s$$

and

$$\text{GAT} = 6:00:00 - 0:09:32.2 = 5:50:27.8$$

On February 1, 1984, at UT=10:28:29.3

$$\text{Eq.T.} = -13^m\ 28.6^s + (-0.36)\ (10.47) = -13^m\ 32.4^s$$

and

$$\text{GAT} = 10:28:29.3 - 0:13:32.4 = 10:14:56.9$$

The calculated GAT is used to calculate hour angle, t, for solution of the hour angle equation. The conversion of GAT to t will be covered in Chapter 6. Let us first complete the explanation of the ephemeris.

5-4 GREENWICH HOUR ANGLE AND DECLINATION OF POLARIS

The extraction of the GHA of Polaris at zero hours is straightforward in the ephemeris tables. The change to GHA at observation time will be covered in Chapter 6 since this does not involve further use of the ephemeris. It is appropriate here to illustrate and explain what this hour angle is. Local Hour Angle, LHA, was defined in Chapter 2, being the angle measured westward along the celestial equator from the observer's meridian to the meridian of the star. The GHA has the same meaning except that the angle is measured from the Greenwich meridian instead of the observer's meridian. The LHA and GHA are measured from the upper transit of the star over the observer's meridian (noon), rather than the lower transit (midnight). This has no significance concerning time reckoning but does need to be realized when visualizing the movement of Polaris, the sun, or any star. For Polaris, GHA has been tabulated in the ephemeris as reckoned from the upper transit over the Greenwich meridian for each date.

The declination of Polaris is listed in the ephemeris, to the nearest second, for each date. Unlike the sun, Polaris' declination changes very slowly; therefore, there is no need for interpolation. The value listed for each date has sufficient precision for third-order accuracy.

For a given date, values for GHA and declination of Polaris change each year due to the leap year shift and other factors related to the difference between star time (sidereal) and sun time (solar), and the gradual change in Polaris' declination each year as the rotation axis of the earth shifts slightly. Just as with the sun, it is important to use the ephemeris for the current year for GHA and declination of Polaris.

CHAPTER 6

CALCULATIONS

6-1 GENERAL PRECAUTIONS IN CALCULATIONS

Unless significant figures are understood, round-off errors or unnecessary work with excess digits can result. The potential problem areas will be mentioned in each step as the examples are discussed.

Algebraic signs must be carefully observed. Some of the arithmetic has already been discussed concerning equation of time and sun's declination. Other problem areas are in the sign of the hour angle, t, and the sign of the final computed astronomic azimuth. These potential problems will be discussed in the examples to follow.

An electronic calculator having at least eight places in trigonometric functions is highly recommended. Hand calculations, using five or six-place tables will surely destroy the precision so carefully designed into the observational procedures. Using storage registers (rather than pausing to record numbers in intermediate steps) is recommended when solving the azimuth equation, to avoid mistakes and round-off errors.

6-2 COMPUTATION OF GREENWICH HOUR ANGLE

Solar Observations

Hour angle can be stated in units of time or angle, and is measured between hour circles, one being through the celestial body and the other through a reference such as Greenwich or the observer's station. (See Section 2-1 for definition and illustration). The local hour angle, LHA is measured westward from the observer's meridian to the hour circle through the body. The Greenwich hour angle, GHA is measured westward from the Greenwich meridian to the hour circle through the body. It is noted that meridian lines are hour circles.

Referring to Section 5-3, it is recalled that Greenwich Apparent time, GAT, was computed from the observed time and equation of time. Apparent time, whether measured from the Greenwich or a local time zone, is referred to zero hours as midnight. Any solar time system, then, uses the zero reference as the place where the sun crosses the meridian on its lower side (lower transit), since noon occurs when it crosses the meridian above (upper transit). But, the hour angle in the spherical triangle to be solved is reckoned from the upper transit (noon). Therefore, to reduce GAT to GHA, a 12 hour shift is required. It doesn't matter whether 12 hours are added or subtracted, since the shift is purely one of achieving a solvable triangle and does not affect the ephemeris date. In most of the United States at

CALCULATIONS

typical times of observation, 12 hours is usually subtracted from computed GAT. If this results in a negative number, add 24 hours back.

$$\text{If } GAT > 12 \quad GHA = GAT - 12^h \quad (6.1a)$$

$$\text{If } GAT < 12 \quad GHA = GAT + 12^h \quad (6.1b)$$

At this stage of the computation GHA is in units of hours. It must be converted to angular units by multiplying it by 15 degrees/hour. It is important that the conversion does not allow loss of precision. It is this stage where the **tenths** of seconds so carefully retained in the time have meaning. Some calculators do not provide for entering hours-minutes-seconds with the conversion to decimal hours retaining the precision of tenths in the seconds. If so, the calculation must be performed without using the automatic conversion. Loss of precision in this step could result in a few seconds error in azimuth.

Example 6.1 Convert 10:20:25.6 GHA to angular units

Solution: $\quad \dfrac{25.6^s}{60} = 0.4267^m \qquad \dfrac{20.4267^m}{60} = 0.340444^h$

$GHA = 10.340444^h \times 15\ °/hr = 155.106666°$

$0.106666° \times 60\ '/° = 6.399996' \qquad 0.399996 \times 60\ '/" = 24"$

$\therefore GHA = 155°06'24"$

Had 10:20:26 been used, the incorrect answer of 155°06'30" would have resulted.

Example 6.2 One example in Section 5-3 gave a solar observation made at UT=10:28:29.3, and after correction for equation of time, GAT=10:14:56.9. Compute the GHA of the observation.

Solution: $\quad GHA = GAT + 12^h \qquad (6.1b)$

$GHA = 10:14:56.9 + 12:00:00 = 22:14:56.9$

$GHA = 22.249139$ (in decimal hours)

$GHA = 22.249139 \times 15\ °/hr = 333.73708$ (in decimal degrees)

$GHA = 333°44'13.5"$

Alternatively,

```
22 x 15   =           = 330°
14 x 15   = 210'      = 3°30'
56.9 x 15 = 853.5"    =   14'13.5"
              GHA     = 333°44'13.5"
```

6-2 COMPUTATION OF GREENWICH HOUR ANGLE

Polaris Observations

The ephemeris lists the GHA for Polaris with no need to consider shifting the time 12 hours. However, the ephemeris listing is for zero hours and must be changed to the GHA at observation time on the date. This can be done by interpolating in the tables between the zero hour value for the date of observation and for the next date, or by using the average change per hour of 15°x1.0027379. An example will illustrate this computation. But, let us first explain the nature of the relationship of time and hour angle in reference to the stars.

There are 366.2422 sidereal (star) days in a year, covering the same time span as 365.2422 solar days. The extra day results because the stars are vastly further away than the sun and because the earth revolves around the sun once a year, in effect gradually subtracting one of the days as it orbits. The earth takes a full 360° to turn and face the stars each rotation, but because if moves 1/365 of its orbit each day, it requires less angle or less time to face the sun. The difference is approximately $3^m 56.5^s$, or more precisely 0.0027379 hours/hour. Since the time observed is on the solar time scale, this scale must be changed to account for the fact that the earth's rotation rate with respect to the stars is relatively different. The average increase of GHA (or any hour angle) of Polaris is

$$\frac{366.2422}{365.2422} = 1.0027379$$

in relation to angular time relationships using the sun. Though the solar time may be 24 hours as read from a watch, this represents 24.065710 hours or 360°59'08" angular change in GHA of Polaris. This value varies a few seconds as the earth revolves around the sun, but using the year's average will never affect a computed azimuth more than 0.5" arc and this would only occur with extreme conditions of Polaris being at culmination, UT approaching 24 hours, and at the time of the year when the variation of this rate of change is maximum.

To compute GHA for a Polaris observation, read the value for zero hours in the table and add it to

$$UT \times 1.0027379 \times 15 \text{ °/hr}$$

or,

$$GHA_{UT} = GHA_0 + UT(15.04107)$$

where GHA_{UT} is the GHA at observation, GHA_0 is the GHA listed in the ephemeris for zero hours, and UT is the observation time as explained in Section 6-1.

Example 6.3 Compute GHA of Polaris at local daylight savings time of 8:05:03 P.M. as observed on the watch, Central Time Zone, USA. The watch is $02^m 12^s$ fast as calibrated with time signals. The date is October 20, 1984.

Solution:

```
                    8:05:03 + 12 hrs = 20:05:03   (on 24 hr. clock)
                    watch correction =   -02:12
            Central Daylight Time = 20:02:51
  Correction, Daylight Saving Time = -1:00:00
            Central Standard Time = 19:02:51
                  Zone Correction = +6:00:00
                              UT = 25:02:51      October 20, 1984
                              UT =  1:02:51      October 21, 1984
                                 = 1.04750       hours
```

$$GHA_0 = 355°20'54''$$
$$1.04750 \ (15.04107) = 15°45'20'' \quad (15.75552 \text{ degrees})$$
$$GHA_{UT} = 371°06'14'' \text{ or } 11°06'14''$$

Example 6.4 Observed UT=22:04:50 using a chronometer calibrated with UT. November 10, 1984 is the date.

Solution:

$$UT = 22:04:50 = 22.080555 \text{ hour}$$
$$GHA_0 = 15°02'51''$$
$$22.080555 \ (15.04107) = 332°06'55'' \quad (332.11517 \text{ degrees})$$
$$GHA_{UT} = 347°09'46''$$

6-3 COMPUTATION OF HOUR ANGLE

This computation is the same for solar and Polaris observations. Local Hour angle is

$$LHA = GHA - \lambda_w \tag{6.2a}$$

or

$$LHA = GHA + \lambda_E \tag{6.2b}$$

In the United States, λ_w is used. Henceforth the subscript, w, will be omitted, it being understood that if longitude is subtracted, it is a west longitude. Astronomic longitude (See Section 3-2) should be used in solar observations.

The hour angle, t, is the angle measured from the observer's meridian to the hour circle through the celestial body and is measured either east or west, depending on which is smaller. Thus, it is the same in definition as LHA, but is reckoned either east or west, as necessary to form a solvable spherical triangle.

If LHA > 180°,   t = 360° - LHA,   t is east and ***positive.***

If LHA < 180°,   t = LHA   ,   t is west and ***negative.***

COMPUTATION OF HOUR ANGLE

It is important to recognize the algebraic sign of t when the azimuth equation is solved.

Example 6.5 In a previous example, the GHA of the sun was found to be 333°44'13.5". The longitude was scaled from a USGS map, corrected to astronomic λ and found to be 72°14'13.5". Find LHA and t.

Solution:
$$\begin{aligned}
\text{GHA} &= 333°44'13.5" \\
\lambda &= \underline{\ 72°14'13.5"} \\
\text{LHA} &= 261°30'00.0" \\
t &= +98°30'00.0" \quad \text{(east)}
\end{aligned}$$

Example 6.6 A Polaris observation was made and a GHA=151°04'50" determined. The longitude was scaled from a map and found to be 81°36'40". Find LHA and t.

Solution:
$$\begin{aligned}
\text{GHA} &= 151°04'50" \\
\lambda &= \underline{\ 81°36'40"} \\
\text{LHA} &= 69°28'10" \\
t &= -69°28'10" \quad \text{(west)}
\end{aligned}$$

Example 6.7 A solar observation was made and GHA=45°12'16" computed. λ=95°45'06". Find LHA and t.

Solution:
$$\begin{aligned}
\text{GHA} &= 405°12'16" \\
\lambda &= \underline{\ 95°45'06"} \\
\text{LHA} &= 309°27'10" \\
t &= +50°32'50"
\end{aligned}$$

These examples illustrate the situations where LHA > 180° and LHA < 180°, and also the situation where GHA is less than λ, necessitating adding 360° to it before subtracting λ. From the values of t, the surveyor can immediately begin to check and analyze results. In the first and third example, the t value reveals that the observation was made in the morning, since t is positive (east). The first example indicates a very early morning summer observation because t is greater than 90° and this could only occur when the days are long and the sun rises in the northeast, north of the prime vertical. Example 6.6 results in a t value that will yield a bearing of Polaris in the northwest quadrant since t is west. Furthermore, Polaris was moving downward and slightly toward the west, approaching west elongation in another 20° or 1.3 hours rotation time. These intermediate inspections and observations aid in understanding the theory and analyzing results. Readers are encouraged to check results in this way. For example if one knew that Example 6.5 represented an afternoon observation in November, there would be little sense in computing the azimuth until the mistake in t was found!

6-4 PROCEDURES FOR AZIMUTH CALCULATION

Individual Computation of Observations

Observations are computed individually for maximum precision and opportunity to analyze results. The simplest way to determine an azimuth would be to average all times and angles, and make one azimuth computation. But, mistakes or gross errors would not be discovered this way, nor would there be opportunity to study data distribution, ranges, and overall precision. Some may feel that at least averaging the time and angle data in each set would be advantageous and save time. But, there are several reasons why it is better to calculate each observation independently of the others.

First of all, it gives more "results" to view, so as to analyze errors and mistakes. If a faulty pointing had been made in one observation, the surveyor wouldn't know whether it was in the direct or inverted observation if they were first meaned and computed in sets. But, by computing separately, all direct and inverted (or all left and all right limb) observations should agree favorably, and the faulty observation can be more directly spotted. The entire set might need to be omitted anyway, but sometimes a new set can be created from a direct and inverted pointing from two sets having one bad observation, if the fault was in a direct observation in one and in the inverted observation in the other. Thus, three sets could be salvaged from four original if any one direct and any one inverted pointing was not agreeing with its counterparts among the other sets. But, had the angle and time data been meaned first, and four azimuth calculations performed, the result would be omitting two sets, leaving only two, since the opportunity to use "halves" of each set would not have revealed itself.

Secondly, the movement of the sun is not in a straight line, but in an arc as it moves across the sky between sunrise and sunset. Meaning angles and times assumes a straight-line, rather than curved movement of the sun and the mean position between pointings would not be as it actually was at the mean time for the set; therefore, an angular error results in the computed azimuth. The error would be small if pointings between left and right limbs were a few minutes apart; but nevertheless, the precision of the method would always be affected when third-order precision is sought. When computed individually, any left limb pointing can be used with any right limb pointing (if the telescope is in the opposite position) in any set-up, no matter how much time has passed between them, and the result is compensated for instrumental error, with no errors from sun's curved movement being involved. The only caution here is in regard to the effect of changing altitude on sun's semi-diameter. This is discussed in Appendix B and is of concern if several minutes of time pass between limb pointings and h is large.

A third reason why observations are not combined into sets to reduce the number of equation solutions is the ever-present possibility that the step of meaning times and angles results in an undetected arithmetic mistake. Then, if not traced as such, field data may be judged faulty and computations discarded unnecessarily.

Seeming simplifications often create more problems, cause more time and effort, reduce chances to properly analyze results, and adversely affect accuracy.

Format

The form used in the examples may vary slightly. There is nothing fixed about the format. The author has found it convenient to squeeze eight observations onto one computation page because it is then easier to make calculations for each variable all at one time, across the page. This helps speed the calculations and check each step as the work progresses. Assuming four sets of observations are being calculated, this means one page contains everything, including an easy visual study of results, final or intermediate. As experience is gained, readers may wish to modify the form to suit their individual preferences. Any change is appropriate as long as it doesn't lead to mistakes or modify the theory. (Note: The forms reproduced herein have been reduced slightly).

Although discussion of each step, vertically down the form, will be done for one observation in the examples, it should be recognized that, in reality, each step was completed for all observations before going to the next one, a procedure that proves to work efficiently and seems to result in fewer mistakes.

Breakdown of Calculations

The calculations can be visualized as consisting of six steps for each observation:

1. Calculation of UT from UTC, DUT1, watch corrections, zone and daylight savings time corrections, as applicable.
2. Calculation of hour angle, t, using ephemeris data and longitude of observation.
3. Determination of declination, δ, from ephemeris.
4. Solution of azimuth equation for astronomic azimuth to celestial body.
5. Calculation of horizontal angle between mark and celestial body.
6. Calculation of astronomic azimuth to ground mark.

After these steps for each observation, they are treated as follows:

1. Investigate agreement of data, possible mathematical errors or mistakes, discard some observations if necessary.
2. Combine acceptable results into sets, D&R, L&R sun limbs, and calculate mean azimuth for each set.
3. Investigate distribution of mean azimuths in sets and, if acceptable, mean all azimuths to arrive at the best value for astronomic azimuth to ground mark.
4. Consider other errors discussed in Appendix B, if necessary, and apply corrections if appropriate.

Angle Calculation

A horizontal angle is the difference between two directions. When defined this way, it is recognized that the angle from the ground azimuth mark to the celestial body is always computed by subtracting the reading (direction) to the ground mark from the reading to the celestial body, regardless of whether a direction or a repetition instrument was used. When using a direction theodolite, the reading to the celestial body is often a smaller number than the one to the ground mark, depending on the circle setting. If so, add 360° to the reading to the celestial body before attempting the subtraction for the angle. Consistency in the way the angle is computed, being always as measured *from* the ground mark *to* the celestial body (and not vice versa) will help avoid mistakes.

Time Used in Computation

For Polaris observations, the time entered into the ephemeris for initial stages of the computation is the UTC read directly from the short-wave radio or chronometer calibrated with UTC, or the UTC time as determined by applying watch and/or zone corrections to observed watch time. It is important to use the correct date. Often, evening observations result in the ephemeris date being the next day. Thus the date of observation in the field notes, locally, may not be the UT time date.

For solar computations, the time should be corrected for DUT1 as explained in earlier sections. As with Polaris, evening observations in some time zones can occur the next day, UTC, so that the local field date may be incorrect.

Latitude and Longitude Used in Computation

Recall that for best accuracy, astronomic position must be used, particularly for solar observations. For Polaris observations, the map-scaled position is accurate enough. (See Section 3-2). For the examples to follow the correction was ignored since it was small and within the map-scaling accuracy and range of the values scaled by the seminar participants.

6-5 SOLAR AZIMUTH COMPUTATIONS

General Comments

The two examples to be discussed are observations made on the campus of Louisiana State University in conjunction with one of the author's workshops. The data were gathered by the author in late afternoon the day before the seminar and at the beginning of the seminar the next morning to demonstrate the observation and calculation techniques. A Polaris observation was also made after the afternoon solar observations which provided additional data to prove the precision of the methods and provide data for the workshop participants to use to compare with their own data gathered later. The first note form excludes complete instrument station descriptions and other refinements so as not to unnecessarily clutter these examples. The essentials are on the forms. The original notes are used. The afternoon observation was made just before sundown, a little later than desired due to a delay in arrival to the site. An attempt for six sets was aborted due to the sun sinking behind trees on the horizon. But, five good sets were completed in less than 20 minutes time.

Example S1

This example includes calculation of the first four sets of observations taken on the afternoon of January 14, 1982. Figures 6.1 and 6.2 are the field notes and calculations. The ephemeris table is at the beginning of Appendix F.

After calculating all GHA's, LHA's, t angles, declinations, and horizontal angles between ground target and celestial mark using procedures thus far discussed, the equation for azimuth was solved. Prior to this, the intermediate steps were examined to see if they seemed logical. For example, such questions as--do all UT's increase, do all GHA's increase, do all LHA's

Figure 6.1 Solar Field Notes, Afternoon Observation

S1 *FIELD NOTES FOR ASTRONOMIC OBSERVATION*

Astronomic Target __Sun__ Latitude of Instr. __LSU 30°24' 36"__
Instrument Station __LSU__ Longitude of Inst. __91° 11' 10"__
near Stadium Mark __WBRZ__ Date __1/14/92__ Observer __RBB__ Rec __Breaux__
Baton Rouge ~ 4:30 PM Local Time
DUT1 = 0.0   Limb

SET D/R	MARK READ DEG	MIN	SEC	STAR/SUN READ DEG	MIN	SEC	TIME UTC HR	MIN	SEC	Limb
1-D	0°	26'	06"	236	~~36~~ 10'	~~14~~ 58"	22	29'	37.0	L
1-R	180°	25'	42"	56°	57	30	22	31	12.0	R
2-R	225	31'	13"	102°	04'	23"	22	35	8.9 (08.9)	L
2-D	~~282'~~ 45°	~~46~~ 31'	~~52"~~ 31"	282	46	52	22	36	13.0	R
3-D	90°	~~38~~	00"	327°	55'	14"	22	40	19.2	L
3-R	270°	37'	34"	148°	37'	45"	22	41	29.7	R
4 R	315	23	09	193°	13'	30"	22	44	13.0	L
4 D	135	23	36	13°	55'	05"	22	45	13.8	R
5 D	0°	43'	34"	239	07	24	22	48	13.4	L
5-R	180°	43'	15	59	47	14	22	49	07.2	R
6-R	270°	42'	33"	149	39	12	22	52	12.0	} Image Into Trees
6							22	53		

DUT1 = 0.0

Figure 6.2 Solar Computation, Afternoon Observation

S1 Solar Observation Computation

LSU Campus

- Sta: **LSU**
- Target: **WBRZ Tower**
- Date: **1/14/82 4:30 PM**
- Inst. type: **T2E (1974)**
- Observer: **RBB**
- Computer: ~~RBB~~ **RBB**
- ϕ **30° 24' 36"**
- λ **91° 11' 10"**

Observation	1D (L)	1R (R)	2R (L)	2D (R)	3D (L)	3R (R)	4R (L)	4D (R)
UT	22:29:37.0	22:31:12.0	22:35:08.9	22:36:13.0	22:40:19.2	22:41:29.7	22:44:13.0	22:45:13.8
Eq. Time, 0	-8:50.9							8:50.9
-0.91^s × UT	20.5	20.5	20.6	20.6	20.6	20.7	20.7	20.7
Eq. Time	-9:11.4	-9:11.4	-9:11.3	-9:11.3	-9:11.3	-9:11.2	-9:11.2	-9:11.2
less 12 hrs.	-12:00:00							-12:00:00
GHA, hours	10:20:25.6	10:22:00.6	10:26:57.6 (25)	10:27:01.7	10:31:07.9	10:32:18.5	10:35:01.8	10:36:02.6
GHA, arc = GHA$_h$ × 15	155 06 24	155 30 09	156 44 24 (29)	156 45 25	157 46 58	158 04 37	158 45 27	159 00 39
λ	91 11 10	91 11 10	91 11 10	91 11 10	91 11 10	91 11 10	91 11 10	91 11 10
LHA	63 55 14	64 18 59	65 33 14 (18)	65 34 15	66 35 48	66 53 27	67 34 17	67 49 29
t	-63 55 14	-64 18 59	-65 33 14 (18)	-65 34 15	-66 35 48	-66 53 27	-67 34 17	-67 49 29
δ, o	-21 23 52							
26.5" × UT	09 56	09 57	09 59	09 59	10 01	10 01	10 03	10 03
δ	-21 13 56	-21 13 55	-21 13 53	-21 13 53	-21 13 51	-21 13 51	-21 13 49	-21 13 49

If LHA > 180°, t = 360° - LHA, t is east and +

$$\tan Z = \frac{\sin t}{\tan\delta \cos\phi - \sin\phi \cos t}$$

If LHA < 180°, t = LHA, t is west and -

Azimuth, Z	238 10 04	238 23 56	238 58 13	239 07 25	239 42 29	239 52 27	240 15 25	240 23 54
Angle, Target to ☉	235 44 52	236 31 48	236 33 10	237 15 21	237 17 14	238 00 11	237 50 21	238 31 29
Astro. Az. to target	2 25 12	1 52 08	2 25 03	1 52 04	2 25 15	1 52 16	2 25 04	1 52 25
Set Average	2° 08' 40"		2 08 33.5		2 08 45.5		2 08 44.5	

Average Astronomic Azimuth to Target = **2 08 40.9**

45^m calculation time

increase, do all t angles increase (or decrease) as viewed from left to right across the page? GHA and LHA must always increase as time passes. The t angle will decrease if east, increase if west. Computational, data entry, or field mistakes would cause reversal or interruption in these trends. Note that the DUT1 correction is zero, resulting in the UT for computation being the same as read in the field. This wouldn't usually be the case.

After calculating all observations down to the δ line for all observations, and entering the horizontal angles on the form, the azimuth equation was solved for each observation using an electronic calculator with storage capacity so that no trigonometric functions or arithmetic steps were written down. Only the final azimuth was entered on the form. For illustration purposes only, the observation 1D will be solved in detail so that readers can check the solution done semi-automatically in the calculator.

$$\sin t = \sin (-63°55'14") = -0.89818535$$
$$\tan \delta = \tan (-21°13'56") = -0.38852157$$
$$\cos \phi = \cos\ \ 30°24'36"\ \ =\ \ 0.86242534$$
$$\sin \phi = \sin\ \ 30°24'36"\ \ =\ \ 0.50618429$$
$$\cos t = \cos (-63°55'14") = 0.43961696$$
$$\tan \delta \cos \phi = -0.33507085$$
$$\sin \phi \cos t = 0.22252720$$
$$\tan \delta \cos \phi - \sin \phi \cos t = -0.55759805$$
$$\tan Z\ \ \ \ = 1.61081150$$
$$\text{Arctan } Z = 58.167779° = 58°10'04"$$
$$Z\ \ \ \ \ \ \ \ = 238°10'04"$$

The reader should note the strict attention to algebraic signs and carrying at least eight places throughout. The tan Z and Arctan Z are positive in this example, indicating that the azimuth must be either in the northeast or southwest quadrant. The fact that the observation was afternoon or that t was west dictates the choice of the two. Thus 180° is added to the acute angle indicated by the calculator and the result is an azimuth of 238°, not 58°. At this stage, one should pause to ask if this is a logical value. It is because it was late afternoon in mid-winter when the sun sets southwesterly by several degrees south of due west. This last check gives confidence that the remainder of the azimuth calculations can proceed.

The equation was solved for Arctan Z for each observation. The horizontal angles were then subtracted from the computed azimuths to the sun. The result in each case is an azimuth affected by the instrumental errors and in error by half the sun's diameter. (See Figure 2.4). Azimuths using the left limb should agree. Likewise, those computated from right limb pointings should agree. Lack of agreement is affected primarily by instrument errors--some left limb pointings were made with telescope direct, some with it inverted; likewise with the right limb pointings. After discovering one mistake in a GHA computation, it is seen that the discrepancies are small, the range being 12" in the left limb pointings and 21" in the right limb pointings. These discrepancies are typical, considering the unknown instrumental error. This check has value primarily to discover computational and field mistakes.

The real check on quality is after the L and R limb observations are meaned into sets. In this example, it is seen that this range is 12" and that no set average deviates more than 7.5" from the overall mean. Some may wish to omit the second set but it is probably as good as the others, statistically, and no real basis exists to omit it without additional data.

Example S2

The field notes of Figure 6.3 are data taken from the same station the following morning. Four sets were completed in 17 minutes time. Computations are illustrated in Figure 6.4. The steps will not be discussed in detail, it being assumed that the reader can check and follow the calculations now. The only arithmetic step not done in the last example was adding 360° to the GHA in order to have a positive LHA after subtracting λ. Note that t is positive in this example since it is east. The sign of tan Z will be negative. The calculator will indicate a value for the first azimuth of -49°55'54". Tan Z is negative only in the southeast and northwest quadrants. Because the observation was made in the morning, this dictates a southeast direction. Therefore, the azimuth is 180° +(-49°55'54")=130°04'06". Its size is about right for a 9 A.M. pointing in the winter, indicating that there are probably no large blunders in the field work or computations.

The results differ from the previous example in that there are somewhat larger discrepancies, the values for left limb and right limb pointings do not match those of the afternoon observation, and the mean is lower. The larger discrepancies and the different values for the individual azimuths can be attributed primarily to instrumental errors. A different instrument, older and not as clean as the previous one, was used. Furthermore, its instrumental error in the axes relationships was opposite that of the other instrument, as is revealed by studying the closures to the ground azimuth mark. Thus, a difference in systematic errors in the instruments is the probable cause of the differences in left and right limb results between the two sets of observations. The range in the final results is not as narrow as in the previous example. Considering all random errors combined, the 18" range is not excessive. Usually the scatter and range are better than in this set. The importance of a clean and well-maintained instrument, of good manufacture, cannot be overlooked.

A summary of the results of all nine sets of solar observations is as follows:

```
        Afternoon January 14 -- 1    2°08'40.0"
                                2         33.5"
                                3         45.5"
                                4         44.5"
                                5         36.0"  (computation not shown)
        Morning   January 15 -- 1         35.0"
                                2         35.5"
                                3         28.0"
                                4    2°08'46.0"
                        Mean = 2°08'38.2"      Range = 18"
                           σ = ±6.2             E₉₀  = ±10.2
                          σₘ = ±2.1            E₉₀ₘ =  ±3.4
```

Figure 6.3 Solar Field Notes, Morning Observation

S2 — FIELD NOTES FOR ASTRONOMIC OBSERVATION

Astronomic Target **Sun**　　Latitude of Instr. **LSU 30° 24' 36"**
Instrument Station **LSU**　　Longitude of Inst. **91° 11' 10"**
(near Stadium)　Mark **W8RZ**　　Date **1/15/82**　Observer **RBB**　Rec **S.V.E.**
　　　　　DUT1 = 0.0　　　　9:00 A.M Local

SET D/R	MARK READ DEG	MIN	SEC	STAR/SUN READ DEG	MIN	SEC	TIME UT HR	MIN	SEC	Limb
1D	0°	41'	21" (48")	128°	19'	48"	14	50	43.6	L
1R	180°	41'	37"	309°	15'	05"	14	52	43.0	R
2R	225°	25'	47"	354°	02'	27"	14	56	15.8	L
2D	45°	25'	37"	174°	48'	54"	14	57	23.5	R
3D	90°	34'	36"	219°	58'	34"	15	00	39.8	L
3R	270°	34'	56"	40°	50'	37"	15	02	13.2	R
4R	315°	30'	17"	85°	56'	06"	15	06 (05)	21.2	L
4D	135°	29'	54"	266°	43'	39"	15	07	31.1	R

Figure 6.4 Solar Computation, Morning Observation

S2 Solar Observation Computation

Sta: **LSU** Inst. type: **T2E** φ **30 24 36**
Target: **WBRZ Tower** Observer: **RBB** λ **91 11 10**
Date: **1/15/82 9AM** Computer: ~~Staff~~ **RBB**

Limb →	L	R	L	R	L	R	L	R
Observation	1D	1R	2R	2D	3D	3R	4R	4D
UT	14:50:43.6	14:52:43.0	14:56:15.8	14:57:23.5	15:00:39.8	15:02:13.2	15:06:21.2	15:07:31.1
Eq. Time, 0	−9:12.7							
-0.88^s × UT	13.1	13.1	13.2	13.2	13.2	13.2	13.3	13.3
Eq. Time	−9:25.8	−9:25.8	−9:25.9	−9:25.9	−9:25.9	−9:25.9	−9:26.0	−9:26.0
less 12 hrs.								
GHA, hours	2:41:17.8	2:43:17.2	2:46:49.9	2:47:57.6	2:51:13.9	2:52:47.3	2:56:55.2	2:58:05.1
GHA, arc = GHA$_h$ × 15	40 400 19 27	400 49 18	401 42 28	401 59 24	402 48 28	403 11 49	404 13 48	404 31 16
λ	91 11 10	91 11 10	91 11 10	91 11 10	91 11 10	91 11 10	91 11 10	91 11 10
LHA	309 08 17	309 38 08	310 31 18	310 48 14	311 37 18	312 00 39	313 02 38	313 20 06
t	+50 51 43	+50 21 52	+49 28 42	+49 11 46	+48 22 42	+47 59 21	+46 57 22	+46 39 54
δ, o	−21 13 15							
27.5" × UT	06 48	06 49	06 51	06 51	06 53	06 54	06 55	06 56
δ	−21 06 27	−21 06 26	−21 06 24	−21 06 24	−21 06 22	−21 06 21	−21 06 20	−21 06 19

If LHA > 180°, t = 360°−LHA, t is east and +

$$\tan Z = \frac{\sin t\,(5)}{\tan δ \cos φ\,(3)(4)(2) - \sin φ \cos t\,(1)(6)}$$

If LHA < 180°, t = LHA, t is west and −

Azimuth, Z	130 04 06	130 24 59	131 02 32	131 14 36	131 49 51	132 06 44	132 52 26	133 05 00
Angle, Target to ☉	127 38 27	128 33 28	128 36 40	129 23 17	129 23 58	130 15 41	130 25 49	131 13 45
Astro. Az. to target	2 25 39	1 51 31	2 25 52	1 51 19	2 25 53	1 51 03	2 26 17	1 51 15
Set Average	2 08 35.0		2 08 35.5		2 08 28.0		2 08 46.0	

Average Astronomic Azimuth to Target = **2° 08' 36.1"**

Figure 6.5 Polaris Observation Field Notes

P1 FIELD NOTES FOR ASTRONOMIC OBSERVATION   LSU Campus

Astronomic Target ★   Latitude of Instr. 30°24'36"
Instrument Station LSU   Longitude of Inst. 91°11'10"
near stadium Mark WBRZ   Date 1/14/82 Observer RBB Rec Breaux
Polaris.   5:30 PM Local

SET D/R	MARK READ			STAR/~~Sun~~ READ			TIME UT (uncorrected)			CORR'N	
	DEG	MIN	SEC	DEG	MIN	SEC	HR	MIN	SEC	MIN	SEC
1-D	0°	45	19	358	54	51	23	25	13	+2	20
1-R	180 ~~178~~	44 ~~54~~	43 ~~00~~	178	54	00	23	27	20		
2-R	225	29	26	223	38	06	23	29	47		
2-D	45	29	56	43	38	09	23	31	33		
3D	90	36	51	88	44	24	23	34	03		
3R	270 ~~268~~	36 ~~43~~	28 ~~39~~	268	43	39	23	35	07		
4R ~~4D~~	315	41	10	313	47	52	23	37	55		
4D	135	41	34	133	48 ~~40~~	00	23	39	15	+2	20

Read 40ᵐ 40ˢ
True 43ᵐ 00
Watch is 2ᵐ 20ˢ slow

T = R + C
C = T − R = +2:20

6-6 POLARIS AZIMUTH CALCULATIONS - EXAMPLE P1

The data of Figure 6.5 were taken soon after the solar observation on January 14, as the observation time reveals. The gap in time between the last solar and first Polaris pointing (33 minutes) was spent performing a preliminary calculation, then going through the field steps of finding the star. The actual observations for four sets required an additional 14 minutes of time. A wrist watch, calibrated to UTC was used. The watch and time signal readings for this calibration are shown in the notes, the watch having read $40^m \ 40^s$ when the announced time was $43^m \ 00^s$. The same instrument and set-up were used as with the afternoon solar observations.

The computations are shown in Figure 6.6. The reader should be able to follow and check the calculations, having been provided earlier examples for each step. The algebraic sign of tan Z is positive for all observations, indicating a northeast bearing, a fact that is also easily seen by the value of LHA. The LHA also shows that the star is moving westerly, approaching upper culmination. The azimuths are given on the form as 360° (plus the computed minutes and seconds) rather than 0° merely to enable subtracting the angle from the ground mark to the star.

The agreement among the individual azimuths is quite satisfactory, as is the agreement among the means in the sets. The overall mean of 2°08'35.6" agrees well with the solar observations, indicating that any of the sets of observations could be used as representing the third-order azimuth to this mark.

It should be noted that five sets of solar and four sets of Polaris observatons were completed from one instrument set-up in approximately 70 minutes observation time, including calculation for finding Polaris and then the time to actually find it. The result is an azimuth with an overall mean of 2°08"37.8", not deviating more than 2 or 3 seconds from the mean for either method alone. This Polaris example, with Example S1, more or less illustrates how one method can be used to check another in a short period of time, with a result having the benefits of the precision afforded by using the mean of a larger number of acceptable observations.

6-7 FINDING POLARIS IN DAYTIME - EXAMPLE PP1

At the conclusion of the field observations for Example S1, a calculation to find Polaris was made. The calculations for this are shown in Figure 6.7. Section 4-3 described field techniques. The following is an explanation of the calculations. First, it can be realized that high precision in the calculation is not required, since an error of a few minutes in final results will not cause Polaris to be outside of the field of view, as discussed in Section 4-3. Thus, the planned time is approximate (within a few minutes), ϕ and λ can be used to the nearest minute, and seconds can be omitted anywhere during hour angle calculations. (Figure 6.7 is on page 80.)

Following the form from top to bottom (lines are numbered for easy reference), Line 1 is the planned time, UTC. Line 2 is the GHA at 0 hours for January 14, 1982, listed in the ephemeris. Line 3 is the increase for 23:00 hours UT being 23.00 times the average 1.0027379x15°/hr (rounded to 4 places). Line 4 is the GHA of the observation, computed by adding lines 2 and 3. Line 6 is the LHA, computed by GHA-λ. The t angle is 360°- LHA

Figure 6.6 Polaris Observation Computation

P1   Polaris Computation Form

Sta: LSU   **Inst. type:** TZE (1974) ϕ = 30°24'36"
Target: WBRZ Tower   **Observer:** RBB λ = 91°11'10"
Date: 1/14/82 5:30 PM   **Computer:** RBB δ = 89 11 11

Observation	1D	1R	2R	2D	3D	3R	4R	4D
UT Time Uncorr.	23:25:13	23:27:20	23:29:47	23:31:33	23:34:03	23:35:07	23:37:55	23:39:15
Zone Corr.								
Watch Corr.	+ 2:20							+ 2:20
GCT (UT)	23:27:33	23:29:40	23:32:07	23:33:53	23:36:23	23:37:27	23:40:15	23:41:35
GHA. 0 hrs.	79°38'54"	79 38 54	79 38 54	79 38 54	79 38 54	79 38 54	79 38 54	79°38'54"
GCT × 1.0027379 × 15°/hr.	352 51 03	353 22 53	353 59 44	354 26 19	355 03 55	355 19 58	356 02 04	356 22 08
GHA	432 29 57	433 01 47	433 38 38	434 05 13	434 42 49	434 58 52	435 40 58	436 01 02
λ	91 11 10	91 11 10	91 11 10	91 11 10	91 11 10	91 11 10	91 11 10	91 11 10
LHA	341 18 47	341 50 37	342 27 28	342 54 03	343 31 39	343 47 42	344 29 48	344 49 52
t	+ 18 41 13	18 09 23	17 32 32	17 05 57	16 28 21	16 12 18	15 30 12	15 10 08

If LHA > 180°, t = 360°−LHA, t is east and positive.
If LHA < 180°, t = LHA, t is west and negative.

$$Z = \tan^{-1} \frac{\sin t}{\tan\delta \cos\phi - \sin\phi \cos t}$$

Azimuth, Z	360 18 17	360 17 47	360 17 12	360 16 46	360 16 11	360 15 55	360 15 15	360 14 56
Angle, Target to ★	358 09 32	358 09 17	358 08 40	358 08 13	358 07 33	358 07 11	358 06 42	358 06 26
Astro. Az. to target	2 08 45	2 08 30	2 08 32	2 08 33	2 08 38	2 08 44	2 08 33	2 08 30
Set Average	2 08 37.5		2 08 32.5		2 08 41.0		2 08 31.5	

Average Astronomic Azimuth to Target =   2 08 35.6

25 min solution time, HP 45

since LHA>180°. The declination is read directly from the ephemeris for January 14. Line 9 is the polar distance, which is 90°- δ.

The latitude, φ, is entered in Line 10 in preparation to correct it for how much Polaris is above or below the elongation positions. If Polaris was *at* east or west elongation, the vertical angle to it would be equal to the latitude. But, in all other positions, a small amount must be added to the latitude to get the true altitude. From the position of Polaris, plotted using the LHA, it is seen that a correction must be added. From a derivation shown in Appendix A,

$$\Delta h = p \cos t \qquad (A.13)$$

The altitude in Line 12 is φ +Δh. Had the star been between t=90° and t=270°, h would be φ -Δh.

Zenith angle is 90°-h, which explains the value in Line 13. Zenith angles, rather than vertical angles (altitude) are read from optical theodolites. With the telescope inverted, the zenith angle reads 360° minus the direct reading, theoretically. Thus, the values in Line 13 and 14 add to 360°. The direct zenith angle reading was set on the theodolite after focusing the objective lens to infinity as explained in Section 4-3. This positions the telescope vertically. The inverted zenith angle reading is used later, when pointing with the telescope inverted.

An approximate equation for Polaris azimuth, derived from the sine law (See Appendix A) is

$$Z = p \frac{\sin t}{\cos h} \qquad (A.12)$$

After solving this using the previously determined values for p, t, and h, it is known that the star is about 24 to 25 minutes from true north. From the value of LHA, it is known that this bearing is northeast. This explains the entries on Lines 15 and 17.

A USGS map was used to plot the instrument station, roughly scale φ and λ, and identify radio tower WBRZ. A protractor was then used to scale the "true" azimuth to WBRZ, this value being 2°10", as entered on Line 16. Using the sketch of the relationship of the star and WBRZ to true north, it is seen that line 17 subtracted from Line 16 yields the angle between the star and WBRZ. This angle is accurate to protractor scale accuracy (perhaps ±15'-30') but is close enough to know where to start looking for Polaris.

The field steps to find the star required measuring the angle to the left of 1°45' from WBRZ and elevating the telescope to 58°52' zenith angle. After a little time and effort, caused primarily by a small amount of haze, and correcting one mistake in the calculation (not shown here), Polaris was found and the previously detailed observation P1 made.

6-8 PROGRAMMED CALCULATIONS

The previous examples, and those to be provided in Appendix C, are all calculated on a form designed for use with a non-programmable calculator. The intermediate computations detailed on the format help to explain the theory, as well as the steps, in that each computed value is included. Reduction of four sets of solar observations requires from 30 to 60 minutes of time; and, four Polaris sets may require 20 to 40 minutes, depending on experience and any mistakes encountered. Although there are advantages in computing using a non-programmable calculator (i.e., checking intermediate results, learning the theory), time will be saved if computations are programmed. After initial programming of the programmable calculator, a complete reduction of four sets of observations should not require more than 10 minutes and there is less chance for mistakes. Since many surveyors may desire to use this method, programs have been provided in Appendix E for popular programmable calculators. These programs follow the same format and sequence of solution as illustrated in the examples elsewhere in this manual. The reader is encouraged to check the examples in this manual to be sure that the same answers (within ±1-second in azimuth) result when using either the programs or a non-programmable calculator.

6-9 SUMMARY AND REVIEW OF PRECAUTIONS

It is hoped that the illustrations and explanations of this chapter have answered the common questions concerning calculations. Questions regarding how to handle increases in equation of time, declination, GHA, and similar interpolation problems are best answered by stating that if attention is paid to algebraic signs and then observing how values increase or decrease in the tables from day to day, mistakes should be avoided.

There are a few steps that require more attention to significant figures than others. Retaining the tenth of a second precision in the time until the last step to determine t is important in solar observations. Using at least 8 places in all trigonometric functions (most calculators carry 10 to 13) is important. Expression of the final azimuth to the nearest second is close enough.

The surveyor should keep in mind the importance of carefully locating and scaling the ϕ and λ of the instrument station since the accuracy of the result is affected, an error of 3" in either component causing approximately the same error in azimuth as an error of 0.2^S in time.

The algebraic sign of t must be carefully observed. If the wrong sign is entered, the numerical value of the azimuth will be correct, but its sign will be wrong since $\sin(-\alpha) = -\sin\alpha$. This is usually no problem except to cause temporary confusion since one can usually determine the quadrant using logic. But, it can result in mistakes. Evidence that a wrong quadrant has been decided usually shows up in the astronomic azimuths decreasing instead of increasing and the ground mark azimuths becoming progressively and uniformly smaller.

If a wrong sign of δ has been used the product of $\tan\delta \cos\phi$ will have the wrong sign since $\tan(-\delta) = -\tan\delta$. Such a mistake will cause the denominator in the equation to have the wrong value after subtracting $\sin\phi \cos t$ (which is always positive). The result, of course, is azimuths which are all wrong.

Other common mistakes are to use the wrong ephemeris date, to omit an hour when a watch is used for Polaris local time and daylight savings time is in effect, to have the wrong sign in the watch correction, to forget the DUT1 for solar observations, to make mistakes in units and decimal places when interpolating for equation of time and declination, to mistake the quadrant of the astronomic azimuth, and to misuse the horizontal angle data in the last step. Mistakes such as these can usually be resolved by close study of the results, but they are time consuming fo find. The best way to deal with them is to avoid them by taking care and by understanding the theory and the procedures.

Appendix C contains additional examples, some with complete solutions, some to be left as student exercises. The reader is encouraged to practice on these examples to check computational skill and help avoid mistakes in actual work which could result in higher costs than the investment in time to study the examples.

The next chapter explains the final refinement to the precise azimuth determination thus far presented--how to convert to geodetic, then grid azimuth so that the surveyor can properly use plane surveying computations and methods.

FIGURE 6.7
Preliminary Calculation to Find Polaris

Date: January 14, 1982 $\phi = 30°24'$
Location: LSU Campus $\lambda = 91°11'$

1. Planned Time, UT = 23:00
2. GHA_0 = 79°38'54"
3. Increase = 15.04 UT = 345°56'40"
4. GHA = 425°35'34"
5. λ = 91°11'
6. LHA = 334°24'
7. t = 25°36'   east
8. δ = 89°11'11"
9. p = 90 - δ = 0°48'49"
10. ϕ = 30°24'
11. Δh = p cos t = 0°44'
12. h = $\phi \pm$ p cos t = 31°08'
13. Zenith angle, direct = 58°52'
14. Zenith angle, inverted = 301°08'
15. Approx. Z = $p \frac{\sin t}{\cos h}$ = 0°24'38"
16. Scaled azimuth from map = 2°10' (to WBRZ)
17. Polaris, approx. azimuth = 0°25'
18. Approx. angle to left, = 1°45'
    WBRZ to Polaris

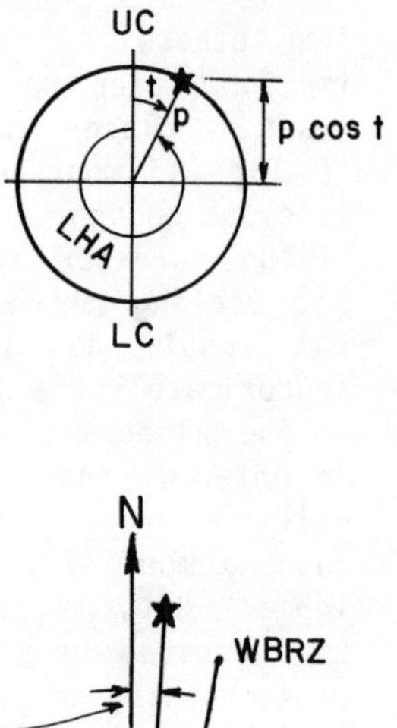

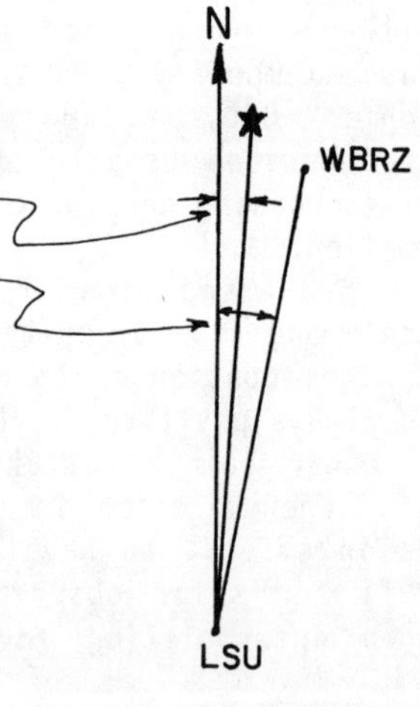

CHAPTER 7

GEODETIC AND GRID AZIMUTH

7-1 THE LAPLACE CORRECTION

The difference between geodetic and astronomic azimuth was introduced in Section 1-2 where reference meridians were defined. The reader may wish to review that Section. The geometry of the deflection of the vertical is illustrated in Figure 7.1.

$$\text{Geodetic Azimuth} = \text{Astronomic Azimuth} + \text{Laplace Correction} \quad (7.1)$$

The Laplace correction is determined by multiplying the difference between the astronomic and geodetic longitude, by the sine of the latitude.

$$\text{Laplace correction} = (\lambda_A - \lambda_G) \sin \phi \quad (7.2a)$$

or,

$$\text{Laplace correction} = \eta \tan \phi \quad (7.2b)$$

The terms η, λ_A, and λ_G were explained in Section 3-2.

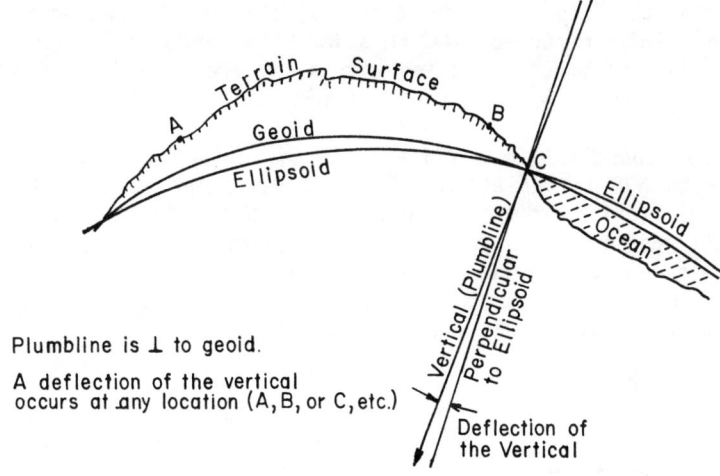

Figure 7.1 Deflection of the Vertical

The surveyor does not determine these deflections during the process of making second or third-order astronomic observations. Deflection stations are geodetic control stations, monumented by the NGS, where observations have been made to determine the deflections. By interpolation, an estimate of the deflections, and thus the Laplace corrections can be made.

In some areas, the deflection stations have been located systematically, providing sufficient density for preparation of an isoline plot (isogram or "contour" plot) of the Laplace corrections. If terrain is fairly flat, and stations are fairly dense, such a plot can be depended upon for at least third-order azimuth accuracy. However, the deflection often does not change uniformly between stations, especially if abrupt changes in topography exist. Isoline plots of Laplace corrections have been prepared by the author for a few states, and are shown in Appendix D. Sparcity of control stations cause uncertainties in some corners of the state of Ohio. Mountains cause uncertainties for the values in Pennsylvania and there was so much irregularity there that the isolines were not even plotted on that map. The plot for Louisiana should be very dependable since the values are small and terrain is flat. In a region where the values are small and do not seem to change abruptly, the map plot might be considered dependable for third-order accuracy, with or without the isolines actually being plotted, a straight line interpolation being used either between lines or as estimated between values for plotted control stations. Uncertainties can be expected to be ±1 to 2".

Appendix D contains a print-out, prepared from Laplace station data provided by NGS for all 50 states in the U.S. The values are given alphabetically by state and with the ϕ and λ position of each station so that the surveyor can plot a Laplace correction map for the region or state where observations are made. Thus, the necessary variables to convert astronomic to geodetic azimuth are included herein for most of the country. At the writing of this manual, the NGS was anticipating preparing astronomic data products for use by private surveyors, including isograms of deflection data and other astrogeodetic data.

Some references, "canned" programs, and other information on azimuth determination ignore the Laplace correction, simply equating geodetic to astronomic azimuth. This is acceptable if azimuths to only ±30" to 1' are being sought, but for the precision being discussed herein, and as a matter of good surveying practice, a systematic error such as the Laplace correction should not be ignored.

In lieu of using Appendix D or isoline plots of Laplace corrections, the surveyor can write to NGS for an estimate of this value and its uncertainty. NGS will use the deflection prediction program discussed in Section 3-2 to arrive at what might possibly be a more accurate value for this correction.

Computation of geodetic azimuth from astronomic is actually only an intermediate step to arrive at grid azimuth. How to compute grid azimuth from geodetic is the objective in the next section.

7-2 MAPPING ANGLE IN SPC SYSTEMS

Introduction

Chapter 8 explains more details on the background and application of the state plane coordinate systems (SPC). For the purposes in this chapter, it is sufficient to say that the SPC provides a way to utilize grid azimuths over large regions within the states, and that these grid azimuths can generally be determined to third-order accuracy using astronomic azimuths, Laplace corrections, and a simple conversion angle to place the azimuths onto the uniform grid for the particular SPC zone.

Two projection systems are used in the United States. The detailed mathematics of the projections are not important to understand. What needs to be known and understood in order to appreciate and properly determine grid azimuths will be explained in this chapter. The cylindrical projection, used in States with generally long north-south dimensions, is called the Transverse Mercator. The conical projection, used in states with generally long east-west dimensions is called the Lambert Conformal.

States are divided into zones along county lines (parish in Louisiana), if necessary. Sometimes a state is comprised of only one zone, but if it extends more than about 158 miles in its narrowest dimension, it will have more than one zone. In each zone is a central meridian with which all other lines are made parallel. Along the central meridian, grid north and geodetic north coincide. For locations east of the central meridian, grid azimuths are less than geodetic azimuths (and vice versa for locations west of the central meridian), because of meridian convergence. See Figure 7.2. The angle of convergence, often called "mapping angle", increases with increased longitudinal distance from the central meridian, and also with increase in latitude. The algebraic sign of this angle is determined by the location of the observation--whether east or west of the central meridian. Examples to follow will illustrate these concepts.

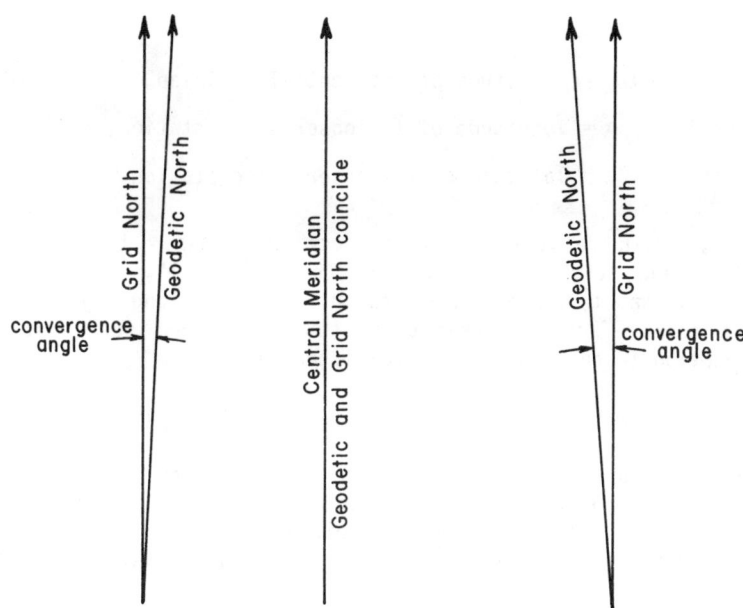

Figure 7.2 Convergence and Mapping Angles

GEODETIC AND GRID AZIMUTH

$$\text{Grid Azimuth} = \text{Geodetic Azimuth} \pm \text{convergence angle} \qquad (7.3)$$

The method of determining this convergence angle will next be explained for the two systems.

Transverse Mercator System

In this system,

$$\text{Grid azimuth} = \text{Geodetic azimuth} - \Delta\alpha - 2^{nd} \text{ term.} \qquad (7.4)$$

The convergence angle is called delta-alpha ($\Delta\alpha$) and it subtracted from the geodetic azimuth. The "2nd term" relates to aspects of the spherical shape of the earth and projection system. It is generally less than 1" arc unless survey lines are approaching five miles in length, are oriented east-west, and are near the zone edges. It can easily be realized that it is negligible for second or third-order work. Thus, let us say that

$$\text{Grid Azimuth} = \text{Geodetic Azimuth} - \Delta\alpha \qquad (7.5)$$

The $\Delta\alpha$ angle is computed as

$$\Delta\alpha = \Delta\lambda \sin\phi + g \qquad (7.6)$$

where

$$\Delta\lambda = \lambda_{CM} - \lambda \qquad (7.7)$$

and

λ_{CM} = longitude of the central meridian,

λ = longitude of the observation station,

ϕ = latitude of the observation station.

The g term, again relates to the projection system. It is a function of the latitude and longitude. It, like the "2nd term", has a maximum value of less than 1" and this maximum occurs only for very long lines at the extreme edges of a zone. Thus, it needs to be considered only for first-order work, and the equation to be used will be rewritten as

$$\Delta\alpha = \Delta\lambda \sin\phi \qquad (7.8)$$

The algebraic sign of $\Delta\alpha$ is determined by the sign of $\Delta\lambda$. The values for ϕ and λ are scaled from a USGS topographic map as part of the astronomic observation data. Values for λ_{CM} are given in Table 7.1 for every zone in the U.S. and other areas where the Transverse Mercator system is employed

$$\text{Grid Azimuth} = \text{Astronomic Azimuth} + \text{Laplace correction} - \Delta\alpha \qquad (7.9)$$

The zone in which the survey lies is probably the one with the central meridian value closest to the λ of the observation station. However, zone boundaries follow county lines and to be sure of the zone, the surveyor should refer to the maps in Figure 7.3 at the end of the chapter. The counties lying in each zone are also listed in the state law covering SPC (for those states having such laws) and are given in SPC projection tables for each state. It is felt that, even though the composite map of Figure 7.3 is small in scale with some county names unclear, a surveyor familiar with an area should be able to identify the county and the zone. Thus, all the information is contained herein to convert an astronomic azimuth to grid azimuth.

Lambert Conformal System

In this system,

$$\text{Grid Azimuth} = \text{Geodetic Azimuth} - \theta + 2^{nd} \text{ term} \qquad (7.10)$$

In this system, the convergence angle is called "the theta angle" (θ), and is subtracted from the geodetic azimuth. The "2nd term" is similar in nature to the one in the TM system. It can be neglected for all but first-order work. Its maximum value is about 3" for a line five miles long, oriented east-west and near the edges of a zone. Therefore, for third-order work and most second-order work,

$$\text{Grid Azimuth} = \text{Geodetic Azimuth} - \theta \qquad (7.11)$$

The θ angle varies with longitude and the north-south location of the SPC zone and other factors. The rate of change of θ per second of longitude is constant in each zone, though it varies among zones, increasing as latitude increases. To determine θ, the surveyor can use the traditional method of interpolating from tables provided in special publications for each state, where it is listed as a function of longitude. However, it is easier to calculate the θ angle from constants furnished in Table 7.2. These constants, defined herein as ℓ, are multiplied by $\Delta\lambda$ to calculate θ.

$$\theta = \ell \cdot \Delta\lambda \qquad (7.12)$$

where $\Delta\lambda = \lambda_{CM} - \lambda$ and λ_{CM} and λ are defined as in the TM system. The algebraic sign of θ is determined by the sign of $\Delta\lambda$.

With the value of λ_{CM} provided in Table 7.1, λ scaled from a map, the values of ℓ in Table 7.2, and the Laplace corrections in Appendix D, the surveyor has the information needed to convert astronomic azimuth to SPC grid azimuth in the Lambert system. The SPC zone must be known, however, in order to know which value of ℓ to use when there is more than one zone in a state. Zones are divided along county lines and do not follow lines of latitude. Figure 7.3 shows the counties and zones of the SPC. Just as with the TM system (see comments there), counties lying in each zone are also listed in applicable state laws and on maps in the SPC projection tables for each state. If the surveyor recognizes the county in Figure 7.3, these outside references need not be consulted.

$$\text{Grid Azimuth} = \text{Astronomic Azimuth} + \text{Laplace Correction} - \theta \qquad (7.13)$$

TABLE 7.1 a
Central Meridians of SPC Zones
Transverse Mercator Zones

State	Zone	λ_{CM} ° '	State	Zone	λ_{CM} ° '
Alabama	E	85 50	Indiana	E	85 40
	W	87 30		W	87 05
Alaska	2	142 00	Maine	E	68 30
	3	146 00		W	70 10
	4	150 00	Michigan (1934)	E	83 40
	5	154 00		C	85 45
	6	158 00		W	88 45
	7	162 00	Mississippi	E	88 50
	8	166 00		W	90 20
	9	170 00	Missouri	E	90 30
Arizona	E	110 10		C	92 30
	C	111 55		W	94 30
	W	113 45	Nevada	E	115 35
Delaware		75 25		C	116 40
Florida	E	81 00		W	118 35
	W	82 00	New Hampshire		71 40
Georgia	E	82 10	New Jersey		74 40
	W	84 10	New Mexico	E	104 20
Hawaii	1	155 30		C	106 15
	2	156 40		W	107 50
	3	158 00	New York	E	74 20
	4	159 30		C	76 35
	5	160 10		W	78 35
Idaho	E	112 10	Rhode Island		71 30
	C	114 00	Vermont		72 30
	W	115 45	Wyoming	1	105 10
Illinois	E	88 20		2	107 20
	W	90 10		3	108 45
				4	110 05

For Zone abbreviations, see page 92.
Source: Reference 13

TABLE 7.1 b

Central Meridians of SPC Zones

Lambert Zones

State	Zone	λ_{CM} ° '	State	Zone	λ_{CM} ° '
Alaska	(Zone 10)	176 00	Nebraska	N	100 00
Arkansas	N&S	92 00		S	99 30
California	1&2	122 00	New York	LI	74 00
	3	120 30	N. Carolina		79 00
	4	119 00	N. Dakota	N&S	100 30
	5	118 00	Ohio	N&S	82 30
	6	116 15	Oklahoma	N&S	98 00
	7	118 20	Oregon	N&S	120 30
Colorado	N,C,S	105 30	Pennsylvania	N&S	77 45
Connecticut		72 45	S. Carolina	N&S	81 00
Florida	N	84 30	S. Dakota	N	100 00
Iowa	N&S	93 30		S	100 20
Kansas	N	98 00	Tennessee		86 00
	S	98 30	Texas	N	101 30
Kentucky	N	84 15		NC	97 30
	S	85 45		C	100 20
Louisiana	N	92 30		SC	99 00
	S&O	91 20		S	98 30
Maryland		77 00	Utah	N,C,S	111 30
Massachusetts	M	71 30	Virginia	N&S	78 30
	I	70 30	Washington	N	120 50
Michigan (1964)	N	87 00		S	120 30
	C&S	84 20	W. Virginia	N	79 30
Minnesota	N	93 06		S	81 00
	C	94 15	Wisconsin	N,C,S	90 00
	S	94 00	Puerto Rico		
Montana	N,C,S	109 30	Virgin Islands, St. Croix		66 26
			Washington D.C.	→ Use Maryland or Virginia	

For Zone abbreviations, see page 92
Source: Reference 13

TABLE 7.2
Theta Angle Conversion Constants, ℓ

State	Zone	Value	State	Zone	Value
Alaska	(Zone 10)	0.79692239	New York	LI	0.65408210
Arkansas	N	0.58189914	North Carolina		0.57717077
	S	0.55969069	North Dakota	N	0.74413340
California	1	0.65388432		S	0.72938260
	2	0.63046797	Ohio	N	0.65695032
	3	0.61223204		S	0.63451954
	4	0.59658714	Oklahoma	N	0.59014707
	5	0.57001192		S	0.56761668
	6	0.54951760	Oregon	N	0.70918602
	7	0.56124321		S	0.68414738
Colorado	N	0.64613348	Pennsylvania	N	0.66153974
	C	0.63068958		S	0.64879317
	S	0.61337805	South Carolina	N	0.56449738
Connecticut		0.66305941		S	0.54465157
Florida	N	0.50252590	South Dakota	N	0.70773818
Iowa	N	0.67774455		S	0.68985196
	S	0.65870102	Tennessee		0.58543973
Kansas	N	0.63271486	Texas	N	0.57953587
	S	0.61452811		NC	0.54539441
Kentucky	N	0.62206727		C	0.51505889
	S	0.60646237		SC	0.48991264
Louisiana	N	0.52870067		S	0.45400685
	S	0.50001270	Utah	N	0.65935549
	O	0.45400685		C	0.64057859
Maryland		0.62763412		S	0.61268734
Massachusetts	M	0.67172866	Virginia	N	0.62411786
	I	0.66109540		S	0.60692482
Michigan (1964)	N	0.72278994	Washington	N	0.74452034
	C	0.70640741		S	0.72639579
	S	0.68052926	West Virginia	N	0.63777297
Minnesota	N	0.74121966		S	0.61819539
	C	0.72338807	Wisconsin	N	0.72137079
	S	0.70092778		C	0.70557663
Montana	N	0.74645181		S	0.68710324
	C	0.73335383	Puerto Rico,		
	S	0.71490124	Virgin Islands,		0.31288823
Nebraska	N	0.67345079	St. Croix		
	S	0.65607640	Washington, D.C. → Use Maryland or Virginia		

For zone abbreviations, see page 92
Source: Reference 21

7-3 GEODETIC AND GRID AZIMUTH CALCULATIONS

Example G1 Determine the grid azimuth at a location in Louisiana where $\phi = 32°00'00"$, $\lambda = 92°00'00"$ and the astronomic azimuth to the mark = $273°14'10"$. This is in Louisiana's north zone.

Solution: From the map of Laplace corrections for Louisiana, the Laplace correction appears to be approximately $+0.3"$ at this location. Therefore

$$\text{Geodetic Azimuth} = (273°14'10") + 0.3" = 273°14'10.3" \quad (7.1)$$

From Table 7.2, $\ell = 0.52870067$. From Table 7.1, $\lambda_{CM} = 92°30'00"$. Therefore,

$$\Delta\lambda = 92°30'00" - 92°00'00" = +0°30'00"$$
$$\theta = 0.52870067 \,(30') = +0°15'51.7"$$
$$\text{Grid Azimuth} = 273°14'10.3" - (0°15'51.7") = 272°58'18.6" \quad (7.11)$$

The value can be rounded to $272°58'19"$, but the tenths should be carried until the last step.

Example G2 Determine the grid azimuth for the observations computed in Chapter 6, the average of all 13 astronomic azimuths being $2°08'37.4"$. $\phi = 30°24'36"$, $\lambda = 91°11'10"$.

Solution: From the Laplace correction map, the Laplace correction is $+0.8"$.

$$\text{Geodetic Azimuth} = 2°08'37.4" + 0.8" = 2°08'38.2" \quad (7.1)$$

As identified by ϕ and λ plotted on the Louisiana map of Figure 7.3, Baton Rouge is in the south zone. From Table 7.1, $\lambda_{CM} = 91°20'00"$ and $\ell = 0.50001270$.

$$\Delta\lambda = (91°20'00") - (91°11'10") = 8'50" = +530"$$
$$\theta = 0.50001270 \,(530") = +265" = +0°04'25"$$
$$\text{Grid Azimuth} = 2°08'38.2" - 0°04'25" = 2°04'13" \quad (7.11)$$

Any example can be illustrated to check the algebraic sign and observe that the convergence is correct. It is seen that since λ is east of λ_{CM}, grid azimuth must be less than geodetic.

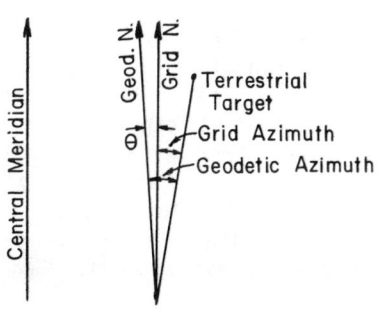

Example G3 Using the station identified as "OHIO2" in Table 3.1 and Section 3.1, compute the Laplace correction at this station and compare it with the value interpolated from the Ohio map in Appendix D.

Solution: From equation 7.2b and Table 3.1

$$\text{Laplace correction} = -5.3" \tan 40°06'20" = -4.5"$$

From the map in Appendix D, the value is approximately -3.7".

Example G4 During a horizontal control survey for a power line project in Illinois, the author made a solar observation for azimuth. Due to suspected problems with the azimuths in the survey, it was decided to check some of them using Polaris observations. Due to minor inconveniences at one point, it was decided to perform this check at the opposite end of the traverse line. The positions of the two traverse courses were determined to be ϕ_{25}=38°13'13", λ_{25}=89°49'49", ϕ_{26}=38°13'13", λ_{26}=89°50'38". The results of the astronomic azimuth computations yielded Z_{25-26}=270°20'02.5", Z_{26-25}=90°19'32.5". The traversing was done using SPC. The data were reduced to grid azimuth as follows:

Solution: From the isoline plot of Laplace corrections it is seen that the Laplace correction is +1.5". From Table 7.1, λ_{CM}=90°10'00", west zone, Illinois.

$$\Delta\lambda_{25} = 90°10'00" - 89°49'49" = +0°20'11" = +1211"$$
$$\Delta\alpha_{25} = 1211" \sin 38°13'13" = +0°12'29.2"$$
$$\Delta\lambda_{26} = 90°10'00" - 89°50'38" = +0°19'22" = +1162"$$
$$\Delta\alpha_{26} = 1162" \sin 38°13'13" = +0°11'58.9"$$

The remainder of the calculations appear in the following table.

Line	Astronomic Azimuth	Laplace Correction	Geodetic Azimuth	$\Delta\alpha$	Grid Azimuth
25-26	270°20'02.5"	+1.5"	270°20'04.0"	0°12'29.2"	270°07'34.8"
26-25	90°19'32.5"	+1.5"	90°19'34.0"	0°11'58.9"	90°07'35.1"

This example illustrates the effect of meridian convergence, and how astronomic or geodetic azimuth cannot be used in surveys if the east-west extent is more than a few hundred feet. Using plane or grid methods, only reference meridians which are parallel can be used. At first glance, one may have thought there was a 30" error in one of the observations or that the methods employed could not check the azimuth any closer than this. The east-west separation of 3970 feet at this latitude caused the 30" discrepancy, not measurement errors or imprecision. After conversion to grid azimuth, there is no discrepancy (to the nearest second). Had the astronomic azimuth been used for a survey such as this, only the line in the survey where the observation was made would be "true". Conversion to grid solves the convergence problem. Furthermore, use of the SPC grid makes it possible for **all** azimuths in a county or region to be compatible, **whether or not actual SPC positions are even calculated.**

Figure 7.4 illustrates the geometry of this example. Other examples employing astronomy, grid azimuth, and SPC combinations are the subject matter of Chapter 8.

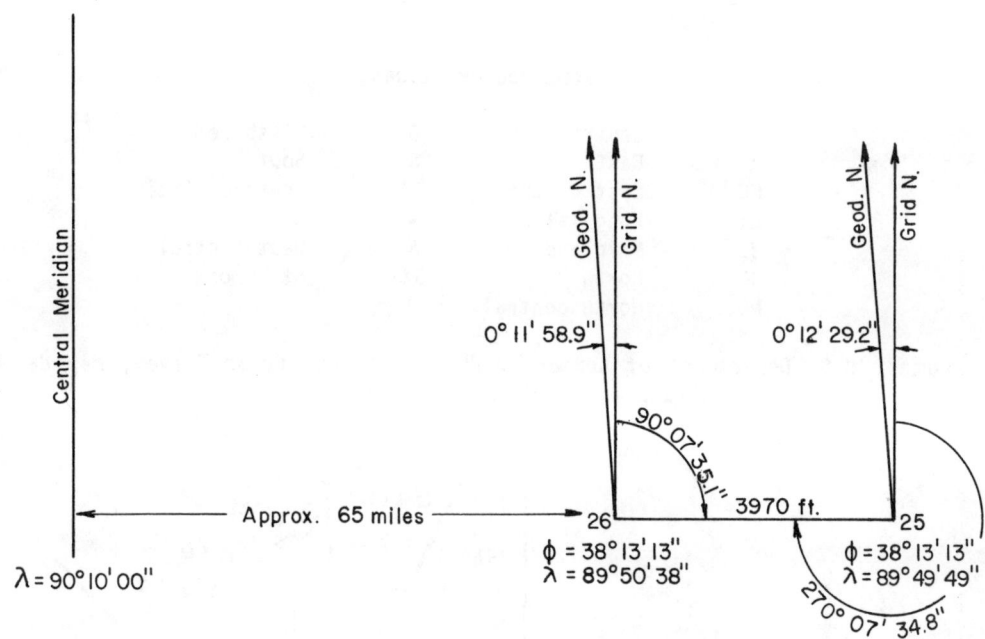

Figure 7.4 Conversions to Grid Azimuth, Example G4

FIGURE 7.3

SPC Zone Boundaries

Zone Code

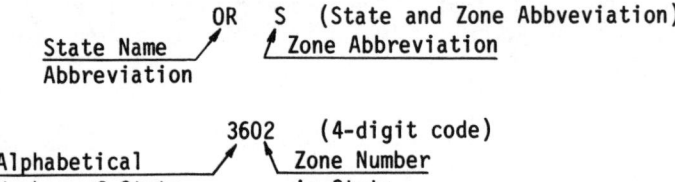

Zone Abbreviations

C	Central	O	Offshore
E	East	S	South
EC	East Central	SC	South Central
LI	Long Island	W	West
M	Mainland	WC	West Central
N	North	SX	St. Croix
NC	North central		

Source: U.S. Department of Commerce, NOAA, National Ocean Survey, Map dated August 1982. Base map by U.S.G.S.

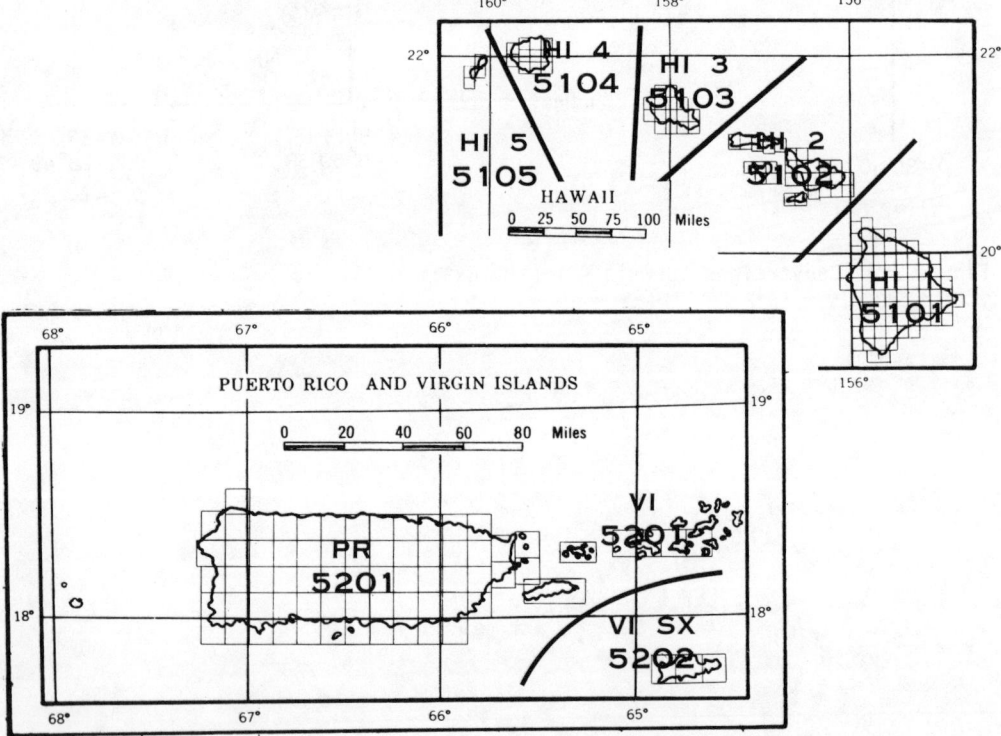

Figure 7.3 SPC Zone Boundaries

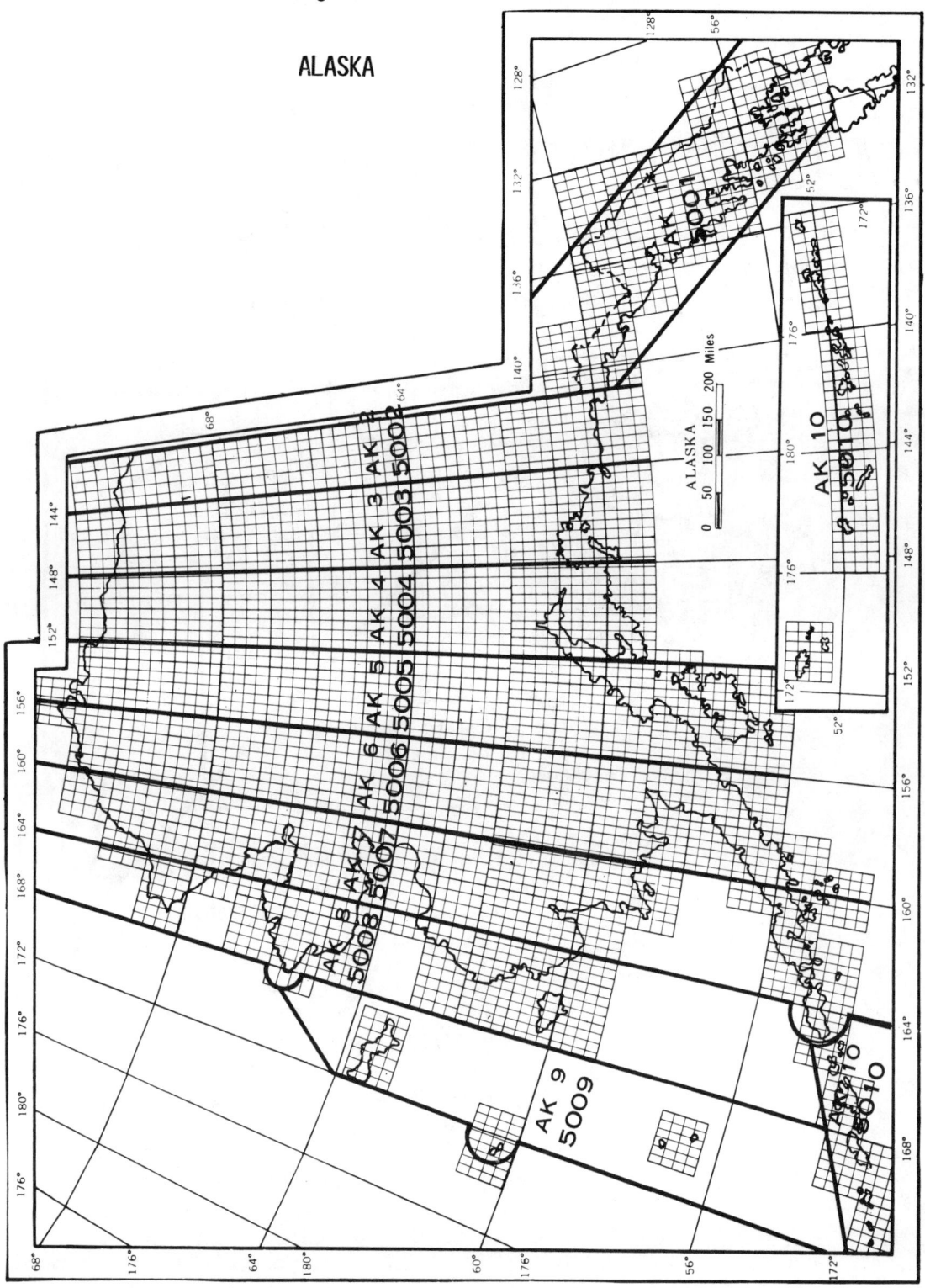

Figure 7.3 SPC Zone Boundaries

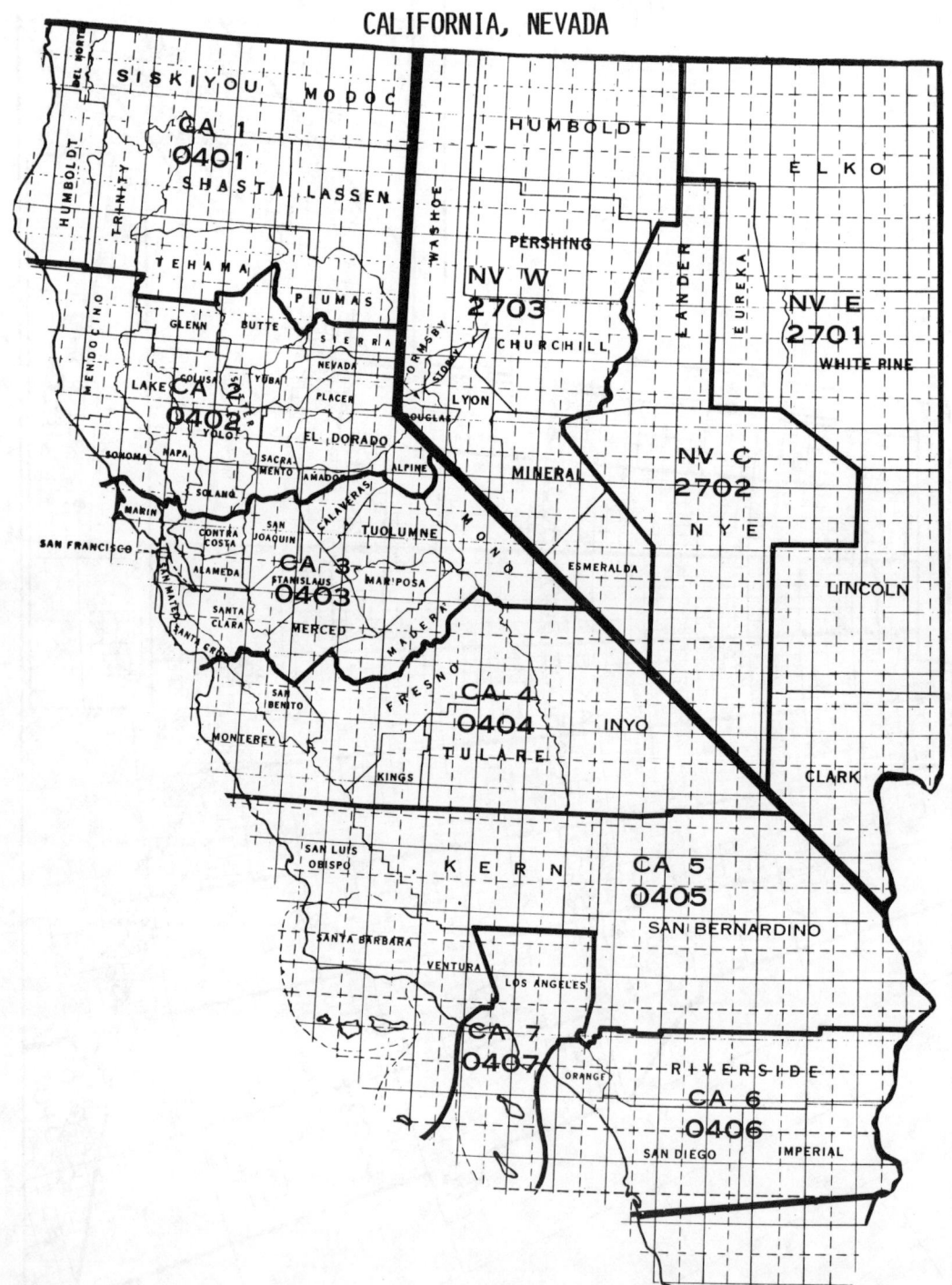

Figure 7.3 SPC Zone Boundaries

WASHINGTON, OREGON, IDAHO

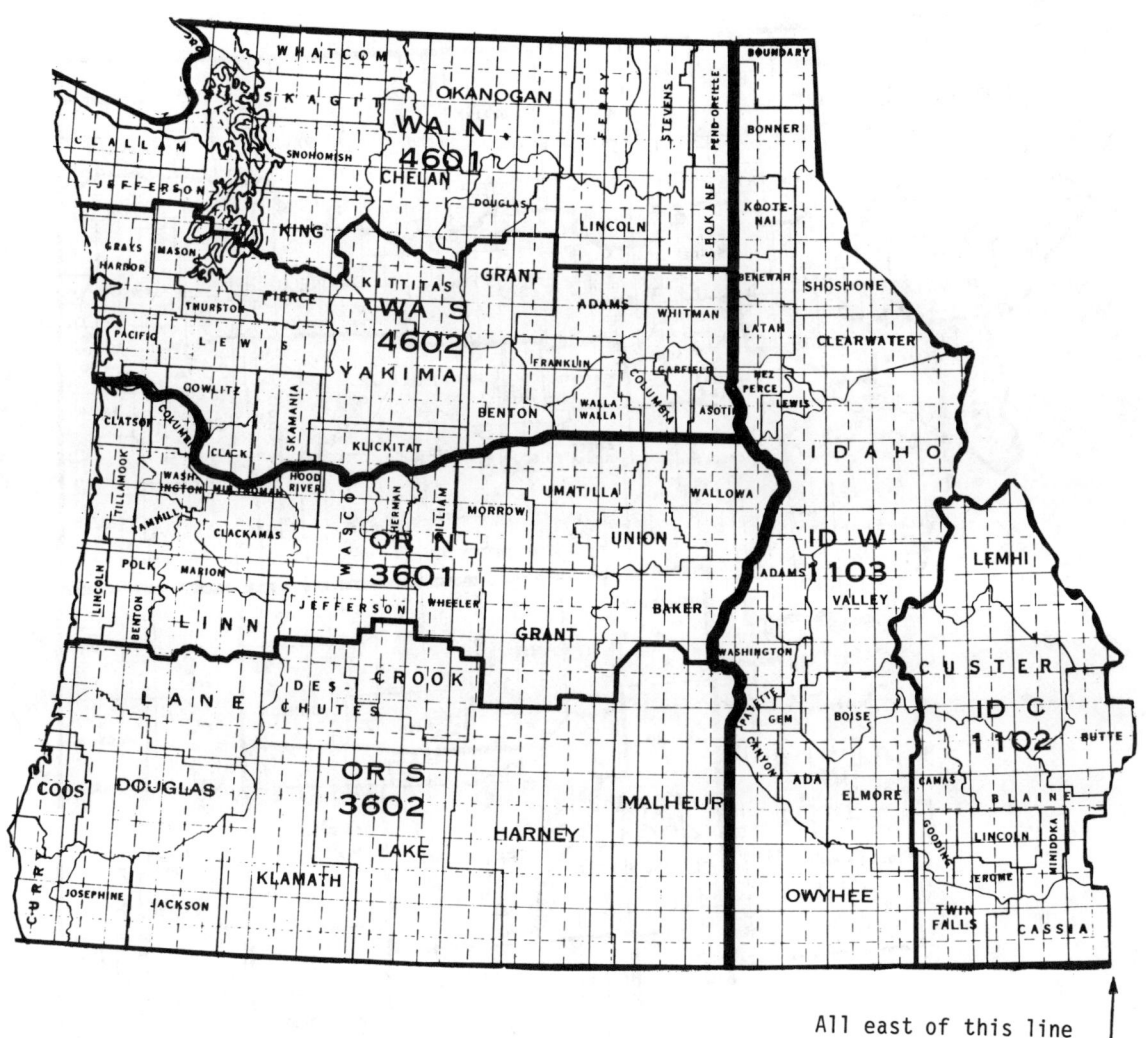

All east of this line are in Idaho, E. Zone. See next page.

Figure 7.3 SPC Zone Boundaries

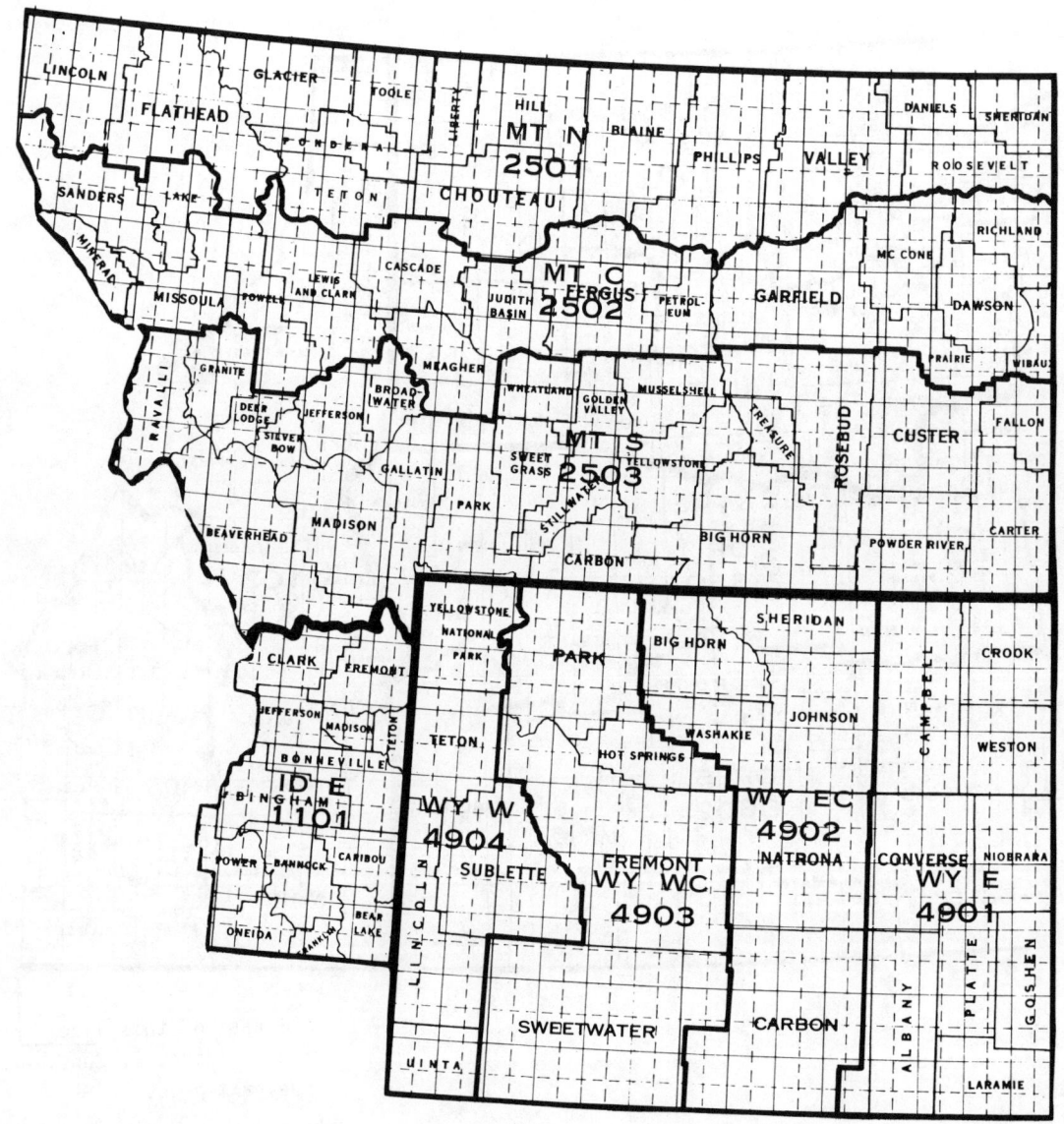

Figure 7.3 SPC Zone Boundaries

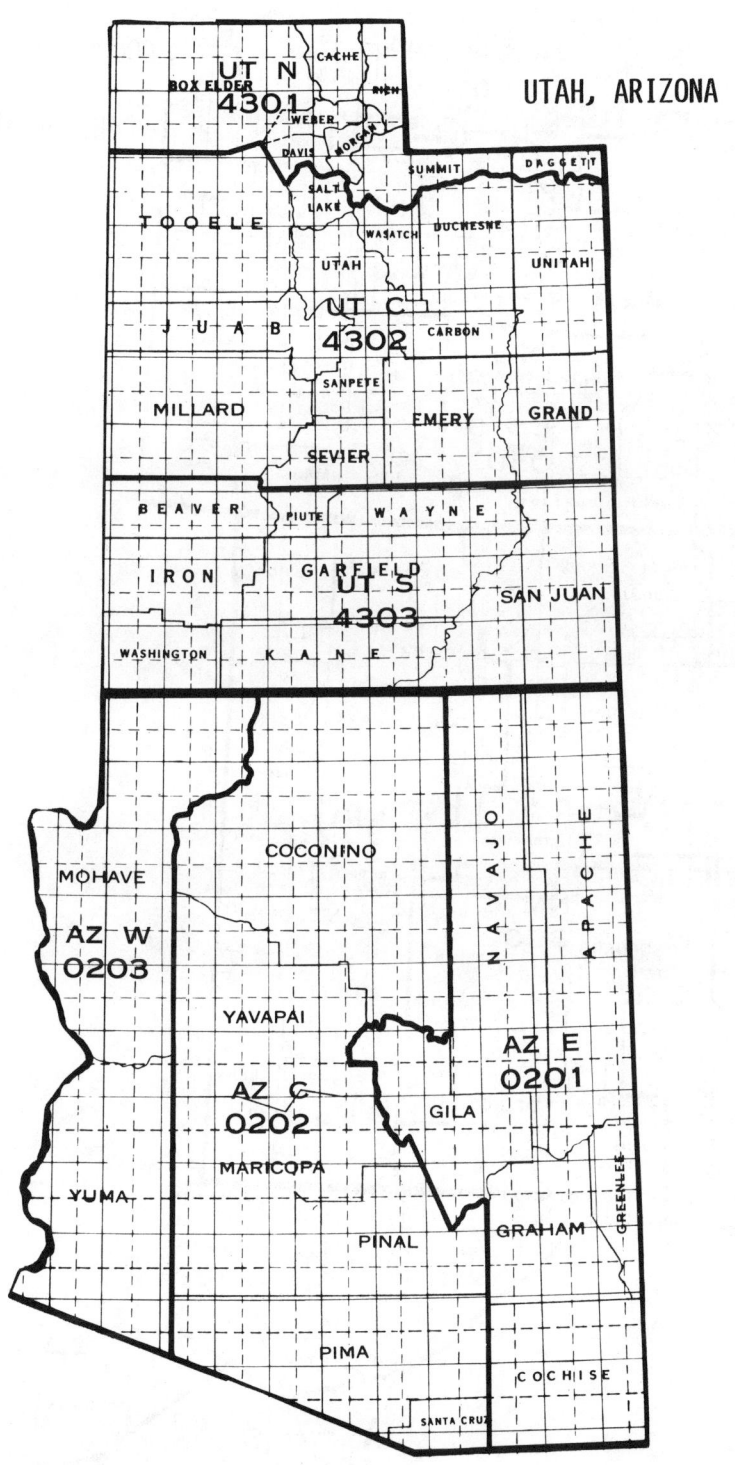

Figure 7.3 SPC Zone Boundaries

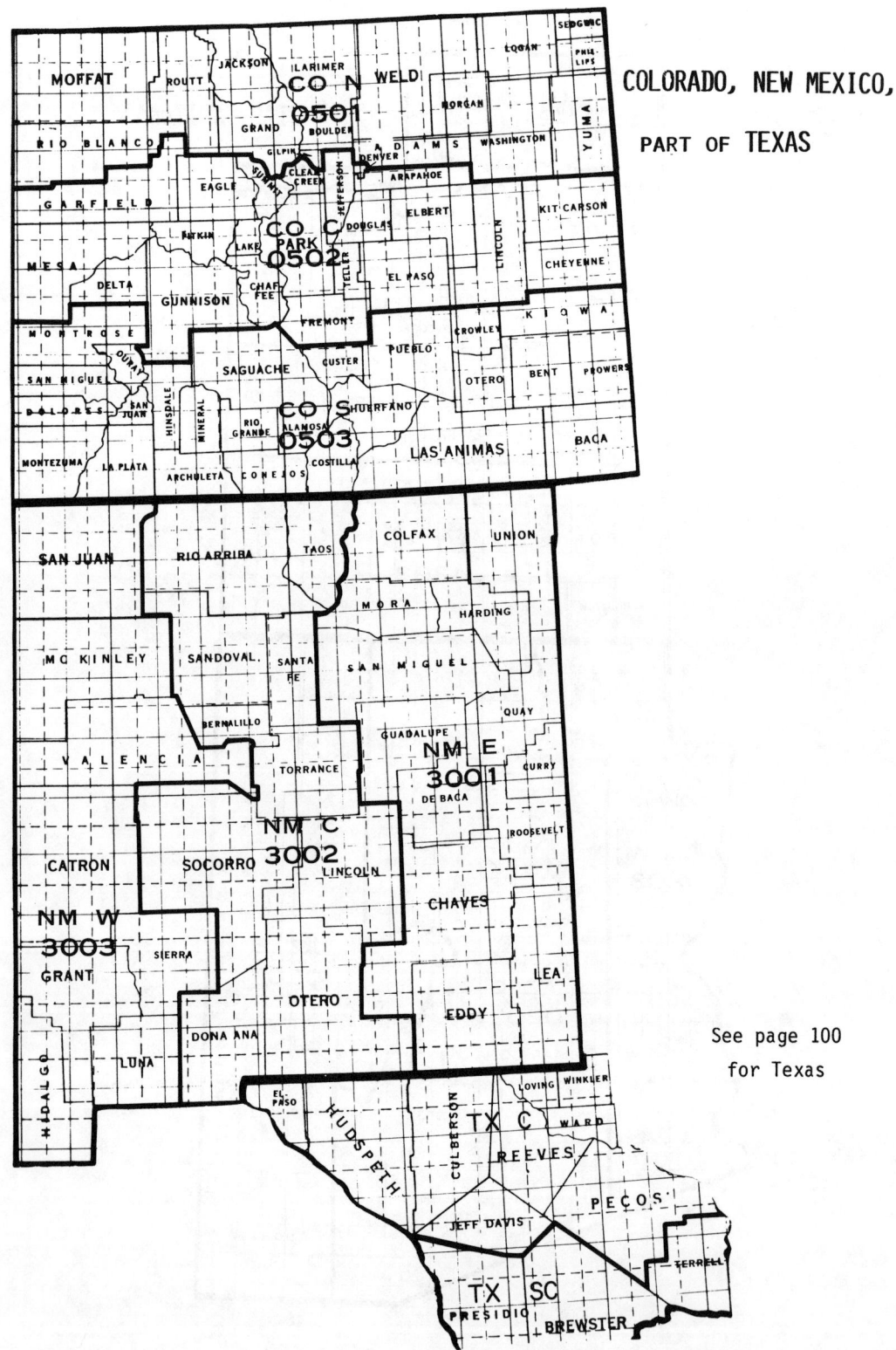

Figure 7.3 SPC Zone Boundaries

All counties north of this line are in N. Zone, N.D.

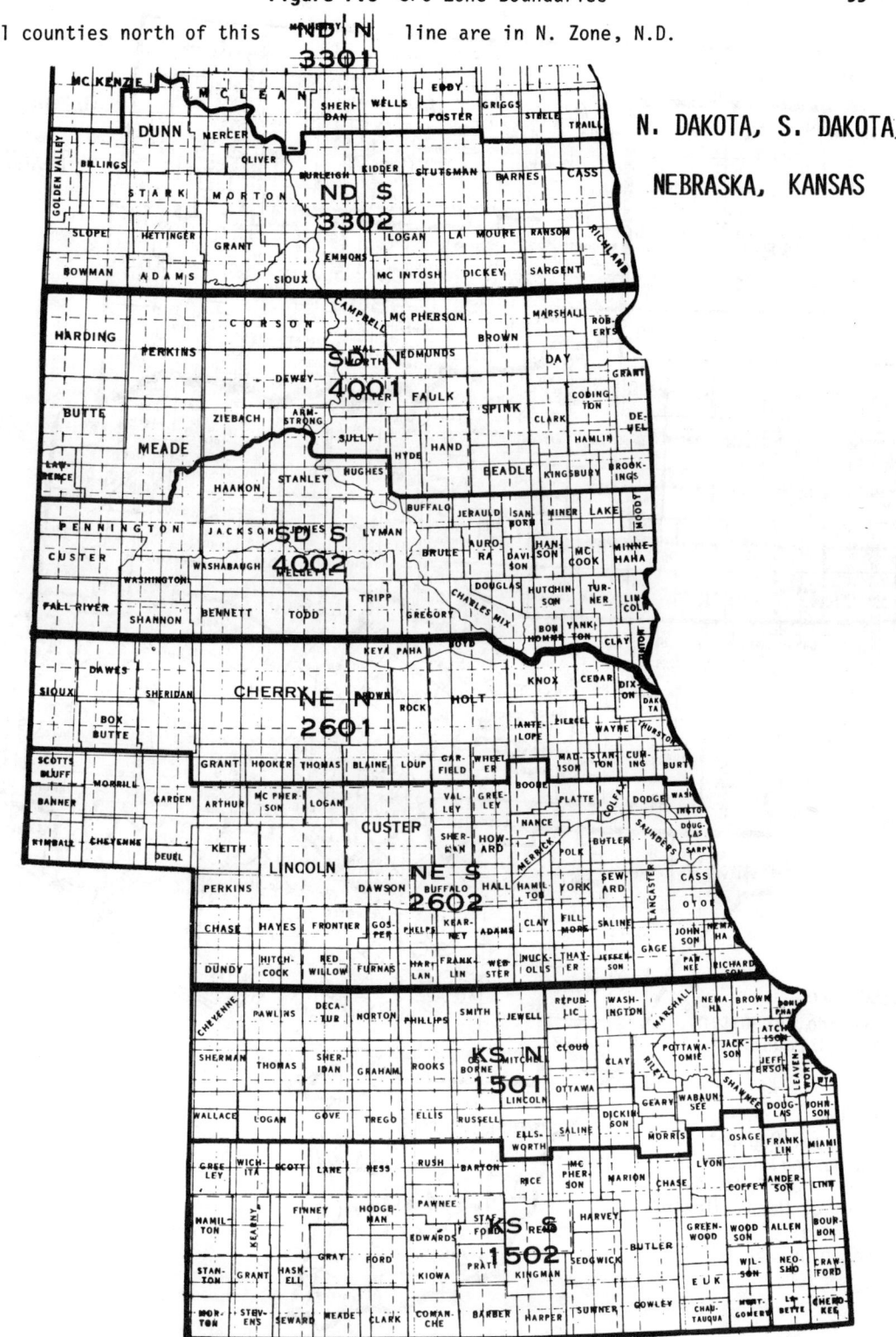

Figure 7.3 SPC Zone Boundaries

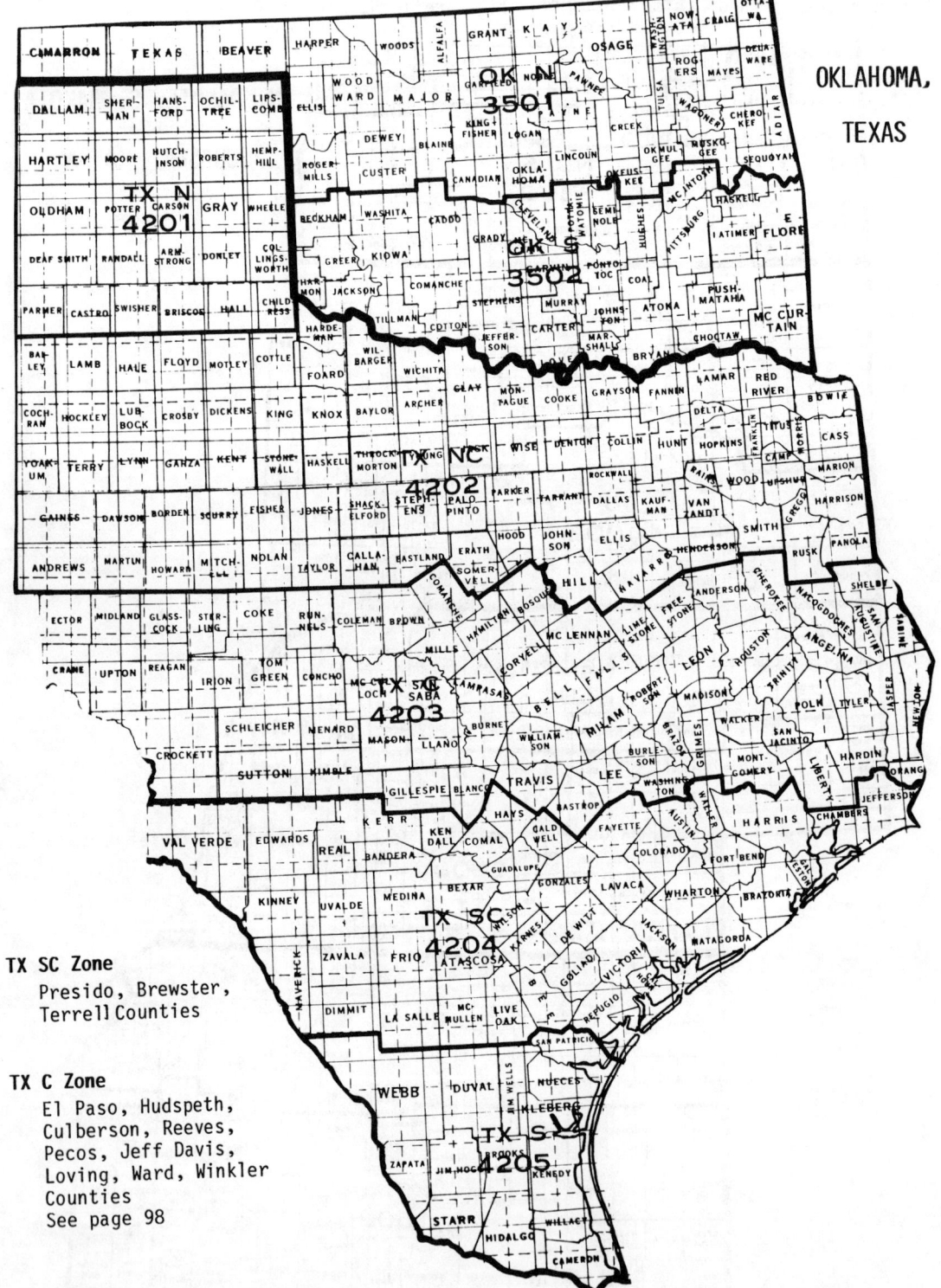

TX SC Zone
  Presido, Brewster,
  Terrell Counties

TX C Zone
  El Paso, Hudspeth,
  Culberson, Reeves,
  Pecos, Jeff Davis,
  Loving, Ward, Winkler
  Counties
  See page 98

Figure 7.3 SPC Zone Boundaries

Figure 7.3 SPC Zone Boundaries

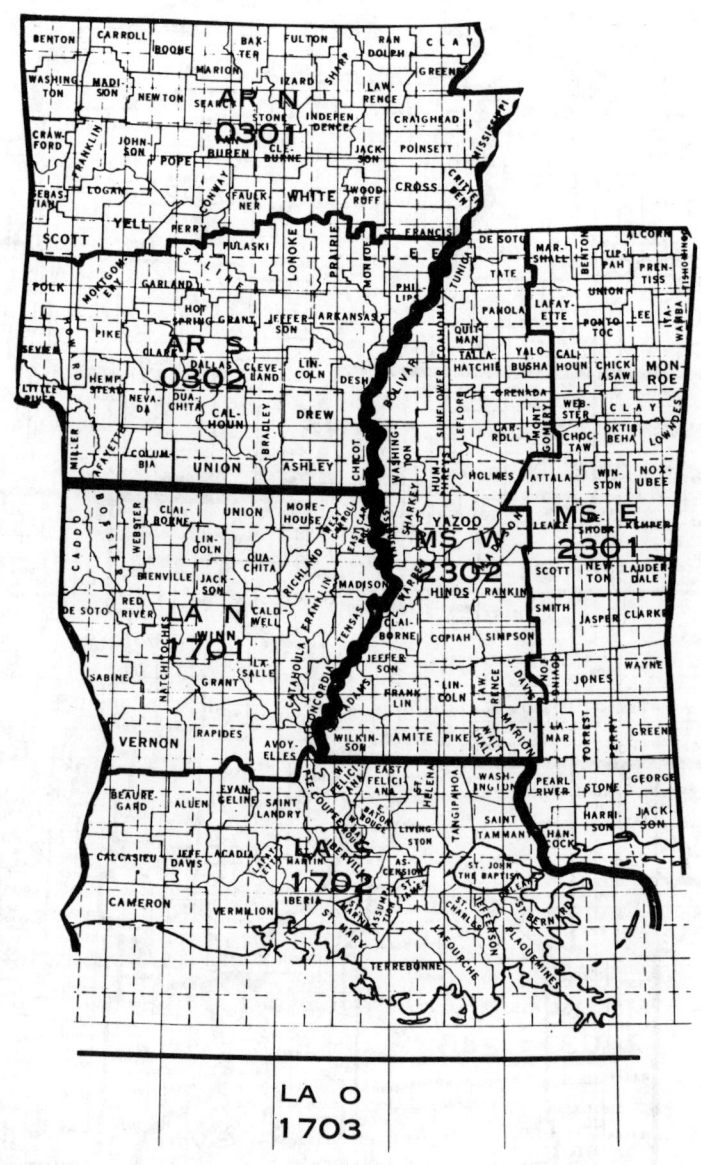

Figure 7.3 SPC Zone Boundaries

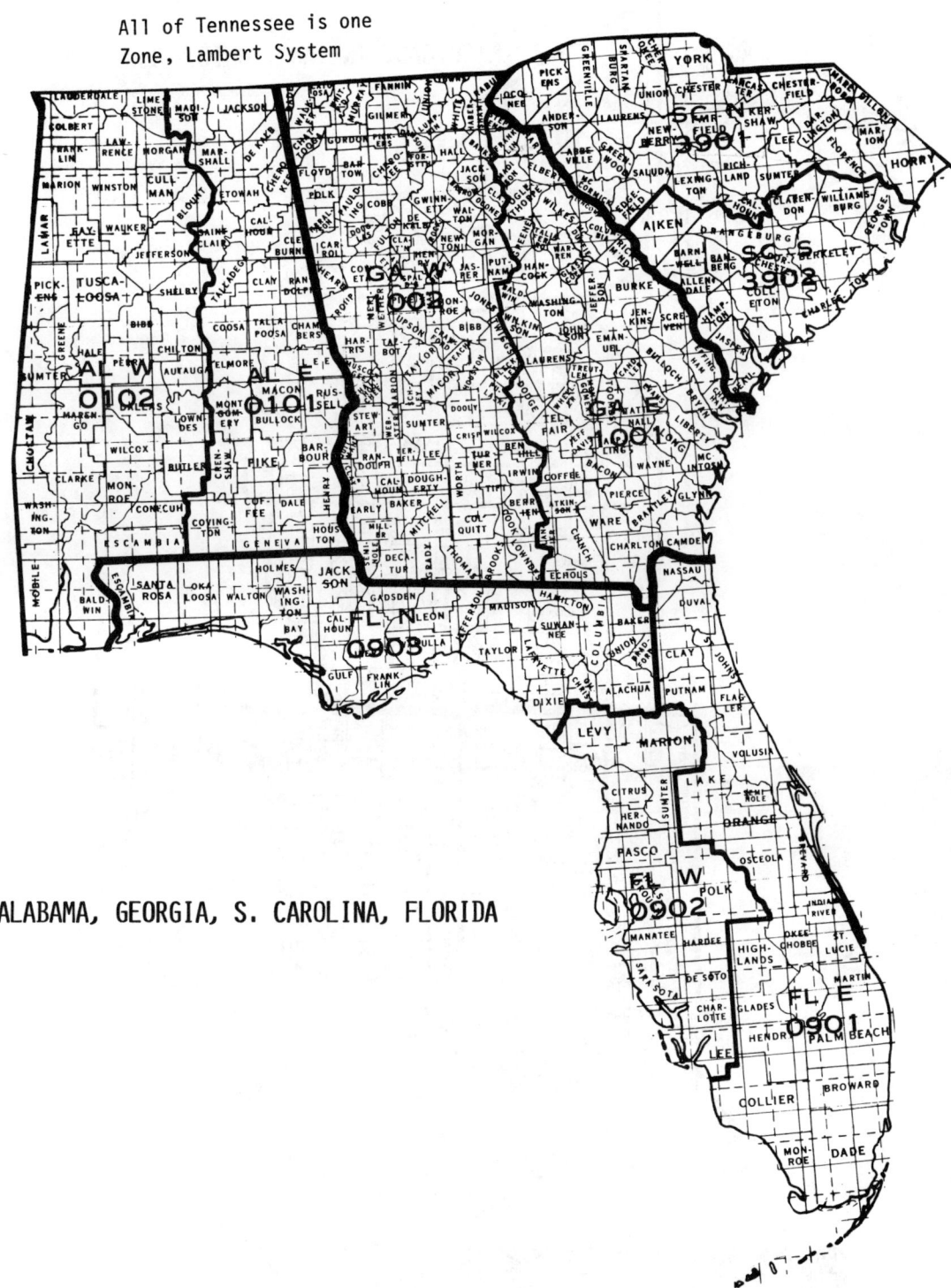

ALABAMA, GEORGIA, S. CAROLINA, FLORIDA

Figure 7.3 SPC Zone Boundaries

MICHIGAN, INDIANA, OHIO, KENTUCKY

Figure 7.3 SPC Zone Boundaries

NEW YORK, PENNSYLVANIA,
VIRGINIA, W. VIRGINIA,
N. CAROLINA, MARYLAND,
DELAWARE

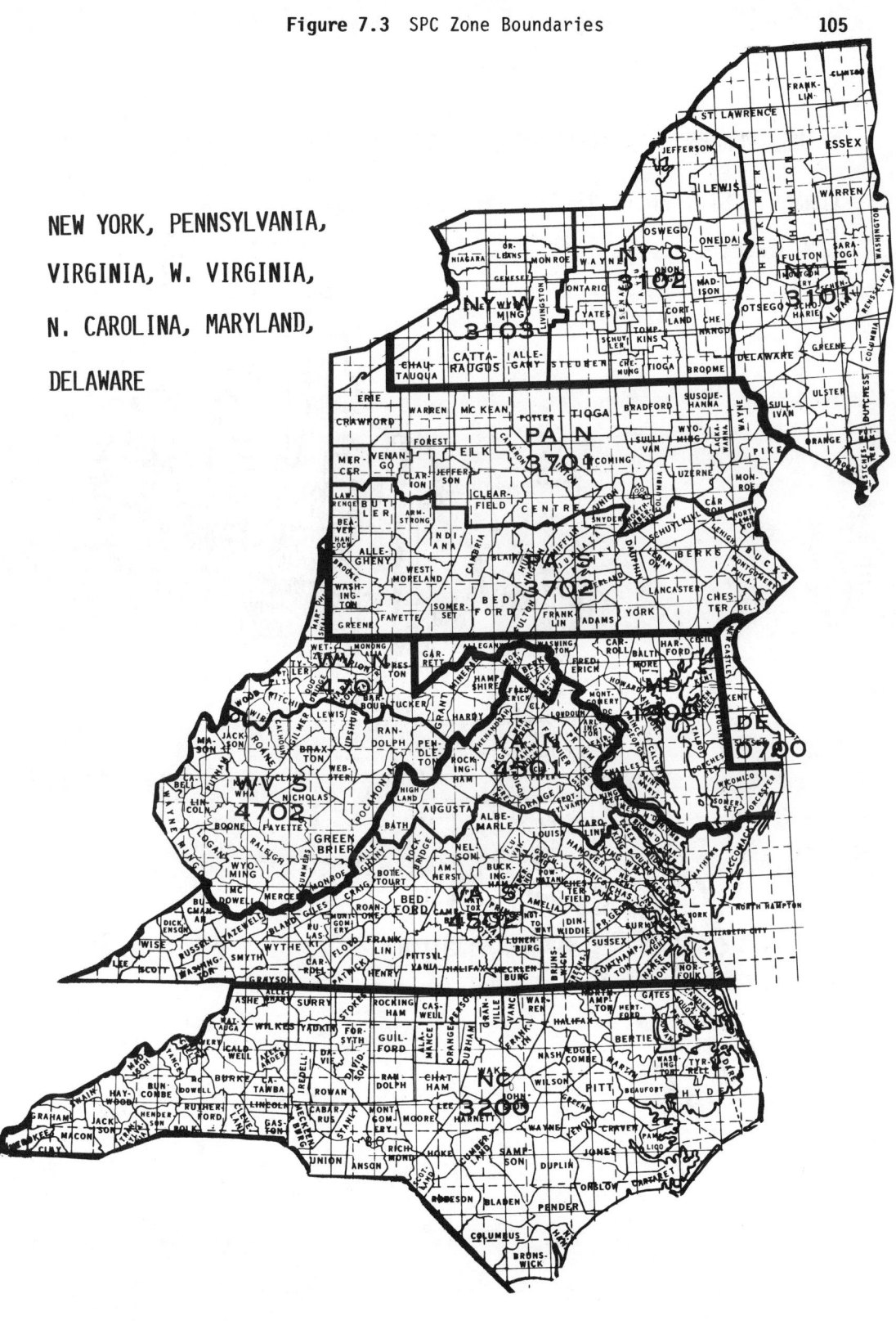

Figure 7.3 SPC Zone Boundaries

MAINE, VERMONT, NEW HAMPSHIRE, MASSACHUSETTS, CONNECTICUT, RHODE ISLAND, NY (LONG ISLAND), NEW JERSEY

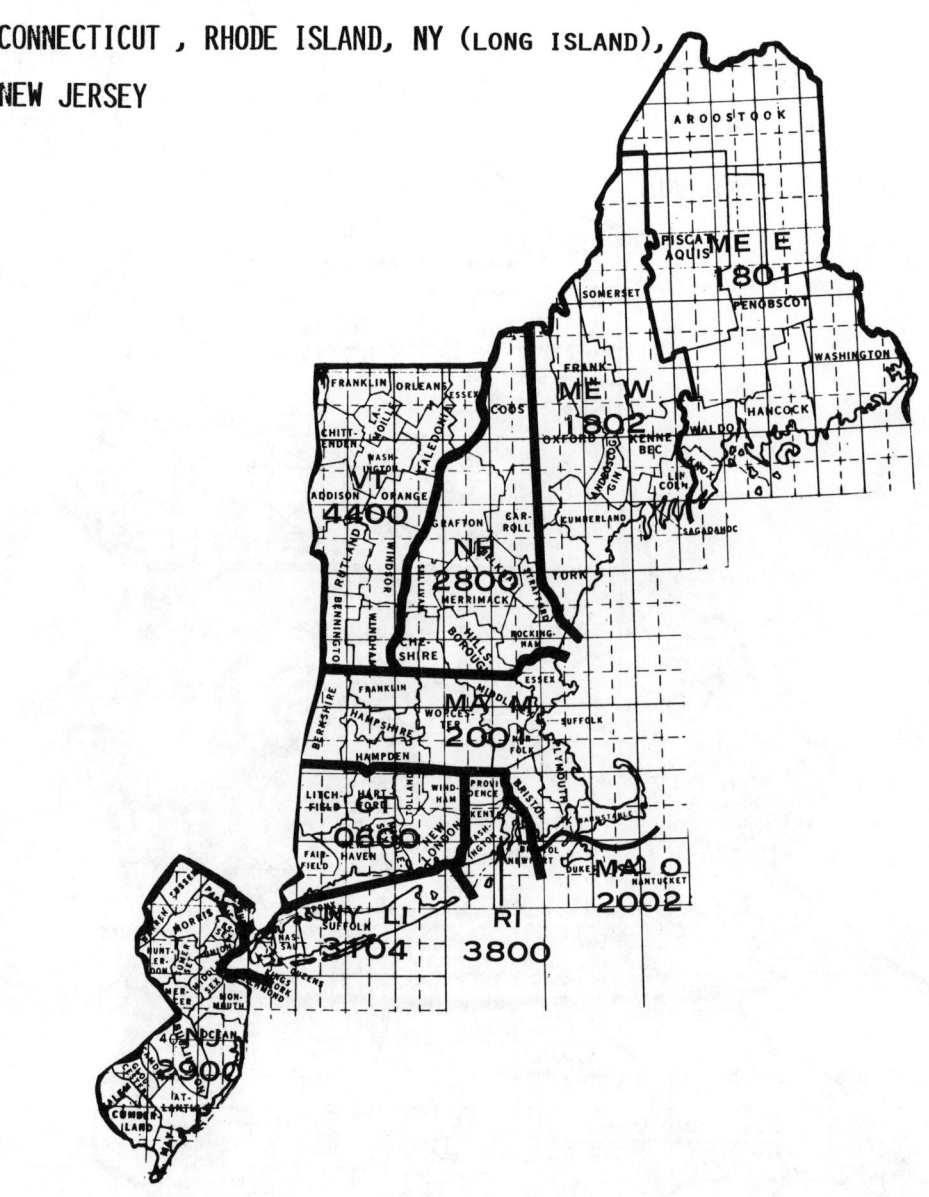

CHAPTER 8

USING STATE PLANE COORDINATES

8-1 INTRODUCTION AND BACKGROUND

Thus far in this manual, the explanation has focused on azimuth determination, culminating in methods to determine grid azimuth on the SPC systems using astronomic azimuth as a starting reference. This chapter is intended to explain how to use SPC in local horizontal surveys. This entails building upon what has already been presented regarding grid azimuth, but adds the necessary formulas and other information needed to convert ground distances to the SPC grid and to reference surveys to SPC monumented control stations.

Use of SPC in local surveys is easy, provided there is a nearby horizontal control station having published SPC to at least second-order accuracy. Too much distance to the nearest control station is the most common obstacle precluding use of SPC by private surveyors. However, using modern methods such as EDM, theodolites, computers, and/or electronic tacheometers, distance to a control monument has become much less of an obstacle than it once was using transit and tape. This means that, in many cases, referencing to a control monument a few miles away may not be impractical for the private surveyor. Of course, if local governments would provide funds and means to densify control, this obstacle would diminish even more in significance.

The calculations and field methods using SPC are simple. The theory should not provide a problem for any surveyor who uses conventional traverse and other simple control methods (i.e., intersection, resection, etc.) and who knows how to balance a traverse and compute plane coordinates. It is assumed that readers understand the fundamentals of traverse calculations. If not, one cannot expect to perform traverse or other coordinate calculations using SPC. The principal theoretical obstacles to learning SPC are not in the understanding of SPC itself, but rather in the understanding of departures, latitudes, and cartesian coordinates on any plane surface. The user does not need to perform observations for latitude and longitude to use SPC. Nor does he need to convert ϕ and λ to SPC since they are given in plane coordinates (X and Y) on control data sheets. This eliminates much of the confusion and misunderstanding about SPC. Conversions between X, Y and ϕ, λ are largely only academic exercises appearing on registration and college exams. Such calculations are rarely needed in practice. This manual will include the information needed to understand and use the systems in everyday surveying work and not confuse the reader with unnecessary theory or operations which are seldom used, the objective being to encourage use of SPC as much as possible.

In 1933, Dr. O.S. Adams, a mathematician in the U.S.C. & GS (now NGS) was asked to develop a system of plane coordinates for the North Carolina State Highway Commission. Soon after, plane coordinate systems were established for all states. The systems developed by Dr. Adams project measured distances onto imaginary surfaces which can be "developed" (rolled out) into plane surfaces. The surfaces used are cones and cylinders. Understanding how this is done mathematically is unimportant regarding use of the system. However, so that an appreciation can be gained of the elementary aspects of the theory, the projection surfaces are illustrated in Figure 8.1 and explained in the next section.

8-2 FUNDAMENTALS OF SPC

Projection Surfaces

Two projection surfaces have been employed for SPC, being the Lambert Conformal Conic Projection and the Transverse Mercator (TM) Projection. The Lambert system projects ground positions onto an imaginary cone, the axis of which coincides with the polar axis of the earth. The Transverse Mercator system projects ground positions onto an imaginary cylinder, the axis of which is perpendicular to the polar axis of the earth. In either system, the imaginary surface is not tangent with the earth, but intersects it slightly. This widens the zone that would be used if it was tangent. Figure 8.1 illustrates these geometrics. The surface used is sea level (the spheroid used in geodetic work). Any ground distance must first be converted to sea level length, then to the projection surface. This, as is seen in the figure, means that grid lengths are less than sea level lengths if a line is between the two lines of intersection and vice versa if outside of these intersection lines. How to correct for sea level and scale will be covered subsequently.

Zones

States with long east-west and north-south dimensions usually employ the Lambert projection since the scale factor remains constant in an east-west direction. States with long north-south dimensions usually use the Transverse Mercator System. Since the maximum width of a zone has been selected to be about 158 miles to keep the scale factor less than 1 part in 10,000, states with widths greater than this must have more than one zone. Illinois for example, being long north-south but wider than 158 miles, has two Transverse Mercator zones. Tennessee has one Lambert zone, but Nebraska has two Lambert zones because it is longer north-south. Both Wisconsin and Minnesota have three Lambert zones. Zones are divided by county and state boundary lines. Zones are illustrated in Figure 7.3.

Axes and Units

The central meridian of a zone is assigned the value of X=500,000 feet in the Transverse Mercator System and usually 2,000,000 feet in a Lambert zone. The choice of these values avoids negative coordinates. A convenient line of latitude to the south of each zone is given the zero Y value. After completion of the readjustment of the North American Horizontal Datum, coordinates will be given in metric, as well as English units, and the numerical value of the central meridian will be different on the metric system.

Figure 8.1 State Plane Projections

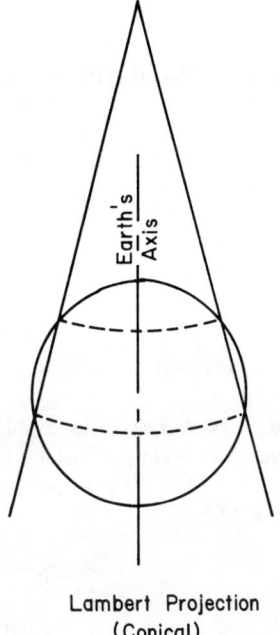

Lambert Projection
(Conical)

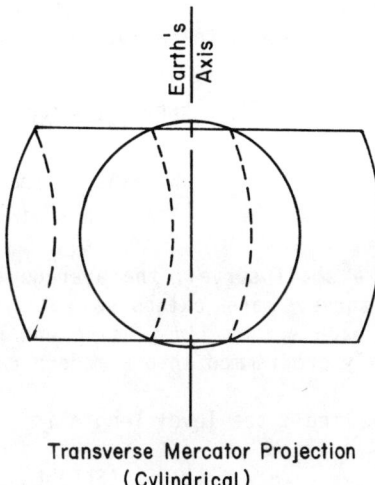

Transverse Mercator Projection
(Cylindrical)

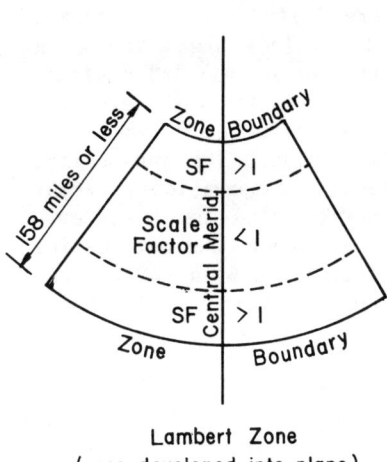

Lambert Zone
(cone developed into plane)

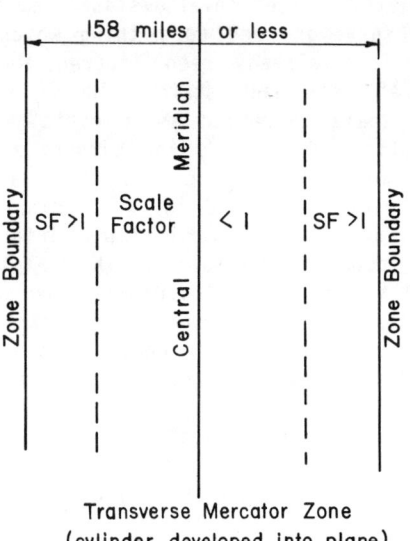

Transverse Mercator Zone
(cylinder developed into plane)

USING STATE PLANE COORDINATES

Except for these changes, the SPC system will not be altered after the aforementioned readjustment (expected to be completed after 1984).

Sea Level Factor

Surveys must be reduced to sea level since the geodetic data of the national control surveys are on this base. The computation is easy. All that is needed is the elevation of the survey line and the earth's radius. Elevations can be interpolated from USGS maps. The earth's radius in the conterminous U.S. is assumed to be 20,906,000 feet.

$$SLF = \left(1 - \frac{h}{20,906,000}\right) \tag{8.1}$$

where SLF = sea level factor
h = elevation in feet, **above** sea level

For a small survey, the average elevation can be used. If terrain is rolling or surveys are extensive, it is more precise to use the average elevation for each survey line, using end point elevations. All of this is, of course, easily programmed into a modern calculator or computer systems.

A line's sea level length is:

$$L_{SL} = (SLF) L_H \tag{8.2}$$

where L_H = ground distance, corrected to horizontal.

Scale Factor and Grid Distance

Next, the sea level distances must be projected onto the projection surface. The surveyor need do nothing which involves sophisticated mathematics since this has already been figured and placed in "Plane Coordinate Projection Tables" for each state. These "scale factors" are listed under a heading of "scale expressed as a ratio" in current tables. They are given as a function of latitude for Lambert zones and as a function of the "x" distance east or west of the central meridian in TM zones. The surveyor must determine the average latitude or X coordinate for the survey (or for each line) from USGS maps and perform simple interpolation in the tables. The map-scaling procedure is the same as that explained for astronomic observations, except that approximate X coordinates are scaled from grid marks using the TM system.

Users of the system need not understand anything about the mathematics of map projections or geodesy--only simple interpolation and when to consider whether to use the average for an area or for each line to avoid slight errors.

A line's grid length is:

$$L_G = (SF) L_{SL} \tag{8.3}$$

where SF = scale factor.

Of course, the grid length can be computed by multiplying SLF x SF x L_H in one step. There is no need for sea level distance as an intermediate step. For a particular line,

$$GF = SLF \times SF \tag{8.4}$$

$$\text{and } L_G = (GF) L_H \tag{8.5}$$

where GF = grid factor

This L_G is then used in all position computations whether employing traverse, triangulation, or any combined horizontal control system.

Grid Azimuth

As was explained in Chapter 7, state plane grid azimuth can be computed from astronomic azimuth. This is appropriate for placing all azimuths on the state grid system when control monuments are not nearby. It must be remembered that actual X and Y coordinates of the SPC system cannot be placed on the local survey points unless the survey is tied into a control monument with published SPC coordinates. If this is done, it is likely that there is also a ground azimuth mark with a published grid azimuth from the control station, making it unnecessary to perform an astronomic observation at all. Such ground azimuth marks, set by the NGS for local surveying work, are located such that surveyors can orient to them without the need of towers. The grid azimuth is generally considered to be of third-order traverse precision. There are occasions when azimuth marks are destroyed, disturbed, or no longer visible, in which case an astronomic azimuth must be made with conversion to grid as explained earlier in this manual. It is recommended that the azimuth to an existing mark be checked by astronomy if the mark appears to have been disturbed or if any uncertainty arises in interpretation of control data. If such a check is made, it is simplified somewhat by the fact that ϕ and λ of the station are published; thus, these values need not be scaled from a map.

In a few areas, modern EDM traverses have been executed by the NGS or other agencies with traverse courses paralleling highways or other routes, all done without towers. In such instances, the intervisible traverse stations, with their published data, provide azimuths and there is no need for additional azimuth marks, terrestrial or astronomic.

Users of any control set by NGS or any agency should carefully inquire and determine that measurements, adjustments, and monumentation have been done in such a way as to assure at least third-order accuracy in the azimuths. If the surveyor must carry the azimuth more than a few traverse courses to tie it into his local survey, the reference azimuth should be to at least second-order accuracy.

Example SP1   A traverse line has been measured as 4324.88 feet, horizontal distance at a location in Pennsylvania. The average latitude and elevation of the traverse course was 40°21'15' and 1025 feet, respectively, as scaled from a USGS map. What is the grid distance of this line?

Solution: $SLF = (1 - \frac{1025}{20,906,000}) = 0.9999510$ \hfill (8.1)

Pennsylvania's plane coordinate projection tables (Special Publication No. 267) lists the scale factor for 40°21' and 40°22' in Table 1, page 17, for the south zone. The SF for 40°21'15" is determined by interpolation as follows:

ϕ	SF
40°21'	0.9999610
40°22'	0.9999606

By interpolation SF = 0.9999609 for 40°21'15"

$GF = SLF \times SF = 0.9999119$ \hfill (8.4)

$L_G = 4324.88 \times GF = 4324.50$ ft. \hfill (8.5)

8-3 TRAVERSING USING SPC

After converting all horizontal ground distances to grid and computing azimuths on the grid system, the rest is no different than any other traverse calculations. Computation of departures and latitudes, adjustment of traverse, and coordinate calculation follow the same procedures used for any plane system. This, of course, is why the SPC system was developed initially--so that positions on our spherical earth could be interrelated using familiar plane surveying methods.

The following steps summarize the procedures to determine SPC using a traverse:

1. Plan the traverse so that it starts and ends on stations of known SPC.
2. Measure the distances between traverse stations, starting and ending on the stations of known SPC and convert distances to horizontal with proper corrections for other systematic errors.
3. Measure horizontal angles between each traverse station, starting and ending the traverse with a sighting to a terrestrial or celestial azimuth mark.
4. Convert astronomic azimuths to grid if necessary, compute the angular closure of the traverse and, if closure is satisfactory, compute adjusted grid azimuths of all traverse lines.
5. Convert distances to the grid by applying the sea level factors and scale factors.
6. Compute the traverse by conventional plane surveying methods and determine the error of closure.
7. If the error of closure is satisfactory, adjust the traverse by any appropriate method.
8. Calculate the SPC of each station using the known SPC and adjusted departures and latitudes.

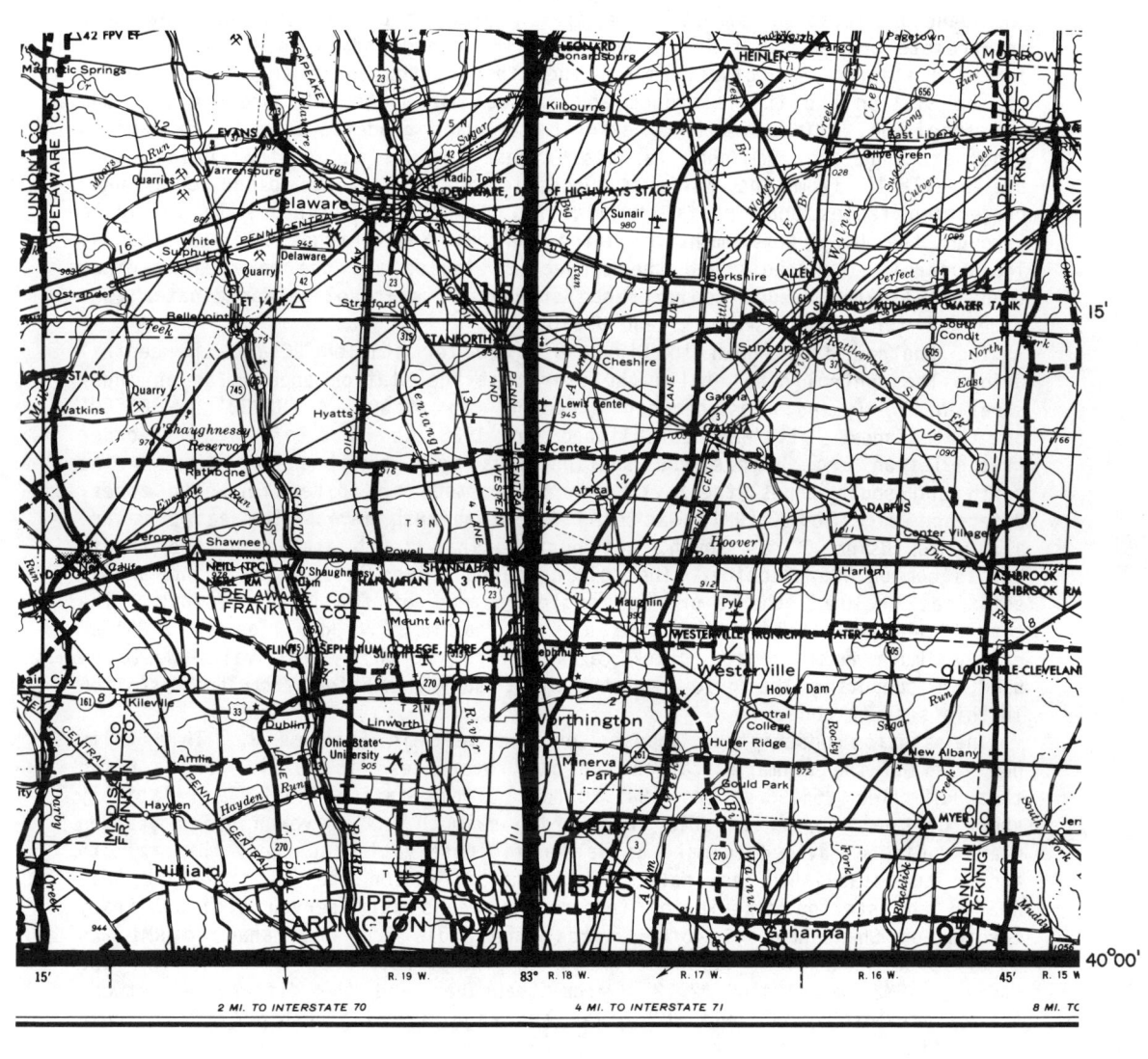

Figure 8.2 Control Station Index Map

MARION, OHIO
NK 17-10

8-4 SPC CONTROL DATA

A portion of an index map and a control data sheet for a horizontal control station are included herein. (Figures 8.2 and 8.3). The index map shows the general location of stations in an area, their names, and how they interrelate on the NGS primary network. The control data sheet is available for each station. It gives, among other things, the location and description of the station and its reference and azimuth mark, the SPC, precise latitude and longitude, grid azimuth to azimuth mark and geodetic azimuth to various other stations and prominent targets. These indexes and control data sheets are available from the National Geodetic Survey. State plane coordinate projection tables and Special Publication 235 (see references) is available through the U.S. Government Printing Office, Superintendent of Documents. The control data sheets and the projection tables are the only data, other than field data, needed to reference surveys to state plane coordinates (See page 253 for addresses of NGS and USGPO).

The control data shown in Figure 8.3 are for station "Clark" in central Ohio. The control sheet tells the user that the station appears on an index sheet NK 17-10 and lies between ϕ =40°00' to 40°30', λ =82°30' to 83°00' on that index. It was established in 1928 as part of a first-order triangulation arc. Since the station lies near the boundary between the north and south zones, the X, Y, and grid azimuth are given for both zones. Reference mark No. 1 (RM1) was used as an azimuth mark. The azimuths are all given using 0° as south, not north. This isn't stated on the sheet, but is common NGS practice. The column headed "plane azimuth, θ or $\Delta\alpha$ angle" sometimes includes the θ or $\Delta\alpha$ angle instead of the plane azimuth. The term "plane azimuth" is the same as "grid azimuth". RM1 lies at 2°36'36" grid azimuth from Station Clark (azimuth from north). This value would be used in computing the grid azimuths in a survey started from this station in Ohio's south zone.

The station description appears on the left in Figure 8.3. This tells how to find the primary station and its references (in 1928). It is noted that RM1 lies 256.55 meters (841.70 feet) from Station Clark, a relatively short distance for a reliable third-order azimuth. A surveyor using station Clark would be advised to make an astronomic observation to check this azimuth and/or to establish a new mark. The term "direction" to Shannahan and RM1, RM2 in the station description information means as observed with arbitrary zeroing on Shannahan. In other words, the angle from Shannahan to RM1 was 15°07'18.90" when it was referenced to the first-order network from the towers. RM2 was 134°00'32.12" from Shannahan and the difference between the directions to RM1 and RM2 gives the angular relationship between these two marks.

As an exercise to understand the data, observe that the geodetic azimuth from Clark to Shannahan (listed under position data) is 167°11'08.79". Adding the 15°07'18.90" to this yields 182°18'27.69". The difference between this and 182°36'36" is the θ angle for the south zone which can be checked by the reader using the formulas of Chapter 7 and Table 7.2.

The note by G.O.W. dated 1934 updates the station information. The more recent recovery note by John P. Apel in 1957 is most valuable in both locating and determining condition of the monuments. His distance of 842.04 feet to RM1 using invar tape and taping bucks suggests that the mark may be

Figure 8.3 SPC Station Description

SEPTEMBER 1959
PUBLISHED AND PRINTED BY:
U.S. DEPARTMENT OF COMMERCE
COAST AND GEODETIC SURVEY
WASHINGTON D.C.

400823 STATION 1003
OHIO DIA NK 17-10
LATITUDE: 40°00'   40°30'
LONGITUDE: 82°30' 83°00'

HORIZONTAL CONTROL DATA
by the
Coast and Geodetic Survey
NORTH AMERICAN 1927 DATUM

ADJUSTED HORIZONTAL CONTROL DATA

NAME OF STATION: CLARK
STATE: Ohio   LOCALITY: Columbus Arc
ORDER: First   TYPE: Triangulation   SOURCE: G-963   YEAR: 1928   FIELD SKETCH: OHIO 12

GRID DATA

	STATE	ZONE N	CODE		x	y
	Ohio	N	3401		1,866,600.13	141,025.27
	Ohio	S	3402		1,866,612.19	748,005.80

GEODETIC DATA

LATITUDE: 40°03'09"894
LONGITUDE: 82 58 35.265

PLANE AZIMUTH θ (ON Δα ANGLE): 182°37'14"   NORTH 0°00'00"10
                                 182°36'36"   WEST

TO STATION	GEODETIC AZIMUTH (From south)	SECONDS IN METERS	LOGARITHM (Meters)	DISTANCE METERS	FEET
UNION	12°22'11.21		FIRST-ORDER		
SHANNAHAN	167 11 08.79		4.012 3175	10,287.68	
DARFUS	222 52 46.16		4.073 9605	11,856.61	
MYER	269 13 03.11		4.264 2241	18,374.86	
HOWELL	308 24 54.00		4.196 9611	15,738.42	
BAKER	356 50 06.21		4.228 8120	16,936.05	
			4.212 3142	16,304.75	

REFERENCE MARK NO 1
REFERENCE MARK NO 1

Card 2 of 2

and is North 56.6° West 203.8 feet from a concrete monument stamped "Reference Mark, U.S.C. & G.S. Clark No. 2, 1928" and is South 83° East 41.4 feet from the centerline of Karl Road; also 15.6 feet in the same direction from a fence running parallel to and along the east side of Karl Road; and South 4.1° West 842.04 feet from a concrete monument with a brass plug stamped "Reference Mark U.S.C. & G.S. Clark No. 1, 1928". For further reference, said "Magnetic Station" is South 55.6° East 224.3 feet from said R.R. spike in said Pignut tree and North 18.1° East 111.7 feet from said "Reference Mark No. 2" and South 7° West 113.5 feet from the fence along the north line of said field; said "Reference Mark No. 2" is located 1.2 feet south of the south fence of said field and 221.7 feet east of the centerline of Karl Road measured along the south fence of said field.

NOTE: Bearings given in the above reference have been determined with a prismatic compass and are not corrected for declination. Directions are merely intended as an aid to recovery in the field and not considered adequate for computation purposes.

HORIZONTAL CONTROL DATA

CLARK (Franklin County, Ohio, W.M., 1928)—About 4 miles east of downtown Columbus, about 70 yards north of the house of James B. Clark, on the east side of Karl Road, and about 1/2 mile south of the intersection of Karl Road with the Sharon-Clinton Township Road. Station is about 8 meters east of the fence on east side of Karl Road, and in the northwest corner of a small lot lying north of Mr. Clark's house.
Mark is set 1 foot below grade.
Reference mark no.1 is north along east fence of Mr. Clark's house.
Reference mark no.2 is near Mr. Clark's house.
Surface, underground and reference marks are standard bronze disks set in concrete as described in notes 1a, 7a and 11a.

OBJECT	DISTANCE meters	DIRECTION
SHANNAHAN	256.55	0°00'00"10
R.M.No.1		15 07 18.90
R.M.No.2	62.06	134 00 32.12

Height of signal above station mark - 100 feet.

CLARK (Franklin County, Ohio, W.M., 1928;G.O.W.,1934).—This description seems to be in error on the general location of the station. Rather than 4 miles east of downtown Columbus the station was reached by proceeding 7 miles north on Cleveland Avenue, then 1-1/4 miles west on the Sharon-Clinton Township Road, better known as Morse Road, to Karl Road, thence south 1/2 mile to station.
The property formerly owned by Mr. James B. Clark is now owned by O.J. Hill. The fence lines have been changed so that the station is no longer #in the northwest corner of a small lot etc., but now measures 16 feet east from the north-south fence paralleling Karl Road, and 126 feet south of a pignut tree which is in the fence line on the north side of this field.
Station in good physical condition as are each of the reference marks which were all visited and the angular measurements between them checked.

U. S. DEPARTMENT OF COMMERCE - COAST AND GEODETIC SURVEY
RECOVERY NOTE, TRIANGULATION STATION
Card 1 of 2

NAME OF STATION: CLARK
ESTABLISHED BY: Wm. Mussetter YEAR: 1928   STATE: Ohio
RECOVERED BY: John P. Apel YEAR: 1957   COUNTY: Franklin

Detailed statement as to the fitness of the original description; including marks found, stampings, changes made, and other pertinent facts:

Station recovered as described excepting distances given in G. O. W., 1934 description as "16 feet east from the north-south fence paralleling Karl Road and 126 feet south of a Pignut tree" were found to be 15.6 feet east from the north-south fence paralleling Karl Road, and 123.7 feet south of a Pignut tree. Distance to Reference Mark No. 1 was found to be 842.04 feet (with invar tape, taping bucks, etc.) instead of 841.60 feet (265.55 meters).
Station in good physical condition as are reference marks, No. 1, No. 2 and magnetic station.
A complete new description follows:
The triangulation station is located 0.54 mile south of the intersection of Morse Road and Karl Road in a field on the east side of Karl Road; said field owned by the Columbus Diocese Catholic Church and in Clinton Township.
The concrete monument with brass plug stamped "Clark - 1928" is located South 76° West 123.7 feet from a spike in the south side of an 18 inch Pignut tree in the north line of said field and is North 87.9° West 201.4 feet from a concrete monument with brass plug stamped "Magnetic Station U.S.C. & G.S. Columbus 1939"

disturbed since this differs from 841.70 feet as converted from the initial 256.55 meters listed in 1928. Note that Mr. Apel made mistakes, however, in converting (He gives 841.60 feet and shows 265.55 meters instead of 256.55). Both of these mistakes may be typing or proofreading mistakes, and it is hoped that his field measurement of 842.04 feet is reliable.

Other data of interest are the ϕ and λ of Station Clark and the geodetic azimuths and distances to other triangulation stations. These stations are shown in Figure 8.2. It is doubtful that the local surveyor could use any of these for azimuth marks since towers were used in the triangulation survey. However, familiarity with the local network should aid in selecting SPC control for referencing nearby surveys.

8-5 PRECAUTIONS IN USING SPC

Surveyors should be aware of the precision needed in the scale and sea level factors and how they change when the variables change. This involves not much more than ordinary consideration of significant figures and round-off errors.

It should also be remembered that the distances computed from SPC are not ground distances. This is one reason why the system is primarily of value as a measuring aid and position preserver, not for describing land. There is no need to show anything but ground distances on a survey plat. If SPC values are given as supplementary evidence, the grid factors should also be given on the plat so that users can account for discrepancies between computed grid distances and ground distances shown.

Land areas computed by SPC must be changed since the desired area is that on the surface of the ground, not on an imaginary surface. There is no need to recompute the traverse using ground lengths. Simply use the SPC for area calculations and divide the result by the square of the grid factor, GF.

Surveyors working near zone boundaries must be certain that they use the coordinates and mapping angle pertaining to the zone in which their survey is to extend. Data sheets for NGS monuments will usually have both given. It should be understood that surveys can commence at a monument inside of one zone and extend well into another zone with no computational problems as long as consistency in use of grid direction and coordinates is exercised, meaning the zone values can't be mixed.

In use of plane coordinate projection tables, caution should be exercised to use the correct zone in states having more than one zone. The zone (east, west, etc.) is stated at the top of the page in the projection tables. A map at the front of the set of tables shows a dividing line between zones for identification of which set of tables is appropriate.

A very common mistake is to misuse the "R" value, listed in projection tables for Lambert zones as a function of latitude, mistaking it for the earth's radius. It is very unfortunate that this value is called "R" since that has connotations of radius. Since these "R" values are usually slightly over 20,000,000, they even approximate the earth's radius, so it is understandable how the mistake occurs. These values are distances from the particular latitude line to the apex of the imaginary cone and are used in making conversions between geographic and grid coordinates. The earth's radius is always to be 20,906,000 feet. It doesn't vary enough to cause concern, considering the purpose for which it is used.

The "2nd term" correction and "g" value mentioned in Section 7.2 should be considered for very long lines near zone boundaries if survey work is to be second-order or better. Other precautions concerning azimuth, mentioned elsewhere in this manual, should be observed as appropriate.

8-6 USES AND ADVANTAGES OF SPC

A Measurement Tool

The principal value of SPC is as a measurement aid. If surveyors have the entire system of land surveys in an area referenced to the system to accuracies approaching second-order, all new land subdivision, photogrammetric mapping, utility and street planning, retracement work, and other surveying and mapping work can be done easier. This is true because coordinate values provide a method of indirect measurements. By inverse calculations for length and direction, the surveyor always has a reliable measurement and this avoids the necessity for direct measurements between unrelated systems. The planning of new subdivisions, calculations of land divisions, and analysis of boundary evidence is greatly enhanced.

Preserving Evidence of Monuments

Mathematical monuments are indestructible. If their accuracy is good enough, they are better than ground reference ties. This isn't to say that a ground monument isn't necessary for a land corner, but that tying this to a reproducible position that cannot be destroyed enhances its preservation and is a highly important part of retracement evidence. Just as the reproducible meridian is important, also important is the position that doesn't depend upon other ground monuments and conflicting evidence for reestablishment.

Referencing Engineering Works

If underground utilities and other public works are referenced to SPC, they can be located easier in the future. This can save money for communities in planning future public works since much design can be done using the advantages of a common coordinate system. Also, much time and money can be saved in locating buried utilities. Concerning certain utilities, such as gas mains, knowing and mapping the position using SPC provides safety precautions to those who may need to excavate in the area.

Integrated Land Systems

If all cadastral, engineering, natural resource, taxation, and other land-related information is coordinated and mapped using a common grid such as SPC, all planning and surveying in the future can be conducted more efficiently.

The state plane system provides a common base for mapping remonumentation. If surveyors use the system now, this is building toward a common future data base. The more it is used, the easier it will become to use and develop modern land systems. If control survey measurements have been properly executed and adjusted, the data will become part of a future integrated system. Procrastination on use of SPC in land surveying will continue to delay modernization of land data records and other systems.

APPENDIX A

MISCELLANEOUS CONCEPTS AND DEFINITIONS

A-1 TERRESTRIAL AND CELESTIAL COORDINATE SYSTEMS

Terrestrial Systems and Spherical Terms

The earth is not a true sphere. It is flattened slightly at the poles. This flattening is unimportant as regards the terminology herein. The celestial sphere can be considered a perfect sphere of infinite radius. Whether speaking of the earth or the celestial sphere, several basic terms have the same meaning, except that one must consider the context (earth or celestial) in which each is used.

A **sphere** is a surface, all points of which are equidistant from its center point. A **great circle** is one on the surface of a sphere, formed by a plane passing through the center of the sphere. A **meridian** is a great circle passing through the poles of the sphere. The **prime meridian** is the meridian used as a reference. In our familiar systems, this is the Greenwich meridian, given 0° longitude, and passing through the Naval Observatory at Greenwich, England. A line of **longitude** is often considered synonymous with a meridian line. Both are north-south directions, known as **cardinal** directions because they point to the poles. As a unit of measurement, **longitude** is the angular distance of a position on the sphere from the prime meridian. Our familiar longitudinal system divides the sphere into 180° west and 180° east of the prime meridian.

The terms zenith, nadir, observer's meridian, horizon, azimuth, altitude, declination, polar distance, hour angle, and the meaning of celestial sphere were defined in Chapter 2. An understanding of some of these terms is necessary in order to comprehend some of the following.

The **prime vertical** is the great circle passing through the observer's zenith at right angles to the observer's meridian. It is an east-west vertical plane. The **equator** is a great circle lying midway between the poles. In our familiar system, **latitude** is the angular distance north or south of the equator and varies from 0° at the equator to 90° at the poles. A **vertical circle** is a great circle passing through the zenith and nadir of an observer. An **hour circle** is a great circle passing through a celestial body and its poles. The word **transit** means the crossing of a meridian by a celestial body. It can be "upper" or "lower".

In a terrestrial system, the position of an observer is given by latitude, ϕ, and longitude, λ, with origins of these coordinates as explained above. Let us next turn to celestial systems as used in observational astronomy.

A-1 TERRESTRIAL AND CELESTIAL COORDINATE SYSTEMS

Celestial Systems

Horizon system. This is the system most easily visualized as we stand on the earth's surface. In this system, the position of a celestial body is given in terms of its altitude and azimuth. As the earth rotates, these values are constantly changing. Measurements are made to the body from the plane of the horizon through the plane of the vertical circle passing through the body, and from the observer's meridian through the horizontal plane.

Hour angle system. In this system, the position of the celestial body is given in terms of hour angle and declination. Thus, angular measurements are made to the body from the observer's meridian through the plane of the equator, and from the equator through the plane of the hour circle through the body. This system is sometimes called the **equator system**.

Combined system. The combination of the horizon and hour angle systems was used to arrive at the equations for astronomic azimuth in this manual. The above concepts merge into one overall view of the celestial sphere as shown in Figure 2.1 and as explained in Chapter 2.

Equatorial system. In this system, the position of a celestial body is given in coordinates of **right ascension** and **declination**. Some new terms must be defined in order to explain this system. The **ecliptic** is the great circle on the celestial sphere defined by projecting the plane of the earth's orbit onto the celestial sphere. The **obliquity of the ecliptic** is the angle between the planes used to define the ecliptic and the plane of the equator. This angle corresponds to the sun's maximum declination; that is, it is approximately 23.5°. These two great circles (ecliptic and celestial equator) cross at two points which are relatively fixed on the celestial sphere. These points are called the **equinoxes.** The **vernal equinox** is the point of intersection of the ecliptic with the celestial equator when the sun crosses the celestial equator from south to north. The **autumnal equinox** is the point of intersection of the ecliptic with the celestial equator when the sun crosses the celestial equator from north to south. The dates March 21 and September 23 are the dates of occurrence of the vernal and autumnal equinoxes, respectively. The relationship between these concepts and declination should be immediately seen. The **solstices** are the points on the ecliptic where maximum declination occurs. The **winter solstice** and **summer equinox** occur on December 22 and June 22, respectively. The **equinoctial colure** is the hour circle passing through the vernal equinox.

The **right ascension** of a celestial body is the angle measured to the body from the equinoctial colure to the hour circle through the body, measured eastward through the plane of the celestial equator. The **sidereal hour angle** is similar to right ascension, except that it is measured westward. Defining right ascension as RA and sidereal hour angle as SHA, RA = 360°-SHA. These concepts are shown in Figure A.1.

The right ascension and declination of a star remain practically constant for years. Therefore, unlike the horizon and hour angle systems, the positions of a star do not change. This system has certain advantages when using the distant stars for observational astronomy. It has not been used in this manual because of the relative simplicity of the combined horizon and hour angle systems and their compatability using either the sun or Polaris for azimuth.

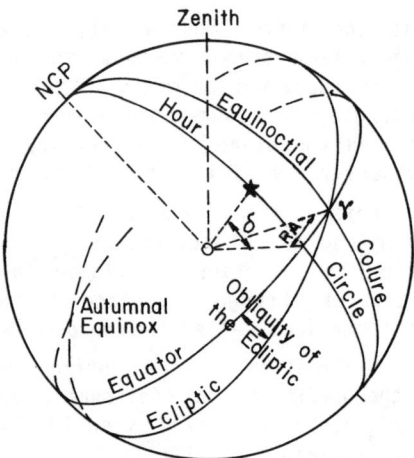

ϓ = Vernal Equinox (First Point of Aries)
δ = Declination
RA = Right Ascension

Figure A.1 Equatorial System

Simple transformations. Relationships between the hour angles involved in the various systems are depicted in Figure A.2. Note that the point of the vernal equinox is defined by the symbol ϓ, since it is also called the **First Point of Aries** and this is the Zodiac symbol for Aries, the Ram.

A-2 TIME SYSTEMS

Several terms related to time were introduced in Sections 5-3 and 6-2. These concepts will be further explained in this section.

A basic assumption in reckoning time is that the earth rotates uniformly on its axis. The period of rotation is the **day**. Because the earth is revolving around the sun, and does so in an elliptical orbit, there are different ways to measure the day, or the time it takes for a celestial reference point to transit the observer's meridian. The three kinds of days with their reference point and characteristics are:

Kind of Day	Reference Point	Characteristics
Apparent Solar Day	Apparent Sun	Non-Uniform
Mean Solar Day	Mean Sun	Uniform
Sidereal Day	Vernal Equinox	Uniform

The following definitions were taken from Reference 16. An **apparent solar day** is the interval of time between two successive lower transits of the real sun over the same meridian. A **mean solar day** is the interval of time between two successive lower transits of the mean sun over the same meridian. A **sidereal day** is the interval of time between two successive upper transits of the vernal equinox over the same meridian.

A-2   TIME SYSTEMS   121

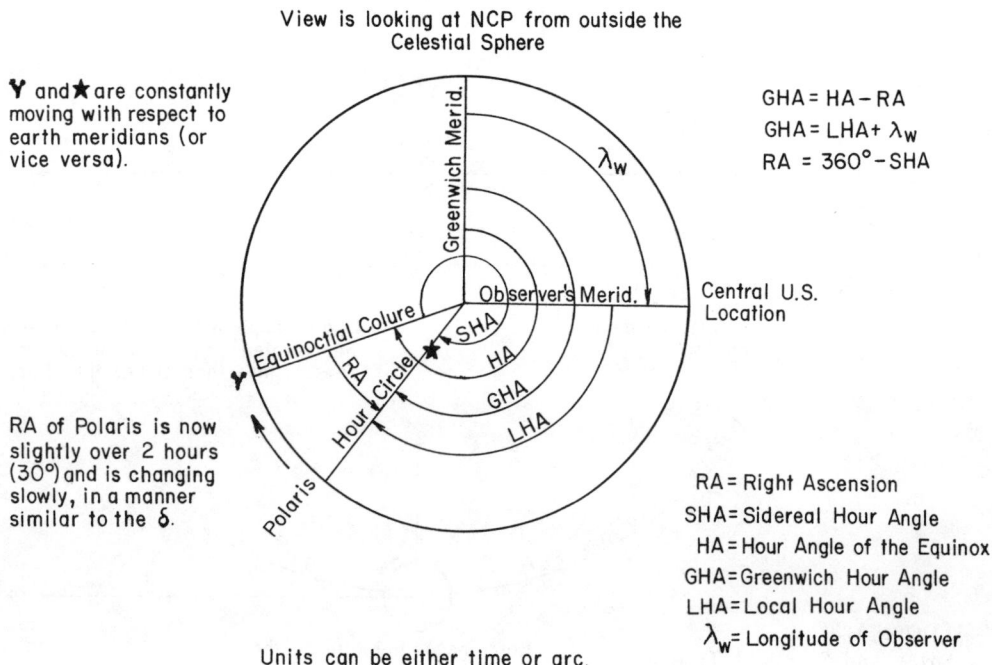

Figure A.2 Hour Angle Transformations

In keeping solar time, the reference is the lower transit over the meridian since our solar day begins at midnight. As between apparent and mean time, they differ by the equation of time which has been explained in Section 5-3.

In keeping sidereal time, a sidereal day is divided into 24 hours. Thus, the length of time unit differs between the solar and sidereal systems, a solar day being longer than a sidereal day. Clocks can be, and are, constructed to read sidereal time, but they would be awkward to use in everyday life since the familiar sun would continually shift as the year passes and it wouldn't rise, set, etc., at watch times which are relatively fixed.

The relationship between solar time and sidereal time is shown in Figure A.3.

$$\text{One Sidereal Day} = 23^h\ 56^m\ 04.091^s \text{ of mean solar time}$$
$$= 0.99726957 \text{ mean solar days}$$

$$\text{One Mean Solar Day} = 24^h\ 03^m\ 56.555^s \text{ of sidereal time}$$
$$= 1.00273791 \text{ sidereal days}$$

$$\text{One year} = 365.242195 \text{ mean solar days}$$

$$\text{One year} = 366.242195 \text{ sidereal days}$$

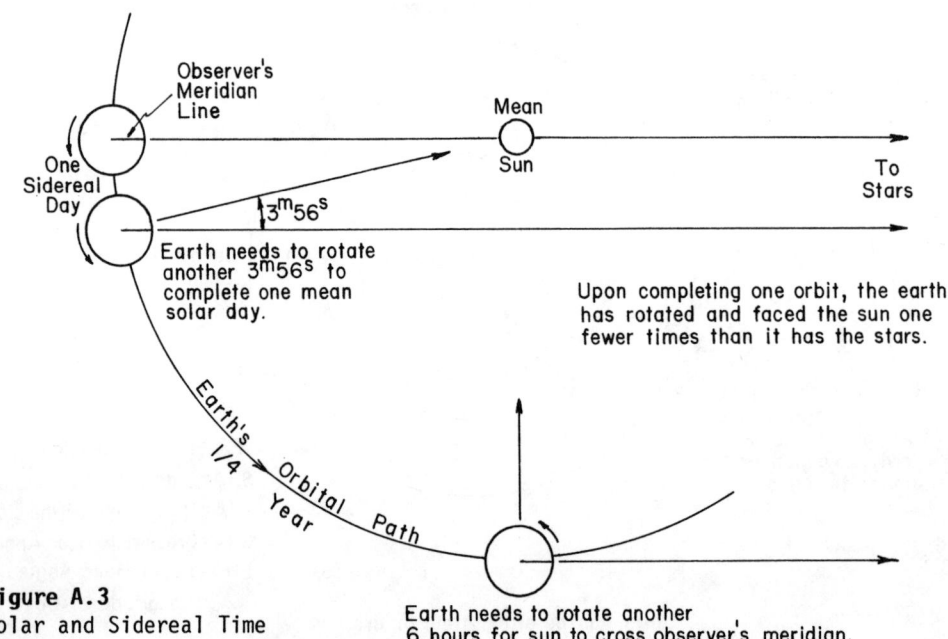

Figure A.3
Solar and Sidereal Time

One relationship based on Figure A.2 and these time concepts is as follows:

$$GHA = \text{Sidereal time} - \text{Right Ascension} \tag{A.1}$$

It is recalled that, using solar time,

$$GHA = GAT - 12^h \tag{6.1}$$

where

$$GAT = UT + \text{Equation of Time} \tag{5.3}$$

Using ephemerides listing sidereal time and right ascension, an azimuth solution will yield the same answer as computed using solar time systems and equation of time since the GHA is the same as computed by either approach. Astronomers use sidereal time, but it wasn't felt practical to expect readers to purchase sidereal clocks or use additional tables to determine third-order azimuth. Thus, solar time was used in this manual. Sideral time simply provides an alternate method to arrive at GHA. It has no other real advantage.

Standard time is the local mean time for a designated meridian, in zones each 15° wide. The center of each zone is the designated meridian. Thus, there are 24 such meridians, at 15° intervals east and west of the 0° meridian, each representing 1^h of time. Daylight savings time, when used, shifts each zone 15° east. The map shown in Figure A.14 at the end of this chapter illustrates time zones used throughout the world. Zone corrections are necessary only when local time is recorded for observations. When UTC is used directly, zone and daylight savings time do not enter into calculations.

A-3 DERIVATIONS OF AZIMUTH SOLUTIONS

Spherical Trigonometry

A spherical triangle is a triangle on a sphere, formed by the intersection of three great circles. The angles of a spherical triangle, are conceptually, the same as in a plane triangle. The sides are arcs of great circles. In spherical trigonometry, the lengths of the sides are not measured in linear units, but in angular units, the angle being that subtended at the center of the sphere by the two end points of the line or side. In a spherical triangle ABC, the angles at its corners are denoted by A, B, C, and the opposite sides by a, b, c. In astronomy, the sphere to be considered is the celestial sphere, having infinite radius. Figure A.4(a) shows a spherical triangle ABC. Figure A.4(b) shows the spherical triangle on the celestial sphere (See Figure 2.1 for this triangle superimposed on the full sphere). This triangle is drawn with a point of view outside of the celestial sphere.

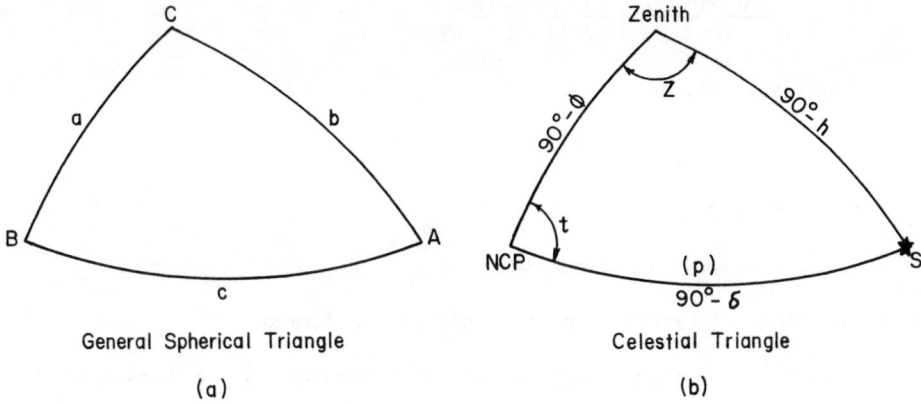

Figure A.4 Spherical Triangle

Basic spherical formulae are:

Sine Law: $\dfrac{\sin A}{\sin a} = \dfrac{\sin B}{\sin b} = \dfrac{\sin C}{\sin c}$ (A.2)

Cosine Law: $\cos c = \cos a \cos b + \sin a \sin b \cos C$ (A.3)

Tangent Law: $\tan \tfrac{1}{2} A = \sqrt{\dfrac{\sin(s-b)\,\sin(s-c)}{\sin s\,\sin(s-a)}}$ (A.4)

where $s = \dfrac{a+b+c}{2}$

Two angles and Three Sides given:

$$\sin b \cos C = \sin a \cos c - \cos a \sin c \cos B \tag{A.5}$$

Identities:

$$\sin X = \cos(90° - X) \tag{A.6a}$$
$$\cos X = \sin(90° - X) \tag{A.6b}$$

Hour Angle Equation

Sine Law:

$$\frac{\sin(90°-h)}{\sin t} = \frac{\sin(90°-\delta)}{\sin Z} \tag{A.7}$$

Cross Multiplying

$$\sin Z \sin(90°-h) = \sin t \sin(90°-\delta) \tag{A.8}$$

From Equation A.6,

$$\sin Z \cos h = \sin t \cos \delta \tag{A.9}$$

From Equation A.5, another relationship can be formed:

$$\sin(90°-h)\cos Z = \sin(90°-\phi)\cos(90°-\delta) - \cos(90°-\phi)\sin(90°-\delta)\cos t \tag{A.10}$$

which, by simplifying using Equation A.6, this becomes:

$$\cos h \cos Z = \cos\phi \sin\delta - \sin\phi \cos\delta \cos t \tag{A.11}$$

Dividing A.9 by A.10

$$\frac{\sin Z \cos h}{\cos h \cos Z} = \frac{\sin t \cos \delta}{\cos\phi \sin\delta - \sin\phi \cos\delta \cos t}$$

Since $\dfrac{\sin\alpha}{\cos\alpha} = \tan\alpha$, this becomes

$$\tan Z = \frac{\sin t}{\tan\delta \cos\phi - \sin\phi \cos t} \tag{2.1}$$

A-3 DERIVATIONS OF AZIMUTH SOLUTIONS

Equations for Finding Polaris

1. Azimuth Equation

Sine Law:

$$\frac{\sin(90°-h)}{\sin t} = \frac{\sin(90°-\delta)}{\sin Z} \tag{A.7}$$

Substituting $p = 90° - \delta$ and $\sin(90°-h) = \cos h$

$$\frac{\cos h}{\sin t} = \frac{\sin p}{\sin Z}$$

Because Z and p are always very small angles, generally less than 1°, the ratios of their sines are approximately equal to the ratio of the angles themselves. Thus,

$$\frac{\cos h}{\sin t} = \frac{p}{Z}$$

Or,

$$Z = p \frac{\sin t}{\cos h} \tag{A.12}$$

The units of Z are the same as the units of p. The sign of Z depends upon the sign of t.

If Polaris is at culmination $\sin t = 0$ and $Z = 0$. If Polaris is at elongation $\sin t = 1$ and $Z = p/\cos h$.

This equation is very convenient for finding Polaris. It is less desirable than the hour angle method (Equation 2.1) for final azimuth computation because of measurement and computational errors associated with determination of a precise value for h and because of the slight round-off error associated with the approximations made. This equation can also be derived from spherical trigonometric equations, as well as from plane trigonometry as shown. This alternate approach to the derivation is left as an exercise for the interested student.

2. Equations for Zenith Angle to Polaris

The altitude, h, is equal to the latitude ϕ when the star is at elongation. In order to know the altitude of the star to bring it into the field of view, a vertical angle must be added to or subtracted from ϕ. From Figure 2.2, it can be seen that,

$$\Delta h = p \cos t \tag{A.13}$$

$$h = \phi + \Delta h \tag{A.14}$$

$$z_1 = 90° - h \tag{A.15a}$$

$$z_2 = 360° - z_1 \tag{A.15b}$$

where

Δh = correction to latitude to determine vertical angle,

h = vertical angle to Polaris,

z_1 = zenith angle, telescope direct,

z_2 = zenith angle, telescope inverted.

The sign of Δh is positive if t is 0° to 90° and negative when t is 90° to 180°, t being either east or west.

Solar Altitude Method

Law of Cosines:

$$\cos(90° - \delta) = \cos(90° - \phi)\cos(90° - h) + \sin(90° - \phi)\sin(90° - h)\cos Z$$

From Equation A.6,

$$\sin \delta = \sin \phi \sin h + \cos \phi \cos h \cos Z$$

Solving for $\cos Z$,

$$\cos Z = \frac{\sin \delta - \sin \phi \sin h}{\cos \phi \cos h} \tag{2.2}$$

It is seen that the hour angle does not appear in the formula, but the altitude does. In this statement lies the advantage and the disadvantage of this method--the time is not critical to know accurately, but the altitude is required, with all of its associated uncertainties and required corrections and field observation problems. Time needs to to be known to only a few minutes and is used merely to determine δ from the ephemeris.

Alternate Solar Altitude Equation

An equation can be derived, based on the tangent law (A.4)

$$\cot \tfrac{1}{2} Z = \sqrt{\frac{\sin(s-\phi)\sin(s-h)}{\cos s \cos(s-p)}} \tag{A.16}$$

where $s = \dfrac{\phi + h + p}{2}$ and $p = 90° - \delta$

It is noted that this equation has the same variables as 2.2. The advantages and disadvantages are the same as for the more common altitude method of Equation 2.2. This equation has been used by the USGS and others.

Equation When Latitude is Unknown

Sine Law:

Using the same formulas used to derive Equation A.12,

$$\sin Z = \cos \delta \, \frac{\sin t}{\cos h} \tag{2.3}$$

This formula gives poor results when Z approaches 90°. For this reason, and because both t and h need to be accurately measured, this formula is undesirable for solar observations, but does provide a crude solution if latitude is unknown. This would be an unusual circumstance since λ would need to be known to determine t; and, if λ could be found, ϕ could probably likewise be found. If ϕ and λ were known there would be no need to use this formula. It is suitable for Polaris, but contains the same problems associated with h as previously discussed for Equation A.12.

A-4 REFRACTION, PARALLAX, SUN'S SEMI-DIAMETER

Atmospheric Refraction

This is of concern **only** if altitude observations are made. Therefore, it is not a consideration when using the hour angle method. **Atmospheric refraction** is the effect on a light ray from a heavenly body as the ray passes through the atmosphere. The ray changes direction, resulting in the celestial body appearing to be higher than it actually is. This results in vertical angle readings being larger than the actual angle. That is, they contain positive errors. See Figure A.5. The lower the star is on the horizon, the greater will be the correction. The correction is also affected by the density of the atmosphere. Atmospheric density is primarily a function of the barometric pressure and air temperature. Pressure is often related to elevation for purposes of applying the correction when pressure readings are unknown, but this gives a poor solution since pressure is influenced by weather, as well as elevation. In order to evaluate the correction, readings of temperature and pressure need to be made at the time of observation. Weather radios, broadcasting these values over frequencies of 162.40 or 162.55 MHz in major cities are useful for these data. Otherwise, a thermometer and calibrated barometer would need to be available.

An empirical formula, developed by Comstock (Reference 16) is

$$r = \frac{983 \, b}{460 + t} \cot h' \tag{A.17a}$$

where

r = refraction error, in seconds,

b = barometric pressure, in inches,

t = temperature, in degrees Fahrenheit,

h' = observed altitude.

If zenith angle is observed,

$$r = \frac{983\,b}{460 + t}\tan z' \qquad (A.17b)$$

where

z' = observed zenith angle.

$$h = h' - r \qquad (A.18a)$$

or

$$z = z' + r \qquad (A.18b)$$

where

h = corrected altitude,

z = corrected zenith angle.

Sun's Parallax

Unlike the stars, the sun is close enough to the earth that the assumption of an observer being at the earth's center for altitude measurements is only an approximation. Figure A.6 illustrates the sun's parallax. The error in a vertical angle measurement varies with the angle itself. It is maximum when the sun is on the horizon and would be zero if the sun was ever straight overhead. The error in a vertical angle is always negative.

$$pc = 8.8'' \cos h' \qquad (A.19a)$$

where

pc = parallax correction, in seconds,

h' = observed altitude.

If zenith angles are observed,

$$pc = 8.8'' \sin z' \qquad (A.19b)$$

where

z' = observed zenith angle.

$$h = h' + pc \qquad (A.20a)$$

or

$$z = z' - pc. \qquad (A.20b)$$

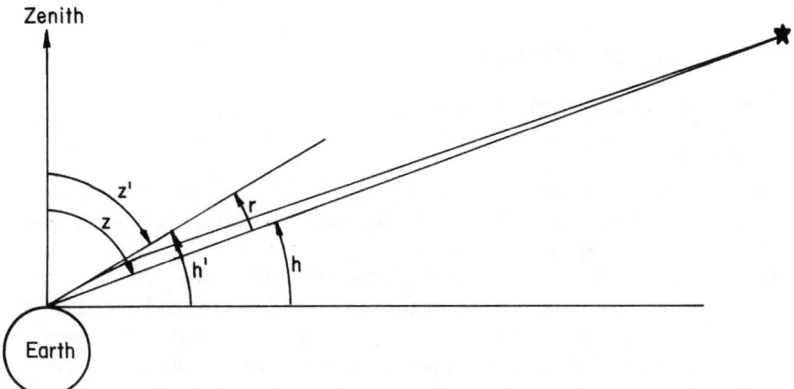

Figure A.5 Effect of Atmospheric Refraction

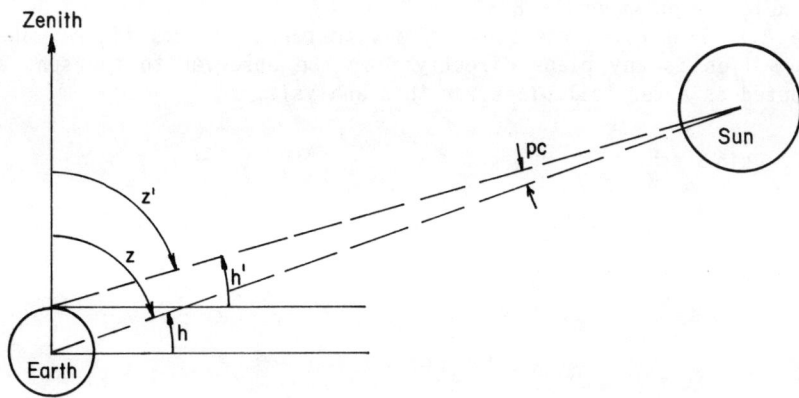

Figure A.6 Sun's Parallax

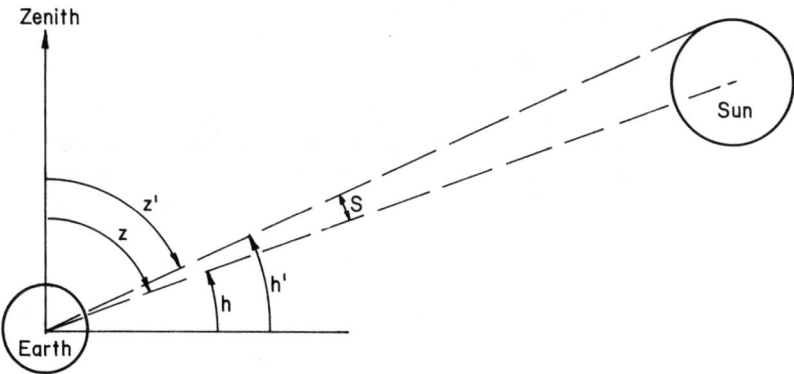

Figure A.7 Sun's Semi-Diameter

where

> h = corrected altitude
>
> z = corrected zenith angle.

Sun's Semi-Diameter

In either the hour angle or the altitude method of solar observations, techniques of observing both limbs of the sun provide means to average the measurements made to the solar target which has an "eccentricity" of half its diameter. Thus, the sun's semi-diameter would not usually need to be considered unless an observer inadvertently sighted to only one limb of the sun for all observations. As a practice, this is not recommended, since additional ephemeris tables for semi-diameter would be required and the technique of sighting both limbs is more "pure". That is, the effects are generally compensated with no risk of round-off errors, misuse of listings of semi-diameter, or mistakes in computations. However, compensation is not exact if too much time passes between pointings in each set. This point will be explained in Appendix B.

Figure A.7 illustrates the sun's semi-diameter. In this figure, the plane in which S lies is any plane directly from the observer to the sun, and can be considered as a vertical plane for this analysis.

$$\sin S = \frac{d}{R}$$

Since S is very small

$$S = \frac{d}{R} \rho \qquad (A.21)$$

where

> S = sun's semi-diameter, in seconds,
>
> d = earth's diameter,
>
> R = distance between earth and sun,
>
> ρ = radians-seconds conversion.

The value of R varies as the earth orbits the sun. Therefore S varies daily. Theoretically, the value of S also changes with elevation above sea level, but this effect is negligible. The semi-diameter, S, is the same as measured in a vertical plane or through any plane passing through the sun's center, since the sun is circular.

$$h = h' + S \quad \text{(lower limb sighted)} \qquad (A.22a)$$

or

$$h = h' - S \quad \text{(upper limb sighted)} \qquad (A.22b)$$

where

> h' = observed vertical angle,
>
> h = corrected vertical angle.

Also,

$$z = z' - S \quad \text{(lower limb sighted)} \tag{A.23a}$$

or

$$z = z' + S \quad \text{(upper limb sighted)} \tag{A.23b}$$

where

> z' = observed zenith angle,
>
> z = corrected zenith angle.

When the lower limb is sighted, the correction to the true center is positive if vertical angles are used and negative if zenith angles are used. These signs are reversed if the upper limb is sighted.

For horizontal angle measurements, a formula which is derived from the sine law in spherical trigonometry, reveals that

$$s = \frac{S}{\cos h} \tag{A.24}$$

where

> s = correction to measured horizontal angle.
>
> $\theta = \theta' + s \quad \text{(left limb sighted)} \tag{A.25a}$

or

$$\theta = \theta' - s \quad \text{(right limb sighted)} \tag{A.25b}$$

where

> θ' = observed horizontal angle from terrestrial target,
>
> θ = corrected horizontal angle.

When the left limb of the sun is sighted, the correction is added to θ'. When the right limb is sighted, it is subtracted. Figure 2.4(b) illustrates this concept.

Since values of S vary daily, a table is convenient to use for semi-diameter. Table A.1 lists values of S. Values in the table are valid for any year to the precision indicated. Any interpolated value should not

TABLE A.1
Sun's Semi-Diameter

Date		Semi-Diameter ′ ″	Date	
Jan.	3	16 17.55	Jan.	3
↓	18	16 17.0	Dec.	19
	28	16 16.0		10
Feb.	5	16 15.0	Dec.	2
↓	11	16 14.0	Nov.	26
	16	16 13.0		21
↓	21	16 12.0	↑	16
Feb.	26	16 11.0		11
Mar.	1	16 10.0	↑	7
↓	5	16 09.0	Nov.	3
	9	16 08.0	Oct.	30
↓	13	16 07.0		26
	17	16 06.0	↑	22
↓	20	16 05.0		18
	24	16 04.0	↑	15
↓	28	16 03.0		11
Mar.	31	16 02.0	Oct.	8
Apr.	4	16 01.0	Oct.	4
	7	16 00.0	Sept.	30
↓	11	15 59.0		27
	15	15 58.0	↑	23
↓	18	15 57.0		19
	22	15 56.0	↑	15
↓	26	15 55.0		12
Apr.	30	15 54.0	↑	7
May	5	15 53.0	Sept.	3
	9	15 52.0	Aug.	30
↓	14	15 51.0		25
	19	15 50.0	↑	20
↓	24	15 49.0		15
May	31	15 48.0	Aug.	8
June	7	15 47.0	July	31
June	17	15 46.0		21
July	5	15 45.35	July	5

Values are accurate for any year ±0.1″. From <u>American Ephemeris and National Almanac</u>

A-5 MISCELLANEOUS ATTACHMENTS FOR CELESTIAL OBSERVATIONS

differ by more than ±0.15" from more precise values given each year in the referenced ephemeris.

A-5 MISCELLANEOUS ATTACHMENTS FOR CELESTIAL OBSERVATIONS

Roelofs Solar Prism

The Roelofs solar prism is an attachment which fits over the objective end of a theodolite. The one shown in Figure A.8 is made for Wild theodolites. It can be adapted to other models. The image seen through the prism, when sighting the sun, is as pictured in Figure A.8. The sun is split into four overlapping suns, geometrically. The cross-hairs are aligned to the sun's center, theoretically, when pointed to the center of the overlapping images (cloverleaf) as shown.

The solar prism is more advantageous for the altitude method than the hour angle method because the sighting to the center makes field manipulation and data computation a little easier. For the hour angle method, individual computations of each direct and each inverted pointing result in the same azimuths theoretically, just as they do with Polaris observations, and this helps to detect mistakes earlier in the computations. This is not an important advantage, however, since when the solar eyepiece method is used the computed azimuths to each respective limb should check closely and thus mistakes can be discovered. Furthermore, the mean in each set (L and R limb) should agree, which is another check. Therefore the intermediate azimuths may appear as more realistic values when using the solar prism, but this isn't an important advantage and doesn't increase accuracy. One small advantage of the solar prism is that it avoids a possible error present in limb pointings which relates to changing diameter of the sun (See Section B-2).

Tests have shown that the pointing error to the center of the prism is comparable to that of sighting a sun's limb. (Reference 19). Therefore, the same precision should result from using either a prism or a solar eyepiece, since the same number of solar pointings are made for the same number of sets. The experience of the author is that the prism image is often not as clear as a limb of the sun, probably because the sun's rays go through a lot more glass in the prism, cutting down the light and increasing fuzziness. For this reason the sun limb with eyepiece method gives more confidence in pointing. However, there is that small, uncertain feeling when sighting the right limb of the sun where the cross-hair suddenly emerges from the blackness and becomes tangent. Considering this extra attention needed here, some may prefer the solar prism with its center image.

There is no more or no less field work with the solar prism. The instrument slips over the objective lens and is hinged so that it can be opened to observe the ground mark and closed to sight the sun. There is probably no more trouble in doing this than in putting the solar eyepiece filter on and taking it off. The prism needs an initial rotation to align the image properly with the cross-hairs but this can be done quickly. There is a little added inconvenience in that the prism weighs enough that the objective lens drops down unless the vertical motion is clamped. Also, the telescope cannot be plunged by passing the objective lens beneath the standards, a minor disadvantage.

The main disadvantage of the solar prism is its price. In 1983, it listed for $634 compared to $32 for a solar eyepiece filter. Unless the surveyor makes solar observations daily or very regularly, and especially enjoys using the prism or sees other advantages to it (or already owns one) the solar eyepiece is the one recommended for purchase as part of the instrument package totalling "less than $100" as promoted by the author.

Prism for High Altitudes

When Polaris observations are made in high latitudes, a prism which delivers the image at an angle of 90° with respect to the line of sight may be needed. This is commonly called a diagonal eyepiece prism. Such a prism would not ordinarily be used for solar observations since the sun would not be sighted at a high vertical angle. Using Polaris observations, it is well to point out that when the altitude is so great that such a prism is needed, the methods described thus far probably would no longer yield third-order azimuth because of bubble centering errors and their effects on azimuths at such steep angles. A diagonal eyepiece prism is shown in Figure A.9, along with an eyepiece prism.

Targets for Night Observations

Night targets for Polaris observations are manufactured in various styles. Some surveyors like to experiment with their own design, saving costs on the more expensive factory made ones. Care should always be made to assure that the design results in the center of the illuminated image being directly vertical over the ground azimuth mark. A commericial target, illuminated by a battery attachment, is shown in Figure A.10.

Striding Level

When the horizontal (trunnion) axis of a theodolite is in proper adjustment (parallel with the plate bubble axis) and the bubble is centered, the line of sight passes through a vertical plane when the telescope is pointed above or below the horizon. If the bubble is allowed to drift off center, the line of sight will be cast to the right or left of the desired vertical plane when the telescope is pointed above or below the horizon. This causes an error in the direction to an elevated or depressed target. A similar error occurs if the horizontal axis is not in proper adjustment, but this is a systematic, instrumental error which is compensated when using the mean of observations made with the telescope direct and inverted. The error in a direction caused by bubble centering errors is not compensated by the measuring procedures. It must be either avoided by keeping the plate bubble in proper adjustment (perpendicular to the vertical axis) and perfectly centered when sighting to an elevated or depressed target, or corrected by taking bubble "readings" and solving an equation.

When high precision is desired, such is the case for Polaris observations, the plate bubble must be very carefully centered. This has been discussed in previous chapters and is also demonstrated by examples in Appendix B. If the observer is incapable of keeping the plate bubble centered with the desired precision, the bubble ends can be "read" and a small correction made for the leveling error observed at the moment of pointing. For even higher precision, a *striding level* can be used and read in a similar manner. A striding level is an instrument which attaches to the theodolite standards, thus being parallel to the plate bubble. (See Figure A.11). It is adjusted so that its axis is perpendicular to the vertical axis of the theodolite. A striding level typically has a sensitivity of 5", as compared to the 20" sensitivity of a plate bubble. Thus, small changes in the orientation

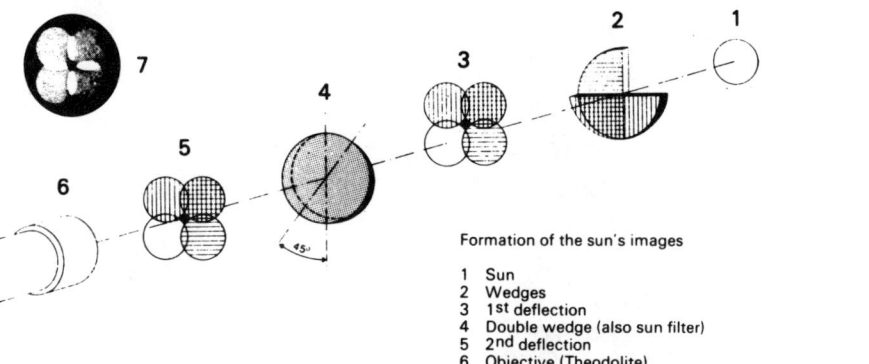

Formation of the sun's images

1 Sun
2 Wedges
3 1st deflection
4 Double wedge (also sun filter)
5 2nd deflection
6 Objective (Theodolite)
7 Field of View

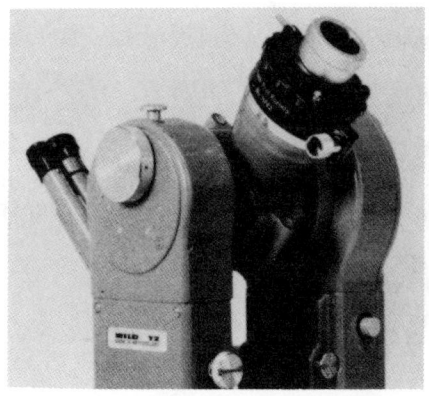

Wild T2 Universal Theodolite fitted with the Wild-Roelofs Solar Prism (note the prong-shaped vane) and diagonal eyepieces.

Solar Prism folded down for sighting a terrestrial mark.

Figure A.8 Roelofs Solar Prism and Solar Image
(Photos courtesy of Wild-Heerbrugg Instruments, Inc., Farmingdale, N.Y.)

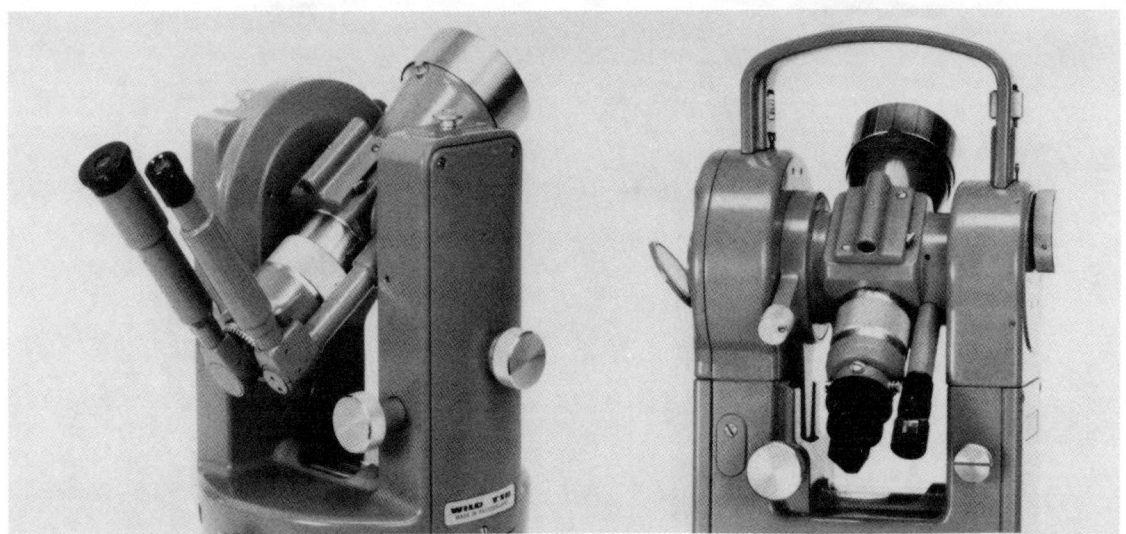

Figure A.9 Diagonal Eyepiece Prism and Eyepiece Prism Set

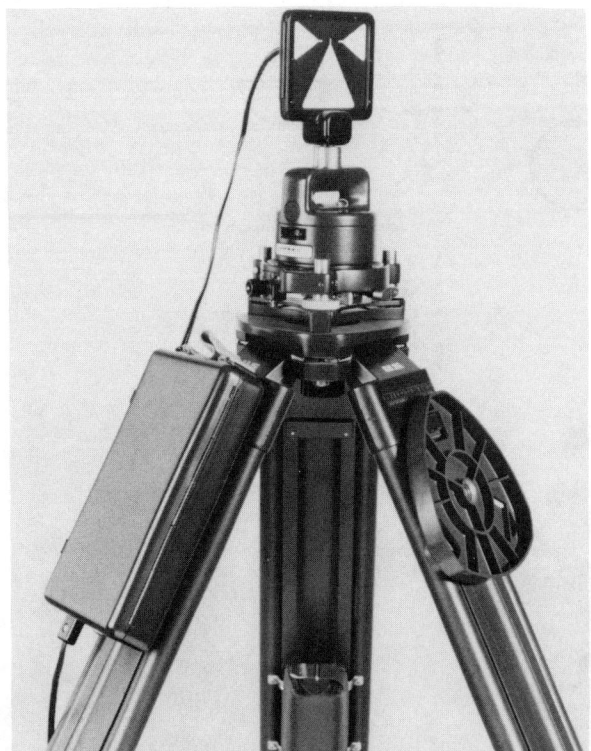

Figure A.10
Target for Night Observations

(Photos courtesy of
Wild-Heerbrugg Instruments, Inc.,
Farmingdale, N.Y.)

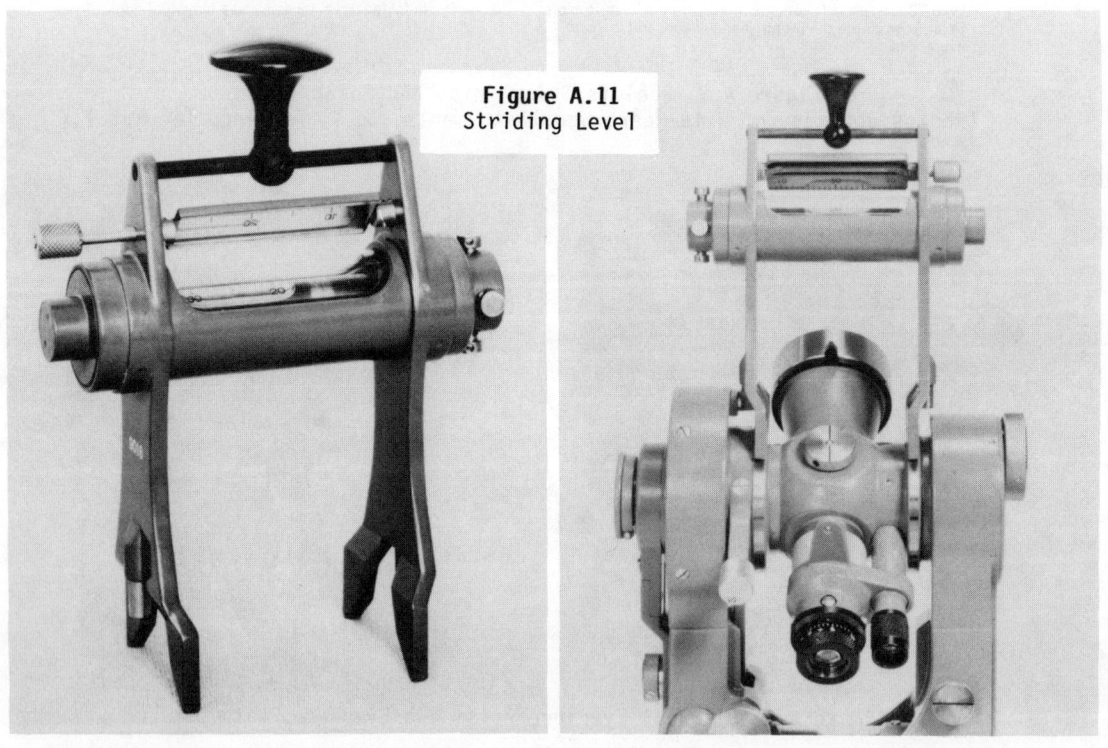

Figure A.11
Striding Level

of the horizontal axis can be more readily observed than when using only the plate bubble.

Using either the striding level or the plate bubble, the correction for a pointing to an elevated or depressed target is a function of the bubble sensitivity, the amount the bubble is off center, and the vertical angle. In the field, the surveyor counts the divisions or marks from the geometric center of the vial to the point where the bubble rests at both ends of the level vial (bubble tube). A striding level generally has "scale readings" which are a count of the divisions left and right of its geometric center. If angle observations are treated individually as is done in the methods described in this manual, the following equations apply:

$$C_{B_D} = (L_D - R_D) \frac{\mu}{2} \tan h \quad (A.26a)$$

$$C_{B_I} = (L_I - R_I) \frac{\mu}{2} \tan h \quad (A.26b)$$

where

C_{B_D} = correction to a direction, telescope direct (D),

C_{B_I} = correction to a direction, telescope inverted (I),

L_D = reading on left side of bubble, telescope direct,

R_D = reading on right side of bubble, telescope direct,

L_I = reading on left side of bubble, telescope inverted,

R_I = reading on right side of bubble, telescope inverted,

μ = bubble sensitivity,

h = vertical angle.

When angles are meaned within sets, which is the usual case when only terrestrial targets are involved (or as some do with astronomic observations), one equation can be used which combines all readings. In this case, the correction to the mean of the direction readings to the particular target is

$$C_B = [(L_D + L_I) - (R_D + R_I)] \frac{\mu}{4} \tan h \quad (A.26c)$$

This is the equation generally given in references describing the striding level.

If, in astronomic observations, the terrestrial target is elevated, a correction for bubble centering might need to be done for pointings, in addition to the correction for pointings to the celestial target. Normally, a correction is not needed for either the terrestrial or astronomic target if vertical angles are less than about 20° and the plate bubble is kept centered better than about $\pm\frac{1}{2}$ division. When conditions exceed this,

MISCELLANEOUS CONCEPTS AND DEFINITIONS

consideration for a correction, preferably using a striding level, should be made if the equivalent of third-order astronomic azimuth is to be maintained. Appendix B includes more discussion on bubble centering care.

Example A.1 A Polaris observation was made. The altitude of Polaris was 42°30'. The terrestrial target's altitude was 0°45'. A striding level of 5" sensitivity was used. Readings were $L_D=5.0$, $R_D=3.5$, $L_I=5.5$, $R_I=6.5$. After subtracting circle readings, the horizontal angles from the terrestrial target to the star, telescope direct and inverted were $\theta_D=36°14'12"$, $\theta_R=36°14'18"$. What are the corrections for bubble centering and what are the corrected angles?

Solution:

$$C_{B_D} = (5.0 - 3.5)\frac{5"}{2} \tan 42°30' = 3.4" \qquad (A.26a)$$

$$C_{B_I} = (5.5 - 6.5)\frac{5"}{2} \tan 42°30' = -2.3" \qquad (A.26b)$$

$$\theta_D = (36°14'12") + 3.4" = 36°14'15"$$

$$\theta_R = (36°14'18") - 2.3" = 36°14'16"$$

Scale readings on the striding level were not observed for the terrestrial pointings since tan 0°45' is so small. At this altitude, a plate bubble of 20" sensitivity would need to be off center 4 divisions to cause even 1" direction error.

The sign of the correction is determined by the scale readings. If L > R, it is positive. If L < R, it is negative. This can be reasoned as follows: when the level bubble is left of center (L > R), the horizontal axis is tilted down on the right side. When the telescope is elevated in this situation, it points too far to the right. Thus, the surveyor must bring it back to a target by moving it counterclockwise, thus causing a horizontal direction reading which is too small.

Although this discussion is intended to describe the use of a striding level, it should be remembered that a plate level can be read in the same manner if desired. For pointings at high altitudes where a striding level is not available this is probably more accurate than trying for perfection in plate bubble centering in order to avoid the error. Plate bubble division counts from its center should be read to at least the nearest 1/4 division, and preferably to 1/8 or 1/10 when the altitude approaches 40°, if equivalent third-order accuracy in astronomic azimuth is to be maintained.

Theoretically, if a plate bubble is off center when pointing to a target with the telescope in one position (either direct or inverted), it will be off center the same amount and in the same direction when the telescope is plunged and again pointed to the target. This assumes that the instrument is unaffected by settlement or any other natural or personal errors and the bubble axis is adjusted perpendicular to the vertical axis. Thus, unless affected by minor natural or personal errors, C_{B_D} and C_{B_I} would be equal. It is hoped that these points will help the reader to realize that these corrections relate only to bubble drift due to unknown causes and are separate from considerations of either horizontal axis or bubble axis adjustment relationship of the theodolite.

A-6 SOLAR ALTITUDE METHOD

Although this method is not recommended, it does need to be explained for comparison purposes, and so that the reader can have the option to use it in instances where only 30" to 1' precision is needed and where accurate time is unavailable, or when the observation station's position in longitude is uncertain. The altitude method was briefly compared with the hour angle method in Section 1-7. The most common solution is

$$\cos Z = \frac{\sin \delta - \sin \phi \sin h}{\cos \phi \cos h} \tag{2.2}$$

The field observations for the altitude method include reading both the vertical and horizontal circles of the theodolite, when either the sun's limbs are tangent to the vertical and horizontal cross-hairs, or when the center of the Roelofs solar prism image is at the intersection of the cross-hairs image. This requires "tracking" the sun's image on one cross-hair, using the fine motion screw of the theodolite, then stopping the motion the instant the other cross-hair is aligned on the sun's image. Using the solar eyepiece, a set of observations would normally include two pointings, one being in the opposite quadrant of the cross-hairs from the other. Figure A.12 illustrates what might be seen for morning and afternoon observations. Before proceeding with solution of Equation 2.2, the two vertical circle readings would need to be averaged, and likewise the two horizontal circle readings. When the solar prism is employed, this averaging is generally also done before proceeding with a solution for each set of observations.

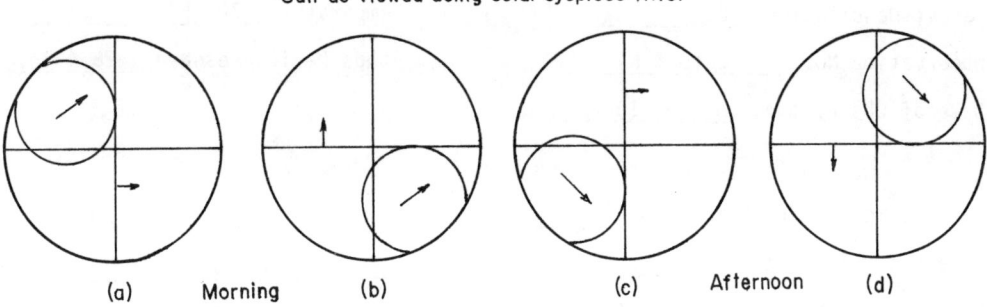

Sun as viewed using solar eyepiece filter

(a) Morning (b)   (c) Afternoon (d)

In (a) and (c), sun is tracked with vertical crosshair until horizontal crosshair becomes tangent, and vice versa in (b) and (d).

Figure A.12 Image of the Sun Using Altitude Method

Time needs to be observed only to the nearest five or ten minutes. It is not used in the solution, but only to interpolate the sun's declination in the ephemeris, and a few minutes error causes a negligible error in results. Latitude is scaled from a map and should be done accurately. Since the method is not as precise as the hour angle method, the conversion from geodetic to astronomic latitude can be ignored unless deflections of the vertical are suspected to approach 60". Longitude is unimportant. As is seen, it isn't used in the solution.

The vertical angle must be corrected for sun's parallax and atmospheric refraction. Thus, field readings of temperature and atmospheric pressure are required. Equation A.17 can be used for refraction. Equation A.19 can be used for parallax.

After the above observations and computations are computed for each set, Equation 2.2 is solved for each and the mean value used to represent the best value, assuming that agreement among the sets is realized. Three sets of data should yield an azimuth with an accuracy of approximately ±30" using a repeating or a direction theodolite.

Figure A.13 shows the computation of one observation set. The field data form is omitted but the important field data are shown on the computation form.

It should be noted that this manual contains all formulae and tables necessary to calculate an azimuth by the solar altitude method. No ephemeris or other tables are needed from other sources.

Figure A.13 Solar Observation Computation, Altitude Method

Instrument Station __A__   Latitude __38°10.1'__

Target Description __B__   Temperature __70° F__

Observation No. __1 (D & R)__   Atmospheric Pressure __28.4 in.__

Date of Observation __May 6, 1967__

Watch Time (24 Hr. Clock) __15:40__   Measured Altitude __34°03.9'__ (Mean of D&R)

Zone and DST Correction __+6:00__   Parallax __0.1'__ (Eq. A.19a)

GCT of Observation __21:40__   Refraction __1.3'__ (Eq. A.17a)

Sun's Declination, 0 hours __+16°16.7'__   True Altitude __34°02.7'__

Diff. for 1 hr., (0.70)xUT __+15.2'__   (See Eq. A.18a and A.20a)

Sun's Declination at Observation __+16°31.9'__

$$\cos Z = \frac{\sin \delta - \sin \phi \sin h}{\cos \phi \cos h} = -0.094289 \qquad Z = 264°35.4'$$

Angle, Mark to Sun = 157°34.8' (mean of D & R)

Astronomic Azimuth to Target = 107°00.6'

Figure A.14 Standard Time Zones and their Relationship to UTC
(Source: Reference 22)

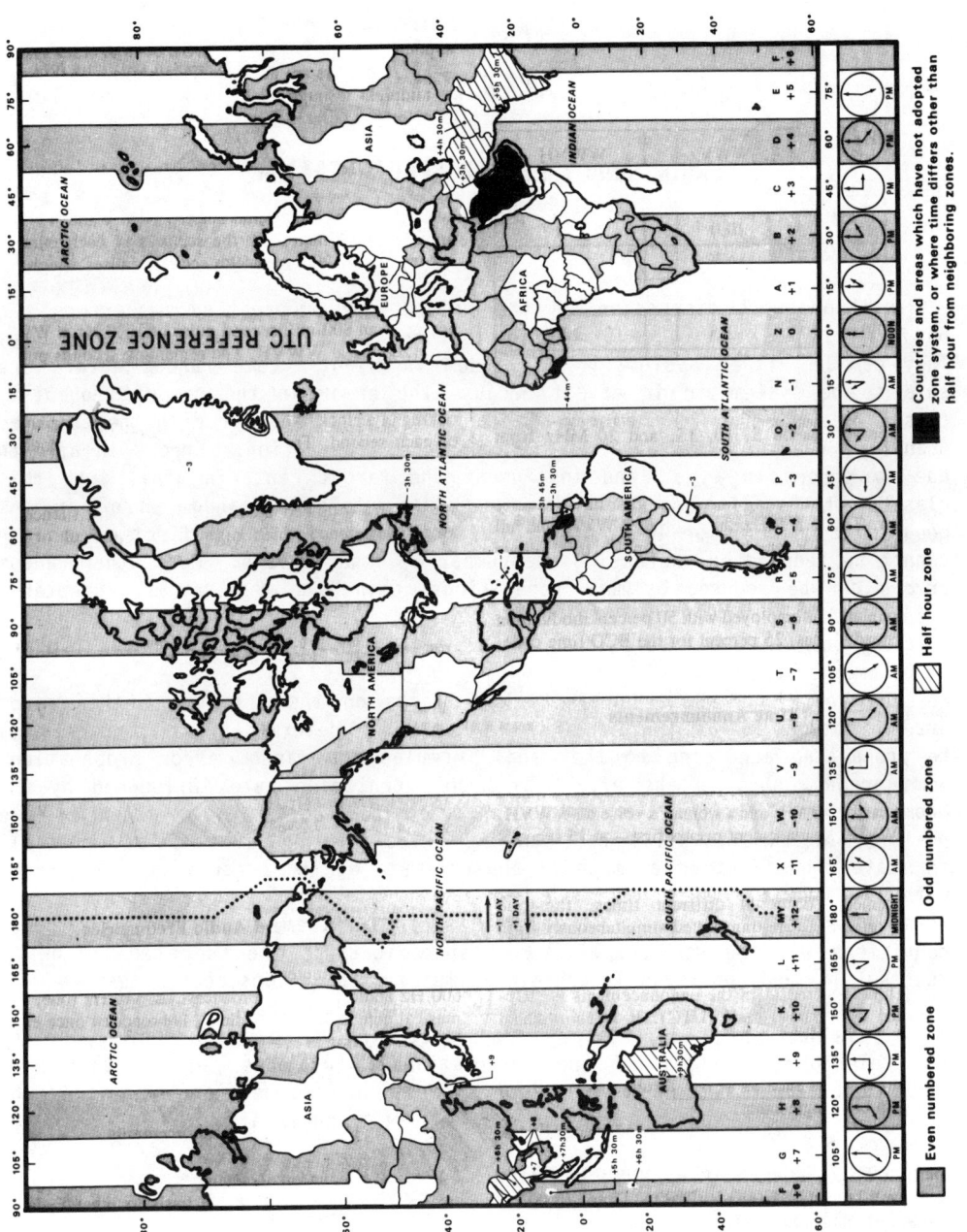

APPENDIX B

ERROR ANALYSES AND SPECIFICATIONS

B-1 ERRORS AND THEIR NATURE

A theoretical discussion of the nature and propagation of errors can be found in Reference 8. Errors are either systematic or random in **nature**. They can be either systematic or random in **effect**. Some random errors are said to be systematic in effect because the error in the computed quantity caused by the error in the variable contributing to the error in the computed quantity, cannot be reduced by the process of repetition. Errors in azimuth due to errors in ϕ, λ, and instrument and target centering fall into this class. The scaling or centering procedure yields a random error in the geographic position or location of the instruments, but the effect on azimuth cannot be reduced by increasing the number of observations. Most other random errors can be reduced by more repetitions, using the mean of all acceptable results. These errors include those due to timing, reading the theodolite circle, pointing the telescope, plate bubble centering, and interpolation in ephemeris tables.

Most of the discussion herein will focus on the effects of the random errors common to astronomic and grid azimuth determination. Errors will be propagated according to the usual formulas for random error propagation. Some errors, such as the effect of bubble centering, are influenced by the geometric relationships in the observations. Others are affected by the position of the celestial body on the celestial sphere. It is impossible to state the effect on azimuth of any estimate of the error from any source, without knowing the positions of the targets and instrument in three-dimensional space. For this reason, extreme or perhaps average conditions will be used to arrive at specifications the user can follow to successfully and consistently achieve azimuths with various accuracies.

Systematic errors are also discussed. Those that are caused by theodolite axes maladjustments are mostly compensated by the procedure of inverting the telescope for half of the observations. For other errors that have systematic effects, methods to correct or compensate for them will be discussed. Some have already been explained in Appendix A.

B-2 ERROR SOURCES IN ASTRONOMIC AZIMUTH

Reading of Theodolite

Several years of study of the reading errors of optical theodolites including several hundred results from the participants in measurement analysis seminars taught by the author has yielded some fairly consistent results from theodolite error tests. Precision of a theodolite varies with manufacturer, observer, lighting, optical cleanliness, understanding of the error test, and the design of the reading system. Using "name" brand instruments which are in good condition, an experienced observer generally observes reading errors of ±0.7" to 1.2" standard deviation for common direction theodolites and ±3" to 5" for repetition theodolites. Details of how to determine these errors for any instrument/observer combination can be found in Reference 8.

The reading error propagates as follows:

$$\varepsilon_{\alpha_r} = \frac{\varepsilon_r \sqrt{2}}{\sqrt{n}} \tag{B.1}$$

where

ε_{α_r} = the error in angle due to ε_r,

ε_r = the error in one reading of the instrument,

n = the number of repetitions of the angle.

Normally, errors would propagate differently for a repeating instrument. But, since the repetition feature isn't used for astronomical observations, the above equation is valid for either a repeating or a direction instrument.

Pointing of Theodolite

The pointing error to the terrestrial targets which are "well-defined", has been proved to be in the range of ±1.5" to 2.5" standard deviation using common theodolites of reputable manufacture, having clean optics and experienced observers. The target range has no effect on this error, provided the target is not obscured or blurred in any way.

This error propagates in the same manner as the reading error.

$$\varepsilon_{\alpha_p} = \frac{\varepsilon_p \sqrt{2}}{\sqrt{n}} \tag{B.2}$$

where

ε_{α_p} = error in an angle due to ε_p,

ε_p = error in one pointing to a well-defined target,

n = the number of repetitions of the angle.

It might be argued that the error in pointing to either the sun or Polaris is greater than to a ground target. In considering the fact that Polaris is a distinct pinpoint of light, barely covering the cross-hair, and that the carefully focussed limb of the sun is as distinct as the edge of any disk near the ground surface, there is no reason to conclude that these targets yield a pointing error greater than that to common, fixed ground targets. The error associated with pointing to the celestial targets is incorporated into the time/motion effects on azimuth, rather than into effects due to optical qualities (definition) of the telescope or the distinctness of the target.

Theodolite and Target Centering

Since the celestial bodies are considered to be at infinity, and with no "centering" error, effects due to erecting instruments and ground targets are all that need to be considered. A target such as a tower or steeple has no centering error. Centering of a theodolite and a target such as a range pole or geodetic target must be considered in its effect on azimuth. The effect is a function of the error itself, the direction of the error with respect to the survey line, and the length of the line. The maximum effect occurs when the error is perpendicular to the line. Naturally, being random in nature, the surveyor has no way of knowing either the magnitude or the direction of this error around the point on the ground. An estimate, based purely on judgment must be made, and then an assumption made that this centering error is perpendicular to the line. Thus, for one error,

$$\varepsilon_{c_1} = \frac{d}{D} \rho \tag{B.3}$$

where

ε_{c_1} = error in direction to the backsight, in seconds,

d = estimate of centering error (same units as D),

D = distance between theodolite and ground target,

ρ = conversion between radians and seconds.

If a ground monument is used rather than a tower, there are two occurrences of this error (one at the target and one at the theodolite) in the line's direction. Assuming d the same for both instrument and target station, the error in the direction due to centering errors at both the theodolite and the target is

$$\varepsilon_c = \frac{d}{D} \rho \sqrt{2} \tag{B.4}$$

following the usual laws of error compensation.

Plate Bubble Centering

The effects on a line's direction, due to bubble centering vary with the care in centering, the bubble sensitivity, and the vertical angle. A derivation from Reference 8 results in

$$\varepsilon_b = \pm f_d \, \mu \, \tan \gamma \tag{B.5}$$

This error propagates as the inverse of the square root of n, where n is the number of repetitions. It must be evaluated and propagated for each of the two targets comprising an angle measurement, considering the vertical angles to each.

ϵ_b = angular error in a direction due to bubble centering,

f_d = the fraction of a division the bubble is estimated to be off (care in centering),

μ = bubble sensitivity, in seconds,

γ = vertical angle (altitude) to target.

Figure B.1 shows errors in a direction due to bubble centering at various estimated uncertainties in centering, as a function of vertical angle.

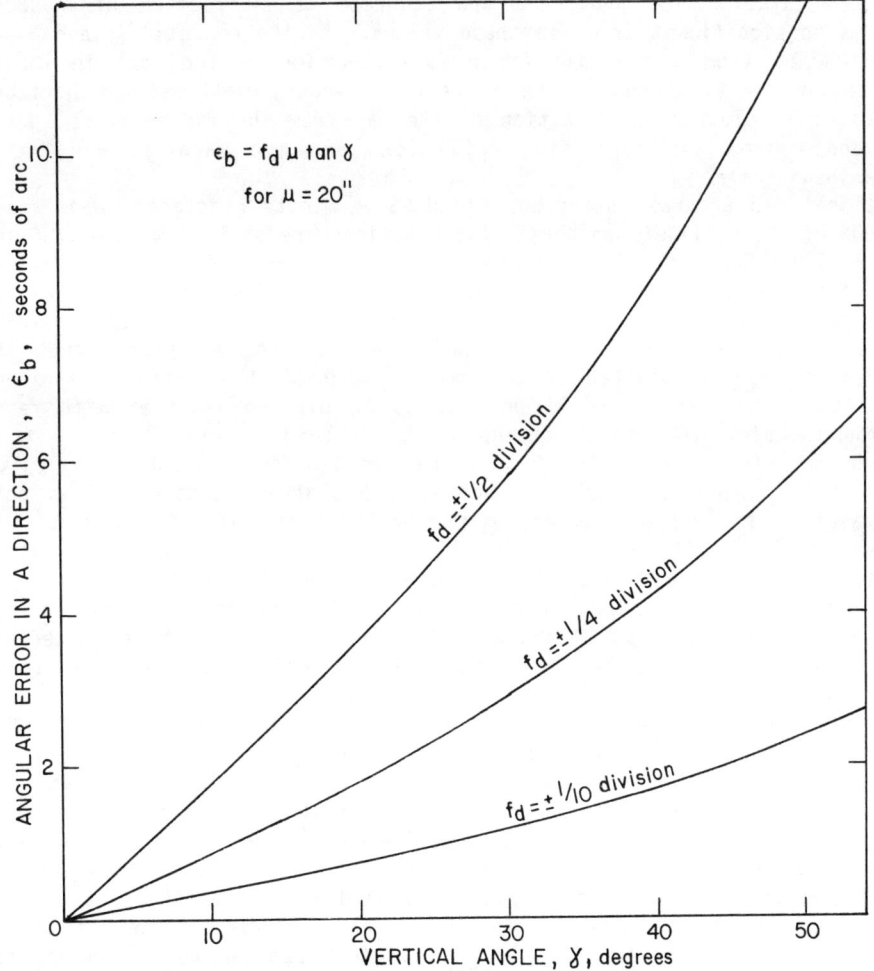

Figure B.1 Effects of Bubble Centering Error

The altitude of the sun is solved by the equation,

$$\sin h = \cos \delta \cos \phi \cos t + \sin \delta \sin \phi . \tag{B.6}$$

Figure B.2 shows the altitude to the sun as a function of t for the different times of the year. This is useful in planning observations to avoid steep vertical angles and thus minimize the effects on azimuth due to bubble centering.

Equation B.5 yields an evaluation of the random errors. Correction for systematic errors when the bubble ends are "read", rather than the observer attempting to keep them "centered", is discussed in Appendix A. Equation A.26 is used to make the corrections.

Latitude and Longitude

Error from map scaling procedures. If the geodetic ϕ and λ of the instrument station are published and accurate to a fraction of a second, there is no significant error in these values. If the geodetic ϕ and λ must be determined from a map, the error is a function of the care in locating the station on the ground with respect to local, well-defined horizontal map features, plotting the station on the map from the map features, scaling of ϕ and λ from the map using grid ticks and map margins, and the map accuracy and scale.

National map accuracy standards for USGS 7½-minute quadrangle maps specify that 90% of the well-defined horizontal positions are plotted within 1/50 inch. At 1:24,000 scale, this is ±40 feet on the ground. Therefore, if local, well-defined features are used to reference the survey station, using distance measurements even as imprecise as pacing, the field location should contribute no significant error in the final position. If it is assumed that these measurements can be plotted on the map to ±0.02 inches using an engineer's scale, then the position might be said to be plotted ±40 feet with respect to ground scale. The final action is to scale from grid lines. If this scaling can also be done to ±0.02 inches or ±40 feet ground equivalent, it might be said that a carefully located and plotted station has an actual uncertainty, with respect to any map coordinate system, of $\pm 40\sqrt{3}$ ft. This assumes first of all that the map sheet is, in fact, produced to National Map Accuracy Standards. It also contains a few assumptions regarding the direction of the positioning, plotting, and scaling errors. The best estimate, yielding an error perhaps slightly on the high side is that the uncertainty in position follows the law of compensation, considering the three effects mentioned.

But, how much of this composite error goes into ϕ uncertainty and how much into λ ? Again, from random error theory, this might be analyzed as $\pm 40\sqrt{3}/\sqrt{2}$ into each component, yielding an estimated uncertainty in determination of approximately ±50 ft. in each of the components ϕ and λ. How this converts to geographic coordinates depends upon the latitude, since lines of longitude vary in separation with latitude. Each second of latitude represents approximately 100 ft regardless of latitude. Each second of longitude measures approximately 80 feet at the mid-latitudes of the U.S. and varies between approximately 70 and 90 feet between the north and south extremes of the 48 states. ($\phi = 25°$ to $50°$).

147

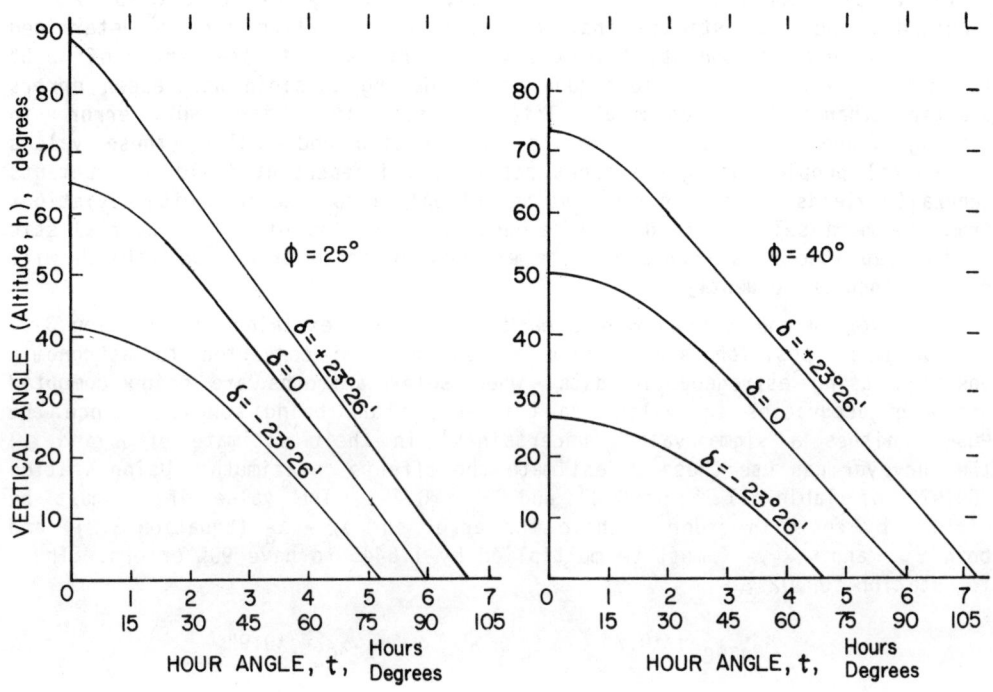

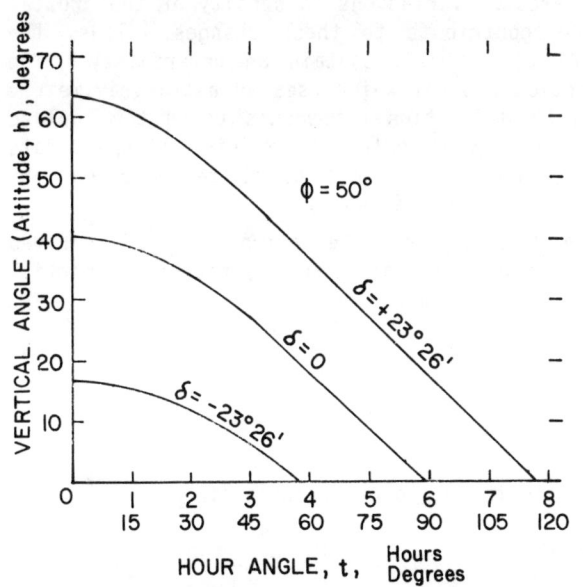

Derived from
$\sin h = \cos \delta \cos \phi \cos t + \sin \delta \sin \phi$

Figure B.2
Vertical Angle to the Sun

Thus, using 100 ft. for 1" of latitude, an average of 80 feet for 1" of longitude, and the estimate that the instrument station can be determined ±50 feet in each component, this translates into an estimated error of ±0.5" in latitude and ±0.6" in longitude. Considering possible unforeseen errors greater than those assumed, let us say that the 90% error in scaling ϕ and λ is ±1.0". Results of plotting and scaling these values by several people, using different scales and independent field measurements generally yields a range in their values of only 1 to 3 seconds with deviations from the mean seldom exceeding 1 second. A precision of ±1" seems realistic if the map itself is accurate. The effects of these errors on azimuth will be explained subsequently.

Converting ϕ and λ to astronomic position. As was explained in Section 3-2, the geodetic position scaled from a map must be converted to astronomic position using astrogeodetic data, when solar azimuths are being computed and when accuracies equivalent to third-order are being sought. Since the NGS furnishes a sigma value (uncertainty) in their estimate of η and ξ, the surveyor can use these to estimate the effects on azimuth. Using station "OHIO2" of Table 3.1, $\varepsilon_\xi = \pm 0.4"$ and $\varepsilon_\eta = \pm 0.4"$. The value of ε_η must be divided by $\cos \phi$ in order to have the error in $\lambda_A - \lambda_G$ (Equation 3.2) and both ε_ξ and $\varepsilon_{\lambda_A - \lambda_G}$ must be multiplied by 1.6445 to have 90% errors. Thus, for station "OHIO2",

$$\varepsilon_{\phi_A - \phi_G} = \pm 0.7" \qquad \varepsilon_{\lambda_A - \lambda_G} = \pm 0.9"$$

If the values of ϕ, δ and t are known, the effect of these errors on azimuth can be estimated using Figures B.3 and B.4, or Equation B.7 and B.8, which will be explained subsequently.

Geodesists warn that there can be abrupt changes in the deflection of the vertical, even in non-mountainous areas. Variations in density of the crustal rock and other natural conditions contribute to these changes. Thus, the values for the astrogeodetic data will always contain an uncertainty. The NGS program is a deflection prediction program which uses an astro-gravimetric collocation technique to interpolate deflection components using the Laplace and other gravity observation stations, with data as complete and up-to-date as possible. Their uncertainty estimates are felt to be reliable and are assumed to consider local conditions as they are known.

Effects of ϕ and λ errors on azimuth. An error equation can be derived for the effect of an error in ϕ on the error in azimuth by taking the partial derivatives of the hour angle formula. The result is

$$\varepsilon_{Z_\phi} = \pm \left[\left(\frac{\sin \phi \tan \delta + \cos \phi \cos t}{\sin t} \right) \sin^2 Z \right] \varepsilon_\phi \qquad (B.7)$$

where

ε_{Z_ϕ} = error in azimuth due to error in latitude,

ε_ϕ = error in latitude.

B-2　ERROR SOURCES IN ASTRONOMIC AZIMUTH

It is seen that the effect on azimuth varies with ϕ, δ, t and Z, as well as the error in latitude. A plot of $\varepsilon_{Z\phi}$ vs. t is shown in Figure B.3 for $\varepsilon_\phi = \pm 1"$ and the extreme values of δ for the sun.

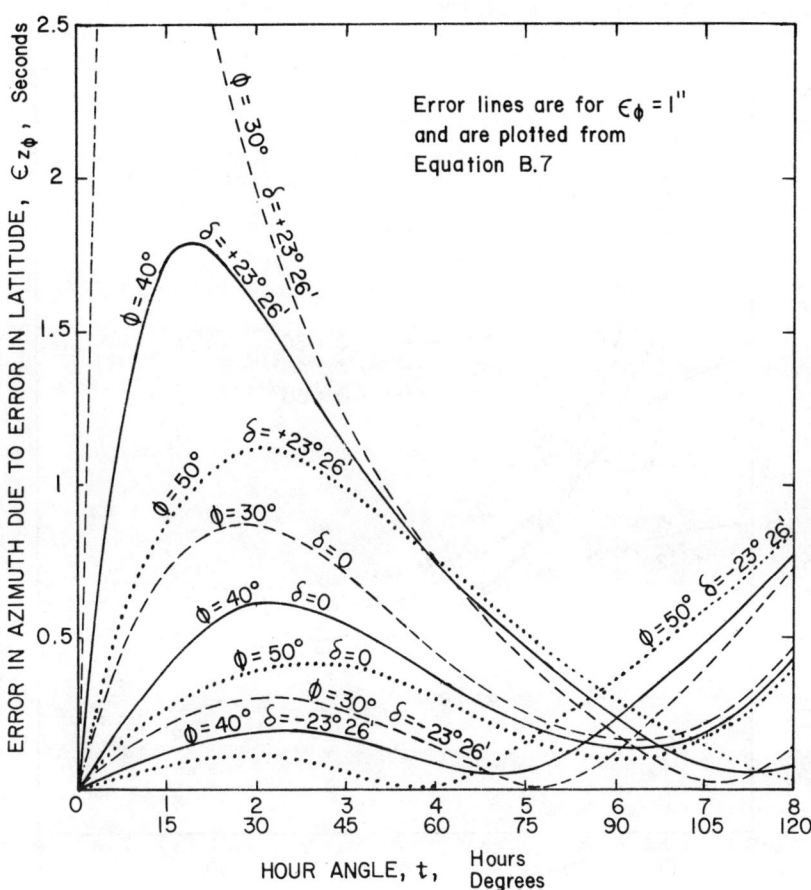

Figure B.3 Latitude Error Effects on Solar Azimuth

An error in λ causes an error in t. An equation for the effect of an error in λ on the error in azimuth is as follows:

$$\varepsilon_{Z_\lambda} = \pm[(\sin \phi \sin Z - \cot t \cos Z) \sin Z] \varepsilon_\lambda \qquad (B.8)$$

where

ε_{Z_λ} = error in azimuth due to error in longitude, in seconds,

ε_λ = error in longitude, in seconds.

The effect of an error on azimuth varies with ϕ, δ t, Z, and the error in scaling λ. Although it isn't apparent from the error formula that δ affects the error, this is understood when it is realized that change in δ changes Z in an hour angle equation solution. Figure B.4 shows ε_{Z_λ} vs. t for the extreme values of δ of the sun.

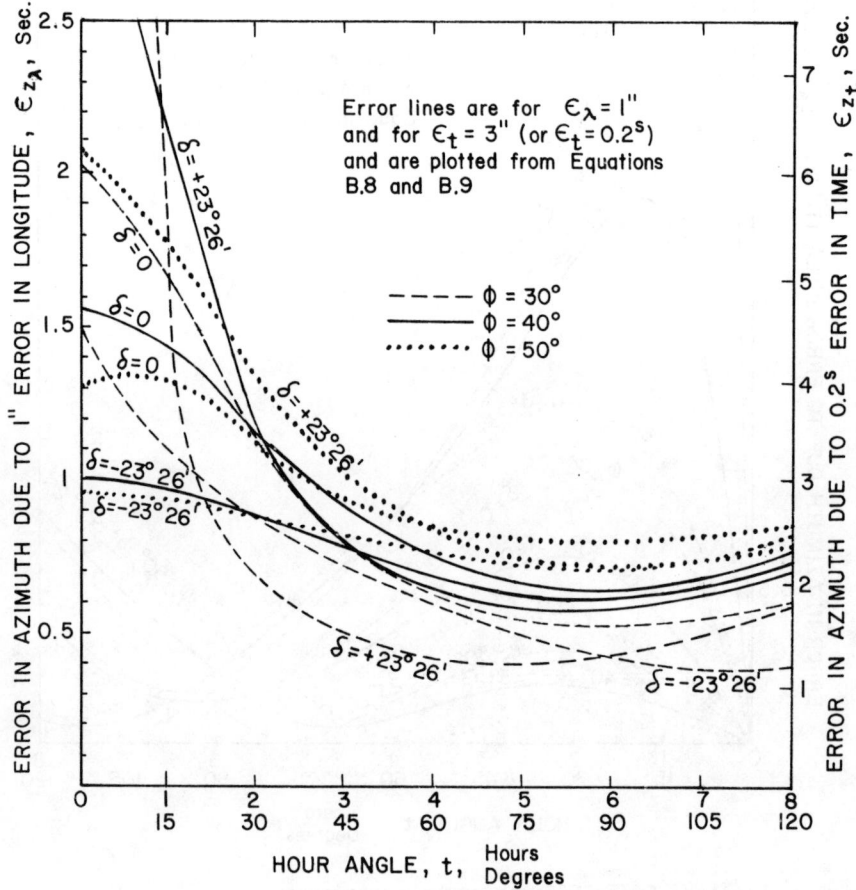

Figure B.4 Longitude and Timing Error Effects on Solar Azimuth

The evaluation of the error equations for Polaris azimuth observations reveals that ϕ and λ can each be in error as much as 50" or approximately a mile in most of the U.S., and not affect the azimuth more than 1". For this reason, if care is taken to scale ϕ and λ the effects of errors in these measurements on azimuths are negligible. Therefore, no further analysis will be presented here concerning these error effects on Polaris azimuths.

Minimizing errors due to ϕ and λ errors. The effects of an error in latitude (whether from scaling or other sources) can be compensated in the solar hour angle solution if observations are made at the same hour angle each side of local noon. For example, an approximate 8 A.M. set of observations and a 4 P.M. set on the same day, when averaged, will contain an equal plus and minus error in azimuth due to any small error in latitude. This fact is realized when it is recognized that sin t is negative in Equation B.7 when t is negative and positive when t is positive. Analysis of Equation B.7 or study of Figure B.3 indicates that the difference in t can be up to one hour between the morning and afternoon sets and the effects of a 1" error in latitude will be compensated within approximately 0.4" azimuth error, maximum. Thus, the morning and afternoon sets need not be taken at exactly the same hour angle from local noon. This technique affords a means to achieve accurate solar azimuths when the position in latitude is uncertain.

The effect of an error in longitude cannot be compensated. It can be minimized by observing at larger hour angles (t=4 to 6 hours) as revealed by Figure B.4. This figure shows that the effect on azimuth of an error in longitude of 1" can be kept well under 1" in azimuth, at any time of year, when observing at t between 4 and 6 hours.

Timing

Solar observations. One of the most significant error sources in solar observations is the timing error. Time affects t and the same error equation is derived for effects of time errors as for λ errors.

$$\varepsilon_{Z_t} = \pm[(\sin \phi \sin Z - \cot t \cos Z) \sin Z] \varepsilon_t \qquad (B.9)$$

where

ε_{Z_t} = error in azimuth due to error in time, seconds,

ε_t = error in time, converted to seconds of t (angle).

Using a stop watch which reads to either hundredths or tenths, tests have shown that the starting and stopping on signals yields a standard deviation of approximately $\pm 0.09^S$, or a 90% error of $\pm 0.15^S$. (Reference 5) Considering other possible unforeseen personal and other factors, let us assume $\pm 0.2^S$ for the 90% error in determining time using the radio/stop watch method. This assumes that the DUT1 correction has been determined and applied and that no other systematic errors exist in timing. The $\pm 0.2^S$ error converts to $\varepsilon_t = \pm 3"$. Thus, this error is approximately three times that assumed for the error in λ. Figure B.4 shows the effects of ε_t on solar azimuth.

Polaris observations. The simplified equation for azimuth of Polaris, derived in Appendix A, is

$$Z = p \frac{\sin t}{\cos h} \qquad (A.12)$$

Differentiating this equation yields

$$\varepsilon_{Z_t} = p \frac{\cos t}{\cos h} \varepsilon_t$$

For the purposes of analysis, we can use $h = \phi$. Thus

$$\varepsilon_{Z_t} = \pm p \frac{\cos t}{\cos \phi} \varepsilon_t \qquad (B.10)$$

where

ε_t = error in time, converted to radians.

It is seen that $\varepsilon_{Z_t}=0$ when t=90° (elongation) and is maximum at t=0° (culmination). The effect increases with latitude. For an assumption of ϕ =40°, p=49', t=0°, and $\varepsilon_t=\pm 1^s$ (=±15"=±0.0000727 radians), ε_{Z_t}=+0.28". From this it can be seen that, in general, the effects of timing errors are negligible if controlled within a few seconds. If Polaris is near culmination, the time error should not exceed 3 seconds if the effect in azimuth is to be kept under 1".

Change in Horizontal Angular Diameter of the Sun

As described in Section A.4, the diameter of the sun changes in horizontal angular value with changes in altitude. If the time lapse between limb pointings becomes too great in a set an error will result in the mean azimuth in each set because of these changes. It is systematic and cumulative. Between sunrise until approximately t=3 hours (A.M.), and from t=3 hours (P.M.) until sundown, the sun is changing approximately 10' arc in altitude for each minute of time (in latitudes from about 25° to 50°). This can be seen from Figure B.2 or Equation B.6. Because of the change in horizontal angular diameter of the sun during the time lapse between left and right limb pointings, the theoretical center of the sun does not lie at the same angular value right and left of the limb. Thus, there is a small, uncompensated error remaining in the azimuth when the left and right limb pointings are meaned. The higher the sun is, or the greater the time lapse is, the larger this error becomes. For example, if h=5°00' for one pointing, it would be about 5°10' for the second pointing one minute later. From Equation A.24 the value of s would change only 0.26" which is negligible. However, for h=40°, the value of s would change 3.1" in one minute's time which would change the mean azimuth for that set by half that amount, or 1.5", which is beginning to be significant. For time lapses greater than 1^m between pointings, this error changes proportionately. It can be shown that, if the same limb is consistently sighted first in each set, the error due to neglecting this change is cumulative and proportional to the inverse cosine of the average altitude and the average time lapse in each set. If the vertical angle is 5°, and there is 2^m average difference in time between L and R pointings in each set, the mean azimuth for any number of sets would be 0.26" higher. If h=40° and this same average time lapse occurred, the overall mean would be 3.1" higher and this would be too much to disregard.

The effects of this error can be kept negligible by minimizing the time lapses between limb pointings, and by making observations when the altitude of the sun is small. The error can be **compensated** by alternating between sighting the left limb, then the right limb first in successive sets. It can be **corrected** by using Table A.1 and Equation A.24 for semi-diameter for each observation. However, in order to correct this error, altitude would need to be known (to perhaps ±1') by either measuring it or computing it using Equation B.6. It is suggested that the error be either compensated as described above or rendered negligible by making observations when the sun is low and making L and R pointings in a set as quickly as possible. This error can be **avoided** by using the Roelofs solar prism instead of the limb pointing method.

Instead of correcting for each observation, it is generally accurate enough to use average conditions unless time lapses vary considerably between observations in the sets. To evaluate this error to see if it is negligible in a particular series of observations, or to see if a correction should be applied, Figure B.5 can be used. This plot gives the error (or correction) for one minute average time lapse between left and right limb pointings. It is a plot of change in s using an assumed increase of 10' in h for each 1^m time change. The systematic error is negative for left to right limb pointing sequence in the morning and positive for the same sequence if the sets are afternoon. To apply a correction, multiply the average time lapse between observations in each set by the value from Figure B.5, being sure to remember that the **correction** is opposite in sign to the **error** as described above.

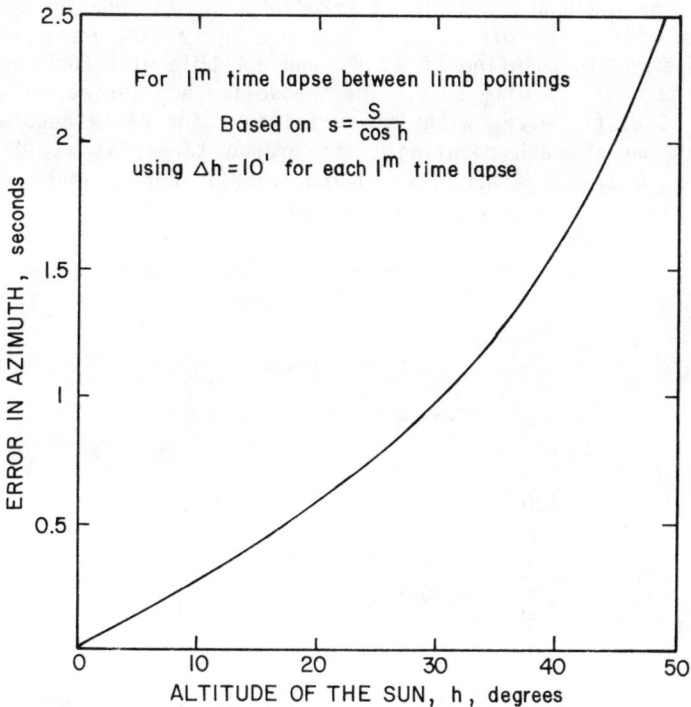

Figure B.5 Error in Solar Azimuth Due to Changing Diameter of the Sun

For example, the January 15, 1982 observations in Baton Rouge, Louisiana, at approximately 9 A.M, had time lapses betwen observations in each set of 2.0, 1.1, 1.6, and 1.2 minutes yielding a mean of 1.5 ^{m}. From Figure B.2, h at average t=50°, ϕ=30°, δ=-21°, is approximately 20°. From Figure B.5 this yields -0.55" azimuth error for 1^m average time lapse. Since the actual average time lapse was 1.5^m, $-1.5 \times 0.55 = -0.8"$. The correction is $+0.8"$. The corrected azimuth would be $2°08'36.9"$. From a separate calculation, adjusting each semi-diameter separately, and computing the azimuth each time, the mean value agrees with the above at $2°08'36.9"$.

B-3 PROPAGATED ERRORS IN ASTRONOMIC AZIMUTHS

Solar Observations

In order to estimate the probable errors in astronomic azimuths, let us use an example where some of the variables are such that maximum effects are realized. Then, this example can be used as a basis to discuss the errors in other situations and how to minimize and control the errors.

Figures B.3 and B.4 reveal that, at t=3 hours, a summer observation yields an error of $±1.1"$ due to error in ϕ and $±0.8"$ due to error in λ. But, a winter observation at t=3 hours has an error of $0.2"$ due to an error in ϕ and $±0.8"$ due to error in λ. The summer observation will be used as an example.

Example B.1 At t=3 hours, $\phi=40°$, $\delta=+23°26'$, an astronomic observation for azimuth was made. The direction theodolite has a 90% error in reading of $±1.5"$, a 90% error in pointing of $±3.0"$, and a bubble with 20" sensitivity. The terrestrial target is ¼ mile away. The theodolite and target are estimated to be centered $±0.01$ ft. each, with 90% certainty. The plate bubble is kept centered $±¼$ division at each pointing. The ground target is at 0° vertical angle. A stop watch is used for timing. DUT1 corrections are made. The ϕ and λ are scaled from a USGS 7½ minute quadrangle map in the usual manner. The estimated sigma in ξ is $±0.5"$ and in η is $±0.4"$. What is the uncertainty (90% estimated error) in the mean of four sets of observations for astronomic azimuth?

Solution:

Reading errors: $\quad \varepsilon_{z_r} = \dfrac{±1.5\sqrt{2}}{\sqrt{8}} = ±0.75$ (B.1)

Pointing errors: $\quad \varepsilon_{z_p} = \dfrac{±3.0\sqrt{2}}{\sqrt{8}} = ±1.50"$ (B.2)

Instrument and target centering: $\quad \varepsilon_{z_c} = \dfrac{±0.01}{1320} 206265\sqrt{2} = ±2.21"$ (B.4)

Bubble centering: Since the bubble is checked and centered with each pointing, it propagates as a random error. This error tends to compensate over the eight observations since the mean value is used as a "best value". The vertical angle is determined to be 49° from Figure B.2.

$$\varepsilon_b = \frac{±0.25\,(20")\tan 49°}{\sqrt{8}} = ±2.03" \quad (B.5)$$

(The same answer is obtained by dividing the value from Figure B.1 by $\sqrt{8}$)

B-3   PROPAGATED ERRORS IN ASTRONOMIC AZIMUTHS

Latitude and longitude effects: For $\varepsilon_\phi = \varepsilon_\lambda = \pm 1"$, the error estimates for azimuth, from Figures B.3 and B.4 are

$$\varepsilon_{Z_{\phi_1}} = \pm 1.1"$$

(due to map scaling)

$$\varepsilon_{Z_{\lambda_1}} = \pm 0.8"$$

For $\sigma_\xi = \pm 0.5"$, $E_{90_\phi} = \pm 0.8"$ and, for $\sigma_\eta = \pm 0.4"$, $E_{90_\lambda} = \pm \dfrac{1.6445\,(0.4)}{\cos \phi} = 0.9"$. From Figures B.3 and B.4,

$$\varepsilon_{Z_{\phi_2}} = \pm 0.9"$$

(due to conversion to astronomic ϕ and λ)

$$\varepsilon_{Z_{\lambda_2}} = \pm 0.7"$$

Timing: Assuming $\pm 0.2^s$ timing error, Figure B.5 yields $\varepsilon_{Zt} = \pm 2.4"$. Since a total of 8 pointings are made, and the mean value used,

$$\varepsilon_{Zt} = \frac{\pm 2.4}{\sqrt{8}} = \pm 0.85"$$

The total error is the square root of the sum of the squares of the separate errors. Thus

$$\varepsilon_Z = \sqrt{0.75^2 + 1.50^2 + 2.21^2 + 2.03^2 + 1.1^2 + 0.8^2 + 0.9^2 + 0.7^2 + 0.85^2}$$

$$\varepsilon_Z = \pm 4.0" \text{ (90\% certainty)}$$

If the assumptions on % certainty of the error estimates are valid, this should theoretically yield an azimuth of third-order, Class I precision. Thus, it is seen that, with 4 sets of observations, a solar observation is reliable under the conditions of the example.

As the time gets closer to t=0°, the error would increase somewhat due to timing errors. Figure B.4 reveals that the ±0.85 error computed would increase by a factor of 3 if t=1 hour. At this time, due to the larger vertical angle, the error would double from the bubble centering causes (if the observer could even see the sun at h=69°). The only error that would decrease would be that caused by latitude determination. All considered, the observer would not want to make a summer observation closer than t=3 hours to local noon.

It could be shown that if t=4 hours in this example, $\varepsilon_Z=\pm 3.4''$ for the mean value.

Obvious ways to further reduce the expected error would be to sight when the sun is even lower on the horizon, at t equal to perhaps 6 hours, and to use an azimuth mark much further away and/or a mark not having a centering error (tower, etc.). If, for example, the sun was at t=6 hours and a tower 2 miles distant was used, ε_Z would be $\pm 2.0''$, theoretically.

Investigation of the variables in the equation shows that the best way to reduce errors is to use longer sight distances, use optical plummets at targets, sight when t is at least 3 hours and the sun as low on the horizon as possible. Other ways are to make more repetitions and/or to use an instrument of better precision.

To illustrate how minor the reading error effects are, however, the initial conditions of the above example could be reworked using a repetition instrument of $\varepsilon_r=\pm 5''$ and $\varepsilon_p=\pm 3''$. This yields $\varepsilon_Z=\pm 4.6''$ in comparison to $\pm 4.0''$ using the 1" direction instrument. This relatively small effect occurs because there are many contributors to the total error other than reading precision. The azimuth error would still have been under $\pm 5''$.

One may wish to consider whether bubble centering can be done better than $\pm\frac{1}{4}$ division, or if this $\frac{1}{4}$ division is unrealistic. This is a large contributor to uncertainty in azimuth. Its effect can be reduced for solar observations by sighting in early morning or late afternoon. For the solar observation example, if the bubble was kept centered only to $\pm\frac{1}{2}$ division instead of $\pm\frac{1}{4}$ division, the resulting mean azimuth would have an error $\varepsilon_Z=\pm 5.3''$ instead of $\pm 4.0''$. And, if bubble centering had been controlled to only ± 1 division, the resulting error in azimuth would have been $\varepsilon_Z=\pm 8.7''$, assuming all other estimates the same as in Example B.1. Thus, the importance of careful bubble centering *and* sighting earlier in the morning or later in the afternoon is evident.

Polaris Observations

If properly controlled, effects due to latitude, longitude, and timing are negligible for azimuths determined by Polaris observations.

Example B.2 Using the instrument of Example B.1 and the same target and instrument location, determine the uncertainty in the mean of four sets of observations.

Solution: From Example B.1,

$$\varepsilon_{Z_r} = \pm 0.75''$$

$$\varepsilon_{Z_p} = \pm 1.50''$$

$$\varepsilon_{Z_c} = \pm 2.21''$$

The bubble centering effects must be calculated, based on $\phi=h=40°$

$$\varepsilon_{Z_b} = \frac{\pm 0.25\ (20")\ \tan 40°}{\sqrt{8}} = \pm 1.48"$$

Propagation of these four errors yields $\varepsilon_Z = \pm 3.1"$

For the repetition instrument having $\varepsilon_r = \pm 5"$, this would have yielded $\varepsilon_Z = \pm 3.9"$. A practical way to reduce the errors in this example would be to improve the sight distance/target centering aspects.

Unlike the solar observations, errors due to bubble centering cannot be controlled by observing with the star lower on the horizon. If $\pm\frac{1}{2}$ division had been assumed for bubble centering, the resulting azimuth error would have been $\varepsilon_Z = \pm 4.1"$ instead of $\pm 3.5"$. And, if bubble centering had been controlled to only ± 1 division, the resulting azimuth error would have been $\varepsilon_Z = \pm 6.5"$ in this example situation. Thus it is seen that bubble centering must *always* be carefully controlled for Polaris observation, or a striding bubble used.

Other Error Sources

There are possible round-off errors in calculating the azimuths if attention is not paid to significant figures in all intermediate steps. These precautions were discussed in Chapter 6.

Ephemerides list variables to varying precisions and the number of digits given can affect the precision in a computed azimuth. The values for δ, equation of time, and GHA have all been given in Appendix F to a precision sufficient for third-order azimuth, assuming that significant figures and other precautions discussed in that Appendix are observed. Use of the Polaris tables for azimuths better than third-order equivalent will be explained toward the end of Section B-4.

Unforeseen personal, natural, and instrumental errors could, and often do, cause azimuth uncertainties higher than what would be propagated by the sources cited herein. It is assumed that knowledgeable instrument operators will realize the effects of wind, thawing ground, faulty instruments, and lack of care in measurements, and take the necessary measures to minimize these effects on azimuths. Naturally, all systematic errors and mistakes must be corrected or avoided.

Some references mention other error sources such as **diurnal abberration** (slight apparent displacement of the star due to the motion of the observer about the earth's axis), **skew normal** (error due to difference in direction of vertical planes when observation and target stations are at different elevations), **polar variation** (small "wobble" in the earth's polar axis which changes ϕ and λ slightly), and the time delay between propagated time signal and receipt of the signal by the observer. The maximum effects of these errors are on the magnitude of a few tenths of a second in azimuth and can be considered negligible for all but first-order azimuth.

This discussion and that in the next section on specifications concern astronomic azimuths. One additional consideration for SPC grid azimuths is the random error in determining the Laplace correction and the mapping angle. These errors are considered separately in Section B-5.

B-4 SPECIFICATIONS FOR ASTRONOMIC AZIMUTHS

General Comments

The examples of the previous section contain assumptions that could be unrealistic in the real world. Likewise, the derived errors are subject to several variables, all of which will alter the expected uncertainty somewhat. It is felt that the examples do represent reasonable expected errors, however. When it is considered that systematic effects must be carefully controlled as discussed, the remainder of the propagated error is purely random and is revealed in the actual field results. Both in practice, in research, and in conducting several seminars, the collective results reveal ranges and deviations indicating that the solar and Polaris results in the "real world" are consistent with the errors derived from theory. Using all of the geometry, error theory, error estimates, tested results, and other experience together, the specifications to follow have been derived. The basis and support of these are found throughout this manual. It must be stated that the claims of the author for "third-order" azimuth are not necessarily shared by the NGS. Their field instructions and opinions vary from those described herein.

The term "error" in the following specifications means 90% error, either as determined by statistical tests or estimates.

Solar Observations - - For 90% Error ±5" (Third-Order, Class I Equivalent)

1. Instrument - - Use an optical theodolite, either direction or repeating, having a circle graduation system yielding an error in reading of at least ±5", a pointing error of at least ±3", with bubble sensitivity at least 20". Instrument should be in good adjustment and clean. The plate bubble should be adjusted so that no noticeable error is observed.

2. Timing - - Use a propagated time signal and a stop watch reading to at least ±0.1^S, starting the watch on a minute time signal and stopping it at each solar limb pointing. Read the watch to at least ±0.1^S. Correct for DUT1.

3. Latitude and Longitude - - Scale from USGS 7½-minute map (or equivalent or larger scale and accuracy), to arrive at uncertainties not exceeding ±1" for each component. To do so, determine the field position to at least ±10 feet with respect to at least two well-defined map features lying generally 90° to each other (cross-roads, for example). Plot the position to at least ±0.02 inch precision on the map and scale the ϕ and λ with at least this same care. Compute the ϕ and λ to the nearest second from the interpolated, scaled measurements. Correct ϕ and λ to astronomic position using NGS data. If the sigma of ξ and η is less than ±1" (which can generally be assumed for non-mountainous areas), the above accuracy can be maintained; otherwise, assume it to be third-order, Class II equivalent. Or, if the NGS statement on sigma of ξ is greater than 1", or if data on the latitude is unavailable or otherwise uncertain, make observations at equal hour angle morning and afternoon and use mean of acceptable sets to compensate for ϕ error.

4. Azimuth Mark - - Select a well-defined azimuth mark not exceeding 5° vertical angle from the instrument station, lying at least ¼-mile from the instrument station if a ground target must be erected over or behind it.

SPECIFICATIONS FOR ASTRONOMIC AZIMUTHS

5. Care in Observing -- Center theodolite and target to at least ±0.01 ft. using optical plummets if possible. Center plate bubbles to at least ¼ division when pointing to all targets. Keep the time interval between limb pointings less than one minute in each set, or alternate between pointing to left limb, then right limb between successive sets, or correct for change in horizontal angular diameter.

6. Reading and Recording -- Read theodolite circle readings to at least the nearest second for direction instruments and to ±2 to 3" for repetition instruments, depending on the style of reading system. Make at least four sets of observations, equal number with telescope direct as inverted.

7. Time of Day -- Observe not closer than 3 hours to local noon and when the vertical angle does not exceed 30° to the sun.

8. Calculations -- Carry at least 8 places in all computations. Use ephemeris data of at least the significant figures as in Appendix F. Compute all azimuths to the nearest second, computing each separately. Use other precautions as outlined in Chapter 6. At least three acceptable sets must be used for the mean azimuth. Correct the δ using Table F.1 if appropriate.

Polaris Observations -- For 90% Error ±5" (Third-Order, Class I Equivalent)

1. Instrument -- same as for solar.

2. Timing -- Calibrate watch with time signals to nearest 1^s. Remove any systematic error due to loss or gain of the watch so that this error source does not contribute more than 1^s to the total time error. Read the watch to the nearest 1^s at each pointing to Polaris. Record times to nearest 1^s.

3. Latitude and Longitude -- Determine ϕ and λ to at least ±20" or approximately ±2000 ft.

4. Azimuth Mark -- same as for solar.

5. Care in Observing -- same as for solar (except as regards limb pointings)

6. Reading and Recording -- same as for solar.

7. Time of Day -- At night or within an hour or two before sundown or just after sunrise.

8. Calculations -- same as for solar.

Observations for 90% Error ±13" (Third-Order, Class II Equivalent)

In general, the same specifications are used as for ±5" precision, except that two or three of the following relaxed procedures can be permitted:

1. A repetition theodolite is appropriate.
2. Vertical angle to the sun can be up to 45°.
3. The observation can be made up to 2 hours from local noon, provided the vertical angle does not exceed 45°.
4. Three acceptable sets are minimum.
5. Time interval between solar limb pointings in each set should be kept to under two minutes when altitude of the sun exceeds 40°; otherwise, it can be up to three minutes. No semi-diameter changes need be considered.

6. When using solar azimuths, the latitude and longitude can be corrected to astronomic position from data for nearby control stations listed in Appendix D. Differences of less than ±2" between geodetic and astronomic position can be ignored (applies to solar observations only).

Observations for 90% Error ±2" (Second-Order, Class II Equivalent)

Solar observations. The author has not become totally convinced that solar observations can be used to achieve this precision, primarily because of effects of uncertainties in ϕ, λ and the effects of bubble centering error. Indications are, however, that such accuracy could be claimed if only a direction theodolite is used, the azimuth mark is on a stable tower at least a mile away, the vertical angle is not allowed to exceed 20° (or, a striding level used if it does), corrections are made for change in horizontal angular diameter of the sun, at least six acceptable sets of observations are used, the ϕ and λ are known accurately (such as would be the case for an NGS station of known position), and the errors in conversion to astronomic position are less than ±0.5" estimate. The effects of latitude error can be compensated if half of the observations are made in the morning, half in the afternoon, meaning that this contributor to the error could be dealt with in this way, even when ϕ is scaled from a map. This would mean that only the error in λ would remain as a concern as regards these two error sources. The suitability of solar observations for an azimuth of this precision lends itself to further research.

Polaris observations. Polaris can be used for this precision, if a striding level is used. In addition, a direction theodolite should be used. The azimuth mark should be on a stable tower at least a mile away. Latitude and longitude should be determined to at least ±10". Time should be determined with a stop watch as done in solar observations. However, it needs to be read and recorded to only the nearest second and DUT1 can be neglected. If Polaris is near elongation, the timing method for ±5" azimuth can be used. Six sets of acceptable observations are recommended.

The δ in Appendix F is given to the nearest second. Values are truncated. Interpolation or estimation should be made to yield δ to the nearest 0.2" in order to assure no round-off errors from this variable. For example, the value for January 8, 1984 is δ =89°11' 47", and likewise for January 16. For January 11, use δ =89°11' 47.3". Since it is increasing from 46" to 48", the interpolation was made to add 0.3" to 47". In this instance, the change was approximately 0.1" per day. In most cases, this rate would be different. For example, use δ =89°11' 37.9" for April 4, 37.6" for April 5, and 37.3" for April 6, 1984. Failure to interpolate for δ would cause a maximum error of 0.7" in an azimuth observed at elongation; therefore, neglecting this refinement would probably not jeopardize the accuracy even for ±2" precision in azimuth. This refinement is definitely not needed for lesser accuracy and the values as read directly from the tables can be used as discussed throughout this manual.

Except as noted above, other procedural specifications are as those for ±5" azimuth by Polaris.

Grid Azimuth Specifications

The above specifications, although derived for astronomic azimuth, should yield a similar uncertainty for grid azimuth after applying the Laplace correction and convergence angle, as is demonstrated by Example B.3, unless the uncertainty in η or λ is very much larger than in that example. The next section adds the considerations for grid azimuth.

B-5 CONVERSION TO GRID AZIMUTH

The two-step process of converting from astronomic to grid azimuth was explained in Chapter 7. Taking partial derivations of Equation 7.2b and propagating the errors according to random error theory yields

$$\varepsilon_{LC} = \sqrt{(\varepsilon_\eta \tan \phi)^2 + (\eta \sec^2 \phi \, \varepsilon_\phi)^2}$$

where

ε_{LC} = error in Laplace correction due to errors in ϕ and η,

ε_η = error in η,

ε_ϕ = error in ϕ, in radians.

It can be shown that, even for very large values of η, the second term in this equation has negligible effects. Thus

$$\varepsilon_{LC} = \pm \varepsilon_\eta \tan \phi \qquad (B.11)$$

The error ε_{LC} is the error in geodetic azimuth after converting from astronomic (Equation 7.1). In addition, the error in $\Delta\alpha$ or in the θ angle (See Equations 7.4, 7.10) needs to be considered. From Equation 7.8 and random error theory,

$$\varepsilon_{\Delta\alpha} = \sqrt{(\varepsilon_{\Delta\lambda} \sin \phi)^2 + (\Delta\lambda \cos \phi \, \varepsilon_\phi)^2}$$

From Equation 7.12 and random error theory,

$$\varepsilon_\phi = \sqrt{(\varepsilon_{\Delta\lambda} \, \ell)^2 + (\Delta\lambda \, \varepsilon_\ell)^2}$$

In either of these equations, the error in $\Delta\lambda$ is affected only by the error in λ of the station since λ_{CM} is fixed. Also, since ℓ is a constant in a Lambert zone, it is error free. Furthermore, even at extreme edges of a TM zone where $\Delta\lambda$ is maximum, any conceivable error in ϕ causes the term $\Delta\lambda \cos \phi \, \varepsilon_\phi$ to be negligible. Thus,

$$\varepsilon_{\Delta\alpha} = \varepsilon_\lambda \sin \phi \qquad (B.12)$$

$$\varepsilon_\theta = \varepsilon_\lambda \, \ell \qquad (B.13)$$

where

$\varepsilon_{\Delta\alpha}$ = error in convergence angle, $\Delta\alpha$,

ε_θ = error in convergence angle, θ,

ε_λ = error in longitude of instrument station.

Example B.3 Using the data from Example B.1 and a Transverse Mercator Zone, compute the uncertainty in the grid azimuth (Assume 90% errors).

Solution:
$$\varepsilon_{LC} = \pm 0.4'' \,(1.645)\, \tan 40° = \pm 0.55'' \qquad (B.11)$$

There are two sources of the ε_λ, one from scaling and one from ε_ξ. From the values given or assumed in Example B.1,

$$\varepsilon_\lambda = \sqrt{1^2 + 0.9^2} = \pm 1.35''$$

Thus,

$$\varepsilon_{\Delta\alpha} = \pm 1.35 \sin 40° = \pm 0.86'' \qquad (B.12)$$

The combined effect on the grid azimuth, according to the law of random error addition, is

$$\varepsilon_{GA_1} = \sqrt{\varepsilon_{LC}^2 + \varepsilon_{\Delta\alpha}^2} = \sqrt{0.55^2 + 0.86^2} = \pm 1.03''$$

But, the actual error in grid azimuth is the error in the astronomic azimuth plus the error caused by conversion from astronomic to grid. The error computed in Example B.1 must be added (according to random error theory) to the ±1.03". Thus,

$$\varepsilon_{GA} = \pm\sqrt{4.0^2 + 1.03^2} = \pm 4.1''$$

It should be noted that an example in the Lambert system would have yielded similar values for ε_θ since ℓ and $\sin\phi$ are nearly equal for the same value of ϕ.

It is seen that the expected uncertainty in the grid azimuth is only slightly larger than in the astronomic when uncertainties in η and λ are reasonably small.

TABLE B.1
Summary of Error Sources

Error	Nature	Effect	Minimizing and Controlling
Theodolite Reading	Random	Random	Use care, clean instrument, added repetitions.
Theodolite Pointing	Random	Random	Use care, good optics, good sight conditions, added repetitions.
Instrument and Target Centering	Random	Systematic	Use longer sight distances, care in adjustment and use of optical plummets, stable target.
Bubble Centering	Random or Systematic	Random or Systematic	Use sensitive bubble, better care in centering, or striding level. Sight with sun lower on horizon. Increase repetitions. Correct if bubble ends are read.
Latitude and Longitude	Random (2 sources)	Systematic	Use larger scale map, more care in plotting and scaling. Use NGS data to correct for deflection of the vertical. Compensate for ϕ error by A.M./P.M. observations in sets.
Timing	Random and Systematic	Random and Systematic	Use stopwatch to minimize random error and DUT1 correction to reduce systematic error for solar. Calibrate time with UT signals for Polaris observations.
Sun's Horizontal Diameter Changes	Systematic	Systematic	Minimize time lapse within sets to keep it negligible. Sight with sun lower on horizon. Correct using equation or graphs. Alternate between L to R – R to L limb pointings. Use solar prism method.
Declination, Equation of Time, Polar distance, GHA	Random	Random	Carry sufficient figures from ephemeris and in calculations. Winter solar observations are more favorable δ than summer.
Nature	Random	Random	Avoid observations when wind, heat waves, clouds, haze, or other conditions hamper observations.
Computational	Random	Random	Avoid round-off errors by using enough significant figures.
Conversion to Grid Azimuth	Random	Random	Careful scaling of λ ; use NGS program data for η.

APPENDIX C

SAMPLE OBSERVATIONS

Introductory Comments

The following examples are solar and Polaris observations made throughout the United States. Some solutions are provided. Other sets of field data are left to be computed by the student. Answers are given in most instances. Each sample includes the original field notes and calculations with minor editing as needed for clarity. In each solar example, it should be assumed that ϕ and λ are astronomic. Each example should be self-explanatory. However, some comments are made in the following paragraphs, for the purpose of directing the reader to some important points concerning each example.

C1 Solar Observation, New Orleans

Four of the five sets of observations are computed. The fifth is left as an exercise for the reader. The results of Set 5 are shown at the bottom of the computation page. The DUT1 correction was disregarded. An analysis of all five sets revealed a range of 16" and a 90% error in the mean value of ±4.6". This observation was performed as a demonstration during one of the author's seminars. Note the late afternoon summer results (NW quadrant).

C2 Solar Observation, State College, Pennsylvania

This observation was made as a demonstration for a workshop on the campus of Penn State University. A possible mistake in observation 1D caused it to be questionable. The fourth set was incomplete because clouds caused the sun's image to become obscured after observation 4D. Since 1R and 4D were taken to opposite limbs of the sun, it was decided to use 1R and 4D as a set since both were free of mistakes. However, the amount of time lapsed between the two observations warrants consideration of the effect of the change in horizontal angular diameter of the sun as it sets. An analysis using Figures B.2 and B.5, with the discussion of page 152 indicates that a correction of +2.6" is appropriate for the 1R/4D set. This solution is as follows: Figure B.2 indicates h = 9°. Figure B.5 indicates 0.2" error per 1^m time lapse at h = 9°. Since there is 13^m time lapse, 0.2 x 13 = 2.6" correction, based on Figure B.5. The other sets, having only about a minute time lapse between L & R points need not be corrected since the correction is less than the significant figures of the solution. The algebraic sign for the computed correction can be determined based on the discussion on page 153 of Appendix B. A more accurate mean for the three sets is 126°11'10.0" instead of 126°11'09.0" as shown on the computation form.

C3 Solar Observation, State College, Pennsylvania

This series of solar observations was made at the same station as the previous example, the next morning. The computation form became somewhat cluttered after re-calculating because of forgetting to consider the DUT1 correction. The changes in t and Z can be observed for the +0.3^S DUT1 correction. Neglecting it would have caused an error of -3" in the azimuths.

The vertical angle was approximately 33° as noted in the field notes. Note that Figure B.2 reveals the same value. Using this and Figure B.5 interpolated value of 1.0^S error per minute of time lapse and average time lapse of 2.7^m between L & R limb pointings, the correction for change in horizontal angular diameter of the sun is +2.7". Therefore the mean is 126°11'10.1" as shown. Corrections to individual sets need not be made.

C4 and C5 Solar Observations at State College

These two sets of field notes were taken by participants of the same seminar in which C2 and C3 data were observed. They have not been checked by the author. They are provided as student exercises. The azimuths may be different than those computed in C2 and C3 since different instrument stations were used, offset from "JW1". The difference is probably only a few seconds or minutes, however, as the same distant water tank light was used as a target. Note that C4 data were taken with a 1" direction theodolite and C5 data with a late model Wild T1 repetition instrument having 6" divisions with estimation to 3".

C6 Polaris Observation, State College, Pennsylvania

Note that this observation was made from the same station and to the same target as data sets C2 and C3, immediately following the C3 data and during the same instrument set-up. A stop watch was used for timing, but the tenths of seconds were ignored. The results agree favorably with the solar observations. The reader should note how LHA and azimuth to Polaris increase with time as the star moves toward east elongation. Comparisons with the solar results, and other points about these data sets will be discussed shortly.

C7 Polaris Observation, State College, Pennsylvania

Like data sets C4 and C5, this observation was made by a seminar participant. It has not been checked by the author. Results should show an azimuth slightly different than the others since "stake 3" was used as the instrument station. The reader should seek agreement among the sets of observations and/or attempt to find mistakes as necessary.

The State College Seminar Results

The observation station is the one identified as "Pennsylvania" in Table 3.1. It is noted that the difference between astronomic and geodetic position was ignored in the computation, the ϕ and λ being as scaled from a map in the examples. If corrected, using the Table 3.1 information and Equations 3.4 and 3.5, the values would be

$$\phi_A = 40°48'36" - 0.1" = 40°48'35.9"$$

$$\lambda_A = 77°51'22" + (\frac{-1.7"}{\cos \phi}) = 77°51'19.8"$$

The effect of ignoring the ξ value is negligible. The effect of ignoring the η value can be evaluated from Figure B.4 or from recalculating the data using λ_A instead of λ_G. Either analysis reveals that approximately 1.0" should be added to the solar azimuths in both sets C2 and C3. A table follows which contains corrections due to both the changing horizontal angular diameter of the sun and the effect of ignoring the η component, and which shows the ultimate agreement of all solar and Polaris observations.

Astro Target	Obs./Set		Observed	Corr. for S	Corr. for η	Corrected
Sun	C2	1	126 11 09.5	+2.6	+1.0	126 11 13.1
Sun		2	126 11 07	-0.3	+1.0	126 11 07.7
Sun		3	126 11 10.5	-0.3	+1.0	126 11 11.2
Sun	C3	1	126 11 10.5	+2.4	+1.0	126 11 13.9
Sun		2	126 11 11	+2.6	+1.0	126 11 14.6
Sun		3	126 11 04	+2.8	+1.0	126 11 07.8
Sun		4	126 11 04	+2.7	+1.0	126 11 07.7
Polaris	C6	1	126 11 11.5	-	-	126 11 11.5
Polaris		2	126 11 14.5	-	-	126 11 14.5
Polaris		3	126 11 10	-	-	126 11 10.0

Mean of Solar C2 = 126°11'10.7"
Mean of Solar C3 = 126°11'11.0"
Mean of All Solar = 126°11'10.9"
Mean of Polaris = 126°11'12.0"
Mean of All 10 Sets = 126°11'11.2"

σ_{solar} = ±3.1" $E_{90_{solar}}$ = ±5.1"

σ_{all} = ±2.8" $E_{90_{all}}$ = ±4.6"

From the Laplace correction map for Pennsylvania in Appendix D, the Laplace correction is approximately 0.0". Therefore, the geodetic azimuth is equal to the astronomic azimuth. From Table 7.1, the central meridian of the SPC system, both north and south zones of Pennsylvania, is 77°45'. From Figure 7.3, it is observed that the observations were made in the north zone. (Location is Center County, which would be known to the observer). From Table 7.2, ℓ = 0.66154.

$$\Delta\lambda = (77°45') - (77°51'22") = -382" \qquad (7.7)$$

$$\theta = 0.66154 \; (-382") = -252.7" = -04'12.7" \qquad (7.12)$$

$$\text{Grid Azimuth} = (126°11'11.2") - (-04'12.7") = 126°15'23.9" \qquad (7.11)$$

The value of 126°15'23.9" or 126°15'24" would be considered the "best value" from astronomic observations used, as corrected for astrogeodetic values and theta angle. Had only one set of solar observations or the one set of Polaris observations been used, the difference would have been a mere 0.8" at most.

As a means of demonstrating the closeness of agreement between grid azimuths computed from astronomic observations and those observed from NGS control, the angles in the traverse shown below were measured.

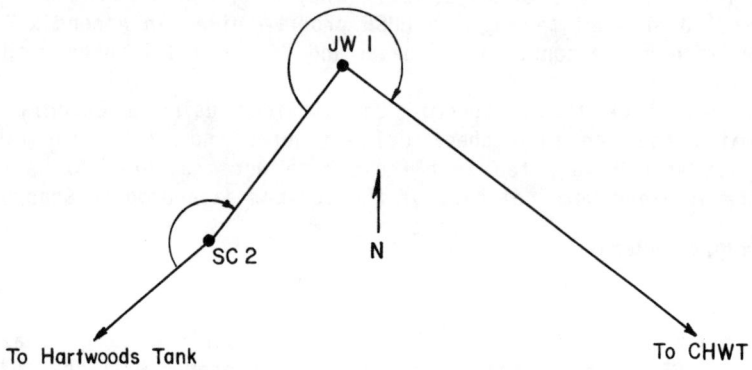

The grid azimuth from SC2 to Hartwoods Tank was published by NGS. Using the angles at SC2 and JW1, the grid azimuth to CHWT was calculated as 126°15'24" using a series of angles with one instrument, and 126°15'21" using another. In either case, the agreement with the azimuth from direct astronomic observations is excellent.

C8 Polaris Observation in Michigan

The data were used for a surveying project in Michigan. The results have not been thoroughly checked by the author. The data are somewhat incomplete and the reader must make some assumptions regarding the time. Working with this set of data may serve to illustrate why all data must be complete in the field notes.

C9 and C10 Solar Observation, Fresno, California

These data sets were taken for a class of sophomore students at California State University. They illustrate results obtainable using a less precise theodolite.

The instrument of Example C.9 had an inverted image telescope. The final azimuth computations are shown. The reader should note that the values for "azimuth to target" are larger in the "R" (right limb) column than the "L" (left limb) column for each set, but this is the opposite for Example C.10 and all other solar observations with an erect image telescope. This is because the left limb as it appeared was actually the right limb, and vice versa, in Example C.9.

In Example C.10 it was decided that a mistake occurred in observation 3D. Therefore, that set was rejected. The angle reading for observation 2D is also suspect, but was accepted, considering the precision of the instrument (which was in need of cleaning and other repairs).

Although there was "play" in the horizontal motion of the Wild T1, the results were more consistent than for the T16. The results illustrate how a repetition instrument can yield results as consistent as when a 1" instrument is used. The T16 results should have been nearly as good. The fault can be traced to an instrument in need of repair and/or observation recorder mistakes.

C11 Solar Observation, Louisiana

These data were furnished by Mr. James M. Theriot and Ronald J. Rodi of Brown and Butler, Inc., Baton Rouge, La. They also furnished the solution shown, as computed and printed by the HP85 program given in Appendix E. The program was written by personnel from Brown and Butler and is used with their permission.

The reader can check the astronomic calculations using a calculator and previous formats, and can also check using figures and tables furnished in this manual. It should be noted that these data were gathered for surveying projects in the area and were not part of the seminar data used in Chapter 6.

Baton Rouge Data, Chapter 6

Before ending the discussion of results, it is well to illustrate once more, the manner in which data agree among themselves for astronomic azimuths and for subsequent grid azimuth calculations in comparison to grid azimuths determined using NGS control. The observations in Chapter 6 will be used.

A study of the time lapses between observations in each set (See Figures 6.2 and 6.4) and Figures B.2 and B.5 reveals that corrections due to change in horizontal angular diameter of the sun are only a few tenths of a second since the sun was low on the horizon in both data sets and observations were made quickly. As was mentioned in Section 6-4, the ϕ and λ scaled from a map were used as astronomic position since effects due to deflections of the vertical are slight in the area. Therefore, no additional corrections to results were needed. A summary of solar results is shown in Section 6-5. When combined with the Polaris results, the overall mean astronomic azimuth is 2°08'37.4", as compared to 2°08'38.2" for all nine solar sets or 2°08'39.9" and 2°08'36.1" for the solar observations grouped into the sets of the two respective dates.

Example G.2 resulted in a grid azimuth of 2°04'13" when all 13 sets were used for the mean astronomic azimuth. To illustrate how closely this azimuth agrees with one determined from NGS control, a tie was made to a nearby NGS first-order traverse with intervisible stations for azimuth control. After three angle points and approximately one-half mile of traverse length, the grid azimuth to the WBRZ tower was calculated to be 2°04'13", which is identical to that determined from all 13 sets of astronomic azimuths and corrections for astrogeodetic differences and theta angle. Had any one of the three observation sets been used, the discrepancy would not have exceeded 2.5". Hopefully, this example gives the reader additional confidence that third-order grid azimuth can be determined astronomically.

C 1   *FIELD NOTES FOR ASTRONOMIC OBSERVATION*

Astronomic Target **Sun**   Latitude of Instr. **29° 57' 19"**
Instrument Station **B**   Longitude of Inst. **89° 58' 46.5"**
New Orleans   Mark **A (WOSU)**   Date **7/31/81**   Observer **BB**   Rec **SE**

Sun's Limb

SET D/R	MARK READ DEG	MIN	SEC	SUN READ DEG	MIN	SEC	TIME HR	MIN	SEC	
1D	0	43	02	169	11	56	22	19	18.7	L
1R	180	43	35	350	05	09	22	21	34.5	R
2R	225	36	33	34	46	14	22	25	12.0	L
2D	45	36	06	215	37	59	22	27	28.9	R
3D	90	29	15	260	14	43	22	30	23.2	L
3R	270	29	31	081	03	27	22	32	09.6	R
4R	315	32	38	125	47	20	22	34	35.5	L
4D	135	32	16	306	33	43	22	36	15.0	R
5D	0	31	16	171	17	10	22	39	12.0	L
5R	180	31	44	352	00	46	22	40	20.4	R

C 1

Solar Observation Computation

St. Bernard Cultural Center, New Orleans — Chalmette Quad. Map

☼ Sta: **B**   Inst. type: **T2** ϕ **29 57 19.0**

Target: **WDSU Radio Tower**   Observer: **BB** λ **89 58 46.5**

Date: **7/31/81 Friday**   Computer: **BB**   page 1 of 2

Observation	1D (L)	1R (R)	2R (L)	2D (R)	3D (L)	3R (R)	4R (L)	4D (R)
UT	22:19:18.7	22:21:34.5	22:25:12.0	22:27:28.9	22:30:23.2	22:32:09.6	22:34:35.5	22:36:15.0
Eq. Time, 0	-0:06:19.7	—	—	—	—	—	—	-0:06:19.7
0.13 × UT	02.9							02.9
Eq. Time	-0:06:16.8							-0:06:16.8
less 12^h	-12:00:00							-12:00:00
GHA, hours	10:13:01.9	10:15:17.7	10:18:55.2	10:21:12.1	10:24:06.4	10:25:52.8	10:28:18.7	10:29:58.2
GHA, arc = GHA$_h$ × 15	153 15 28.5	153 49 25.5	154 43 48.0	155 18 01.5	156 01 36.0	156 28 12.0	157 04 40.5	157 29 33.0
λ	89 58 46.5							89 58 46.5
LHA	63 16 42	63 50 29	64 45 02	65 19 15	66 02 50	66 29 26	67 05 54	67 30 47
t	− "	− "	− "	− "	− "	− "	− "	− "
δ, o	18 20 44.0							
37.15" × UT	13 49	13 51	13 53	13 54	13 56	13 57	13 59	14 00
δ	+18 06 55	18 06 53	18 06 51	18 06 50	18 06 48	18 06 47	18 06 45	+18 06 44

$$\tan Z = \frac{-\sin t}{\tan\delta \cos\phi - \sin\phi \cos t}$$

	1D	1R	2R	2D	3D	3R	4R	4D
Azimuth, Z	273 46 26	274 02 03	274 27 11	274 42 52	275 02 46	275 14 53	275 31 26	275 42 42
Less ∡	168 28 54	169 21 34	169 09 41	170 01 53	169 45 28	170 33 56	170 14 42	171 01 27
Astro. Az. to target	105 17 32	104 40 30	105 17 31	104 40 59	105 17 18	104 40 57	105 16 44	104 41 15
Set Average	104 59 01		104 59 15		104 59 07		104 58 59	

Set 5 Results
104°59'05"
(p. 2 not shown)

Average Astronomic Azimuth to Target = **104 59 05**
of 5 Sets   Range = 16"

$\sigma_1 = \pm 6.2"$ $\sigma_s = 2.8"$ All 5 Sets

$E_{90} = \pm 10"$ $E_{90_s} = 4.6"$ " " "

C2 FIELD NOTES FOR ASTRONOMIC OBSERVATION

171

Astronomic Target **Sun**   Latitude of Instr. **40° 48' 36"**
Instrument Station **JW1**   Longitude of Inst. **77° 51' 22"**
State College   Mark **CHWT**   Date **9/3/91**   Observer **RBB**   Rec **HW**
Pennsylvania   Tk   PM

Sun Limb

SET D/R	MARK READ DEG	MIN	SEC	STAR/SUN READ DEG	MIN	SEC	TIME HR	MIN	SEC	~~CORR~~ ~~MIN~~	~~SEC~~
?1 ~~D~~	2~~0~~	10 ~~02~~	10	146	09	00	22	46	09.4	L	
1R	182	10	48	327	05	13	22	48	35.0	R	
2D	0	35	38	145	44	39	22	53	21.0	L	
2R	180	36	08	326	29	39	22	54	37.9	R	
3R	225	49	27	11	38	10	22	57	25.4	L~~R~~	
3D	145	49	04	192	20	30	22	58	31.8	R~~L~~	
✓4D	90	40	25	237	12	11	23	01	50.4	L	
R										clouds	
							DUT1 correction = +0.3^s				
				146	09	00	270	25	59		
				2	10	10	143	58	50		
				143	58	50	126	27	09		

C 2

Solar Observation Computation

Sta: __JW1__ Inst. type: __T2 (New)__ φ __40° 48' 36"__
Target: __CHWT__ Observer: __RBB__ λ __77° 51' 22"__
Date: __9/8/81 Tues. PM__ Computer: __RBB__

Observation	OMIT L / 1D	R / 1R	L / 2D	R / 2R	L / 3R	R / 3D	L✓ / 1D	1R
UT	22:46:09.7	22:48:35.3	22:53:21.3	22:54:38.2	22:57:25.7	22:58:32.1	23:01:50.7	
Eq. Time, 0	2:10.7						02:10.7	
0.86 × UT	19.6	19.6	19.7	19.7	19.7	19.8	19.8	
Eq. Time	+2:30.3	2:30.3	2:30.4	2:30.4	2:30.4	2:30.5	+2:30.5	
less 12^h	-12:00:00						-12:00:00	
GHA, hours	10:48:40.0	10:51:05.6	10:55:51.7	10:57:08.6	10:59:56.1	11:01:02.6	11:04:21.2	
GHA, arc = GHA$_h$ × 15	162 10 10	162 46 24	163 57 55	164 17 09	164 59 01	165 15 39	166 05 18	
λ	77 51 22						77 51 22	
LHA	84 18 48	84 55 02	86 06 33	86 25 47	87 07 39	87 24 17	88 13 56	
t	-84 18 48	-84 55 02	-86 06 33	-86 25 47	-87 07 39	-87 24 17	-88 13 56	
δ, o	+5° 48' 42"						5 48 42	
-56.4" × UT	21 24	21 26	21 31	21 32	21 35	21 36	21 39	
δ	+5° 27' 18"	5° 27' 16"	5° 27' 11"	5° 27' 10"	5° 27' 07"	5° 27' 06"	5° 27' 03"	
	↑ OMIT	tan Z = sin t / (tanδ cosφ − sinφ cos t)					↓ Use as 1D	
Azimuth, Z	270° 25 59	270 49 36	271 36 08	271 48 39	272 15 51	272 26 39	272 58 54	
Less ∡	143 58 50 / ~~146 06 50~~	144 54 25	145 09 01	145 53 31	145 48 43	146 31 26	146 31 46	
Astro. Az. to target	126 27 09 / ~~+24 19 09~~ OMIT	125 55 11	126 27 07	125 55 07	126 27 08	125 55 13	Use 126 27 08	
Set Average	126 11 09.5		126 11 07		126 11 10.5			

Average Astronomic Azimuth to Target = __126 11 09.0__

Note: correction for Sun's semi-diameter change is appropriate

C3 *FIELD NOTES FOR ASTRONOMIC OBSERVATION* 173

Astronomic Target **SUN** Latitude of Instr. **40° 48' 36"**
Instrument Station **JWL** Longitude of Inst. **77° 51' 22"**
State College Mark **CHUT** Date **9/9/81** Observer **RBE** Rec **JW**
Pennsylvania Wednesday A.M. SIDE **(Limb)**

SET D/R	MARK READ			STAR/SUN READ			TIME UT			~~CORR'N~~	
	DEG	MIN	SEC	DEG	MIN	SEC	HR	MIN	SEC	~~MIN~~	~~SEC~~
D	0	27	23	333	43	54	12	31	11.6	L	
R	180	27	35	154	43	41	12	33	37.0 ~~7.0~~	R	
✓R	225	23	00	199	47	30	12	37	38.0	L	
D	45	22	40	20	49	02	12	40	15.6	R	
D	90	37	43	66	24	44	12	45	21.0	L	
R	270	38	17	247	29	59	12	48	11.0	R	
R	315	28	36	292	56	14	12	54	36.5	L	
✓D	135	28	10	114	00	23	12	57	21.0	R	
							Add	0.3^s	DUT1		
				TOTAL OBSERVATION	Time						
							26^m				
				Set-up, take down			8^m				
						TOTAL	34^m				
				Some cloud interference							
✓R	Zenith Angle			~~242~~ 292	53		Vert. ∡ ~	33°			
D				65	56		Vert ∡ ~	34°			

C 3

Solar Observation Computation

Sta: __JW1__ Inst. type: __T2 (New)__ φ __40°48'36"__
Target: __CHWT__ Observer: __RBB__ λ __77°51'22"__
Date: __9/9/81 Wed. AM__ Computer: __RBB__

Observation	L 1D	R 1R	L 2R	R 2D	L 3D	R 3R	L 4R	R 4D
UT	12:31:11.6	12:33:39.3/17.0	12:37:38.0/3	12:40:15.6/9	12:45:21.3/3	12:48:11.0/3	12:54:36.3/8	12:57:21.3/3
Eq. Time, 0	+2:31.3s							+2:31.3
0.87 × UT	10.9	10.9	11.0	11.0	11.1	11.1	11.2	11.3
Eq. Time	2:42.2	2:42.2	2:42.3	2:42.3	2:42.4	2:42.4	2:42.5	2:42.6
less 12ʰ GHA, hours	0:33:54.1/53.8	0:36:19.5/59.2	0:40:20.6/3	0:42:58.2/57.9	0:48:03.7/03.4	0:50:53.7/53.4	0:57:19.3/14.0	1:00:03.9/03.6
GHA, arc = GHAh × 15	8 28 27/31	8 59 48/9 04 52	10 05 05/08.5	10 44 28.5/32.5	12 00 51/55	12 43 21/25	14 19 45/49	15 00 54/58
λ	77 51 22	77 51 22	77 51 22	77 51 22	77 51 22	77 51 22	77 51 22	77 51 22
LHA	290 37 05/09	291 08 26/13 30	292 13 43/47	292 53 07/11	294 09 29/33	294 51 59/52 03	296 28 23/27	297 09 32/36
t (east) +	69 22 55/51	68 51 34/46 30	67 46 17/13	67 06 53/49	65 50 31/27	65 08 01/07 57	63 31 37/33	62 50 28/24
δ, o	+5°26'08"							5°26'08"
-56.7" × UT	11 50	11 52	11 56	11 58	12 03	12 06	12 12	12 15
δ	+5°14'18"	5°14'16"	5°14'12"	5°14'10"	5°14'05"	5°14'02"	5°13'56"	+5°13'53"

$$\tan Z = \frac{\sin t}{\tan\delta \cos\phi - \sin\phi \cos t}$$

Azimuth, Z	99°44'43"/46	100 06 39/10 42	100 52 37/40	101 20 31/34	102 15 04/07	102 45 41/44	103 55 49/52	104 26 04/07
Less λ	333 16 31	334 16 06	334 24 30	335 26 22	335 47 01	336 51 42	337 27 38	338 32 13
Astro. Az. to target	126 28 12/15	125 50 33/54 36	126 28 07/10	125 54 09/12	126 28 03/06	125 53 59/54 02	126 28 11/14	125 53 51/54
Set Average	126 09 14/22.5 10.5		126 11 11		126 11 04		126 11 04	

Average Astronomic Azimuth to Target = __126° 11' 07.4"__
Correction due to S __+ 02.7"__
__126° 11' 10.1"__

C 4 — FIELD NOTES FOR ASTRONOMIC OBSERVATION

Astronomic Target **SUN** Latitude of Instr. **40° 48' 36"**
Instrument Station **14** Longitude of Inst. **77° 51' 22"**
Mark **WT** Date **9/10/81** Observer **JSA** Rec **JWA**

SET D/R	MARK READ DEG	MIN	SEC	STAR/SUN READ DEG	MIN	SEC	TIME HR	MIN	SEC	LIMB
1D	0	29	58	331	16	56	12	13	~~06.6~~	L
1R	180	31	27	153	26	13	12	22	19.8	R
2R	225	31	32	199	28	01	12	31	25.5	L
2D	45	30	01	20	28	50	12	34	10.1	R
3D	90	29	57	66	00	04	12	39	18.3	L
3R	270	31	22	247	20	29	12	44	29.2	R
4R	315	31	18	293	44	20	12	55	19.9	L
4D	135	30	03	115	02	51	12	59	23.9	R
5D	180	30	03	161	10	47	13	08	34.9	L
5R	0	31	20	343	18	56	13	16	27.6	R

Astronomic Target **SUN** Latitude of Instr. **40° 48' 36"**
Instrument Station **5** Longitude of Inst. **77° 51' 22"**
C 5
Mark **CHWT** Date **9-10-81** Observer **GWM** Rec **LSH**

T-1 # 248200

SET D/R	MARK READ DEG	MIN	SEC	STAR/SUN READ DEG	MIN	SEC	TIME (U.T.) HR	MIN	SEC	LIMB
1D	0	00	00	333	45	54	12	32	24.4	L
1R	180	00	06	155	14	18	12	37	33.3	R
2R	225	00	00	201	59	42	12	50	33.1	L
2D	44	59	48	23	47	03	12	57	11.8	R
3D	90	00	00	70	06	15	13	07	21.0	L
3R	270	00	03	251	29	45	13	11	32.4	R
4R	315	00	00	297	03	57	13	17	33.5	L
4D	134	59	54	118	01	21	13	19	25.6	R

C6 — FIELD NOTES FOR ASTRONOMIC OBSERVATION

Astronomic Target **Polaris**   Latitude of Instr. **40° 49' 36"**
Instrument Station **Jw 1**   Longitude of Inst. **77° 51' 22"**
State College   Mark **CHWT**   Date **9/8/81** Observer **REB** Rec **HW**
Pennsylvania   Tue: P.M (7:30 DST)
                                  (stop watch used)

SET D/R	MARK READ DEG	MIN	SEC	STAR/SUN READ DEG	MIN	SEC	TIME UT HR	MIN	SEC	CORR'N MIN	SEC
1D	0	25	47	235	04	24	23	35	06	0	
1R	180	26	14	55	05	23	23	37	04	0	
2R	240	29	57	115	10	24	23	45	10	0	
2D	60	29	30	295	10	13	23	46	41	0	
3D	120	45	45	355	27	25	23	51	13	0	
3R	300	46	15	175	28	02	23	53	10	0	

	1D	1R	2R	2D	3D	3R
GCT (UT)	23:35:06	23:37:04	23:45:10	23:46:41	23:51:13	23:53:10
GHA, 0 hrs.	313 24 09					313 24 09
GCT × 1.0027379 × 15°/hr.	354 44 37	355 14 12	357 16 01	357 38 51	358 47 02	359 16 21
GHA	308 08 46	308 38 21	310 40 10	311 03 00	312 11 11	312 40 30
λ	77 51 22					77 51 22
LHA	230 17 24	230 46 59	232 48 48	233 11 38	234 19 49	234 49 08
t	+129 42 36	129 13 01	127 11 12	126 48 22	125 40 11	125 10 52

$$z = \tan^{-1} \frac{\sin t}{\tan\delta \cos\phi - \sin\phi \cos t}$$

	1D	1R	2R	2D	3D	3R
Azimuth, z	360 49 54	360 50 15	360 51 42	360 51 57	360 52 44	360 53 03
Less x	234 38 37	234 39 09	234 40 27	234 40 43	234 41 40	234 41 47
Astro. Az. to target	126 11 17	126 11 06	126 11 15	126 11 14	126 11 04	126 11 16
Set Average	126 11 11.5		126 11 14.5		126 11 10	

Average Astronomic Azimuth to Target = **126° 11' 12.0"**

C7 FIELD NOTES FOR ASTRONOMIC OBSERVATION

Astronomic Target **POLARIS** Latitude of Instr. **40° 48' 36"**
Instrument Station **STK 3** Longitude of Inst. **77° 51' 22"**
Mark **WT** Date **9/10/81** Observer ____ Rec ____

SET D/R	MARK READ DEG	MIN	SEC	STAR/SUN READ DEG	MIN	SEC	TIME HR	MIN	SEC	CORR'N MIN	SEC
1D	0	37	15"	235°	26'	28"	0	42	56	-0ᵐ	18ˢ
1R	180°	37	36"	55°	27'	19"	0	46	56		
1D	0°	07	05	235	02'	29"	0	56	18	-0ᵐ	18ˢ
1R	180	07	18	55°	03'	37"	0	59	11	"	"
2D	45	05'	11"	280	01	15	1	05	06	"	"
2R	225	05	35"	100	02	17	1	06	57		
3D	90	06	19	325	03	33	1	15	13		
3R	270	06	10	145	03	49	1	19	59		
4D	135	06	25	10°	04	41	1	36	45		
4R	315	06	30	190	05	22	1	41	30		

C8 FIELD NOTES FOR ASTRONOMIC OBSERVATION

Astronomic Target **Polaris** Latitude of Instr. **43 13 48**
Instrument Station ____ Longitude of Inst. **84 03 35**
Mark ____ Date **7/21/82** Observer ____ Rec ____

SET D/R	MARK READ DEG	MIN	SEC	STAR/SUN READ DEG	MIN	SEC	TIME HR	MIN	SEC	CORR'N MIN	SEC
1D	0	00	18	158	14	31	10	08	12	-0	50
1R	180	00	17	338	14	48	10	10	55		
2D	45	00	17	203	15	48	10	14	40		
2R	235	00	23	23	16	08	10	15	03		
3D	90	00	19	248	17	06	10	19	40		
3R	270	00	18	68	17	17	10	20	56		
4D	135	00	19	293	18	37	10	26	20		
4R	315	00	20	113	19	02	10	27	46	-0	50

C 9 *FIELD NOTES FOR ASTRONOMIC OBSERVATION* Campus Calif. State U.

Astronomic Target **Sun**   Latitude of Instr. **36° 48' 44.5'**
Instrument Station **NE Cor**   Longitude of Inst. **119° 44' 30"**
Wild T1   Mark **Water Tower CSUF**   Date **10/5/91**   Observer **RBB**   Rec **Heather**
Inverted (20")

SET	MARK READ			STAR/SUN READ			TIME UT			Limb
D/R	DEG	MIN	SEC	DEG	MIN	SEC	HR	MIN	SEC	~~MIN SEC~~
1D	0	00	00	186	06	30	17	11	28.8	L
1R	179°	58	20	6	19	25	17	15	12.2	R
2D	0	00	00	187°	52	00	17	18	39.7	L
2R	180°	00	00	7°	49	45	17	21	13.0	R
3D	0	00	00	189	02	25	17	23	22.6	L
3R	179°	59	40	8	35	00	17	24	16.0	R
4D	0	00	00	189	51	00	17	26	34.2	L
4R	179°	59	40	9	34	25	17	28	09.8	R

Clovis Quadrangle

C9

Solar Observation Computation

Sta: NE Corner **Inst. type:** Wild T1 (20") ϕ 36°48'44.5"
Target: CSUWT Antenna **Observer:** RBB λ 119°44'30"
Date: Oct. 5, 1981 **Computer:** RBB (Inverted Image)

Sun's Limb → L R L R L R L R

Observation	1D	1R	2D	2R	3D	3R	4D	4R
UT	17:11:28.8	17:15:12.2	17:18:39.7	17:21:13.0	17:23:22.6	17:24:16.0	17:26:34.2	17:28:09.8
Eq. Time, 0	+0:11:25.4	→						0:11:25.4
$0.75^s \times$ UT	12.9	12.9	13.0	13.0	13.0	13.1	13.1	13.1
Eq. Time +	0:11:38.3	0:11:38.3	0:11:38.4	0:11:38.4	0:11:38.4	0:11:38.5	0:11:38.5	0:11:38.5
less 12^h	-12:00:00	→						-12:00:00
GHA, hours	5:23:07.1	5:26:50.5	5:30:18.1	5:32:51.4	5:35:01.0	5:35:54.5	5:38:12.7	5:39:48.3
GHA, arc = GHA$_h \times$ 15	80°46'46"	81 42 37	82 34 31	83 12 51	83 45 15	83 58 37	84 33 10	84°57'04"
λ	119 44 30	→						119°44'30"
LHA	321 02 16	321 58 07	322 50 01	323 28 21	324 00 45	324 14 07	324 48 40	325 12 34
t	38 57 44	38 01 53	37 09 59	36 31 39	35 59 15	35 45 53	35 11 20	34 47 26
δ, o	-4°36'45"	→						-4°36'45"
$-57.7" \times$ UT	16'32"	16 36	16 39	16 41	16 44	16 44	16 46	16 48
δ	-4°53'17"	-4 53 21	-4 53 24	-4 53 26	-4 53 29	-4 53 29	-4 53 31	-4 53 33
			$\tan Z = \dfrac{\sin t}{\tan\delta \cos\phi - \sin\phi \cos t}$			Sun's Limb sighted as it appeared (inverted image)		
Azimuth, Z	130 21 32	131 15 29	132 06 21	132 44 21	133 16 49	133 30 16	134 05 18	134 29 43
Less ∡	186 06 30	186 21 05	187 52 00	187 49 25	189 02 25	188 35 20	189 51 00	189 34 25
Astro. Az. to target	304 15 02	304 54 24	304 14 21	304 54 56	304 14 24	304 54 56	304 14 18	304 55 18
Set Average	304° 34' 43.0"		304° 34' 38.5"		304° 34' 40.0"		304° 34' 48.0"	

Average Astronomic Azimuth to Target = **304° 34' 42.4"**

C 10 — FIELD NOTES FOR ASTRONOMIC OBSERVATION

Astronomic Target **Sun** Latitude of Instr. **36° 48' 44.5"**
Instrument Station **NE Cor** Longitude of Inst. **119° 44' 30"**
Mark **CSU WT** Date **10/14/81** Observer **RBB** Rec _____

Wild T16 — Direct Reading to 1', estimate to 0.1"

Sun's Limb

SET D/R	MARK READ DEG	MIN	SEC	STAR/SUN READ DEG	MIN	SEC	TIME HR	MIN	SEC	~~CORREN~~ ~~MIN~~ SEC
1D	0°	00		183	42.5		16	50	33.4	L
R	180	00.2		4	50.7		16	52	47.5	R
2D	0	00		184	59.0		16	56	14.1	L
R	180	00.2		5	52.3		16	57	25.1	R
3D	0	00		185	45.6		16	59	27.5	L
R	180	00.1		6	44.2		17	01	12.7	R
4D	0	00		186	46.5		17	04	08.4	L
R	180	00.2		7	39.8		17	05	13.5	R

C 11 — FIELD NOTES FOR ASTRONOMIC OBSERVATION

Astronomic Target **SUN** Latitude of Instr. **30° 19' 18.75"**
Instrument Station **24A007** Longitude of Inst. **91° 07' 44.75"**
Mark **Pt 1 (½" I.P.)** Date **12-21-82** Observer **GH** Rec **GWM**

1" Theodolite

Limb

SET D/R	MARK READ DEG	MIN	SEC	STAR/SUN READ DEG	MIN	SEC	TIME HR	MIN	SEC	~~CORRN~~ ~~MIN~~ SEC
1D	0°	00'	15"	147°	03'	38"	14	07	19.1	L
1R	180°	00'	13"	328°	08'	25"	14	10	45.6	R
2R	225°	00'	12"	13°	27'	26"	14	16	19.5	L
2D	45°	00'	07"	194°	43'	43"	14	20	45.2	R
3D	90°	00'	12"	239°	46'	44"	14	24	31.9	L
3R	270°	00'	08"	61°	11'	12"	14	29	41.2	R
4R	315°	00'	05"	106°	18'	00"	14	33	46.3	L
4D	135°	00'	10"	287°	39'	15"	14	38	23.6	R

C 10

Solar Observation Computation

Sta: NE Corner **Inst. type:** Old Wild T-16 (0.1') ϕ 36°48'44.5"
Target: CSUWT Antenna **Observer:** RBB λ 119°44'30"
Date: Oct. 14, 1981 **Computer:** RBB Erect Image

Sun's Limb →	L	R	L	R	L	R	L	R
Observation	1D	1R	2D	2R	3D	3R	4D	4R
UT	16:50:33.4	16:52:47.5	16:56:14.1	16:57:25.1	16:59:27.5	17:01:12.7	17:04:08.4	17:05:13.5
Eq. Time, 0	+0:13:51.9	→						+0:13:51.9
0.58^s × UT	09.8	09.8	09.8	09.8	09.9	09.9	09.9	09.9
Eq. Time	+0:14:01.7	+0:14:01.7	+0:14:01.7	+0:14:01.7	+0:14:01.8	+0:14:01.8	+0:14:01.8	+0:14:01.8
less 12^h	-12:00:00	→						-12:00:00
GHA, hours	5:04:35.1	5:06:49.2	5:10:15.8	5:11:26.8	5:13:29.3	5:15:14.5	5:18:10.2	5:19:15.3
GHA, arc = GHA_h × 15	76°08'46"	76 42 18	77 33 57	77 51 42	78 22 19	78 48 37	79 32 33	79 48 49
λ	119 44 30	→						119 44 30
LHA	316 24 16	316 57 48	317 49 27	318 07 12	318 37 49	319 04 07	319 48 03	320 04 19
t	43 35 44	43 02 12	42 10 33	41 52 48	41 22 11	40 55 53	40 11 57	39 55 41
δ, o	-8°01'43"							-8°01'43"
-55.8" × UT	15'40"	15 42	15 45	15 46	15 48	15 50	15 52	15 53
δ	-8°17'23"	-8 17 25	-8 17 28	-8 17 29	-8 17 31	-8 17 33	-8 17 35	-8 17 36

$$\tan Z = \frac{\sin t}{\tan\delta \cos\phi - \sin\phi \cos t}$$

Azimuth, Z	128 36 24	129 05 59	129 52 01	130 07 58	130 35 39	130 59 35	131 39 54	131 54 56
Less $\angle$	-183 42 30	184 50 30	184 59 00	185 52 36	185 45 36	186 44 06	186 46 30	187 39 36
Astro. Az. to target	304 53 54	304 15 29 ✓	304 53 01 ?	304 15 22 ✓	304 50 03 ?	304 15 29 ✓	304 53 24	304 15 20 ✓
Set Average	304°34'41.5"		304°34'11.5"				304°34'22.0"	

Average Astronomic Azimuth to Target = 304° 34' 25"
(Note: Instrument in need of cleaning & adjustment)

C 11

****

SOLAR OBSERVATION COMPUTATION

****

HOUR ANGLE METHOD AS PRESENTED
IN DR.R.B.BUCKNER WORKSHOP

```
PROJECT NAME:WOODSTOCK
STATION:      24A007
TARGET:       PT. 1
OBS. DATE:    12/21/82

LONGITUDE=        91   07   45.01
LATITUDE =        30   19   18.54
EQ. OF TIME=      0    02   14.7
DELTA EQ. TIME=-  0    00   01.3
DECLINATION=    - 23   26   09.0
DELTA DECL.=    - 0    00   00.8

UNIVERSAL TIME=   14   07   19.1
AZIMUTH=          126  36   47.0
SUN READING=      147  03   38.0
MARK READING=     0    00   15.0
MARK TO SUN=      147  03   23.0
TAR.ASTRO.AZ.=    339  33   24.0

UNIVERSAL TIME=   14   10   45.6
AZIMUTH=          127  08   42.0
SUN READING=      328  08   25.0
MARK READING=     180  00   13.0
MARK TO SUN=      148  08   11.0
TAR.ASTRO.AZ.=    339  00   30.0

UNIVERSAL TIME=   14   16   19.5
AZIMUTH=          128  01   06.0
SUN READING=      13   27   26.0
MARK READING=     225  00   12.0
MARK TO SUN=      148  27   14.0
TAR.ASTRO.AZ.=    339  33   52.0

UNIVERSAL TIME=   14   20   45.2
AZIMUTH=          128  43   30.0
SUN READING=      194  43   43.0
MARK READING=     45   00   07.0
MARK TO SUN=      149  43   36.0
TAR.ASTRO.AZ.=    338  59   54.0

UNIVERSAL TIME=   14   24   31.9
AZIMUTH=          129  20   10.0
SUN READING=      239  46   44.0
MARK READING=     90   00   12.0
MARK TO SUN=      149  46   32.0
TAR.ASTRO.AZ.=    339  33   38.0

UNIVERSAL TIME=   14   29   41.2
AZIMUTH=          130  10   59.0
SUN READING=      61   11   12.0
MARK READING=     270  00   03.0
MARK TO SUN=      151  11   04.0
TAR.ASTRO.AZ.=    338  59   55.0

UNIVERSAL TIME=   14   33   46.3
AZIMUTH=          130  51   53.0
SUN READING=      106  18   00.0
MARK READING=     315  00   05.0
MARK TO SUN=      151  17   54.0
TAR.ASTRO.AZ.=    339  33   58.0

UNIVERSAL TIME=   14   38   23.6
AZIMUTH=          131  38   52.0
SUN READING=      287  39   15.0
MARK READING=     135  00   10.0
MARK TO SUN=      152  39   05.0
TAR.ASTRO.AZ.=    338  59   47.0

SET AVERAGE-1=    339  16   57.0
SET AVERAGE-2=    339  16   53.0
SET AVERAGE-3=    339  16   47.0
SET AVERAGE-4=    339  16   53.0
```

IF SET AVG. OK,ENTER 5.

TO EXCLUDE ONE SET AVG.
ENTER THAT SET NUMBER.

EXCLUDE SET 3

AVG.ASTRO.AZ.=   339 16   54.0

CENTRAL MER.=   91   20   00.0
MAPPING ANGLE=   0   06   07.0
LAPLACE COR.=   0   00   00.8

GRID BRG N 20 49 12 W

APPENDIX D

ASTROGEODETIC CONVERSIONS AND LAPLACE CONVERSIONS

The Laplace Correction maps were plotted from data furnished by the National Geodetic Survey. The theory has been explained in Sections 3-2 and 7-1.

Laplace corrections are listed in the last column, in seconds of arc--LAPLACE(S). The columns headed GLAT and GLON are geodetic latitude and geodetic longitude for the station named, in degrees, minutes, seconds. These values can be used to plot positions on a map for possibly plotting isolines or for interpolation of Laplace corrections or differences between astronomic and geodetic position.

The columns headed ALAT-GLAT(S) and ALON-GLON(S) are for converting geodetic to astronomic position (See text Section 3-2). They are differences between astronomic and geodetic position, in seconds. For example, a geodetic position scaled from a USGS topographic map near Station "Connerly 1929" in Arkansas should be corrected as follows:

ϕ_G = 33° 07' 30"	and	λ_G = 91° 14' 22"	(Map scaled)
Correction = −1.6"		Correction = −1.8"	(From tables)
ϕ_A = 33° 07' 28"		λ_A = 91° 14' 20"	(Use in Solar Computation)

Be cautious to observe the algebraic sign of the correction.

If asterisks appear (******) in the ALAT-GLAT column, this means that the astronomic latitude was not determined. If they appear in the ALON-GLON column, this means that the astronomic longitude was not determined, in which case it is not possible to compute the Laplace correction, since it depends on having ALON.

Values interpolated from the isoline plots of Laplace corrections are probably accurate to ±1 to 2 seconds.

LAPLACE CORRECTIONS — ILLINOIS

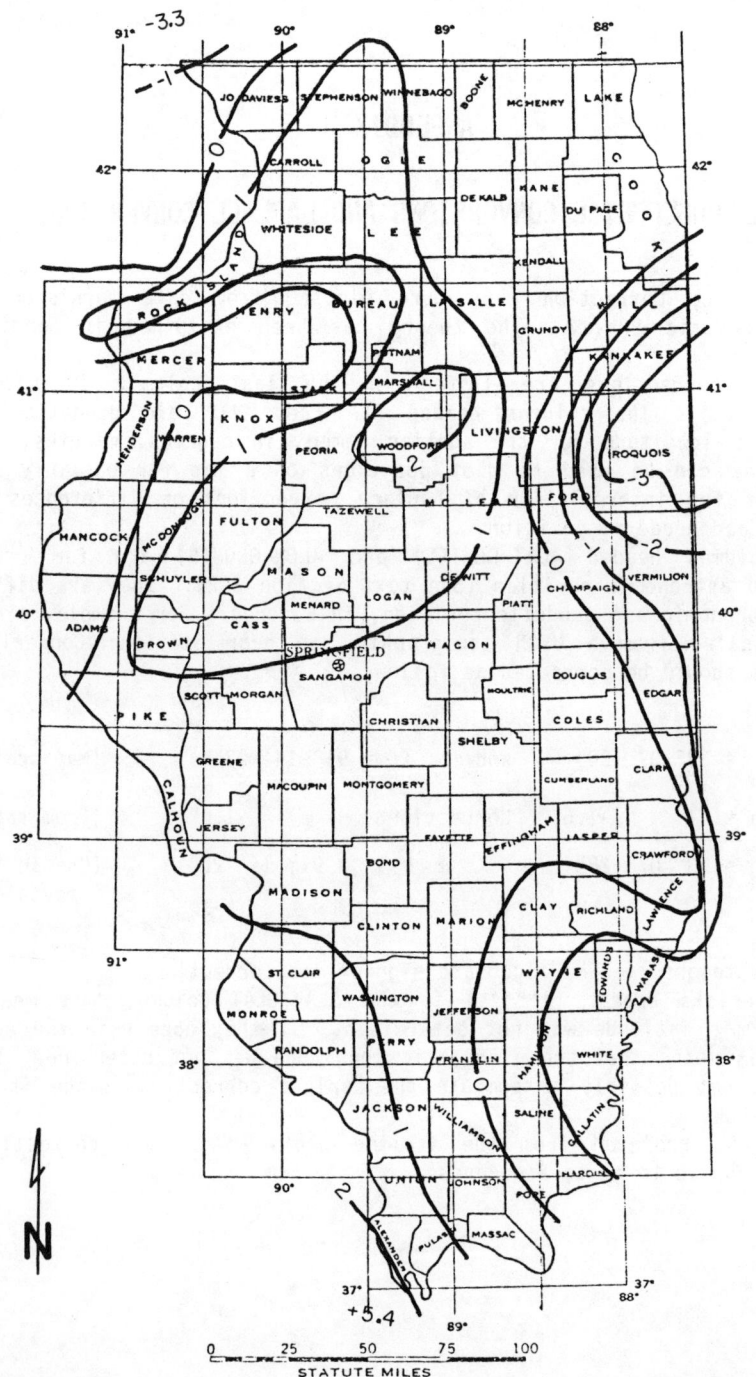

USGS Base Map Isolines are at 1" Intervals Plotted by Dr. Ben Buckner

LAPLACE CORRECTIONS — INDIANA

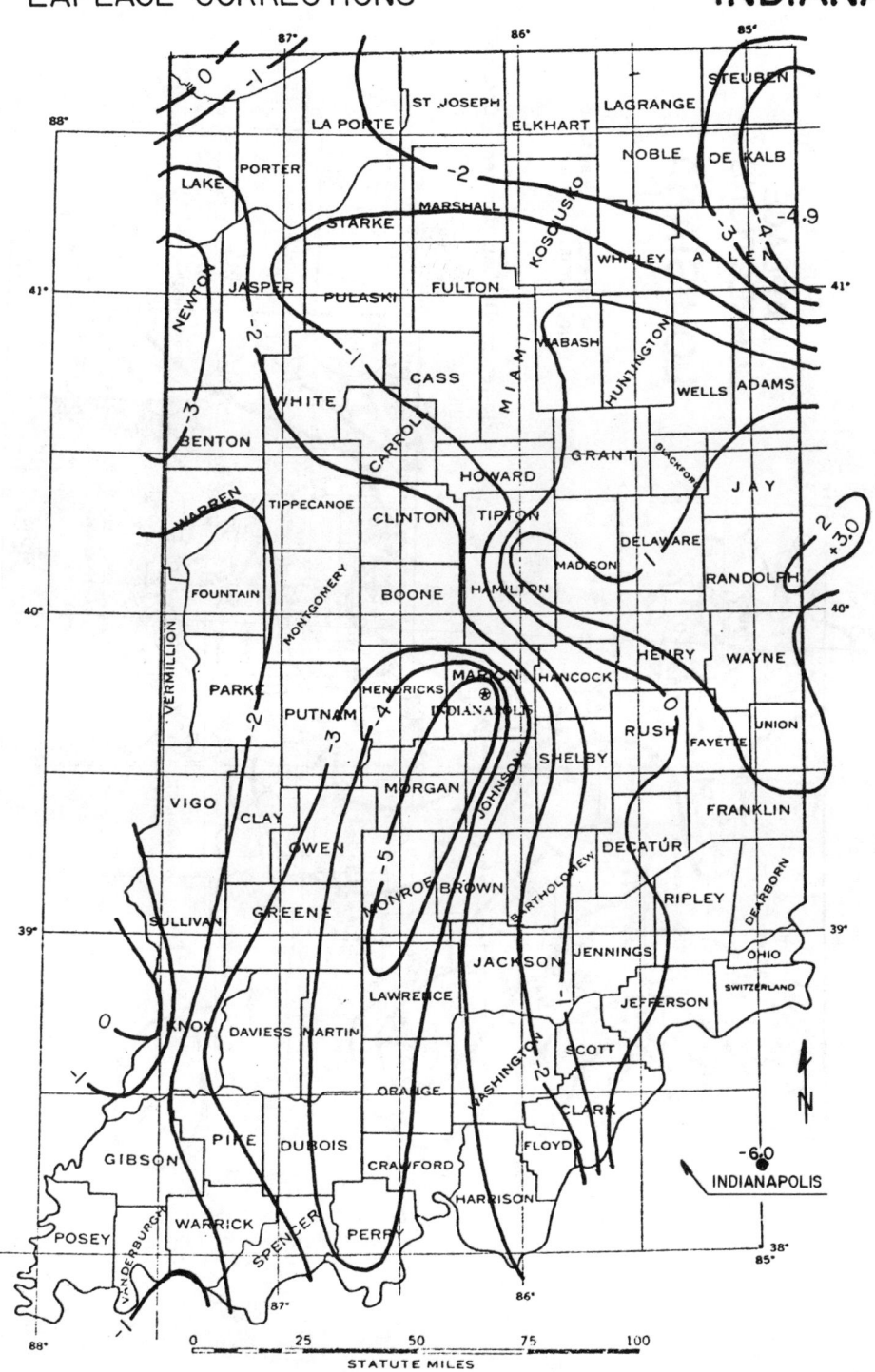

USGS Base Map — Isolines are at 1" intervals — Plotted by Dr. Ben Buckner

LAPLACE CORRECTIONS
LOUISIANA

Isolines are in 1-second intervals.

Plotted by Dr. Ben Buckner, Dec. 1981

USGS Base Map

LAPLACE CORRECTIONS
OHIO

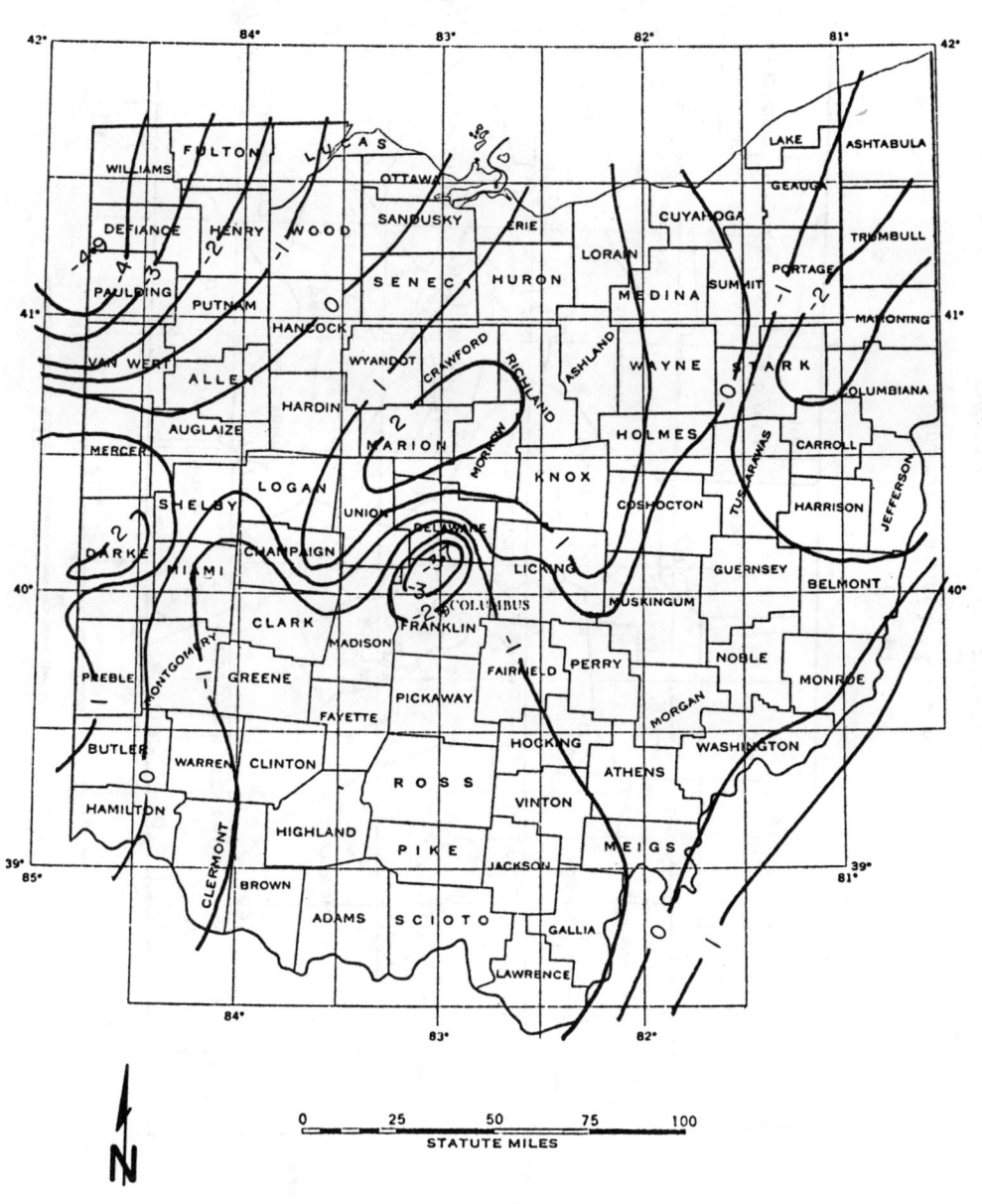

USGS Base Map Isolines are at 1" intervals Plotted by Dr. Ben Buckner

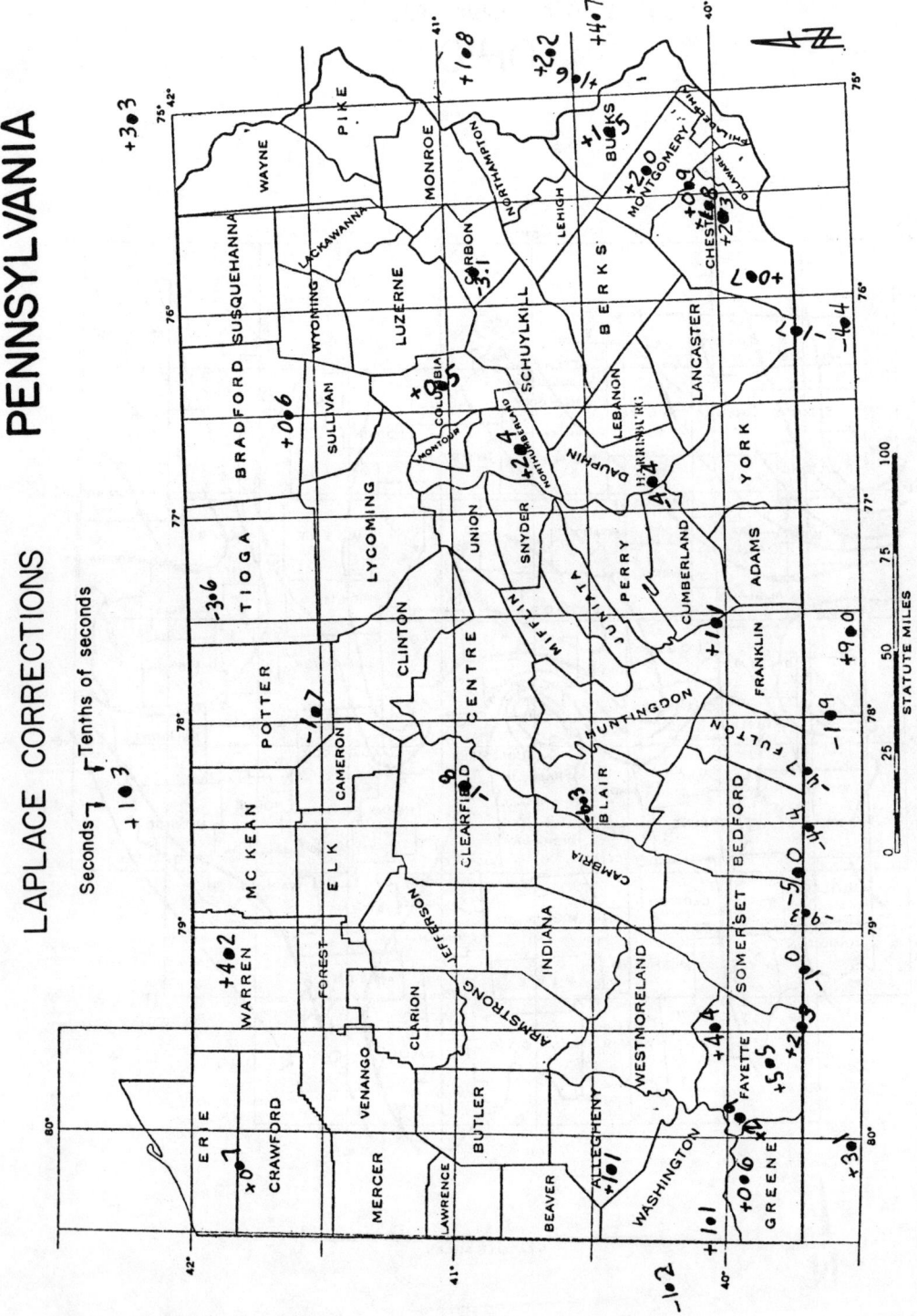

STATE	STATION NAME	GLAT(D,M,S)	GLON(D,M,S)	ALAT-GLAT(S)	ALON-GLON(S)	LAPLACE(S)
	BEAVER CREEK ASTRO 1943	62 27 21.76	140 51 0.29	-2.690	-1.730	-1.53
	STE MARIE USLS 1893	46 31 4.57	84 17 52.32	-57.730	179.840	25.05
	TURNERS ASTRONOMIC STA 1890	67 25 3.74	140 59 36.13	1.370	********	******
AK	AS 1147 USLM 1950	55 59 28.04	160 34 25.37	8.800	19.850	16.45
AK	ASTLEY LAPLACE 1523	57 42 59.62	133 39 0.34	0.140	17.410	14.72
AK	ASTRO 1913	59 3 50.71	161 45 28.55	0.890	********	******
AK	ASTRO 2 1934	52 12 19.34	174 11 9.88	-5.270	5.960	4.71
AK	ASTRONOMIC STATION	59 11 48.87	135 27 22.66	-7.900	********	******
AK	BAKER USN 1953	69 22 20.12	152 6 39.21	4.240	7.010	6.56
AK	BARTER ASTRO 1948	70 7 15.39	143 36 59.55	1.270	18.400	17.31
AK	BASE 1946	58 39 58.97	156 37 17.06	-0.420	12.500	10.68
AK	BEALES SOUTH BASE 1942	63 54 24.48	145 47 12.32	-6.870	8.580	7.72
AK	BETHEL WEST BASE 1949	60 46 49.18	161 43 6.08	-0.470	10.510	5.17
AK	BETTLES ASTRO STA 1955	66 55 10.80	151 30 57.88	-3.500	17.030	15.88
AK	BISHOP 1952	55 11 7.06	162 43 19.10	5.960	12.840	10.54
AK	BOUNDARY ASTRO STA 1906	64 40 53.55	141 0 7.49	-2.290	2.280	2.05
AK	BURROUGHS BAY ASTRO STA	56 2 16.23	131 5 45.46	-4.660	********	******
AK	CAIRN ASTRO 1953	61 12 50.76	154 44 34.34	1.610	11.240	9.85
AK	CHANCE 1952	63 15 24.22	153 32 41.90	0.410	12.540	11.19
AK	CITY 1900	64 30 3.34	165 21 25.49	-5.670	14.470	13.06
AK	COPPER CENTER ASTRO STA 1911	61 57 17.80	145 18 5.98	-6.930	9.680	8.54
AK	COPPER RIVER DELTA N BASE 1858	60 18 41.52	145 4 41.28	-9.370	********	******
AK	DIG ASTRO USN 1944	51 23 41.46	180 44 0.57	-22.510	-3.400	-2.66
AK	EAST BASE ASTRO STATION 1915	56 19 44.01	133 16 52.17	-0.650	********	******
AK	ELD 1922	58 58 16.08	135 13 8.97	-9.080	16.580	14.21
AK	EMEL 1952	62 41 21.63	164 36 35.37	-1.420	13.710	12.18
AK	FAIRBANKS ASTRO STATION 1900	64 50 54.26	147 43 26.28	-1.250	7.350	6.64
AK	FOX 1917	56 53 36.59	132 54 49.07	-4.410	-6.770	-5.67
AK	FT WRANGELL ASTRO 1893	56 28 19.95	132 23 8.83	-4.030	********	******
AK	GAMBELL 1950	63 46 25.74	171 44 7.45	-0.140	-0.570	-0.51
AK	GLOBE B 1 E USE 1961	61 17 1.98	149 49 22.54	2.440	55.570	48.74
AK	HAR 1916	56 14 10.55	132 39 28.48	1.020	5.680	4.73
AK	HOLY CROSS 1949	62 12 25.91	159 46 8.79	-2.050	1.680	1.50
AK	HOPER 1951	61 31 64.91	166 5 50.20	-0.260	12.170	10.70
AK	ILIAMNA EAST BASE 1946	59 45 50.11	154 49 6.08	-5.330	5.950	5.15
AK	ISLAND LATITUDE STATION 1899	64 27 6.11	162 52 30.23	5.190	********	******
AK	KANAK 1952	59 54 37.20	158 16 12.77	-0.010	19.170	16.59
AK	KIPNUK 1949	59 56 21.50	164 2 26.98	-0.270	11.220	9.71
AK	KIVALINA SE 1950	67 43 36.40	164 31 47.49	-0.550	14.310	13.23
AK	KODIAK ASTRO STATION 1907	57 47 25.31	152 24 2.53	-2.520	********	******
AK	KUH KLUX LATITUDE STA OF 1869	59 23 57.41	135 53 20.92	-16.360	********	******
AK	LEE 1944	57 26 11.57	156 15 50.50	-12.800	-4.710	-3.98
AK	MAGNETIC STATION 1897	56 36 3.98	169 32 36.57	6.320	********	******
AK	MARY ISLAND ASTRO STA 1895	55 5 37.68	131 13 15.35	-4.460	********	******
AK	MATTHEWS ASTRO 1944	60 21 16.26	172 41 28.38	-1.960	8.730	7.59
AK	MELLICK ASTRO 1951	61 41 4.20	157 10 35.72	1.440	1.400	1.22
AK	MESHIK 1949	56 54 59.69	158 40 47.81	-0.630	15.130	12.68
AK	MILLER ASTRONOMIC-RM 3 1906	56 57 34.13	154 7 24.17	-5.480	********	******
AK	MILLER NORTH BASE 1941	63 25 59.04	145 46 36.92	-0.810	22.600	20.22
AK	NEVAI 1922	70 14 20.23	161 20 28.07	-2.490	8.170	7.68
AK	NOTCH 2 1964	60 55 7.89	147 7 44.97	-8.490	********	******
AK	ORCA ASTRONOMICAL STA 1898	60 34 50.93	145 42 56.87	-10.240	********	******
AK	OTTER COVE ASTRO STATION	54 45 36.73	163 20 12.11	8.190	********	******
AK	PETE ASTRONOMICAL STA 1898	60 22 54.42	145 26 37.02	-11.740	********	******
AK	PETERS EAST BASE 1922	61 28 9.25	149 22 30.87	1.150	51.970	45.65
AK	PLATINUM BASE A 1948	59 1 17.17	161 49 39.96	1.440	11.440	9.81
AK	POINT HOPE 1950	68 20 46.72	166 48 14.66	-0.850	20.530	19.07
AK	POP ASTRO 1954	60 5 32.69	159 40 13.16	0.180	15.670	13.77
AK	PT AGASSIZ S BASE 1887	56 54 52.29	132 51 22.45	-7.630	********	******
AK	PT BARROW ASTRO 1945	71 20 18.17	156 38 24.70	1.170	11.230	10.63
AK	PT BARROW S BASE 1945	71 19 27.11	156 36 31.07	0.860	10.040	9.50
AK	RANGE ASTRO 1941	64 45 0.89	147 38 42.71	1.880	8.950	8.10
AK	RAPID SOUTH BASE 1953	67 18 36.52	141 34 5.29	-2.820	8.050	7.41
AK	REDCLBT ASTRO 1944	60 29 13.79	152 18 54.82	-10.910	-61.620	-53.63
AK	RUBY ASTRO 1942	64 44 37.28	155 28 8.48	0.700	11.590	10.47
AK	SEAL 1907	54 55 57.48	131 35 32.71	-8.600	9.700	7.93
AK	SEWARD ASTRO STA 1905	60 6 9.28	149 26 59.14	-3.360	********	******
AK	SHAKAN 1886	56 8 57.08	133 37 57.34	-0.580	********	******
AK	SHEA ASTRO 1953	70 38 32.52	160 0 48.40	********	17.940	16.94
AK	SHEEP ASTRO 1943	61 48 0.96	147 34 24.06	-8.540	23.540	20.74
AK	SHIP 1915	55 35 56.54	132 12 5.58	-5.430	12.790	10.56
AK	SHISHMAREF ASTRO 1949	66 15 21.20	166 4 35.99	3.290	14.750	13.50
AK	SOUTHWEST PT ASTRO 1944	57 9 49.47	170 24 10.98	-3.080	-26.250	-22.07
AK	ST MICHAEL ASTRO STA 1891	63 28 39.47	162 2 18.98	-2.040	********	******
AK	ST PAUL ASTRO STATION 1897	57 7 18.85	170 16 34.24	-1.590	********	******
AK	STEESE WEST BASE 1944	65 33 45.74	145 1 27.16	10.700	-7.970	-7.26
AK	SUMMIT ASTRO 1942	63 22 42.93	149 0 44.60	-7.480	-1.960	-1.76
AK	TANACROSS ASTRO 1943	63 22 23.02	143 20 24.61	0.430	-5.980	-5.35
AK	TELLER SOUTHEAST BASE 1943	65 16 37.66	166 20 23.06	********	28.710	22.86
AK	THEODORE ASTRO 1944	52 45 4.94	187 5 1.52	-24.760	17.590	15.98
AK	UNALAKLEET SOUTH BASE 1942	63 52 21.74	160 47 14.22	-1.120	16.800	15.09
AK	UNALASKA ASTRO STATION 1896	53 52 33.55	166 31 56.56	2.900	11.150	9.00
AK	VALDEZ ASTRO STA 1905	61 7 6.08	146 16 4.86	-16.920	28.770	25.20
AK	VALDEZ NORTH BASE 1901	61 7 23.33	146 16 46.67	-18.570	28.400	24.90
AK	VILLAGE ASTRO 1904	56 36 4.07	169 52 34.45	1.570	-35.200	-29.39
AK	WILLOW CREEK SOUTH BASE 1941	61 50 20.55	145 13 31.24	11.410	-1.400	-1.23
AK	YOUNG 1917	58 11 42.24	134 33 24.11	-1.840	1.400	1.19
AL	ALLISON 1938	32 9 42.78	88 9 16.38	5.900	-0.960	-0.51
AL	ANDERSON SOUTHEAST BASE 1943	31 44 1.27	85 50 33.56	********	3.700	1.55
AL	AURORA 1877	34 8 47.27	86 18 1.04	0.440	********	******
AL	BOESCHEN 1964	30 51 51.80	87 44 1.51	-0.650	3.500	1.79
AL	CCCN 1895	31 14 50.08	88 5 43.78	-2.290	********	******
AL	CREOLA 1964	30 54 5.19	88 0 31.00	-2.200	0.0	0.0
AL	ESCM 1941	33 57 29.79	85 23 52.03	1.990	-2.420	-1.35
AL	ETHRIDGE 1892	32 4 45.68	87 3 29.50	-0.220	0.960	0.51
AL	FATAMA 1895	31 53 32.84	87 14 13.47	-2.690	2.090	1.11
AL	FORT MORGAN 1846	30 13 42.02	88 1 23.70	-5.870	2.390	1.21
AL	GREEN 1939	33 28 23.63	87 41 10.15	-4.450	-0.770	-0.42
AL	HAMMER 1939	33 17 26.18	87 32 50.14	-4.190	-1.300	-0.71
AL	KAHATCHEE 1887	33 13 38.04	86 21 36.92	1.650	0.730	0.40
AL	LANIER 1941	32 53 9.10	85 12 39.09	20.370	0.470	0.26
AL	LEON 1944 RM 3 1973	31 13 16.25	85 23 30.51	-0.740	1.340	0.70
AL	LOKER PEACH TREE TELEG LONG	31 50 20.25	87 32 43.91	0.540	-2.170	-1.14
AL	MADISON 1914	34 42 13.01	86 44 42.15	-1.700	-9.200	-5.24
AL	MCGOAN 1943	32 4 4.10	85 56 27.31	5.040	-1.360	-0.72
AL	MONTGOMERY CAPITOL	32 22 39.32	86 18 1.92	-6.310	-1.360	-0.78
AL	NEWVILLE 1943 RM 3 1973	31 25 49.65	85 13 1.68	-1.050	-1.110	-0.58
AL	PERRY 1890	32 45 32.92	86 57 21.47	-1.800	2.840	1.54
AL	ROBERTSON 1934	35 0 47.97	87 31 18.18	0.040	0.130	0.07
AL	SEMMES 1942	30 46 41.36	88 15 11.73	-0.370	1.570	0.80
AL	SMITHERS 1878	34 49 11.14	86 36 58.67	-3.930	********	******
AL	TANYARD 1878	34 39 11.14	88 7 32.67	-1.850	-4.630	-2.41
AL	TURNER 1938	31 24 44.42	87 32 38.25	3.340	1.230	0.66
AL	WALTON 1917	32 42 17.30	85 39 50.53	3.760	3.820	2.19
AL	WELDON 1934 RM 3 1972	34 56 13.37	92 57 4.03	2.050	5.980	3.43
AR	APLIN 1940	34 58 48.55	94 2 39.33	-0.380	1.080	0.60
AR	ASHDOWN SOUTHEAST BASE 1930	33 35 8.66	92 16 17.46	-0.200	3.090	-1.71
AR	BANKS 1952	33 36 6.95	90 23 11.61	-0.200	-1.410	-0.81
AR	BROWNS 2 1954	35 5 30.23	94 10 24.50	-8.770	3.280	1.91
AR	BURGETT 1935	35 30 52.83	92 50 7.04	-0.650	-2.200	1.21
AR	CAMDEN 1941	33 34 49.06	91 16 0.59	-1.560	-1.760	-0.96
AR	CONNERLY 1929	33 8 45.74				

STATE	STATION NAME	GLAT(D,M,S)	GLON(D,M,S)	ALAT-GLAT(S)	ALON-GLON(S)	LAPLACE(S)
AR	DAWSON 1940	34 59 34.92	93 11 44.71	4.540	8.690	4.98
AR	DERMOTT 1939	33 31 20.89	91 26 18.68	0.710	-3.720	-2.05
AR	DOBBS 1952	34 58 19.19	94 12 33.75	-3.320	2.070	1.19
AR	EGORE 1941	33 36 20.28	92 54 25.37	0.980	3.190	1.77
AR	ELNICE 1965	33 32 15.79	91 13 53.50	0.880	-3.510	-1.94
AR	FCLATAIN 1939	33 22 8.61	91 51 2.36	0.590	2.730	1.50
AR	FLRGESON 1934	35 34 3.47	91 47 7.67	-0.100	1.380	0.81
AR	HARDY 1965	33 32 33.79	91 46 0.59	-0.210	1.350	0.75
AR	HARTFORD 1919	35 1 14.41	94 23 12.44	-2.600	3.950	2.26
AR	HAYES 1941	33 3 19.95	93 0 55.67	2.110	0.730	0.40
AR	LITTLE ROCK LONGITUDE CBS PIER	34 44 52.06	92 16 25.18	4.110	*******	*******
AR	LITTLE ROCK LONGITUDE 1885	34 44 51.45	92 16 25.43	4.750	0.690	0.40
AR	LITTLE ROCK NM BASE 1916	34 45 27.63	92 13 44.11	*******	0.800	0.45
AR	LONCON 1940	35 22 3.18	93 15 0.07	-4.240	2.900	1.68
AR	LCST 1954	35 2 28.56	91 25 56.28	1.260	1.810	1.04
AR	MAYFLOWER 1940	34 57 31.12	92 28 48.65	5.980	7.160	4.10
AR	MINTON 1934	34 58 40.73	91 22 36.24	2.880	1.100	0.64
AR	MURPHY 1941	33 14 20.65	92 56 26.83	0.620	1.210	0.66
AR	NUTT 1953	35 56 2.19	90 27 5.43	2.130	-1.920	-1.12
AR	PACE 1965	33 24 18.59	92 55 17.50	0.750	2.440	1.34
AR	PALESTINE 2 1953	34 58 16.07	90 54 31.48	1.610	1.640	0.94
AR	RED 1628	36 12 6.22	93 16 36.27	3.400	-1.260	-0.74
AR	ROVER 1956	34 56 47.04	93 24 8.34	4.320	6.490	3.72
AR	SCHOOL 1929	35 56 39.82	89 43 57.99	4.720	2.760	1.62
AR	SPRINGDALE 1935 RM 4 1971	36 9 51.14	94 9 47.71	0.220	2.130	1.26
AR	SUMPTER 1939	33 29 23.25	92 3 2.02	-1.100	2.460	1.35
AR	SUNNY 1962	35 16 8.03	91 48 1.46	0.550	2.440	1.42
AR	TURNER 1934	34 59 26.86	91 44 2.74	2.650	0.950	0.55
AR	UNION 1952 RM 4	33 32 11.01	92 32 46.11	-0.660	2.160	1.19
AR	WALTREAK 1956	34 57 8.85	93 39 36.22	2.550	3.440	1.97
AR	WARSAW USGS 1940	34 56 37.17	92 8 50.22	4.630	3.200	1.83
AR	WHITE OAK 1916	34 58 43.77	93 56 39.27	2.330	7.790	4.46
AR	WICENER 1916	35 1 42.70	90 40 19.85	1.820	1.650	0.55
AR	WILLS 1935	34 26 41.24	94 28 43.60	4.270	3.540	2.10
AR	WYE 1940	34 55 59.51	92 41 33.73	9.380	3.780	2.17
AR	ZENT 1954	34 59 55.56	91 9 18.18	1.860	3.720	2.14
AZ	A 2 1959	34 9 0.89	112 8 52.09	-5.590	11.530	6.48
AZ	ADAMSON 1970	31 57 42.48	110 10 47.73	-0.600	9.830	4.77
AZ	ACOBE 1935	31 26 40.08	110 58 11.46	0.640	9.400	5.13
AZ	AIRPORT NPS 1945	36 5 53.23	114 2 36.41	2.660	8.530	5.02
AZ	ALKALI 1960	32 12 46.31	109 55 11.95	-1.300	4.910	2.61
AZ	APACHE 1936	34 15 40.64	110 26 41.19	1.110	4.980	2.85
AZ	ART 1945	32 35 54.58	105 42 3.90	5.260	-9.260	-5.00
AZ	ASGP 1955	35 32 17.5	112 8 41.82	-0.530	8.160	4.74
AZ	AWK 1960	34 43 39.85	113 44 55.38	-0.410	6.640	3.59
AZ	AZTEC 1938	33 48 44.50	110 54 17.13	-12.660	0.510	0.28
AZ	B GLC 1934	33 20 54.01	112 49 56.45	-3.300	6.430	3.53
AZ	BABY ROCK 1951	36 46 19.87	110 0 56.48	4.930	0.820	0.49
AZ	BART 1960	32 19 0.34	109 29 0.45	4.700	-0.040	-0.02
AZ	BEACON 2 1971	32 4C 25.74	114 20 12.61	-0.630	5.490	2.96
AZ	BEAUTIFUL 1951	36 4 12.94	109 36 31.98	0.930	7.140	4.21
AZ	BEEHIVE NORTH BASE 1947	36 58 7.37	111 29 54.36	0.000	0.760	0.45
AZ	BEEHIVE SOUTH BASE 1947	36 56 55.84	111 30 25.44	-0.580	1.830	1.10
AZ	BENCH MARK F 35 1936	34 51 26.17	109 59 42.66	-0.170	4.020	2.30
AZ	BENCH MARK K 47 1936	34 14 54.00	109 23 28.97	4.400	2.920	1.64
AZ	BENCH MARK K 65 1936	34 54 25.53	110 8 44.60	-2.700	3.560	2.01
AZ	BENCH MARK Z 101 USFS 1935	34 32 1.83	112 40 59.97	-7.530	13.010	7.38
AZ	BIG LAKE 1958	33 51 43.85	109 23 26.91	-4.880	4.620	2.57
AZ	BLACK PEAK 1931	34 7 28.84	114 12 38.30	-C.380	8.860	4.97
AZ	BM 4179 USGS 1935 RM 3	32 5 22.15	109 54 10.31	0.560	0.540	0.50
AZ	BCNIIA 1935	32 3 58.7C	110 30 23.35	-1.600	9.690	5.14
AZ	BOUNDARY MON 179 US MEX 1920	32 22 46.40	113 22 66.40	-1.660	9.400	4.99
AZ	BOUNDARY MON 189 US MEX 1920	32 14 22.67	113 58 33.64	2.640	7.160	3.82
AZ	BCX G 1936	32 54 31.82	111 0 39.76	2.640	8.900	4.83
AZ	BUILDER 1935	35 50 6.17	114 33 54.71	-1.270	6.590	3.86
AZ	CABEZAS 1970	32 18 52.26	109 45 17.90	1.050	10.030	5.36
AZ	CAMPBELL 1971	34 27 11.47	109 55 37.32	-10.050	10.000	8.56
AZ	CANE EAST BASE 1952	36 50 46.69	112 55 1.25	-1.090	3.950	2.37
AZ	CANE WEST BASE 1952	36 50 45.16	112 59 38.01	0.030	5.910	3.54
AZ	CANE 1936	32 6 6.43	112 45 12.73	1.320	15.250	8.10
AZ	CARDENAS USGS 1934	36 13 8.48	111 46 35.42	-9.310	5.050	2.99
AZ	CASCA 1946	32 16 42.03	110 21 14.15	1.600	6.340	4.45
AZ	CHAVES 1936	34 46 28.86	111 7 22.34	1.740	-5.570	-3.21
AZ	CINDA 1958	35 11 45.51	111 52 14.55	-3.510	10.450	6.03
AZ	CINDER 1958	35 31 20.68	111 32 42.55	5.770	-6.810	-3.95
AZ	CCAL MINE 1951	36 32 44.60	110 29 35.69	0.500	9.810	5.84
AZ	COCO 1936	35 54 43.55	111 41 23.81	3.510	-5.060	-2.96
AZ	CCOY 1950	34 14 34.59	112 20 33.78	0.090	7.220	3.75
AZ	CONTROL 1971	32 48 39.83	113 31 6.97	-0.960	6.930	3.79
AZ	CCRNFIELD 1960	31 15 59.84	110 23 24.60	4.900	2.780	1.61
AZ	COVERED 1936	32 10 0.75	112 7 34.16	-0.490	2.200	1.17
AZ	CCHLIC 1935	31 48 25.01	111 55 14.62	-C.380	7.840	4.14
AZ	CRIST 1960	32 46 14.25	113 36 25.80	-1.940	8.790	4.76
AZ	CEER 1946	32 54 46.41	110 25 38.45	-3.860	9.520	5.17
AZ	CELUGE 1950	34 53 7.89	113 65 55.94	-5.37C	-3.400	-1.55
AZ	CYKE 1934	34 59 6.34	114 25 38.73	-5.140	15.400	8.83
AZ	EHRENBURG USGS 1935	33 36 11.94	114 29 23.54	-0.280	10.700	5.93
AZ	FARGO 1946	31 44 40.16	109 55 58.36	-2.450	8.280	4.36
AZ	FIG SPRING 1948	32 2 20.59	114 16 20.34	-2.270	8.320	5.06
AZ	FLATO 1935	32 4 24.71	110 54 51.56	-1.650	3.670	1.97
AZ	FLITE 1935	32 23 10.15	112 52 19.05	1.650	3.670	1.97
AZ	FLUTED ROCK USGS 1951	35 53 9.57	109 14 54.71	-3.220	5.340	3.13
AZ	FORT 1958	33 45 51.53	109 57 41.39	0.230	12.390	6.89
AZ	FCUR COR GLO STA 1 1934	36 59 56.30	109 2 4C.24	-4.210	10.000	1.10
AZ	FRAZIER 1936	35 49 11.11	111 1 54.32	1.330	1.540	4.71
AZ	FREEZOUT 1960	33 28 55.5C	109 45 20.89	-5.090	8.400	4.64
AZ	GANADO NORTHWEST BASE 1951	35 53 39.90	109 43 50.44	-2.790	4.250	2.49
AZ	GANADO SOUTHEAST BASE 1951	35 47 24.92	109 39 48.64	-3.060	5.090	2.98
AZ	GAS 1958	35 15 56.36	112 19 54.03	-3.540	14.830	8.56
AZ	GERONIMO USGS 1936	34 54 39.56	109 41 23.81	-0.050	7.770	4.45
AZ	GILLARD 1945	33 0 10.10	109 17 9.17	-0.160	4.280	2.47
AZ	GLO 23 1936	32 21 18.14	112 18 20.64	-C.580	4.820	2.59
AZ	GLO 35 1935	35 4C 29.81	113 13 37.15	1.190	6.830	3.98
AZ	GLO 6 1938	31 56 5.78	109 2 54.10	-C.780	1.950	1.03
AZ	GLCBE 1938	33 24 12.12	110 46 16.37	-4.470	9.290	5.11
AZ	GOLD BASIN 1950	35 47 41.26	114 7 53.01	-6.970	15.190	8.86
AZ	HA WHI 1951	35 30 43.36	113 3 5.17	-17.360	9.180	5.34
AZ	HANK 1962	35 18 59.43	112 56 36.43	-6.130	7.610	4.40
AZ	HARQUAHALLA 1910	33 48 42.23	113 20 46.13	-3.520	8.990	5.00
AZ	HERD 1936	35 33 57.65	113 59 55.83	-1.730	5.330	3.10
AZ	HEREFCRD 1938	31 27 49.01	110 9 7.34	-0.020	4.270	2.23
AZ	HILLTOP 1958	33 34 29.76	110 22 40.74	-11.430	9.210	6.60
AZ	HORSE 1971	32 50 50.04	112 20 16.01	-1.040	5.150	2.79
AZ	HWY 60 1938	33 35 42.03	110 38 24.61	-5.470	12.800	7.08
AZ	JAYNES 1935	32 19 24.86	111 3 1.44	-1.590	10.100	5.40
AZ	JIM 1960	32 54 26.19	112 32 1.42	-1.840	9.050	4.51
AZ	JOHNSON 1944	34 50 49.81	113 40 40.40	-4.610	2.950	2.26
AZ	JOLLY 1944	34 43 18.55	113 40 56.79	-7.150	12.460	7.13
AZ	JUNCTION 1935	33 0 8.53	111 31 18.73	-3.530	8.420	4.59
AZ	KEAMS 1951	35 50 38.53	110 5 30.81	-5.380	8.660	5.07
AZ	KEITH 1971	32 42 42.66	111 40 47.80	-2.620	5.070	2.74
AZ	KNOB 1920	36 4 17.26	112 9 14.00	12.360	4.450	2.62

STATE	STATION NAME	GLAT(D,M,S)	GLON(D,M,S)	ALAT-GLAT(S)	ALON-GLON(S)	LAPLACE(S)
AZ	KOFA NORTH BASE 1947	33 27 45.47	114 12 59.89	-5.230	7.760	4.28
AZ	KOFA SOUTH BASE 1949	33 22 37.03	114 12 59.62	-3.390	9.960	5.47
AZ	KOFA 1911	33 21 33.15	114 4 55.58	-5.460	******	******
AZ	LINE 1960	32 49 50.11	111 59 50.64	-1.480	2.510	1.36
AZ	LCRAC EAST 1956	32 52 35.21	113 1 6.44	-2.690	6.350	3.45
AZ	LCRAC NORTH 1956	33 51 44.50	113 56 19.72	-3.740	6.870	3.82
AZ	LCRAC WEST 1956	32 54 12.75	114 20 12.09	-1.200	8.860	4.82
AZ	LOST CABIN 1948	35 26 40.56	114 30 14.61	-10.260	20.090	11.65
AZ	LOST 1955	35 16 20.48	109 20 17.60	-8.690	7.780	4.49
AZ	MAC KENZIE 1948	34 51 45.08	114 1 56.26	-8.000	19.840	11.34
AZ	MANUEL 1970	31 59 55.49	110 30 4.02	-1.560	8.400	4.45
AZ	MARBLE CANYON 1947	36 48 58.15	111 38 12.26	-0.850	-3.960	-2.37
AZ	MARICOPA E PIER LONG STA 1910	33 3 33.46	112 2 59.67	-4.400	4.680	2.55
AZ	MARIN 1964	32 39 10.83	114 39 27.15	-4.180	4.410	2.38
AZ	MC EUEN 1936	32 25 10.37	111 46 1.44	1.060	5.120	2.74
AZ	MC NEAL 1938	31 36 0.01	109 40 8.48	-0.580	10.020	5.25
AZ	MERRIAM 1936 AZ MK	35 18 58.29	111 14 31.82	4.050	-1.770	-1.03
AZ	MESA 1935	33 25 16.50	111 49 47.67	-2.040	3.820	2.11
AZ	METEOR 1936	35 1 45.90	111 1 42.38	-3.210	-3.000	-1.72
AZ	MICA 1950	34 55 0.34	113 29 25.52	-6.120	17.420	9.98
AZ	MIDDLE USGS 1934	36 1 32.18	112 4 59.11	13.170	0.800	0.47
AZ	MILLER 1962	35 11 19.19	113 24 22.90	-4.090	11.500	6.86
AZ	MOHAWK 1910	32 35 22.23	113 38 49.40	-4.370	******	******
AZ	MOSBY 1953	36 33 50.05	113 57 45.60	-7.250	3.700	2.20
AZ	NEHI 1960	36 55 55.14	113 52 50.42	-2.840	20.080	15.67
AZ	NOGALES ASTRONOMIC STA 1893	31 20 1.47	110 56 21.17	2.990	7.660	3.98
AZ	NORTH ROAD USGS 1953	36 54 48.76	112 12 26.52	4.840	6.980	4.19
AZ	OCOTILLA 1960	32 41 8.42	111 27 4.28	-0.770	13.780	7.44
AZ	OLGA 1960	32 17 48.71	109 20 54.37	2.460	2.780	1.49
AZ	ORAIBI 1951	35 54 21.56	110 41 55.32	-4.720	12.550	7.36
AZ	OX BOW 1946	34 10 7.31	111 20 16.01	-2.510	10.150	5.70
AZ	PA 7 1959	31 50 32.39	110 59 57.37	0.640	4.780	2.52
AZ	PALO 1924	33 54 59.28	113 0 17.35	-2.840	6.390	3.56
AZ	PARKER MWD 1932	34 9 20.72	114 17 0.47	-2.470	7.530	4.23
AZ	PARKER SW BASE MWD 1932	34 7 33.84	114 16 15.77	-1.940	7.650	4.29
AZ	PASS 1969	32 38 36.69	111 23 39.92	-0.410	6.960	3.72
AZ	PEPPER 1947	32 34 3.51	110 40 5.43	4.720	-3.540	-2.12
AZ	PGT NO 2 AMS 1971	32 55 38.04	114 18 23.63	-1.970	7.570	4.11
AZ	PGT NO 3 AMS 1971	33 14 21.76	114 15 25.33	-3.470	7.090	3.89
AZ	PGT NO 4 AMS 1971	33 39 7.42	114 13 46.65	-0.920	6.230	3.46
AZ	PGT NO 5 AMS 1971	33 51 42.69	114 13 1.56	-2.990	9.450	5.27
AZ	PHOENIX 1958	34 16 4.57	110 31 8.15	-4.390	3.110	1.75
AZ	PIKE 1944	34 31 1.58	113 8 42.15	-9.160	7.620	4.32
AZ	PINE 1910	34 46 8.02	113 51 10.93	-9.210	******	******
AZ	PIPE 1953	36 51 44.06	112 43 54.32	-4.920	4.960	2.98
AZ	PLAINS 1936	35 43 27.06	111 25 28.09	3.680	-1.500	-1.11
AZ	PLATEAU 1920	35 50 27.15	112 43 43.87	-1.570	7.810	4.57
AZ	PORTAL 1970	32 13 0.63	109 10 32.58	-0.040	7.360	3.52
AZ	POWER 1936	34 36 21.82	113 29 17.92	-8.290	15.380	8.74
AZ	PRESCOTT NORTH BASE 1919	34 50 17.65	112 27 58.41	-0.920	6.810	3.89
AZ	PRINCE 1924	33 46 30.49	112 21 56.42	-6.450	6.380	3.55
AZ	PROM 1938	34 22 2.27	110 36.63	-11.720	8.820	4.98
AZ	PUERTECITO US ARMY 1935	31 31 0.13	111 29 27.83	-0.350	5.850	3.07
AZ	PUSCH USGS 1935	32 22 18.56	110 56 18.41	-2.320	******	******
AZ	RADAR 1971	32 41 0.74	114 3 4.60	-0.720	7.650	4.13
AZ	RAIL 1924	33 59 59.57	112 48 34.31	-5.900	4.810	2.69
AZ	RANE 1949	33 33 53.91	113 46 0.56	-0.610	4.230	2.34
AZ	RANGE 1948	34 35 23.40	114 4 57.78	-0.610	-0.680	-0.39
AZ	REESE 1936	34 59 36.90	109 24 44.59	-0.750	7.270	4.17
AZ	RENO 1946	33 52 5.93	111 18 44.55	-7.220	0.920	0.51
AZ	ROADSIDE 1938	33 22 50.56	111 28 58.21	0.800	10.090	5.55
AZ	ROSE 1934	33 26 18.36	109 22 12.91	-0.130	7.800	4.30
AZ	SAN PEDRO 1936	32 4 29.56	111 31 7.10	-3.530	6.620	3.52
AZ	SAN RAFAEL USGS 1938	31 27 57.28	110 34 38.77	-3.630	9.450	4.93
AZ	SANDHILL USGS 1936	35 8 34.27	109 44 6.94	-3.620	6.870	3.96
AZ	SCRABBLE 1946	34 22 1.58	111 31 54.41	-13.420	16.160	9.12
AZ	SCRUB 1936	34 56 43.47	109 6 38.06	-1.800	5.600	3.21
AZ	SEC CORNER 1948	34 15 41.29	113 32 16.01	-1.250	5.890	3.32
AZ	SENT 1971	32 50 32.16	113 11 41.93	-0.390	6.750	3.66
AZ	SHOT 1962	33 23 47.44	112 20 22.70	-4.060	6.500	3.58
AZ	SHOW LOW 1958	34 15 46.01	110 0 7.44	-5.210	8.230	4.63
AZ	SILVER 1938	31 27 48.05	109 21 48.73	-3.730	-0.060	-0.03
AZ	SNAP 1934	36 10 23.08	113 47 59.43	-1.200	33.170	19.58
AZ	SOGA 1945	33 14 18.23	110 30 49.37	-4.740	-3.000	-1.64
AZ	ST JOHNS 1936	34 27 35.18	109 11 35.86	-4.160	5.410	3.06
AZ	STEVE 1960	32 27 37.91	112 40 46.96	-3.620	13.680	7.35
AZ	STONE TANK 1936	31 54 21.67	112 22 56.11	-2.420	6.840	3.61
AZ	STOVAL 1934	32 45 29.51	113 38 22.25	-1.840	6.060	3.28
AZ	SUNDT 1970	32 7 7.00	110 46 17.53	-1.970	15.180	8.07
AZ	SUNSET EAST BASE 1945	32 33 27.28	110 6 37.06	-1.770	4.590	2.47
AZ	SUNSET WEST BASE 1945	32 33 26.21	110 12 32.65	-1.390	3.850	2.08
AZ	SUPERIOR 1946	33 18 20.65	111 6 30.17	-7.770	16.280	8.94
AZ	TANK 1958	34 16 46.14	109 43 8.20	4.770	7.030	3.98
AZ	TELEPHONE 1953	36 32 45.84	112 10 24.56	2.360	-1.460	-0.87
AZ	TELL 1959	34 45 4.08	111 40 5.48	-6.920	19.000	10.83
AZ	TENDERFOOT 1934	36 11 7.04	112 49 17.59	-2.580	9.530	5.63
AZ	TESHBITA 1951	35 16 7.80	110 4 13.43	-6.380	4.380	2.52
AZ	THEBA 2 1971	32 56 12.33	112 53 10.42	-1.130	4.530	2.68
AZ	THOMAS 1945	33 2 52.82	109 59 11.97	-5.520	8.490	4.63
AZ	TOOFER 1935	33 12 5.33	114 40 12.03	-1.920	8.860	4.85
AZ	TRACKS 1936	34 59 30.45	111 28 23.63	3.890	1.750	1.01
AZ	TRAIL 1949	33 9 27.48	110 56 5.79	-6.100	11.070	6.06
AZ	TRENCH 1953	36 35 11.33	112 52 23.08	-1.130	2.720	1.63
AZ	TRUCK 1960	32 54 34.01	112 35 54.62	-0.420	10.830	5.88
AZ	TUBA CITY 1951	36 8 0.83	111 13 23.04	-5.010	6.880	4.06
AZ	VAIL 1935	32 2 51.67	110 45 2.71	0.460	15.260	8.10
AZ	VALLEY 1948	33 45 8.56	112 51 54.13	-5.500	5.280	3.01
AZ	WALAPAI NORTHEAST BASE 1947	35 21 0.77	113 52 33.53	-0.370	13.670	7.91
AZ	WALAPAI SOUTHWEST BASE 1947	35 16 0.52	113 57 32.56	-0.890	7.500	4.33
AZ	WASP 1953	36 29 58.08	113 26 32.88	0.020	8.420	5.01
AZ	WELT 1960	32 40 21.49	114 8 59.86	0.110	7.940	4.29
AZ	WHITETANK 1910	33 31 1.65	112 33 27.50	-4.010	******	******
AZ	WHITMORE 1953	36 10 54.08	113 15 38.34	2.260	6.160	1.34
AZ	WILDCAT 1946	33 50 54.45	111 50 6.92	-11.980	3.190	3.45
AZ	WILLIAMS LAPLACE 1920	35 15 8.22	112 11 24.48	-0.680	11.910	6.87
AZ	WINSLOW AIRPORT BEACON	35 1 19.09	110 42 54.27	1.270	-2.000	1.15
AZ	WOODLAND 1936	34 59 56.65	111 16 59.58	-5.940	-5.990	-3.44
AZ	WRIGHT 1953	36 59 12.02	113 17 49.65	4.100	21.000	12.66
AZ	YA 2 1959	34 27 58.05	112 1 43.97	-6.100	7.190	4.07
AZ	YCLO 1944	34 44 42.10	112 56 2.39	-6.760	9.660	5.52
AZ	YOUNG 1938	34 6 5.36	110 57 46.22	-7.110	10.190	5.71
AZ	YUMA LATITUDE STATION 1911	32 43 37.26	114 37 19.48	-2.510	4.860	2.63
AZ	ZENIFF 1936	34 41 13.13	110 28 13.78	-2.600	3.570	2.03
AZ	ZIBI DUSH 1951	36 13 37.34	109 27 16.55	-10.590	8.150	4.82
CA	ABALONE 2 1933	34 27 37.91	120 27 18.76	-4.910	7.870	4.48
CA	AC 1962	33 55 52.41	118 23 39.87	-3.840	6.620	3.69
CA	ACADEMY 1950	36 52 53.93	119 33 22.49	-3.870	18.980	11.39
CA	ADOBE 1932	35 39 7.81	121 13 31.20	-12.600	14.350	8.36
CA	AIRPORT BN APT SAN LUIS OBISPO	35 14 23.22	120 38 19.22	-6.360	10.680	6.16
CA	AIRPORT 1952	35 20 53.61	118 5 24.69	-6.380	9.530	5.36
CA	ALAMO 1925	34 41 54.85	120 15 20.86	-4.090	9.390	5.36
CA	ALMOND 1932	35 33 6.46	120 27 7.50	-0.580	6.050	3.52

STATE	STATION NAME	GLAT(D,M,S)	GLON(D,M,S)	ALAT-GLAT(S)	ALON-GLON(S)	LAPLACE(S)
CA	ALICS 1957	37 22 33.32	122 6 58.54	1.140	2.190	1.33
CA	AMADOR 1958	36 16 52.77	120 15 50.27	-1.080	-5.590	-3.30
CA	AMERICAN 1947	37 17 15.84	121 51 53.43	-0.450	7.890	4.78
CA	ANTONIO 1923	34 14 54.53	117 40 27.27	-23.770	********	******
CA	APPLE 1965	34 31 4.52	117 13 5.37	-0.580	9.000	5.10
CA	ARGUELLO 1875	34 34 58.03	120 33 37.64	-4.160	********	******
CA	ARGUELLO 2 1959	34 34 58.02	120 33 37.65	-4.670	11.530	6.55
CA	ARMADA 1929	33 52 14.89	117 12 5.56	-2.030	10.360	5.77
CA	ASH 1934	36 23 26.48	118 1 30.69	-1.780	-29.410	-17.45
CA	ATE 1964	34 57 26.06	118 51 39.22	14.850	12.740	7.30
CA	ATOLIA 1933	35 18 47.39	117 36 30.54	-6.620	0.640	0.37
CA	AUTONETICS A 1962	33 51 19.79	117 51 15.84	-2.720	13.870	7.73
CA	AUTONETICS B 1962	33 51 29.04	117 50 39.27	-2.390	14.160	7.89
CA	BAILEY 1971	35 1 54.87	122 3 21.88	-3.270	7.440	4.69
CA	BALCH 1934	35 1 20.66	116 0 5.55	2.340	7.180	4.12
CA	BALD MTN 1939 RESET 1974	39 57 9.35	121 28 53.16	-13.830	23.760	15.26
CA	BALD MTN 2 USE 1956	35 51 48.08	121 2 23.76	-1.140	0.790	0.46
CA	BARSTOW 2 1965	35 5 26.79	116 56 18.20	-3.500	-0.150	-0.08
CA	BASS 1939	40 43 58.62	122 21 57.57	-8.230	0.750	0.49
CA	BAY CITY 1943	37 13 54.24	120 54 13.08	-5.760	6.400	3.87
CA	BDRY MON 120 CA NV 1934	35 47 16.97	115 37 13.19	-5.350	********	******
CA	BDRY MON 129 CA NV 1898	35 26 39.80	115 11 5.20	-3.480	********	******
CA	BDRY MON 2 CA NV 1894	38 57 33.70	119 56 28.10	4.050	7.050	4.43
CA	BDRY MON 28 CA NV 1894	38 37 44.22	119 27 45.74	2.690	********	******
CA	BDRY MON 41 CA NV 1899	38 16 22.99	118 57 19.57	2.690	********	******
CA	BDRY MON 74 CA NV 1895	37 29 45.44	117 52 29.43	-4.580	********	******
CA	BDRY MON 88 CA NV 1895	37 3 16.26	117 16 36.50	-3.850	********	******
CA	BDRY MON 97 CA NV 1895	36 36 39.68	116 41 13.75	-0.740	********	******
CA	BEACH 1951	34 6 25.75	119 8 20.55	-8.900	11.180	6.27
CA	BENCH MARK J 162 1965	34 58 57.91	116 11 23.28	-2.720	5.390	3.09
CA	BENCH MARK N 166 1948	36 17 16.72	117 36 53.61	-4.300	-2.570	-1.52
CA	BENCH MARK S 672 1950	35 53 58.56	116 39 19.32	-11.240	14.780	8.66
CA	BERKLEY 1965	34 30 22.87	117 35 0.82	4.700	1.760	1.00
CA	BETA 1959	33 14 48.89	119 31 9.01	-7.590	6.830	3.75
CA	BICDLE 1932	35 9 40.52	120 28 26.26	-6.510	10.000	5.76
CA	BILL 1963	35 15 25.53	116 48 1.71	-1.770	0.450	0.26
CA	BITTER 1538	35 0 3.65	119 27 55.61	11.380	2.790	1.60
CA	BLACK FOX 1920	41 20 47.54	121 53 24.87	-7.410	12.010	7.94
CA	BLACK OAK 1932	35 9 42.92	118 41 47.04	4.300	28.250	16.27
CA	BLACK 1929	33 48 22.48	116 42 4.73	4.050	18.810	10.47
CA	BLACK 1933	35 3 44.45	120 36 14.22	-4.970	8.750	5.03
CA	BLANK 1958	35 42 47.90	119 43 4.58	-1.710	6.160	3.60
CA	BLUE 1962	36 17 12.92	118 50 19.63	-14.200	30.280	17.92
CA	BLUFF 1893	34 54 58.90	114 38 1.23	0.0	-0.600	-0.34
CA	BODEGA 1855	38 18 22.57	123 2.45	-2.360	********	******
CA	BORCN 1558	30 52.75	117 39 35.60	-2.430	6.590	3.78
CA	BOUCHER 2 1975	33 20 4.86	116 55 6.23	-7.830	21.060	11.57
CA	BOUNDARY MON 2C8 US MEX	32 42 55.90	114 45 58.81	-5.190	8.290	4.48
CA	BOUNDARY MON 254 ECC	32 32 41.98	116 59 59.06	-5.370	16.460	8.85
CA	BOUNDARY MON 59 CA NV 1894	37 54 15.54	118 26 17.66	-1.370	********	******
CA	BOWL 1934	34 2 51.15	120 25 1.41	-1.330	10.530	5.50
CA	BRENTWOOD NO 2 LAC C 1933	34 2 28.74	118 28 28.95	-17.620	3.920	2.20
CA	BRIDGE 1935	34 43 40.76	115 29.49	-3.800	5.920	3.37
CA	BROWNING 1933	33 37 7.01	117 50 42.71	-2.270	17.190	9.51
CA	BUCK USGS 1949	39 56 17.66	121 5 4.36	-5.490	22.190	14.24
CA	BUENAVISTA 1870	34 3 17.87	118 14 33.12	********	7.840	4.39
CA	BUTE 1910	33 33 40.69	117 20 38.31	-1.700	6.020	3.59
CA	CALAVERAS 1931 RM 6 1971	36 31 50.23	120 16 45.00	-1.890	0.300	0.16
CA	CALEXICO 1934 RM 5 1954	32 39 51.98	115 30 20.16	0.550	-6.340	-4.03
CA	CALIFORNIA NEV IRON MON 1897	39 31 28.48	120 0 2.93	-1.480	********	******
CA	CAMPBELL 2 1972	34 9 22.62	116 1 11.06	2.740	-1.190	0.67
CA	CANTIL 1952	35 16 51.68	117 59 5.80	-5.270	-2.560	-1.71
CA	CAR 1972	34 14 31.82	116 31 55.92	0.520	********	-1.97
CA	CASTLE MOUNT 1885	35 56 20.35	120 20 21.67	-0.560	********	******
CA	CASTRO 1898 RM 3	34 5 8.76	118 47 5.87	-11.900	6.920	3.88
CA	CASTRO 2 1967	36 57 57.75	121 33 40.79	-5.170	0.250	0.15
CA	CATTLE 1932	35 58 15.75	121 27 7.28	-15.220	31.650	18.59
CA	CEMETERY BLUFF 1887	32 41 23.55	117 14 43.74	-5.740	12.230	6.61
CA	CENTER 1926	34 59 27.86	119 55 35.29	-2.500	5.490	3.14
CA	CENTER 1972	34 3 20.35	115 11 26.57	-0.540	5.000	2.80
CA	CHAFFEE 2 1923	34 18 2.20	119 19 47.98	-14.680	9.080	5.12
CA	CHALK TPC 1971	37 9 36.66	122 18 9.22	-11.810	19.550	11.81
CA	CHEWS 1946	36 18 43.90	121 34 4.71	2.790	4.600	2.73
CA	CHICA 1933	33 42 37.18	118 2 55.61	1.720	7.020	7.02
CA	CHINA 1932	35 43 36.08	121 18 38.83	-12.020	22.260	13.00
CA	CHIPS 1948	36 9 29.83	117 30 24.12	0.850	9.560	5.88
CA	CLAREMONT USGS 1949	39 53 7.44	120 55 47.93	5.350	3.620	2.32
CA	CLEMENTE 1933	33 25 52.44	117 35 48.30	-6.790	15.650	8.62
CA	CLIFF 1933	34 53 31.46	120 38 26.55	-2.520	12.560	7.15
CA	COFFIN 1947	37 23 19.01	121 44 43.61	-1.110	4.430	2.49
CA	COHASSET 1939	39 53 0.65	121 44 43.61	-5.390	19.440	12.67
CA	CONCEPTION 1933	34 26 59.68	120 28 10.94	-11.680	10.800	6.11
CA	COOKE 1931	37 3 40.49	122 14 11.26	-16.490	23.940	14.42
CA	CORNING TOWER	39 55 39.74	122 10 42.88	-1.230	********	******
CA	CORNING 1939 RM 3 1971	39 55 53.75	122 9 43.34	-0.070	15.660	10.05
CA	CORRAL 1943	37 37 34.21	121 27 13.32	7.840	-4.520	-2.76
CA	CORT 1971	35 13 33.07	122 11 22.97	-4.830	5.110	3.23
CA	COWLES 1939	32 48 45.96	117 1 51.63	-6.270	17.030	9.23
CA	COXCOMB WEST BASE MWD 1931	33 56 6.72	115 16 36.21	-4.100	-1.960	-1.10
CA	CREEK 1885	35 33 51.08	121 6 26.19	-9.590	14.090	8.19
CA	CROSS 1944	36 37 16.35	121 41 23.65	-3.050	10.790	6.64
CA	CUATE 1933	33 21 16.12	117 31 55.56	-11.830	19.450	10.70
CA	CUT TPC 1971	35 15 9.32	120 45 46.97	-6.410	10.150	5.86
CA	CUYAMACA 1898	32 56 48.10	116 26 21.13	-2.580	11.880	6.46
CA	DALE 1934	34 9 43.82	115 43 5.93	-5.330	4.620	2.59
CA	DAM USE 1957	34 54 16.16	117 37.60	1.800	4.870	2.84
CA	DANCE 1933	35 19 52.51	120 50 32.34	-5.830	16.040	9.27
CA	DAYTON HARRIS GRAVITY STA 1950	36 12 24.94	116 52 8.59	-4.240	-3.910	-2.31
CA	DEER 1932	35 5 8.83	118 30 24.56	11.200	12.830	7.38
CA	DELANO 2 1960	35 46 7.67	119 14 57.62	-4.350	15.260	8.92
CA	DELILAH 1950	36 47 20.20	119 7 3.98	-8.010	22.550	13.51
CA	DENIS 1960	34 37 22.68	118 8 33.10	5.110	0.500	0.28
CA	DIMAS 1922	34 6 49.80	117 49 3.20	-16.320	********	******
CA	DOME 1942	35 16 34.74	119 36 5.10	7.340	-0.530	-0.30
CA	DOMINGUEZ HILL 1870	33 51 54.93	118 14 10.34	0.260	********	******
CA	DOS PALOS 1931	36 41 41.29	120 38 54.37	-3.630	6.540	4.18
CA	DRAIN 1958	35 34 25.82	119 57 7.49	-1.600	8.730	5.08
CA	DUMONT 1952	35 40 47.08	116 9 42.67	-5.960	12.500	7.29
CA	ECHO 1963	35 17 58.22	118 48 18.24	-3.170	0.790	0.45
CA	EDWARDS 1932	34 27 3.49	119 59 9.73	-18.280	11.130	6.30
CA	EGDIR-EGDIR 2 1933	34 38 8.69	120 18 18.14	-1.020	12.620	7.17
CA	ELLIS 2 1971	36 24 1.98	120 11 36.23	-3.020	4.790	2.85
CA	ELSINORE 1939	33 7 18.86	117 29 13.90	1.130	2.780	1.54
CA	ESCONDIDO 1960	33 8 44.86	117 5 13.90	-1.640	16.780	9.17
CA	FIBRE 1958 RM 3 1971	35 23 17.83	119 23 52.18	-1.480	11.380	6.59
CA	FIELD 1913	36 13 0.17	118 45 28.22	-25.730	37.420	22.11
CA	FIGUEROA 1941	34 44 36.50	119 59 2.27	-8.060	20.910	11.92
CA	FILIPPONI TPC 1971	35 28 5.59	120 56 23.21	-9.640	12.140	7.05
CA	FISH 1971	37 47 1.91	118 32 29.57	-2.600	1.980	1.22
CA	FISHER 1891	39 3 58.56	123 35 10.57	-3.840	15.990	10.08
CA	FLASH 1940	34 49 26.28	117 0 56.96	1.900	5.320	3.04
CA	FLYNN 1934	34 59 28.32	115 45 11.70	-8.120	11.580	6.64

STATE	STATION NAME	GLAT(D,M,S)	GLON(D,M,S)	ALAT-GLAT(S)	ALON-GLON(S)	LAPLACE(S)
CA	FRAZIER 1941	34 46 29.90	118 58 5.43	-4.050	-3.120	-1.78
CA	GARRISON 1944	35 50 15.03	120 45 0.74	-3.050	8.650	5.07
CA	GAVIOTA 1873	34 30 6.54	120 11 52.05	-11.550	10.860	6.15
CA	GAZELLE ASTRONOMICAL STA 1904	41 31 35.19	122 31 7.12	6.070	-9.080	-6.02
CA	GIANT 1965	34 13 22.90	116 11 8.53	0.330	-1.990	-1.12
CA	GILROY 1930	36 58 47.40	121 36 54.85	-6.380	3.280	1.97
CA	GLO 9 1926 RM 1	35 27 39.37	118 54 4.05	-6.450	18.760	10.88
CA	GLO 1932	34 57 40.45	118 16 55.38	-7.030	-0.240	-0.74
CA	GORDA 2 CASLC 1959	34 21 20.04	119 26 28.97	-19.650	11.880	6.71
CA	GORGE 1932	35 2 53.03	118 45 55.72	-4.100	25.070	14.40
CA	GOVERNMENT ROAD 1934	35 6 46.50	114 49 41.71	-5.830	5.730	3.30
CA	GREENFIELD 1944	36 18 58.35	121 14 18.08	-2.440	3.930	2.33
CA	GRENADA NORTH BASE 1952	41 35 0.07	122 31 10.57	2.040	-2.020	-1.34
CA	GRENADA SOUTH BASE 1952	41 33 18.27	122 31 29.06	3.140	-6.220	-4.12
CA	GRINELL USGS LA CO 1940	34 45 59.82	117 49 6.52	-0.850	6.330	4.75
CA	GRIZZLY 1920	41 8 41.25	121 58 38.52	-10.460	6.770	4.46
CA	HANS 1958	34 59 8.20	118 1 55.35	2.320	4.740	2.72
CA	HAPPY 1959	34 21 28.67	118 50 57.12	-5.070	11.340	6.40
CA	HARBOR 1860	32 59 54.06	118 33 39.44	3.550	******	******
CA	HATCH 1957	37 7 46.86	117 36 7.29	-3.140	5.700	3.44
CA	HAWES 1958	34 56 53.49	118 38 13.80	0.130	1.860	1.07
CA	HAYWARD 2 1957	37 38 42.30	122 9 15.76	-2.680	8.050	4.91
CA	HEPSEDAM 1885	34 48 52.55	120 45 25.05	-10.170	******	******
CA	HINKLEY 1958	34 54 59.69	117 14 3.45	-0.770	9.340	5.35
CA	HIRZ 2 USGS RM 3 1971	40 53 49.62	122 14 40.84	-9.530	-0.440	-0.29
CA	HOLDE 1949	36 38 16.51	117 15 10.0	-10.100	11.590	6.92
CA	HORSE MOUNTAIN 1949	39 18 24.09	122 58 15.04	-6.100	14.680	9.30
CA	HOT SPRING 1935	32 37 2.87	116 10 48.46	4.810	0.130	0.07
CA	HUCKLEBERRY HILL 2 1932	36 30 43.72	121 55 10.61	3.870	28.540	16.98
CA	INDEPENDENCE SE BASE 1933	36 42 29.69	118 7 47.15	-13.370	-1.790	-1.99
CA	ISLAND 1958	35 8 4.00	119 4 24.77	1.100	9.990	5.75
CA	JAYNE 2 1971	36 16.86	120 9 32.48	-0.020	-0.770	-0.45
CA	JENSEN 1943	37 16 9.03	121 7 46.67	-0.080	-6.130	-3.71
CA	JPL ARIES 1 1975	34 12 16.96	118 10 12.15	-22.100	13.850	7.79
CA	JUNCTION 1934	42 32.90	115 3 40.07	-2.660	5.110	2.76
CA	KING CITY N BASE 1944	36 17 9.77	121 9 3.35	-5.350	9.600	5.68
CA	KING 1935	35 1 20.24	122 10 22.93	-6.840	-0.950	-0.55
CA	KIRKWOOD 1971	35 50 41.06	119 9 40.70	-0.680	14.900	9.55
CA	KIT 1958	35 12 48.08	118 44 3.46	-1.400	8.770	5.06
CA	KLEIN 1952	41 21 1.07	123 58 25.17	-8.900	21.320	12.49
CA	KNEELAND USGS 1941	40 43 36.37	124 54 36.68	-2.310	24.460	15.96
CA	KNOB 1972	34 2 26.94	116 2 11.59	-0.100	5.780	3.24
CA	LA NW BASE AUX 3 LAC	33 56 59.57	118 3 14.37	-6.530	12.280	6.85
CA	LA POSTA USN 1971	32 40 55.85	116 26 4.37	-5.450	-0.650	-0.35
CA	LAB 1960	34 12 10.87	118 10 37.09	-21.860	13.720	7.71
CA	LAGUNA 2 1951	34 6 30.76	119 3 51.14	-10.560	15.230	8.54
CA	LAMB 1934	34 1 18.35	120 18 56.63	-2.880	5.730	3.21
CA	LARD MWD 1931	33 56 40.63	116 48 24.63	-13.920	6.750	3.77
CA	LEMCCRE 1944	36 18 14.83	119 46 52.58	-5.370	10.480	6.20
CA	LEON 1931	36 56 46.12	121 52 22.31	-11.110	9.410	5.66
CA	LEROO 1958	35 29 49.93	119 31 44.61	-1.860	10.390	6.03
CA	LIVERMORE EAST BASE 1947	37 40 41.70	121 47 21.05	0.540	10.180	6.22
CA	LIVERMORE WEST BASE 1947	37 40 14.26	121 51 28.10	-1.550	7.030	4.29
CA	LODGE 1952	35 41 52.71	118 28.34	-2.840	0.600	0.35
CA	LGNE TREE 1930	36 2 17.16	121 17 36.23	-2.840	16.390	9.85
CA	LOOK TPC 1971	35 40 31.11	121 3 9.94	-5.370	5.760	3.36
CA	LOS ANGELES NORMAL SCHOOL	34 3 1.53	118 15 13.42	-17.270	8.360	4.68
CA	LOS ANGELES NW BASE 1889	33 55 4.95	118 3 22.50	-6.880	11.550	6.47
CA	LOS ANGELES SALT LAKE AY BN 15	35 4 24.83	118 23 11.72	0.650	4.140	2.38
CA	LOS BANOS 1943 RM 3 1971	37 3 25.78	120 62 48.27	-1.730	2.030	1.22
CA	LOSPE 1875	34 53 37.53	120 36 18.60	-4.240	7.620	4.35
CA	LUCERNE SOUTH BASE 1928	34 26 22.18	116 52 51.84	7.100	4.630	2.62
CA	MACOCEL 1948	41 49 49.79	122 0 11.54	-2.360	6.050	4.04
CA	MADERA 1931	36 57 27.98	120 3 16.06	-1.450	7.850	4.72
CA	MAGIC 1958	34 23 10.26	118 19 2.70	-1.760	18.680	10.55
CA	MAGNETIC 1908	35 16 27.46	119 16 17.58	-0.260	******	******
CA	MAIN 1958	35 18 44.30	122 9 37.45	-3.830	8.750	5.06
CA	MAJORS TPC 1971	36 58 46.68	121 11.17	-17.930	21.190	12.75
CA	MANIX 1952	35 0 39.21	116 31 49.32	-5.250	4.950	2.62
CA	MARE 1928	33 39 30.48	117 52 46.03	-6.330	15.720	8.72
CA	MARS 1963	35 25 39.62	120 33 46.84	-1.720	5.850	3.39
CA	MARSHALL 1933	34 44 19.41	115 53 19.03	-1.220	7.130	4.15
CA	MASTER CAMERA STATION 1967	32 55 44.25	115 41 58.62	-2.820	-0.300	-0.16
CA	MAYHOOD 1971	38 8 32.16	121 42 21.81	-5.330	5.330	3.00
CA	MC CLURE SOUTHWEST 1932	35 47 19.67	120 6 41.50	3.460	-11.480	-6.71
CA	MCPHERSON RM 3 TPC 1971	34 53 19.44	118 34 3.95	4.120	3.720	2.13
CA	MELLC 3 1967	32 47 55.03	115 37 52.26	-0.920	3.240	1.75
CA	MENDCIA 1944	36 41 26.80	120 21 4.52	-3.480	8.070	4.93
CA	MERRITT 1944	36 2 48.42	121 7 16.87	-2.300	4.310	2.53
CA	MESA USGS-YCLO COUNTY 1939	38 45 11.44	121 53 22.08	-6.440	7.640	4.64
CA	MESQUITE 1950 AZ MK	36 59 52.03	117 21 57.14	-4.310	7.400	4.36
CA	MIDDLE 1972	34 7 7.37	115 28 24.64	-1.620	1.450	0.82
CA	MIDWAY 1965	34 54.98	115 42 2.17	-2.130	-0.030	-0.01
CA	MILLER 1931	36 48 24.92	120 26 22.04	-3.980	7.390	4.43
CA	MINDEGO 1947	37 19 12.40	122 11 55.02	-1.380	10.620	6.57
CA	MINER 1931	38 14 58.33	121 39 19.14	-6.870	7.160	4.17
CA	MIRAGE 1950	35 36 5.14	120 35 15.47	-3.180	5.390	3.17
CA	MITH 1944	36 4 44.12	120 35 41.95	-1.670	6.380	3.81
CA	MOCHO 1887	37 28 30.61	121 33 17.53	-2.420	4.090	2.30
CA	MOLE MWD 1931	34 7 23.45	114 40 31.20	-3.430	1.690	1.17
CA	MONTICELLO 1880	38 39 49.55	122 11 21.12	-5.300	3.770	2.16
CA	MONUMENT USGS 1963	35 22 46.46	116 57 54.08	-1.820	11.710	7.00
CA	MORAN 1944	36 40 19.21	119 53 39.17	-4.730	******	******
CA	MOUNT CONNESS 1890	37 58 54.99	119 19 3.07	-8.420	4.610	2.83
CA	MOUNT DIABLO 1876	37 52 46.03	121 54 47.11	-1.230	13.860	8.66
CA	MOUNT HELENA 1876	38 40 29.59	122 37 56.60	2.240	-6.750	-4.11
CA	MOUNT LOLA 1879	39 25 59.01	120 21 50.48	-7.340	8.910	5.09
CA	MOUNT OSO 1931	37 30 30.32	121 22 26.02	******	******	******
CA	MOUNT PINOS 1941	34 48 46.29	119 5 40.03	6.800	5.390	3.31
CA	MOUNT TAMALPAIS 1882	37 55 24.70	122 35 43.73	6.800	6.380	3.80
CA	MOUNT TAMALPAIS 2 1951	36 55 24.79	122 35 41.73	-7.410	19.010	10.42
CA	MOUNT TORO 1885	36 31 33.66	121 36 31.02	-6.920	6.940	4.13
CA	MOUNT 1933	33 11 27.12	117 20 16.30	1.760	10.930	6.52
CA	MT TCRO RM 4 TPC 1971	36 31 33.93	121 36 30.68	-25.400	9.980	5.61
CA	MT WILSON ASTRONOMICAL STA	34 13 2.26	118 3 42.06	-25.240	11.570	6.51
CA	MTW E-10A LA CO 1971	34 38 2.84	119 17 58.48	-8.980	12.700	7.22
CA	MLASON TPC 1971	38 1 44.36	122 11 47.34	-1.660	6.820	4.20
CA	NADEEN 1955	34 26 21.36	116 44 42.72	-4.430	0.650	0.37
CA	NAKE 1972	38 21 30.84	121 14 14.40	-5.160	5.330	3.31
CA	NAVY 1971	34 50 12.85	114 36 14.95	-2.160	-2.100	-1.20
CA	NEEDLES LONGITUDE STA 1889	34 50 22.85	120 23 8.03	-1.100	******	******
CA	NEW SAN MIGUEL 1873	37 20 13.21	121 1 41.86	-3.550	3.420	2.07
CA	NEWMAN N BASE RM 4 1971	37 15 15.49	120 59 43.86	-3.120	14.250	1.72
CA	NEWMAN SE BASE RM 4 1971	33 30 44.83	117 43 59.82	-4.930	15.140	9.79
CA	NIGUEL 1884	40 47 29.14	117 42 21.06	-6.800	17.190	9.57
CA	NINE MILE 1939	40 47 29.14	117 42 21.06	-6.800	17.190	9.57
CA	NCHL 1958	36 57 19.20	115 14 28.23	-25.120	10.750	6.09
CA	NORDHOFF RM 4 TPC 1971	41 0 58.12	122 6 34.86	-11.780	6.440	4.02
CA	NORTH FORK MTN LSGS 1953	33 16 28.79	117 26 17.49	-10.620	16.440	10.55
CA	OCEANSIDE NORTH BASE 1933	32 40 41.35	115 18 44.50	-0.430	1.930	1.04
CA	OFFSET 217 1934	32 38 52.94	115 43 28.61	4.320	-0.630	-0.34
CA	OFFSET 225 1934					

STATE	STATION NAME	GLAT(D,M,S)	GLON(D,M,S)	ALAT-GLAT(S)	ALON-GLON(S)	LAPLACE(S)
CA	OFFSET 256 1975	32 32 17.39	117 4 41.61	-1.860	15.930	8.57
CA	OILER USGS-RANKIN USGS 1952	35 21 10.55	118 35 11.54	-8.790	18.070	10.46
CA	ORLAND SOUTH BASE 1939	39 46 6.95	122 11 28.42	-1.510	14.990	9.59
CA	ORLAND 1939	39 44 44.60	122 9 2.16	-1.330	14.920	9.54
CA	ORR MOUNTAIN 1948	41 40 3.52	121 58 24.11	0.590	4.740	3.15
CA	OWL 1952	35 36 43.91	116 41 2.71	-2.420	3.110	1.81
CA	OXALIS 2 1971	36 54 51.04	120 33 11.01	-4.190	8.080	4.85
CA	OXNARD WEST BASE 1951	34 11 49.72	119 9 52.63	-2.530	11.400	6.41
CA	PAHRUMP SOUTHEAST BASE 1926	36 0 51.59	115 54 53.68	-7.210	12.290	7.23
CA	PALO CEDRO 1939	40 33 58.86	122 17 52.04	-7.140	5.200	3.38
CA	PALOS VERDES ARIES3 1975	33 44 35.95	118 24 13.71	-13.660	21.000	11.66
CA	PANOCHE MTN USGS 1962	36 43 32.07	120 45 49.71	6.170	-10.000	-5.98
CA	PARACISE TPC 1971	36 49 2.28	121 42 24.93	-7.750	15.860	9.51
CA	PARAL 2 1969	35 47 22.45	119 47 0.07	-3.590	3.750	2.19
CA	PASADENA EAST BASE 2 1922	34 7 14.10	117 41 16.26	-19.860	8.530	4.78
CA	PASADENA WEST BASE 1922	34 7 49.84	118 8.50	-18.520	8.470	4.75
CA	PASQUAL 1939	33 7 3.19	116 53 3.09	-7.550	14.250	7.78
CA	PAUMA 1939	33 17 45.45	117 2 43.86	-3.900	19.680	10.75
CA	PELONA 1932	34 33 39.52	118 21 18.45	-7.350	11.940	6.77
CA	PETALUMA 1930	38 20 45.26	122 34 43.75	-3.750	7.680	4.77
CA	PILLAR POINT 3 CASLC 1962	37 29 53.44	122 29 50.25	-7.490	20.480	12.47
CA	PINE HILL 1876	38 43 10.06	120 59 21.77	-5.370	*******	*******
CA	PINTO USGS 1950	36 23 1.63	117 13 31.86	-1.460	7.210	4.27
CA	PIONEER 1963	35 23 20.97	116 50 51.07	-1.270	7.230	4.18
CA	POINT A NEAR AEROSPACE 1962	33 54 52.54	118 22 28.09	-3.380	6.750	3.77
CA	POINT ARENA LATITUDE STA 1870	38 55 17.90	123 43 35.73	-7.560	*******	*******
CA	POINT ARENA LONGITUDE STA 1889	38 54 35.13	123 41 23.25	*******	23.190	14.56
CA	POINT DELGADA 2 1973	40 1 18.61	124 4 3.46	-4.970	15.560	10.01
CA	POINT PINOS LAT STA 1851	36 37 58.18	121 55 30.35	-1.300	*******	*******
CA	POINT PINOS LAT STA 2 1930	36 38 6.86	121 55 29.02	1.730	16.070	9.59
CA	PORTAIR 1962	36 1 49.17	119 3 27.93	-5.440	15.260	8.98
CA	POWAY 3 CADH 1964	32 56 26.14	116 59 4.65	-5.560	16.970	9.23
CA	PROBERTA 1939	40 4 5.40	122 11 21.12	-1.400	15.670	10.09
CA	QUINCY STA 7051 1974	39 58 24.94	120 56 18.43	-2.530	10.040	6.46
CA	R 2 OFFSET 1951	33 14 5.48	115 52 9.42	-2.410	0.470	0.26
CA	RAMONA 1939	33 1 29.29	116 49 38.37	-6.080	14.800	8.07
CA	RANCHO 1972	34 25 47.19	117 7 55.68	9.860	8.850	5.00
CA	RATEL 1961	37 23 47.87	122 11 19.69	-2.720	1.160	0.70
CA	RED ROCK 1932	34 41 28.44	120 15 57.81	-1.770	8.310	4.73
CA	REDDING ASTRONOMICAL STA 1908	40 34 18.30	122 23 37.50	-10.360	*******	*******
CA	RES K-9A LA CC 1971	34 7 43.44	118 30 43.02	-4.510	3.830	2.15
CA	RIO 1933	34 55 57.19	117 27 33.40	-10.030	10.170	5.83
CA	RIZ 1959 RM 3 1971	39 27 3.99	122 11 27.48	-6.010	11.170	7.10
CA	ROAD 1931	37 14 12.47	122 24 9.35	-6.370	16.750	10.14
CA	ROAD 1951	34 15 10.47	119 30 45.85	-6.080	5.940	3.25
CA	ROCKY BUTTE-ROCKY BUTTE 2 1884	35 39 55.02	121 3 30.79	-10.900	*******	*******
CA	ROCA 1959	34 40 8.58	120 34 50.60	-2.320	7.790	4.43
CA	ROGERS 1958	35 0 17.82	117 48 36.29	-3.250	5.200	2.98
CA	RON 1964	33 17.08	118 59 44.72	18.170	7.770	4.46
CA	ROSA 1933	34 34 11.85	120 1 1 8.59	-1.900	5.460	3.10
CA	ROSS MOUNTAIN 1851	38 30 19.43	123 7 7.98	-5.430	12.530	7.80
CA	ROUND TOP 1879	38 39 49.29	119 59 59.99	-2.390	*******	*******
CA	ROUND USE 1952	35 38 29.64	118 28 34.27	-3.900	6.480	3.78
CA	RP 1 USFS 1956	33 42 27.78	117 32 0.01	-4.020	11.320	6.28
CA	SACRAMENTO STATE CAPITOL DOME	38 34 36.26	121 29 33.36	*******	6.760	4.22
CA	SADDLE PEAK ARIES 3 1975	34 4 41.50	118 39 18.40	-15.250	-6.070	-3.40
CA	SALINE 1950	36 45 35.41	117 35 35.84	-2.560	-0.210	-5.51
CA	SAN BENITO 1931	36 22 11.04	120 38 36.80	2.640	0.390	0.24
CA	SAN DIEGO LATITUDE STA 1851	32 42 3.35	117 14 29.74	-5.560	10.160	5.49
CA	SAN DIEGO LATITUDE STA 1892	32 43 29.85	117 9 27.22	-9.090	14.210	7.69
CA	SAN JACINTO 1858	33 48 52.91	116 40 42.92	13.560	*******	*******
CA	SAN LUCAS 1944	36 7 4.21	121 4 13.60	-5.470	-4.770	2.81
CA	SAN PEDRO LATITUDE STA 1852	33 43 20.93	118 17 0.84	-1.530	*******	*******
CA	SAN PEDRO 3 1974	33 44 48.56	118 20 3.95	-5.610	9.400	5.22
CA	SANDY 1 1959	36 14 36.00	115 15 51.15	-5.570	8.410	4.74
CA	SANTA ANA 1852	36 54 18.29	117 13 56.47	-2.480	*******	*******
CA	SANTA BARBARA ISLAND 2 1940	33 28 19.35	119 2 25.18	0.200	3.350	1.84
CA	SANTA BARBARA 1857	34 24 16.72	119 42 51.74	-17.490	*******	*******
CA	SANTA BARBARA 2 1956	34 24 14.93	119 42 54.68	-17.500	8.100	4.58
CA	SANTA CATALINA IS LAT STA	33 26 29.13	118 29 48.91	6.060	*******	*******
CA	SANTA CRUZ WEST 1874	34 4 23.16	125 1 3.35	4.780	*******	*******
CA	SANTA LUCIA 1885	36 8 44.30	121 26 4.68	-5.810	*******	*******
CA	SANTA MARIA MUNICIPAL TANK	34 56 47.73	120 26 1.38	-8.940	10.740	6.16
CA	SANTA MARIA 2 1934	34 58 3.06	120 38 5.55	-5.700	7.820	4.48
CA	SANTIAGO 1899	33 42 27.89	117 31 59.90	-3.460	11.620	6.44
CA	SANTIAGO 1899 RM 8 1975	33 42 37.01	117 32 0.01	-3.460	11.610	6.44
CA	SATELLITE TRI STA 111 1965	34 22 54.53	117 40 50.50	6.380	-0.820	-0.47
CA	SATELLITE TRI STA 134 1970	34 22 44.44	117 40 50.73	-2.560	-0.820	-0.47
CA	SAWMILL 1932	34 41 35.37	118 33 77.71	-2.560	-2.990	1.70
CA	SCOTT 1957	37 21 8.20	121 57 34.66	-0.660	6.220	3.78
CA	SEARLES 1926	35 30 27.91	117 39 15.98	-2.820	-0.230	-0.14
CA	SECO 1941	37 37 31.71	119 26 26.28	-5.530	12.030	6.84
CA	SEGO NORTHEAST BASE 1949	33 24 7.94	115 28 1.18	-5.720	11.390	7.54
CA	SHEPPARD 1927	32 37 59.39	115 54 4.84	-25.880	13.340	7.54
CA	SHOCKEY 1935	32 36 3.91	116 24 4.27	-0.390	-2.820	-1.52
CA	SHOSHONE 1952	35 58 21.34	116 16 11.51	-9.200	3.380	1.98
CA	SHRUB 2 1951	35 5 17.64	118 21 12.31	-4.640	-3.420	-1.97
CA	SIERRA MORENA 2 1963	37 24 37.23	118 26.73	-2.350	8.470	5.15
CA	SIGNAL HILL 2 1974	33 47 57.26	118 9 55.74	0.130	6.840	3.88
CA	SIMMLER 1926	35 21 5.58	120 0 15.03	0.400	4.090	2.36
CA	SOLEDAD 1872	32 57 3.08	120 6 16.89	-0.500	4.140	2.31
CA	SOLEDAD 1887	32 50 23.32	117 15 5.81	-2.020	16.980	9.21
CA	SOLEDAD 1944	36 25 24.53	121 19 23.61	-6.100	8.820	5.24
CA	SPENCE 1944	36 35 43.89	121 32 43.71	-1.450	12.330	7.35
CA	SPOONER 1939	39 34 56.08	122 8 30.85	-4.180	14.070	8.97
CA	SQUARE 1935	34 50 1.94	115 31.77	-3.120	0.820	0.47
CA	STANDARD 1935	35 22 24.02	115 35 55.18	-3.860	12.010	6.96
CA	STEPHENS BU 1971	41 30 51.56	121 52 51.02	-2.470	5.280	3.51
CA	STIL 1956	33 54 58.13	118 22 32.64	-3.660	6.920	3.86
CA	STOCKTON NORTHEAST BASE 1954	38 1 40.46	121 15 29.99	-5.560	9.280	5.72
CA	STOCKTON SOUTHWEST BASE 1954	37 59 31.65	121 18 35.51	-4.810	8.900	5.48
CA	STONE E 1967	36 38 1.48	121 14 2.37	-4.670	1.020	0.61
CA	STRECH 2 1970	35 58 9.41	119 56 50.83	-1.770	-0.470	-0.28
CA	SLB TPC 1970	34 34 58.50	120 33 37.42	-4.060	11.080	6.29
CA	SULPHUR PEAK 1860	38 45 43.15	122 50 39.01	-8.570	*******	*******
CA	SUNSET 1946	37 57 34.27	119 7 7.47	-0.330	3.650	2.25
CA	SUNSET 2 1971	37 57 21.52	121 38 56.66	-0.530	4.200	2.58
CA	SW 6 1952	36 55 51.44	118 13 53.18	-1.750	-9.230	-5.55
CA	TANK 1933	34 28 16.05	120 13 1.74	-16.450	7.870	4.46
CA	TARGET 1947	34 53 27.27	118 0 36.67	-6.810	3.040	1.74
CA	TECATE 2 1935	32 34 45.84	116 41 16.60	-3.700	13.300	7.16
CA	TEJON 1941	34 48 13.09	118 48 52.63	-6.740	-2.910	-1.66
CA	TEMBLOR 1926	35 6 46.09	119 33 14.54	8.810	-3.080	-1.77
CA	TEMPLETON 1944	35 33 28.66	120 42 0.77	-1.280	2.580	1.50
CA	TEPUSQUET RM 7 TPC 1971	34 54 36.41	120 11 8.25	-10.410	17.350	9.93
CA	TEPUSQUET 1875	34 54 36.52	120 11 8.34	-10.260	*******	*******
CA	TERRACE 1935	34 52 36.26	114 48 47.12	-2.030	-0.630	-1.16
CA	THEO TPC 1970	33 59 41.05	119 37 51.27	-10.890	-0.320	-0.18
CA	TOOMEY 1957	34 55 0.63	116 46 17.98	-3.860	-0.420	-0.24
CA	TOWN 2 1969	32 55 5.06	117 14 55.56	-4.920	18.940	10.30
CA	TRACY 1943 RM 4 1971	37 44 32.77	121 25 10.64	0.880	2.420	1.48
CA	TRANSIT 1893	38 57 15.68	119 56 39.90	4.080	6.280	3.95

STATE	STATION NAME	GLAT(D,M,S)	GLON(D,M,S)	ALAT-GLAT(S)	ALON-GLON(S)	LAPLACE(S)
CA	TRAVER 1931	36 27 12.25	119 25 12.73	-7.510	13.600	8.08
CA	TRIGGER 1935	34 56 29.95	115 10 13.82	-11.230	3.970	2.27
CA	TRONA 1950	35 48 44.21	117 19 42.49	-8.290	-1.010	-0.59
CA	TURK 1948 RM 1	37 48 21.97	122 27 10.14	-2.600	8.010	4.51
CA	TURQUOISE 1934	35 26 8.91	115 55 20.38	-6.320	10.490	6.08
CA	UKIAH ZENITH TELESCOPE	39 8 14.50	123 12 38.13	-2.390	3.780	2.39
CA	UNION B 1965	37 35 52.36	122 0 43.69	-4.420	9.350	5.71
CA	VACA 1880	38 22 32.72	122 0 0.76	-8.920	-1.900	-1.18
CA	VAIL 1939	33 31 18.06	117 15 31.60	-3.140	6.240	3.45
CA	VENTURA 1932	36 19 38.68	121 53 38.35	-8.910	32.910	19.50
CA	VENUS 1963	35 14 50.07	116 47 36.53	-2.920	-0.350	-0.20
CA	VIDAL MWD 1932	34 9 37.73	114 30 6.12	-3.260	5.390	3.03
CA	VINEYARD USGS 1944	33 47 6.58	120 40 22.87	-3.200	9.330	5.46
CA	VIXEN 1949	38 11 24.09	122 51 44.85	-3.430	12.160	7.52
CA	VON SCHMIDTS IRON MONUMENT	39 13 24.82	120 0 16.89	-12.030	-1.400	-0.88
CA	WAUKENA 2 1961	36 8 15.43	119 29 33.08	-7.900	11.880	7.00
CA	WAYS 1957	32 50 15.00	116 47 28.06	-4.570	20.680	11.22
CA	WELL 1898	35 35 22.95	115 25 2.17	-5.550	-0.280	-0.16
CA	WELLS 1935	32 44 57.48	115 67 46.87	3.550	-2.860	-1.54
CA	WESTLEY 1943	37 31 33.05	121 10 17.17	-0.660	0.630	0.38
CA	WHEELER 2 1948	35 0 38.22	119 0 47.99	20.100	8.170	4.69
CA	WHITE 1933	33 4 48.15	117 17 16.11	-6.940	19.520	10.66
CA	WHY DMA 1974	39 46 9.38	120 26 6.89	-2.380	0.160	0.11
CA	WILDER 1931	36 58 29.51	122 5 20.11	-17.860	12.800	7.70
CA	WILLIAMS USGS 1944	39 57 5.60	121 0 2.13	-5.130	4.050	2.38
CA	WORKMAN HILL LAFC LAC C 1932	33 59 30.10	118 0 7.11	-10.920	11.600	6.49
CA	WYLIE 1927	34 27 34.02	119 56 24.28	-18.110	10.120	5.72
CA	YOLO NORTHWEST BASE 1880	38 40 43.72	121 51 27.35	******	******	******
CA	YOLO SOUTHEAST BASE 1880	38 31 41.10	121 47 57.31	-7.150	6.870	4.28
CA	ZAM 1959	38 47 23.11	121 52 15.62	-6.640	7.410	4.65
CA	ZEBRA 1948	41 53 19.49	121 53 19.49	-2.180	5.260	3.52
CO	ADOBE 1881	38 40 40.55	103 33 16.32	-3.010	******	******
CO	BENCH MARK H 12 1954	38 53 4.03	102 21 5.53	0.990	-1.730	-1.09
CO	BETHEL 1954	37 47 54.06	102 57 57.09	-0.230	-2.750	-1.68
CO	BLACK	39 31 4.33	105 21 7.71	-6.200	-22.490	-14.31
CO	BOUNDRY MON 1900 CO NM OK 1922	37 0 0.47	103 0 6.63	-1.450	-8.940	-5.38
CO	BRIGHTON BENCH MARK ECC 1912	40 0 1.24	104 48 58.18	2.830	******	******
CO	BUCKLEY 1960	39 38 54.22	104 41 18.37	5.050	-2.530	-1.62
CO	COLORADO SPRINGS LAT AND LONG	38 50 2.87	104 49 35.06	4.470	-1.130	-15.93
CO	COLORADO SPRNGS ASTRO USE 1873	39 48 20.03	104 54 8.03	-1.870	******	******
CO	COMMERCE 1968			2.380	-7.820	-5.01
CO	CONDON USGS 1963	40 47 19.09	102 42 37.19	-1.160	-0.050	-0.07
CO	DEWEY USGS 1912	40 30 25.50	104 33 16.16	-5.320	-9.420	-5.92
CO	EL PASO EAST BASE 1879	38 57 21.99	104 27 41.92	-5.000	-10.710	-6.73
CO	EL PASO WEST BASE 1879	38 58 42.84	104 35 19.26	-0.620	******	******
CO	ELBERT 1912	39 14 2.58	104 34 33.15	3.910	-20.980	-13.47
CO	ENERGY	39 54 15.78	105 9 59.25	7.950	-1.210	-0.73
CO	GOLF 1945	37 9 6.84	104 30 46.08	-3.030	11.820	7.18
CO	GOOD TPC 1972	37 21 39.74	108 46 27.57	4.780	0.920	0.52
CO	GRAND JCT TELEGRAPH LONG STA	39 3 54.61	108 33 52.92	-1.120	-2.070	-1.29
CO	GREEN 1954	38 46 10.69	102 16 30.43	-1.740	4.070	2.54
CO	GUNNISON AZIMUTH 1894	38 32 46.15	106 55 26.63	-2.660	-4.890	-3.13
CO	HENDERSON USGS 1568	39 52 4.72	104 49 19.66	2.280	12.480	7.93
CO	HOOSIER 1953	39 29 6.66	106 57 14.74	3.000	-3.200	-1.97
CO	HUEY 1922	38 2 20.58	103 14 55.25	-4.710	7.230	4.59
CO	IGNATIO 1936	37 8 3.16	107 38 26.10	-4.240	-3.510	-2.24
CO	INDIAN USGS 1912	39 39 18.45	104 35 5.82	-3.020	-0.620	-0.40
CO	JOHNSON 1934	40 35 24.68	103 26 32.27	-5.000	-0.370	-0.24
CO	KIRKHAM 1934	40 7 9.59	103 28 50.81	-2.010	-5.920	-3.81
CO	LAIRD 1933	40 2 48.07	102 3 19.03	-0.900	0.0	******
CO	LAKE DMWW 1951	39 36 49.87	106 4 1.04	-4.100	-5.020	-3.17
CO	LE BOW 1934	39 16 30.55	103 30 35.37	-2.740	******	******
CO	LONG 1922	37 31 6.43	103 19 46.55	-5.110	******	******
CO	LOVELAND TALL WHITE CHIMNEY	40 24 10.40	105 3 35.91	3.430	-3.390	-2.16
CO	LOW 1960	39 59 51.00	104 29 35.10	3.170	-2.920	-1.86
CO	LOW 1960 AZ MK	39 38 34.68	104 29 34.02	-1.040	-7.610	-4.73
CO	LS 40 USGS 1954	38 14 40.13	102 57 29.61	-1.240	-2.770	-1.80
CO	MARTIN 1922	37 21 30.35	103 6 21.44	-2.220	******	******
CO	MESSIE 1934	39 32 34.19	105 15 9.10	3.330	******	******
CO	MORRISON USGS 1895	39 40 9.27		-3.790	-6.120	-4.55
CO	MOUNT OURAY 1854	38 25 21.78	106 13 27.02	-2.560	-6.370	-4.55
CO	OVAL 1935	37 33 15.77	106 13 27.82	-6.510	******	******
CO	OVERLAND 1881	39 7 20.06	103 10 15.63	-0.320	9.290	5.63
CO	PARK PT 2 RM 3 TPC 1972	37 16 50.47	108 27 39.31	1.960	******	******
CO	PIKES PEAK 1879	38 50 25.93	105 2 37.19	0.230	-1.680	-1.05
CO	PITTS 1954	38 50 40.63	102 48 46.83	-1.750	-8.620	-5.24
CO	POTATO 1922	38 10 26.46	106 6 26.25	-0.520	-1.490	-0.93
CO	QUAIL 1954	38 55 7.54	102 47 15.82	-2.040	-4.800	-3.06
CO	RANGE 1960	35 35 7.76	104 27 45.23	2.140	0.980	0.59
CO	ROMEO NORTH BASE 1935	37 15 51.39	105 57 42.74	2.800	1.240	0.75
CO	ROMEO SOUTH BASE 1935	37 10 47.56	105 59 7.13	-9.820	2.430	1.50
CO	SAN LUIS 1935	38 5 35.11	106 7 3.99	0.250	-2.160	-1.39
CO	SAND 1934	39 50 49.28	103 27 52.06	-4.580	16.230	10.53
CO	SAW 1960	39 18 51.76	104 33 54.88	-2.000	19.720	12.77
CO	STEAMBOAT NORTH BASE 1939	40 25 26.16	106 49 33.93	0.430	-3.130	-2.38
CO	STEAMBOAT SOUTH BASE 1939	40 21 23.29	106 49 41.02	1.330	******	******
CO	SUTTON 1934	39 34 36.13	103 35 49.39	-6.550	-2.830	-1.83
CO	TAVAPUTS 1891	39 32 23.28	109 0 18.72	-2.950	-10.930	-7.01
CO	TAYLOR 1935	40 10 36.13	102 49 2.37	-2.360	******	******
CO	THORNTON 1968	39 52 14.51	104 58 23.01	-2.570	-3.250	-1.99
CO	TREASURY MOUNTAIN 1893	39 0 51.43	107 5 44.44	-1.300	******	******
CO	TURTLE 1953	37 36 33.40	103 1 26.01	-1.300	-7.530	-4.62
CO	UNCOMPAHRE 1855	38 4 17.58	107 27 41.31	0.500	-8.790	-5.38
CO	UTE NORTHWEST BASE 1952	37 48 14.23	104 43 8.85	-1.690	6.430	4.00
CO	UTE SOUTHEAST BASE 1952	37 44 3.40	104 38 39.32	******	-4.520	-2.89
CO	W MTN 1958	38 31 27.30	104 54 23.02	8.010	-5.030	-3.07
CO	WATKINS ASTRONOMIC 1912	39 44 43.43	104 36 18.93	-7.390	-5.330	-3.56
CO	WOLF 1936	37 38 12.49	106 30 57.30	-2.040	-2.750	-1.83
CT	BALD HILL 1861 RM 3	41 48 26.74	72 11 55.38	-4.930	-5.120	-3.39
CT	BOX HILL 1861	41 47 59.31	72 27 21.49	-0.600	2.030	1.35
CT	FORT WOOSTER 3 1882	41 16 55.40	72 51 16.36	-2.270	******	******
CT	RAYMOND HILL CTGS 1942	41 25 46.48	72 15 32.88	-1.100	-7.100	-4.72
CT	ROXBURY 1942	41 33 2.05	73 15 32.88	-3.610	-0.860	-0.74
CT	SANDFORD 1862	41 27 42.35	72 56 59.40	-6.460	-0.110	-0.04
CT	SCHOOLHOUSE HILL 1875	41 39 30.86	72 40 52.98	-1.850	-4.210	-2.79
CT	STERLING 1937	41 38 48.81	71 50 57.85	-3.750	-2.660	-1.77
CT	STORRS 1936	41 48 53.00	72 15 34.69	-2.580	******	******
CT	WEST PEAK CTGS 1538	41 33 43.72	72 50 40.38	-3.100	******	******
CT	4351 CTGS 1575	41 43 42.78	72 9 24.00	-3.810	-4.180	-2.62
DC	CAUSTEN 1851	38 55 34.96	77 4 23.98	-3.510	-7.120	-4.47
DC	FOUR AND ONE HALF STREET OBS	38 53 33.80	77 1 2.50	1.110	-0.210	-0.11
DC	GEORGETOWN COLLEGE OBSERVATORY	38 54 29.61	77 3 39.21	0.550	-0.240	-0.12
DC	OLD NAVAL OBS DOME PIPE 1863	38 53 41.93	77 3 8.28	-0.610	-1.280	-1.24
DC	SEATON ASTRONOMICAL STA 1850	38 53 23.62	76 59 59.70	-0.350	-1.900	-0.90
FL	CAPE HENLOPEN LIGHTHOUSE	38 46 38.40	75 5 10.74	-0.340	34 35.00	
FL	ALLISON 1932	30 24 54.40	83 5 23.43	-0.960	-2.790	-1.33
FL	ALTHA 1938	30 32 12.27	85 8 6.58	0.170	-2.190	-1.04
FL	ARTESIA 1953	28 25 15.36	80 34 54.85			
FL	ATHENA 1933	29 57 51.44	83 28 22.36			
FL	AUX THEOD PIER 29 A	28 25 46.68	80 34 35.00			
FL	BANK 1956	28 24 41.46	80 35 35.35			
FL	BAREA 1956	28 26 52.45	80 34 54.53			

STATE	STATION NAME	GLAT(D,M,S)	GLON(D,M,S)	ALAT-GLAT(S)	ALON-GLON(S)	LAPLACE(S)
FL	BARTH 1938 RM 3	30 45 21.02	87 20 42.58	1.890	-2.260	1.15
FL	BC 4 FC 1 1960	28 40 22.41	81 22 26.23	1.520	-1.950	-0.94
FL	BC 4 1.1 1956	28 30 2.29	80 33 48.94	0.500	-2.730	-1.30
FL	BEAU 1965	28 28 1.63	80 33 32.83	-2.110	-2.920	-1.39
FL	BELLEVIEW 1906	27 56 33.90	82 48 43.16	********	-4.060	-1.90
FL	BETHEL 1932	30 16 30.95	81 46 14.28	0.910	-3.180	-1.60
FL	BETTS 1964	30 32 26.44	85 24 29.27	-0.060	-1.620	-0.82
FL	BORDER 1971	26 10 20.74	80 52 46.42	3.420	4.150	0.51
FL	BOUFFORD 1934	27 24 41.95	80 24 15.73	2.860	-4.760	-2.19
FL	BOUNDRY 1958	28 34 4.53	80 34 14.46	1.070	-3.070	-1.46
FL	BOWMAN 1955	28 28 29.18	80 34 40.10	0.670	-2.720	-1.30
FL	BRADFORD 1933	30 33 43.05	81 13 7.91	0.940	-0.160	-0.08
FL	BROOKSIDE 1934	27 46 2.44	80 31 10.89	4.410	-3.900	-1.82
FL	BUIE 1933	29 5 10.27	82 33 7.76	2.470	-3.230	-1.81
FL	BURBANK 1933	29 16 53.66	82 0 5.17	-2.780	-3.680	-1.58
FL	CAMERA 1959	28 26 52.96	80 32 10.38	-0.410	-1.680	-1.75
FL	CENTRAL 1950	28 29 32.37	80 34 38.77	0.590	-6.410	-0.80
FL	CHANGE 1971	26 13 43.81	80 17 55.33	2.110	-3.270	-2.83
FL	CHESTER 2 1964	28 36 43.40	80 35 57.00	1.940	-0.540	-1.57
FL	CHRISTIE 1963	30 16 46.74	82 37 7.77	1.030	-7.200	-0.27
FL	CLINT 1962	26 24 31.46	80 12 13.84	2.760	-2.530	-3.20
FL	COCOA BEACH 1957	28 21 3.60	80 36 21.13	-1.700	-0.870	-1.20
FL	COON 1936	27 45 11.30	81 4 30.95	8.420	-0.190	-0.41
FL	COUSINS 1962	29 10 12.18	80 46 46.16	2.250	-0.870	-0.09
FL	CRAWFORD 1917 RM 3	30 30 29.74	80 58 55.70	1.140	-2.880	-0.44
FL	CREWS 1956 RM 3 1970	26 33 43.57	80 50 24.05	0.470	-1.860	-1.28
FL	DANIELL 1935	30 39 47.11	90 34 24.78	0.450	-2.970	-0.95
FL	DAREA 1956	28 27 34.46	81 0 59.34	0.930	-0.870	-1.41
FL	DAYTONA 1906	29 12 35.35	81 3 28.47	********	-11.190	-0.42
FL	DELRAY NORTH BASE 2 1970	26 32 34.08	80 6 21.72	3.460	-9.520	-5.00
FL	DELRAY SOUTH BASE 1934	26 26 48.61	82 42 51.58	-3.380	-3.700	-4.52
FL	DESOTO 1973	27 37 23.11	80 32 58.62	1.270	-3.270	-1.72
FL	DOLITE 1956	28 3 2.00	80 57 22.86	4.220	-0.530	-1.54
FL	DUFFIE 1962	28 58 41.29	85 39 1.70	-3.530	-1.800	-0.26
FL	DUMA 1964	30 31 17.48	80 43 15.37	-0.660	-4.090	-0.91
FL	DUMMIT 1934	28 41 46.80	81 20 52.75	2.140	-1.360	-1.97
FL	DUPONT 1934	29 44 49.39	81 37 32.27	0.590	-2.000	-0.67
FL	DUVAL NORTH BASE 1932	30 29 24.10	85 54 6.91	2.540	-1.440	-1.06
FL	EBRO 1934	30 25 54.25	80 35 42.42	-0.880	-2.300	-0.73
FL	EE 1956	28 27 16.62	82 24 50.10	-0.220	-4.870	-1.09
FL	EGGS 152 1960	27 40 29.50	81 24 54.87	6.450	-3.610	-2.26
FL	ELFER 1960	28 12 59.99	81 24 53.45	-1.150	-3.020	-1.71
FL	ELZEY 1933	29 55 5.03	81 5 38.23	3.640	-3.650	-1.50
FL	EMERSON 1960	27 31 21.03	80 0 0.74	2.410	********	-1.69
FL	ESPERANZA 1873	29 22 15.15	81 4 3.51	1.740	-1.180	-0.61
FL	EVERGREEN 1932	30 43 54.58	81 11 6.81	1.400	-1.510	-1.51
FL	FALSE BC 4 1960	28 35 7.20	81 5 5.23	-2.320	-1.090	-0.53
FL	FAVORETTA 1934	29 22 4.40	81 27 42.05	6.480	-6.840	-2.91
FL	FERGAN 1934	26 40 47.99	82 28 13.83	********	-1.410	-0.72
FL	FERNANDINA NASSAU CTHSE CUP	30 40 14.32	81 30 17.35	2.380	-3.930	-1.83
FL	FISH 1934 HILLSBOROUGH	27 50 41.90	86 41 49.40	4.150	-4.460	-2.08
FL	FLEMING 1960	27 48 46.37	80 34 57.79	1.760	0.080	0.04
FL	FLOROSA 1934	30 24 38.37	80 34 37.35	1.730	-2.860	-1.37
FL	FORTY ONE 1965	28 34 59.02	80 37 51.19	0.840	-2.570	-1.22
FL	FORTY 1964	28 33 41.73	80 33 55.07	1.300	-2.980	-1.42
FL	GADGET 1956	28 29 16.56	80 48 0.00	1.350	-2.070	-0.98
FL	GATOR 1960	28 32 20.01	80 6 57.22	0.770	-0.650	-1.01
FL	GEMINI RM 1 1964	28 31 24.57	80 35 3.85	1.800	-2.070	-0.99
FL	GENE 1935	28 6 1.86	81 15 45.20	0.010	-4.010	-1.99
FL	HANGAR AJ 1964	29 42 51.06	84 25 19.98	-0.320	-1.730	-0.88
FL	HASTINGS 1933	30 36 46.10	80 13 54.41	-0.110	-7.910	-3.58
FL	HAVANA 1933 RM 4	30 26 32.17	80 12 10.18	8.750	-6.880	-3.04
FL	HAWK 2 1954	26 9 61.21	80 20 25.02	6.730	-0.790	-2.95
FL	HAWKINS 2 1962	25 45 42.10	81 16 51.53	0.350	-1.410	-0.70
FL	HECTOR 2 1961	29 33 6.98	83 47 28.07	2.810	-0.160	-0.08
FL	HENDERSON 1934	28 26 11.60	80 35 49.84	1.700	-1.780	-0.84
FL	HIGH 1953	30 18 18.82	83 26 4.31	-0.150	-1.380	-0.70
FL	HOLMES 1934	30 31 36.40	82 43 33.28	0.310	-2.920	-1.57
FL	HOLSTON 1935	28 29 28.59	80 33 55.61	5.050	-2.870	-1.39
FL	HONEY 1967	27 53 35.18	85 56 19.15	0.560	-1.350	-1.47
FL	HUB 1953	30 36 53.35	81 42 14.84	-2.110	-2.410	-0.69
FL	HUDSON 1938	30 51 25.90	81 50 57.08	1.990	-1.220	-1.13
FL	ITALIA 1932	28 1 34.41	80 52 57.08	7.320	-8.230	-1.04
FL	JACKSON 1936	27 36 4.13	80 56 32.23	6.070	0.970	-0.62
FL	JANE 1960	30 31 20.67	80 8 45.91	0.180	-4.140	3.63
FL	JASPER 1935	26 13 18.46	85 18 28.55	1.780	-11.730	0.48
FL	JILL 1934 RM 5 1971	29 49 12.04	82 41 35.30	3.430	-10.330	-1.93
FL	JOE 1934	27 55 65.76	80 4 29.25	4.280	-0.150	-5.31
FL	JORDAN 1934	26 48 10.12	80 58 17.13	8.590	********	-4.66
FL	JUPITER INLET LH 1883	27 50 41.85	80 9 21.89	5.840	-3.020	-4.94
FL	KELSEY 2 1959	25 39 64.46	80 33 22.37	10.000	-3.930	-1.45
FL	KENAN 1936	28 31 24.33	80 29 42.46	-1.160	-2.070	-1.83
FL	KEY BISCAYNE S BASE 1849	27 41 20.18	80 36 36.34	4.810	-2.540	-0.98
FL	KIMBALL ECC 1934	28 29 9.23	81 38 59.08	0.720	-3.480	-2.96
FL	KINGSBURY 2 1961	27 53 29.39	81 40 38.30	7.030	-0.460	-1.28
FL	LAB 1960	28 18 37.12	82 11 38.84	2.860	3.340	-1.63
FL	LAKE WALES SE BASE 1935	27 49 41.58	83 2 13.40	4.360	0.440	-0.23
FL	LANCASTER 2 USE 1926	30 19 9.80	80 26 2.13	1.700	-8.060	0.22
FL	LITHIA 1937	25 14 10.12	80 42 44.08	-1.270	-0.700	-3.43
FL	LIVE OAK EAST BASE 1932	27 49 41.58	84 15 34.47	4.850	-0.230	-3.43
FL	LLOYD 1973	30 19 9.80	80 45 34.12	0.560	0.390	0.11
FL	LONG SOUND 1961	28 2 6.86	81 11 42.89	********	-8.870	0.20
FL	MARION 1960	25 14 10.12	80 35 47.14	4.560	-2.590	-3.86
FL	MERIDIAN 1933	28 2 58.49	80 48 12.88	-2.250	-6.130	-1.21
FL	METTS 1964	3C 40 27.08	80 35 47.14	0.940	-6.290	-2.83
FL	MIAMI LONGITUDE STA 1905	30 33 15.99	82 51 2.28	0.650	-2.720	-0.15
FL	MICCO 1934	25 46 28.47	86 31 0.83	1.740	-0.880	-1.17
FL	MILK 1961	27 52 26.08	80 58 20.25	0.600	-1.170	-0.45
FL	MILLER 1932 RM 3	27 26 58.49	82 22 36.82	1.000	-3.540	-0.55
FL	NIBRUD 1962	30 18 5.26	80 19 8.35	3.430	-7.240	-3.54
FL	NICEVILLE 1934 RM 5	30 5 5.58	82 23 36.82	1.040	-0.350	-1.77
FL	NORTH 1955	30 30 37.00	80 19 8.35	4.440	-6.190	-2.83
FL	OKEECHOBEE W BASE 1935	28 13 40.84	87 10 51.35	1.360	3.170	1.62
FL	OLUSTEE 2 1963	27 43 35.80	81 12 21.63	0.920	-2.990	-1.63
FL	OWEN 1934	3C 15 22.56	80 51 27.22	2.480	-1.120	-0.54
FL	PACE 1938 RM 4	27 9 40.89	85 48 21.79	6.830	-1.430	-2.44
FL	PAD 1950	30 39 18.94	80 36 22.08	-0.620	-2.690	-0.23
FL	PADGETT 1935	28 26 6.87	80 34 22.18	-0.350	-2.880	-1.27
FL	PADRICK 1960	27 54 19.79	81 27 7.56	-1.780	-4.010	-1.36
FL	PARISH 1938	30 39 5.27	80 17 28.21	-0.670	-6.270	-1.91
FL	PATRICK N BASE 1953	28 16 17.63	82 38 29.51	-0.350	-6.270	-2.89
FL	PATRICK S BASE 1953	28 6 33.34	82 6 51.68	2.990	-6.830	-2.09
FL	PERSHING ASTRO 1959	28 26 48.29	86 57 59.99	7.820	-2.270	-3.11
FL	PIERCE 1934	27 28 11.73	80 41 38.37	0.450	-4.170	-1.15
FL	PINELLAS 3 1967	27 42 16.06	80 38 58.41	-2.250	-2.600	-2.12
FL	PLANT CITY 1937	28 1 0.30	80 38 23.65	4.740	-5.220	-2.12
FL	PCLK 1964	3C 26 6.57	80 16 40.25	-0.110	-2.850	-1.24
FL	PORTLAND 1934 RM 2	30 30 33.63	80 39 52.27	0.310	-2.240	-1.07
FL	RACK 1935	24 52 49.23				
FL	RANGE 1964	30 50 58.41				
FL	RED TOP 2 1962	25 56 21.92				
FL	REED RM 2 1964	28 25 28.25				

STATE	STATION NAME	GLAT(D,M,S)	GLON(D,M,S)	ALAT-GLAT(S)	ALON-GLON(S)	LAPLACE(S)
FL	REGIS 1963	30 21 1.13	81 59 41.04	1.250	0.770	0.39
FL	RESIN 1963	30 18 53.67	82 11 2.21	0.840	0.690	0.34
FL	ROCKS 1934	27 14 30.01	82 31 59.65	4.540	-2.160	-0.99
FL	RCSE 1934	26 16 5.36	81 47 30.72	5.110	-4.680	-2.07
FL	SAND KEY 1849	24 27 12.13	81 52 39.68	-2.860	*******	******
FL	SAND MTN 1964	30 33 28.97	85 31 21.82	-0.280	-2.000	-1.02
FL	SEBASTIAN 2 1907	27 49 0.17	80 28 13.21	4.260	-5.290	-2.47
FL	SEVENTEEN A 1957	28 26 49.16	80 33 55.01	-0.020	-2.290	-1.09
FL	SKIPPER 1937	28 4 48.11	82 26 59.95	6.120	-4.190	-1.97
FL	SCTTILE 1960	27 54 30.28	80 42 18.18	5.230	-2.900	-1.36
FL	SQUPAD 1957	28 28 56.86	80 34 8.26	0.730	-2.650	-1.26
FL	SOUTHSIDE 1962	30 17 7.46	81 32 35.29	2.560	-0.120	******
FL	ST AUGUSTINE LIGHTHOUSE 1882	29 53 6.12	81 17 19.45	-0.790	*******	******
FL	ST MARKS LONGITUDE STA 1907	30 9 27.41	84 12 17.53	*******	0.930	0.46
FL	STANLEY 1937	28 13 39.06	82 17 14.37	5.730	-6.270	-2.96
FL	STARLING 1964	30 32 22.01	83 31 18.98	1.220	-1.390	-0.71
FL	TALLAHASSEE 1933	30 27 21.57	84 16 59.45	0.630	-0.540	-0.28
FL	TAM 1971	25 45 40.68	80 37 6.27	2.550	-4.900	-1.91
FL	TAMIAMI W BASE 1934	25 45 38.27	80 36 57.94	2.760	-4.230	-1.84
FL	TECH 1961	28 13 34.67	80 36 0.13	1.380	-2.250	-1.06
FL	THEOD 1,40 1963	28 27 58.23	80 35 21.70	-0.110	-2.470	-1.18
FL	TIFTS CBSERVATCRY 1853	24 33 33.76	81 48 27.41	-1.570	-1.660	-0.69
FL	TIMER 1962	25 36 49.36	80 23 3.43	-1.960	-8.330	-3.60
FL	TITUSVILLE SOUTHEAST BASE 1934	28 37 27.13	80 49 23.16	1.110	-2.030	-0.97
FL	TCRREYA 1964	30 33 2.70	84 56 36.48	1.300	4.370	2.23
FL	TUTTLE 1971	25 48 35.42	80 8 45.79	-0.440	-9.980	-4.34
FL	TWELVE NN 1956	28 28 50.95	80 32 39.64	0.810	-2.940	-1.41
FL	VALKARIA 1960	27 57 22.65	80 33 29.03	4.310	-4.130	-1.94
FL	VOLUSIA 1934	28 47 28.98	80 53 3.20	1.180	-3.380	-1.63
FL	WALDIN 1962	25 30 25.25	80 23 17.52	1.290	-8.520	-3.67
FL	WALNUT 1938	30 53 8.13	87 30 24.38	0.530	3.030	1.56
FL	WATER 1971	26 6 7.96	80 11 58.83	2.290	-8.080	-3.56
FL	WELCH 1933	30 28 56.53	83 18 1.09	-0.510	-1.200	-0.61
FL	WEST MARTELLC TWR 1905	24 32 47.85	81 47 10.27	-1.790	-1.770	-0.74
FL	WHEAT 1938	30 31 47.45	84 38 38.25	-0.700	*******	1.77
FL	WILBER 1934	29 7 39.52	80 57 18.21	2.420	-1.350	-0.66
FL	WILLIAMS 1957 RM 4 1960	28 26 59.44	80 45 45.74	0.310	-0.900	-0.43
FL	WILLIS 1935	30 46 6.52	81 56 56.02	0.660	-2.490	-1.27
FL	WILSEN RM 5 1959	28 38 34.94	80 41 56.75	1.850	-3.840	-1.84
FL	WINDHAM 1934	30 30 49.53	86 20 37.63	0.180	-2.010	-1.02
FL	WRIGHT 1965	28 34 57.66	80 37 57.60	1.360	-3.160	-1.51
FL	Y TRANSPCNDER 1953	28 29 33.28	80 32 53.38	1.070	-2.650	-1.26
FL	1OKS PIER 3 1962	27 55 46.65	80 32 56.12	4.550	-4.110	-1.92
FL	1OKN PIER 4 1962	27 56 54.57	80 35 8.39	4.520	-3.480	-1.63
GA	ABBEVILLE 1935	31 55 5.12	83 18 32.37	2.080	-3.410	-1.80
GA	ALBANY 1917	31 34 20.07	84 7 23.51	-1.590	-4.350	-0.99
GA	ALLIGCOD 1935	32 26 5.19	83 0 15.16	-2.490	-5.160	-2.33
GA	ARABIA 1935	31 7 46.71	82 55 32.51	-4.910	-0.850	-2.67
GA	ASHBURN ALUMINUM MUN TANK	31 42 22.58	83 39 17.32	0.680	0.930	-0.44
GA	ATLANTA MIDDLE BASE 1873	33 54 21.26	84 16 37.95	0.960	-12.800	0.52
GA	AUGUSTA LCNGITUDE STA 1890	33 28 23.73	81 58 9.74	-3.210	*******	-7.06
GA	AVON 1934 RM 5	32 56 37.07	83 0 56.47	-1.540	-7.730	-4.24
GA	BARNEY 1932	33 17 22.93	81 54 54.07	2.590	-4.160	-2.21
GA	BC SAV USA 1963	32 0 6.33	81 9 12.86	1.870	-6.470	-3.47
GA	BENCH MARK 128 USGS 1935	32 27 34.58	81 26 1.18	1.040	-4.030	-2.10
GA	BENNETT 1935	31 26 31.72	82 18 57.16	1.880	2.050	-1.07
GA	BERRY 1972	34 21 35.34	85 12 0.08	-4.950	-6.680	-3.61
GA	BLAKELY BLACK MUNICIPAL TANK	31 22 43.86	85 55 59.63	-2.010	1.410	0.76
GA	BLUNDALE 1935	32 44 55.50	82 23 26.12	6.190	-2.650	-1.57
GA	BONITA 1917	32 13 32.51	84 9 48.35	-2.090	-4.360	-2.32
GA	BOMDGN 1941	33 32 12.13	85 18 54.82	1.160	6.670	0.37
GA	BRCWN 1932	32 3 22.53	81 16 6.67	4.890	-2.660	-1.38
GA	BRUCE 1935	34 14 57.78	83 7 31.78	8.840	-0.850	-0.45
GA	BRUNSWICK SOUTHEAST BASE 1917	31 9 26.18	81 29 55.34	2.000	-2.020	-1.04
GA	BUENA 1935	32 17 19.74	84 30 53.72	1.280	2.470	1.33
GA	BURNTFORT 1935	30 56 48.38	81 53 58.49	5.960	2.590	1.42
GA	BUTLER 1917	32 33 23.83	84 13 50.15	5.840	4.930	0.33
GA	BUTTRILL 1930	33 17 32.37	84 0 41.71	0.370	2.040	1.10
GA	BV 058 200 GAHD 1972	34 14 11.87	84 9 39.11	0.290	-5.010	-2.68
GA	BYRD 1974	32 39 23.23	84 7 40.32	0.030	-1.400	-0.79
GA	CANDLER 1935	32 16 9.61	83 30 6.67	-0.620	-3.040	-1.73
GA	CANTON 1935	34 11 13.71	84 29 34.40	2.100	-6.680	-3.59
GA	CARNES 3 1941	33 59 35.11	85 0 52.35	0.030	-0.190	0.11
GA	CARR 1935	32 31 19.26	82 28 6.12	0.530	-1.300	-0.73
GA	CASH 1972	34 29 3.55	84 44 45.16	0.620	0.500	0.28
GA	CASS 1972	34 17 48.03	83 45 53.90	3.950	-1.820	-1.02
GA	CEDAR 1935 RM 4 1975	34 1 19.20	82 52 21.99	4.430	1.670	0.87
GA	CELESTE 1935	33 45 38.54	83 23 25.18	5.080	-0.680	0.37
GA	CHESTNUT 1935	33 28 19.42	85 13 28.27	1.440	-6.280	-3.34
GA	CLARK 1941	33 6 35.66	81 53 21.50	0.430	-3.390	-1.87
GA	CLAXTCN 1935	32 9 42.27	82 29 29.65	-0.160	-4.040	-2.13
GA	CLEM 1941	33 31 9.08	82 53 38.18	4.140	-4.830	-2.58
GA	CLEMENS 1935	31 51 31.83	82 10 27.69	2.850	3.470	3.10
GA	COBB 1935	32 17 34.23	83 21 34.66	1.920	4.260	-3.51
GA	COCHRAN 1917	32 23 19.57	82 35 34.82	1.090	-6.710	3.17
GA	COHUTTA 1874 RM 4 1972	34 53 23.35	84 58 56.39	8.320	5.900	1.16
GA	CCLE 1935	31 36 15.58	84 0 10.22	2.880	2.080	-2.40
GA	COLUMBUS 2 1964 RM 4 1974	32 27 54.91	81 30 19.00	4.300	-4.600	-1.41
GA	CCNYERS 2 1964 RM 4 1974	33 39 46.21	84 13 58.77	0.220	-2.760	2.38
GA	CREDIT 1963	31 31 9.25	82 7 43.81	0.850	4.390	-4.65
GA	CRESCENT 1932	30 56 46.31	84 20 25.76	6.490	-8.200	-4.44
GA	CREWS 1935	32 54 0.85	83 22 33.58	-0.430	-8.200	-4.47
GA	CRYSTAL HILL 1930	34 31 44.69	83 22 33.58	-0.430	-4.500	-2.34
GA	CURRAHEE 1874	34 31 44.86	83 22 33.58	7.770	-4.500	-2.34
GA	CURRAHEE 1874 RM 4 1975	34 31 44.86	83 35 6.51	*******	-4.500	-2.34
GA	CUT 1932	31 19 14.36	84 25 7.75	-1.550	-2.770	-1.46
GA	DARIEN LONGITUDE PIER 1907	31 43 23.18	84 50 0.37	-2.010	-0.380	-0.20
GA	DAVIS 1933	31 0 5.92	83 53 34.44	5.100	0.700	0.39
GA	DAVIS 2 1938 RM 3 1973	33 56 5.21	83 23 15.60	-3.660	0.000	1.13
GA	DOC 1935	31 2 53.10	82 55 45.23	0.260	-0.340	-0.18
GA	DCNALDSCNVILLE 1938	30 34 5.85	82 12 53.76	4.070	-0.910	-0.52
GA	ELLICOTTS MOUND 1935	34 52 57.19	83 6 53.70	-0.640	-2.250	-1.15
GA	ERNEST 1972	30 41 16.66	82 33 58.55	-0.120	-7.170	-3.88
GA	FARGO 1917	32 47 25.95	81 30 9.14	2.070	-1.450	-0.76
GA	FIFER 1935	31 42 58.65	83 14 43.77	3.410	-1.000	-0.55
GA	FITZGERALD ALUMINUM STANDPIPE	33 40 55.64	83 2 43.46	*******	-2.320	-1.19
GA	FLUKER 1935	30 50 17.52	82 0 37.79	-1.390	12.050	6.87
GA	FOLKSTON 1917	34 45 2.16	83 53 3.96	0.570	-3.280	-1.78
GA	FORT 1972	33 3 11.27	81 47 58.03	4.400	-1.670	-4.01
GA	FORTVILLE 1935	31 32 20.44	81 39 48.71	4.290	-2.070	-1.07
GA	GARDI 1917	31 3 52.27	81 40 40.28	2.530	-0.660	-0.49
GA	GEORGE 1932 RM 5	33 1 25.58	84 19 56.37	2.780	-0.660	-3.48
GA	GIRARD 1932	31 28 2.53	84 19 56.37	-4.980	-0.660	-0.37
GA	GRACE 1935	34 29 10.10	85 45 26.63	5.010	4.450	2.43
GA	GRASSY 1874	33 10 10.11	85 9 48.07	-1.210	-3.220	-1.78
GA	GREENVILLE 1941	33 38 39.29	84 16 9.36	5.660	6.900	3.62
GA	GREIG 1973	32 44 19.16	84 13 10.44	5.890	2.200	2.00
GA	HAMILTON EAST BASE 1930	32 33 21.77	81 35 50.55	0.970	-6.670	-3.51
GA	HARMCN 1917	31 49 53.23	84 0 2.01	7.360	2.940	1.60
GA	HINES 1935	33 3 19.73	82 36 59.93	0.940	-6.980	-3.82
GA	HCLLIS 1930	33 12 5.82				
GA	HOPKINS 1934					

STATE	STATION NAME	GLAT(D,M,S)	GLON(D,M,S)	ALAT-GLAT(S)	ALON-GLON(S)	LAPLACE(S)
GA	HOUSE 1972	34 40 36.87	84 14 11.60	-0.640	2.370	1.35
GA	HUMP 1917	32 47 34.63	83 28 36.99	*******	-1.630	-0.88
GA	JESUP 1817	31 36 27.87	81 52 42.65	*******	-7.750	-4.06
GA	JOHNS 1874	34 37 22.71	85 5 54.03	2.170	-1.410	-0.80
GA	JOINT 1517	31 52 5.74	82 35 58.79	4.490	-3.020	-1.59
GA	JONES 1917	31 30 49.18	83 44 16.21	0.930	0.620	0.32
GA	JONESBORO 1918	33 31 58.51	84 18 10.84	3.250	2.100	1.16
GA	JONESBORO 1918 RM 5 1974	33 31 58.52	84 18 11.42	3.250	2.110	1.17
GA	KIMBROUGH 1917	32 0 13.61	84 39 2.35	0.320	-1.250	-0.66
GA	KINGSLAND 1932 RM 5	30 48 45.52	81 41 36.02	1.710	-0.960	-0.49
GA	KITCHENS 1934	33 3 7.89	82 31 28.48	1.430	-5.930	-3.23
GA	LANCE 1933 RM 3 1972	34 58 54.87	84 0 14.15	-0.240	1.800	1.03
GA	LAVENDER 1873	34 19 19.04	85 17 18.62	-1.990	*******	******
GA	LINCOLN 2 1962	33 44 31.56	82 31 28.44	2.070	-3.640	-2.02
GA	LULA 1935	34 11 11.79	83 26 51.27	2.980	-4.000	-2.07
GA	LUMPKIN 1975	33 56 9.98	83 23 14.59	5.310	1.360	0.76
GA	MASON 1941	33 14 58.24	85 15 26.60	-1.870	-2.560	-1.40
GA	MC CUTCHEN 1934	34 41 54.49	85 23 3.88	6.610	-2.040	-1.16
GA	MC KINNON 1935	31 25 39.40	82 54 10.32	-0.680	-4.320	-2.25
GA	MELVIN 1933	34 54 54.28	84 33 43.65	-3.730	-1.260	-0.65
GA	MERCER U 1935	32 49 46.14	83 39 7.89	-0.360	-1.790	-1.79
GA	MILL 1935	32 32 7.52	81 49 53.44	2.220	-5.320	-2.86
GA	MILLWOOD 1917	31 15 58.51	82 40 18.80	0.080	-4.740	-2.46
GA	MCATE 1935	33 13 1.68	84 4 39.77	-0.340	-5.680	-3.12
GA	MONTICELLO RM 3 1974	33 17 47.03	83 41 54.00	8.060	4.630	2.54
GA	MOULTRIE ALUMINUM MUN TANK	31 11 16.17	83 47 25.19	-0.830	0.080	0.04
GA	MUD 1917	32 11 15.40	83 10 17.74	*******	-3.030	-1.62
GA	NAPIER 1935 RM 3	32 39 7.09	83 31 14.76	-0.920	-2.990	-1.61
GA	NEYAMI 1935	31 49 29.09	84 12 56.38	-1.590	-1.320	-0.70
GA	NORRIS 1918	32 30 31.57	82 30 10.45	0.540	-4.610	-2.48
GA	NORWOOD 1976	33 27 57.68	82 42 25.86	1.430	-5.360	-2.96
GA	OGEECHEE S BASE 2 1963	31 52 51.24	82 9 16.20	0.570	-3.800	-2.01
GA	OTTER 1935	31 19 28.80	82 8 3.74	-2.890	-4.310	-1.79
GA	PAULK 1935	31 33 37.37	83 4 32.67	-0.290	-0.500	-0.26
GA	PENIA 1935	31 57 53.44	83 41 19.44	0.790	-0.810	-0.43
GA	PERRY 1917	32 34 43.55	84 33 6.31	4.640	3.040	1.54
GA	PERRY 1935 HOUSTON	32 22 25.51	83 8 0.47	1.170	-5.450	-2.93
GA	PINE 1917	31 48 43.40	83 27 2.47	1.000	-0.370	-2.83
GA	PISGAH 1973	34 0 21.57	84 17 16.89	-0.370	-5.040	-2.68
GA	POWELL 1935	32 9 58.69	83 2 59.61	-0.890	-5.040	-0.89
GA	PUCKETT GAGS 1935	34 38 28.85	83 49 56.76	-0.160	-1.580	-0.50
GA	PUTNEY 1935 RM 3 1973	31 28 25.07	84 7 45.63	-1.710	-0.960	-7.31
GA	RABUN 2 1933	34 57 55.78	83 17 59.66	-1.180	-12.750	-2.33
GA	RAY 1935	30 37 40.13	82 28 42.37	1.100	-4.230	-2.12
GA	RECOVERY 1938	30 42 29.87	84 46 1.77	1.100	4.150	-2.15
GA	RENFROE 1934	33 3 38.99	82 54 29.02	-0.980	-3.960	-2.48
GA	RICE 1932	31 42 4.82	81 24 36.71	2.070	-4.720	2.37
GA	RILEY 1917	32 33 18.37	84 12 0.96	7.460	4.400	0.75
GA	ROSE HILL 1902	34 5 0.71	82 45 33.71	5.420	1.340	0.91
GA	ROSE HILL 1902 RM 4 1975	34 5 0.88	82 45 33.62	5.400	1.340	0.53
GA	SALEM 1935	33 43 1.79	83 23 36.78	3.460	0.950	-0.67
GA	SAWNEE 1873	34 14 11.65	84 9 29.06	-0.560	*******	******
GA	SAXON 1976	33 48 3.97	82 47 59.03	3.790	-1.200	-2.16
GA	SCARBORO 1935	32 46 24.12	81 53 3.69	0.410	-3.980	-3.53
GA	SMITH 1935 RM 2 1974	32 0 55.05	82 57 9.54	-0.180	-3.210	-3.84
GA	SOUTHERN 1935	31 43 28.69	82 12 2.61	1.710	-7.260	-0.59
GA	STATE 1935	31 58 54.75	82 9 59.23	1.980	-1.160	0.79
GA	SWILLEY 1935	34 37 16.49	83 5 28.07	0.240	1.520	-0.74
GA	SYLVESTER D 1917	31 32 48.89	83 54 14.34	-0.140	-1.440	-0.46
GA	THELMA 1937	30 49 26.62	82 6 49 24.73	0.350	-0.900	0.44
GA	TRAIL 1935	30 39 17.10	82 6 13.23	0.100	0.800	-2.75
GA	TURIN 1941	33 19 41.69	84 38 40.12	3.600	-5.110	-0.19
GA	TURNER 1918	32 26 20.05	84 17 4.57	0.200	-0.370	-0.05
GA	VADA 1933	31 4 44.23	84 26 58.93	-2.160	-0.100	-0.88
GA	VALDOSTA 1917	30 45 47.85	83 40 40.08	-1.760	16 57.47	-2.25
GA	VALDOSTA 1935	30 51 35.44	83 53 21.34	-4.190	-4.180	-0.71
GA	VALLEY 1917	32 32 57.46	82 28 42.06	*******	-1.360	0.50
GA	WARESBORO 1917	31 14 55.74	84 51 37.53	2.210	0.870	-3.32
GA	WARR 1934	33 28 30.37	82 25 47.99	0.990	-6.020	-4.11
GA	WAYNE 1935	33 4 3.47	82 2 39.48	1.670	-7.530	-1.38
GA	WEBB 1941 RM 2 1973	33 40 13.28	84 43 17.65	1.160	-2.490	-2.01
GA	WELLS 1976	33 46 29.47	82 24 11.99	0.180	-3.610	0.21
GA	WESTON 1943	31 57 8.18	85 1 19.99	3.110	0.400	-0.73
GA	WHITEHEAD 1935	33 56 15.83	83 4 12.87	3.360	-1.310	-1.49
GA	WILLACOOCHEE 1917	31 20 30.41	83 5 5.13	*******	-2.870	-2.83
GA	WILLIE 1935	34 25 20.07	83 17 24.00	1.450	-5.330	-0.06
GA	WILLINGHAM F 1917	31 33 7.10	83 57 7.01	-1.330	-0.110	-0.85
GA	WINGATE 1935	31 3 42.14	83 29 13.66	-2.920	-1.650	-2.02
GA	WRIGHT 1935	32 5 44.26	82 44 40.06	2.580	-3.800	-2.19
GA	ZOMBIE 1935	32 49 18.50	81 59 21.55	3.980	-4.030	2.74
GA	10E-62 ECC 1935	33 31 27 56.89	83 2.02	-0.160	5.240	1.25
HI	ANTENNA 1958	21 33 21.69	158 14 29.56	-0.540	6.110	-1.77
HI	DIAMOND HEAD 1872	21 15 46.33	157 48 22.18	-27.330	-4.880	-20.76
HI	HALAI HGS 1877	19 43 10.83	155 5 57.57	6.360	-61.540	-15.61
HI	HANA SOUTHEAST BASE 1950	20 47 16.36	156 1 37.17	24.800	-44.000	-0.78
HI	HAWAII ASTRO IGY 1957	20 20 49.81	157 55 8.23	-17.490	-2.140	-0.30
HI	HONOLULU LONGITUDE MON 1903	21 18 23.72	157 51 55.44	*******	-0.830	0.78
HI	KAHE 1875	21 21 17.80	158 7 50.22	-25.000	10.380	-6.53
HI	KAHUKU NEW 2 1565	21 42 46.31	157 58 36.37	-24.710	-17.660	3.78
HI	KAHUKU SOUTHEAST BASE 1948	19 4 16.72	155 43 51.68	-66.770	11.560	-1.13
HI	KAPU ASTRO 1961	21 24 20.34	158 6 3.00	-21.390	-3.100	19.81
HI	KAPUKAMAA 1884	19 31 25.27	155 54 36.27	-34.810	60.260	16.65
HI	KAUPULEHU 1906	19 43 6.99	155 54 40.55	-2.810	49.350	5.29
HI	KANAIWAI 2 1948	20 2 43.69	155 36 7.27	-1.710	15.450	0.78
HI	KEALAHEHA 2 1948	20 16 4.16	155 52 14.45	21.040	-2.250	-15.16
HI	KEOPAHEO 1927	21 56 38.84	159 21 55.06	-30.940	-40.560	-2.98
HI	KOLEKOLE HGS 1876	20 42 37.79	156 15 32.37	-16.360	-8.420	-8.78
HI	LAAUKAHI 1910	20 56 32.51	157 27 42.50	-37.410	-23.500	-2.04
HI	LAHAINA SOUTHEAST BASE 1950	20 50 26.73	156 39 7.39	-3.170	5.720	-2.15
HI	LANI 1910	21 57 10.74	159 35 21.95	-38.640	-6.750	-7.18
HI	LUKE 1912	20 53 0.68	156 29 56.71	8.080	-20.130	-3.72
HI	MAUNA LOA 1872	21 27 26.78	157 9 35.12	2.690	-10.320	-9.48
HI	MOKAPU 1872	21 17 2.43	154 4.67	11.690	-25.930	-6.18
HI	NIU LONGITUDE 1927	22 5 1.80	159 40 9.42	-28.550	-17.040	-0.27
HI	PELE 1910	21 56 44.04	159 32 38.10	-6.000	-0.570	3.57
HI	POHAKEA 1910	19 58 36.32	155 49 32.75	-39.620	29.980	10.24
HI	PUAKO NEW 1948	21 53 45.13	159 36 26.35	6.580	-4.950	-1.85
HI	PUOLO 1910	21 7 35.88	158 13 18.87	-44.330	20.990	7.22
HI	PUU AIEA HITS 1948	21 34 20.23	158 15 55.33	11.880	5.550	2.04
HI	RAO 1958	21 45 50.18	159 29 59.97	-5.680	13.580	4.04
IA	AUDUBON 2 1968	41 37 59.53	91 35 57.13	-0.310	-0.630	-0.02
IA	AVIATION 1930 RM 4 1969	41 48 2.57	94 28 28.43	-4.080	-3.780	-2.52
IA	BAGLEY 1930	41 36 2.69	93 9 3.41	-2.940	-3.120	-2.08
IA	BAXTER 1969	40 50 57.10	93 2 55.66	-1.710	-1.510	-0.99
IA	BURGUS 1931	41 39 32.46	91 48 14.48	-2.820	-2.640	-1.75
IA	CELLMAN 1930	41 42 8.08	92 0 15.05	-2.820	-1.110	-0.74
IA	CONROY 1930 RM 5 1969	40 43 34.31	93 21 44.59	-0.510	-1.470	-0.96
IA	CROYDON SOUTH BASE 1928	42 35 0.93	96 9 16.26	-2.320	-2.740	-1.85
IA	DAHLMAN 1934	40 49 41.04	92 23 13.90	-0.750	-2.140	-1.40
IA	DAVIS 1934	42 50 32.46	95 19 43.19	1.170	-7.300	-4.96
IA	DIVIDE 1940					

STATE	STATION NAME	GLAT(D,M,S)	GLON(D,M,S)	ALAT-GLAT(S)	ALON-GLON(S)	LAPLACE(S)
IA	DURANT 1930 RM 3 1969	41 34 13.52	90 52 52.69	-1.290	0.650	0.43
IA	FAIR 1930	41 32 10.41	90 37 20.75	-2.710	-0.280	-0.19
IA	GARDEN 1969	41 54 10.96	93 47 49.50	-15.860	-14.500	-9.68
IA	GILMAN 2 1969	41 50 51.49	92 50 1.33	-1.440	-1.320	-0.88
IA	HARDY 1938	42 49 48.65	94 5 23.10	4.180	3.600	2.45
IA	HARTFORD 1969	41 45 47.95	93 17 46.90	-3.100	-0.490	-0.32
IA	IRWIN 2 1968	41 45 39.57	95 12 16.91	8.190	10.300	6.86
IA	JOICE 1926	43 21 34.42	93 24 29.98	8.490	11.820	8.11
IA	KUBIK 1934	43 17 34.04	92 17 42.28	-2.460	1.470	1.01
IA	LIBERTY EAST BASE 1930	41 33 1.79	91 18 32.84	-1.600	-0.100	-0.06
IA	LIFE 1934	43 16 3.27	94 51 7.41	-3.540	-0.980	-0.68
IA	MAGNOLIA 1968	41 40 16.92	95 53 8.00	-3.580	-7.870	-5.23
IA	MC CLEARY 1928	41 41 3.10	93 25 53.70	-3.240	-4.450	-2.96
IA	MEFFORD 2 1968	41 42 12.72	95 38 38.82	-0.120	-3.710	-2.47
IA	MCSCCM 1969	41 34 29.63	91 7 11.94	-1.620	0.020	0.01
IA	OSAGE 1971	43 16 59.46	92 48 51.12	-6.680	-10.560	-7.24
IA	PATTEN 1930	41 45 0.52	95 23 15.50	-2.310	-15.670	-10.45
IA	PERRY 1969	41 49 55.56	94 8 24.02	-11.790	2.910	1.93
IA	PHALIA 1968	41 42 11.10	95 23 50.19	7.490	-7.380	-4.93
IA	RICHESON 1928	41 52 36.01	93 27 38.04	-7.450	-10.910	-7.41
IA	SCHRAGE 1938 RM 3 1971	42 49 7.08	92 53 3.31	-4.640	-5.160	-3.53
IA	SHELDON 1934	43 11 4.41	95 51 0.19	-1.890	1.520	1.01
IA	SHERIDAN 1969	41 50 52.76	92 33 40.74	-2.350	-4.450	-3.10
IA	SPENCER 1971	43 9 40.95	95 8 54.72	-3.030	-4.450	-3.05
IA	TIICNKA 1971	43 14 5.39	94 2 54.13	0.210	0.130	0.09
IA	TJEBBEN 1934	42 30 9.78	94 36 26.74	2.220	8.000	5.33
IA	VIOLA 1934	41 46 10.60	94 46 21.69	-0.630	3.290	2.29
IA	WESTGATE 1934	42 47 17.06	91 56 56.28	-41.160	8.600	5.99
ID	ASHTON NORTHEAST BASE 1946	44 5 41.48	111 25 19.80	-2.800	9.120	6.34
ID	ASHTON SOUTHWEST BASE 2 1958	44 1 57.79	111 50 31.49	-10.550	10.480	7.05
ID	BATTLE 1963	42 12 55.05	116 18 35.13	-2.470	6.000	4.44
ID	BUMBLEBEE 1940	47 40 33.42	116 9 44.41	10.010	-11.640	-8.58
ID	BUNKER 1968	47 31 25.56	113 55 37.73	-6.280	-13.590	-9.58
ID	CHIPPS 1968	45 10 42.76	113 20 40.27	-5.210	4.540	3.13
ID	CINDER 1960	43 37 7.54	117 3 8.61	-6.510	4.500	3.14
ID	EATON EAST BASE 1945	44 16 3.38	117 6 38.49	-7.260	3.640	2.54
ID	EATON WEST BASE 1945	44 17 4.46	116 10 54.49	4.520	4.050	2.91
ID	FENN EAST BASE 1945	45 57 41.33	116 17 51.86	-0.560	7.320	4.95
ID	FENN WEST BASE 1945	45 57 41.38	116 34 8.74	1.090	8.720	5.90
ID	FILER EAST BASE 1946	42 34 15.61	114 37 25.17	1.270	5.270	3.90
ID	FILER WEST BASE 1946	42 34 22.89	116 57 58.08	-2.890	10.710	7.93
ID	HAUSER 1968	47 45 9.40	116 35 48.25	-1.370	6.560	4.80
ID	HUCKLEBERRY 1940	47 44 51.94	111 55 30.34	-0.050	9.570	3.49
ID	IDAHO FALLS NE BASE 1946	43 39 1.48	112 1 26.94	-1.570	5.550	4.03
ID	IDAHO FALLS SW BASE 1946	43 30 35.86	113 41 35.24	-0.780	2.900	2.06
ID	MOUNTAIN HOME 1915	43 7 43.34	113 28 15.25	-2.490	11.810	10.40
ID	RUPERT EAST BASE 1950	42 37 9.07	113 52 3.67	0.960	12.800	8.23
ID	SALMON NORTHWEST BASE 1945	45 9 58.56	113 45 44.31	2.710	10.870	7.28
ID	SALMON SOUTHEAST BASE 1945	45 6 45.04	114 56 34.07	-1.400	5.530	9.45
ID	STANLEY 1939	44 13 35.33	115 1 1.05	3.700	-0.330	3.94
ID	SUNSET 1958	47 33 49.27	115 6 5.74	3.660	5.530	-0.24
ID	THREE 1947	42 5 11.37	116 9 44.90	-4.390	-3.550	-2.30
ID	VAN 1945	45 26 12.01	116 19 24.35	1.810	1.140	0.96
ID	WHITE BIRD 1945	45 44 36.96	87 47 49.46	-0.120	1.540	1.02
IL	ASH GROVE USLS 1879	40 35 18.15	89 37 21.40	0.570	-1.510	-0.98
IL	BAER 1939 RM 3 1969	40 54 56.40	88 48 21.72	1.550	******	******
IL	BELLEVILLE EAST BASE 1930	38 31 57.68	89 20 25.71	5.510	1.540	1.02
IL	BORDING 1882	38 36 45.22	89 11 17.85	-2.560	-1.510	-0.98
IL	CHERRY 1930	41 25 17.61	88 21 12.16	1.040	1.340	0.86
IL	EDWARDS 1969	40 45 16.70	91 1 25.13	0.710	******	******
IL	EL DARA 1931	39 37 27.82	87 50 48.27	0.470	3.280	2.14
IL	FAIRMOUNT USLS 1879	40 1 35.50	89 17 19.40	0.290	-4.790	-3.11
IL	GINGRICH 1939 RM 4 1969	40 49 20.34	87 35 45.52	0.960	2.690	1.76
IL	GREER 1955	40 30 51.27	89 37 34.41	0.390	2.090	1.36
IL	JANSSEN 1969	40 50 43.17	88 54 35.14	3.140	0.790	0.44
IL	JOSE 1934	40 17 20.13	88 46 26.35	-0.790	******	-1.71
IL	MOUND 1883	39 4 5.32	87 7 46.17	2.400	0.290	0.18
IL	NEW HAVEN 1957	37 54 14.38	88 50.46	-0.100	******	******
IL	NEWTON 1883	38 55 28.47	88 2 34.78	-2.770	0.020	-0.07
IL	OAKLAND USLS 1879	39 42 31.65	88 6 8.52	-0.010	-1.550	-2.45
IL	OLNEY WEST BASE USLS 1879	38 51 38.45	90 21 45.12	0.760	-3.760	0.37
IL	ORICK 1930	41 20 48.10	88 12 49.14	-0.240	0.590	-0.77
IL	PARKERSBURG USLS 1879	38 34 51.40	88 12 29.83	0.540	-1.170	-2.45
IL	PIPER 1939	40 40 31.41	89 55 25.58	1.350	3.750	2.45
IL	RICE 1934	39 42 46.52	88 1 53.14	-0.140	-5.730	-3.74
IL	RUSSEL 1930 RM 5 1969	41 15 36.97	88 59 18.43	1.010	1.600	0.93
IL	SCHLOSSER 1939	40 52 14.84	87 59 29.60	-0.440	-1.040	-0.68
IL	SPRING CREEK USLS 1879	40 35 29.60	89 39 17.36	-1.650	0.170	0.11
IL	SPRINGFIELD STATE CAPITOL DOME	39 47 54.10	87 43 15.50	-0.800	1.660	1.10
IL	ST ANNE USLS 1879	41 1 19.27	89 52 7.35	4.800	-0.830	-0.39
IL	STARK 2 1969	41 2 43.24	88 38 5.89	0.130	2.111	1.42
IL	SWYGERT 1939	40 54 8.16	90 41 4.18	-0.290	1.360	0.88
IL	TAYLOR 1930 RM 3 1969	41 23 11.43	88 20 58.36	1.840	-6.780	-4.50
IL	THOMAS 1961	38 38 14.47	84 49 0.93	1.180	-4.740	-2.42
IL	VOGHT 1935	42 25 22.30	88 42 22.96	1.230	-6.370	-0.33
IL	WATKINS 1934	40 16 55.46	87 51 5.79	-1.190	3.590	2.29
IL	WILLOW SPRINGS USLS 1874	41 43 36.79	85 30 34.65	1.330	6.200	0.18
IL	ALVARADO 1932	41 33 50.52	86 29 38.17	-0.930	******	******
IN	BANGS 1879 RM 4 1970	38 19 37.00	87 19 50.91	0.620	-4.050	-2.36
IN	BAUGH 1939	40 18 56.49	86 1 47.31	5.180	1.480	0.98
IN	BOGUE 1939 RM 4 1970	39 49 31.53	86 39 42.90	-0.550	-6.450	-3.96
IN	BOSTON 1932	39 42 40.49	85 27 40.86	-0.150	2.620	1.69
IN	CAMMACK 1934 RM 3 TPC	40 12 51.76	85 53 20.68	-2.440	0.150	0.10
IN	CARLISLE USLS 1877	41 40 5.03	85 30 28.75	-0.140	0.340	0.21
IN	CEDAR 1955	40 28 34.56	87 19 50.91	-1.350	-0.620	-0.45
IN	COX 2 1934	40 14 40.41	86 1 47.31	0.810	-5.110	-3.40
IN	DEER 1946	37 57 47.59	86 39 42.90	1.840	-0.910	-0.60
IN	DERROW 1939	35 59 58.20	85 27 40.86	0.050	-0.330	-0.21
IN	EEL 1947	40 54 42.10	85 53 20.68	2.380	-0.610	-0.39
IN	FENLEY 1947	39 22 49.45	85 30 28.75	4.250	-3.560	-2.31
IN	FOREST 2 1968	40 31 54.02	86 7 36.43	-4.250	-3.950	-5.98
IN	FRANKTON 1934	40 14 6.96	85 45 32.81	0.560	1.190	0.77
IN	FREMONT USLS 1878	41 42 31.98	84 57 58.38	-2.310	-3.850	-2.49
IN	GOODELL 1947	41 2 35.27	86 52 38.10	-0.780	-0.370	-0.23
IN	GREEN 1890	39 5 9.42	85 30 10.30	4.160	-5.250	-3.52
IN	HANCOCK 1920	40 0 2.87	86 0 31.94	2.400	-2.770	-0.48
IN	HILLSBURG 1934	39 46 6.87	86 9 43.56	1.470	2.710	1.75
IN	INDIANAPOLIS CAPITOL DOME FLAG	39 46 40.21	84 48 8.65	0.590	-1.510	-0.98
IN	INDOH 2 1970	40 39 40.21	84 36 13.88	1.130	-0.010	-0.01
IN	JEFFERSON 1934 RM 3 1970	40 18 4.72	85 40 11.68	2.270	-6.460	-4.24
IN	JOHNSON 1939	39 34 43.32	86 1 27.52	-1.420	-6.460	-4.24
IN	KINNICK 1939	39 37 25.78	84 48 12.84	-0.300	-4.340	-2.78
IN	MARYSVILLE 1914	38 35 20.79	86 3 7.55	3.330	-3.940	-2.43
IN	NIXON 1934	40 0 16.19	85 38 52.26	-0.090	******	******
IN	NORMANDA 1934 RM 2	40 18 7.12	85 8 24.20	1.190	2.050	1.33
IN	O ANO M 1884	38 29 0.13	85 40 11.68			
IN	OAKLEY 1948	40 42 24.43	86 1 27.52			
IN	PAYNE 1932 RM 3 1970	41 2 56.18	84 48 12.84			
IN	PENDLETON 1920	39 48 48.63	86 3 7.55			
IN	POTTS 1880	38 5 23.63	85 8 24.20			
IN	REIZIN 1889	39 2 53.49	85 40 11.68			
IN	SELMA 1934	40 11 43.42	85 16 20.77	1.190	2.050	1.33

STATE	STATION NAME	GLAT(D,M,S)	GLON(D,M,S)	ALAT-GLAT(S)	ALON-GLON(S)	LAPLACE(S)
IN	SNYDER 1934	38 36 59.18	87 22 4.17	0.850	-4.650	-2.90
IN	STOUT 1890	38 51 12.35	85 34 42.30	1.410	-0.650	-0.41
IN	SHANEY 2 1969	40 17 45.65	87 9 51.25	1.220	-2.590	-1.67
IN	TERRE NORTH BASE 1947	39 28 27.16	87 15 58.98	2.350	-2.860	-1.82
IN	VINCENNES CTHSE CENTER CUPOLA	38 40 35.11	87 31 37.48	1.180	-0.080	-0.05
IN	WALTING 1932	39 54 39.05	84 51 22.61	-1.570	1.320	0.84
IN	WEED PATCH 1889	39 10 0.49	86 13 0.93	0.060	*******	******
KS	ABI 1961	38 56 58.51	97 15 28.92	-7.730	-11.490	-7.23
KS	ADAMS 1888	39 2 41.08	96 4 24.34	0.560	*******	******
KS	ASCO 1961	39 24 49.55	97 40 52.29	0.860	3.460	2.19
KS	BADGER 1956	38 24 44.55	96 2 45.39	-0.250	-2.910	-1.81
KS	BATES 1967	38 27 33.70	98 32 41.47	-0.570	-2.500	-1.59
KS	BENNET 1934	37 58 4.20	102 2 37.21	-0.480	-3.270	-2.01
KS	BIG CREEK 1893	38 55 39.32	99 54 22.44	-1.060	-0.530	-0.33
KS	BROWN 1898	39 31 33.55	95 44 24.33	-3.610	-5.100	-3.24
KS	BULLARD USGS 1923	39 46 37.14	98 42 16.70	-1.890	*******	******
KS	BURLINGTON 1948	37 52 35.29	100 52 51.11	*******	0.890	0.55
KS	BUSHBY 1950	38 33 28.24	95 18 14.54	0.350	-1.460	-0.99
KS	CALKIN 1968	39 7 28.98	100 42 32.72	-0.730	-1.670	-0.99
KS	CAR 1961	37 29 10.32	98 8 4.67	-0.580	-0.690	-0.60
KS	CARLTON 1961	38 45 9.23	98 3 8.39	-1.780	-4.740	-2.96
KS	CHAP 1961	38 40 22.53	97 19 20.29	-1.840	-8.770	-5.48
KS	CHEYENNE 2 1951	38 55 51.08	97 3 56.56	-4.890	-9.020	-5.67
KS	CLEAVES 1934	39 36 16.81	101 56 14.25	-1.290	-1.670	-1.07
KS	COOPER 1858	37 26 23.35	101 51 8.16	-1.590	-6.330	-3.85
KS	CROFT 1938	39 58 41.96	98 35 23.95	1.290	-0.880	-0.57
KS	CUNNINGHAM 1953	37 31 41.81	99 3 3.71	-4.590	-2.500	-1.53
KS	DIRKS 1934	38 41 43.98	97 27 35.36	-2.860	-3.100	-1.89
KS	ELLSWORTH ASTRO STATION 1885	37 25 36.50	95 16 21.38	3.690	-2.200	-1.34
KS	ERWIN 1935	38 43 48.75	98 13 44.16	-1.180	-7.470	-4.68
KS	FREDRICK 1952	37 55 17.14	96 52 13.24	1.330	-3.930	-2.42
KS	GILMORE 2 1953	38 28 42.57	98 17 36.88	-2.250	-6.710	-4.17
KS	GLOW 1960	38 10 1.08	97 57 12.07	-1.800	-2.950	-1.82
KS	GREGG 1934	38 41 43.96	95 52 45.77	-0.520	-1.990	-1.24
KS	HACK 1963	37 13 29.00	99 9 54.35	-3.490	-1.650	-1.00
KS	INGTON 1960	38 13 35.33	100 9 39.18	-0.230	-0.280	-0.17
KS	JANDA 1967	39 6 2.47	97 32 35.29	-3.860	2.330	1.47
KS	KEENE 1960	38 43 56.79	98 25 51.11	0.900	-5.810	-3.63
KS	KINSLER 1967	38 58 35.08	96 2 55.47	-0.180	-2.930	-1.84
KS	KOONS 1967	37 43 26.45	98 10 59.21	-2.190	-0.330	-0.20
KS	LIBERAL 1933	38 16 5.23	98 16 31.04	-2.450	-6.270	-3.88
KS	MACK 1961	37 6 43.41	100 54 16.56	-2.670	-0.400	-0.23
KS	MC ALLISTER 1951	38 28 31.26	97 38 19.94	-0.660	-2.220	-1.38
KS	MEADES RANCH 1891	39 13 26.686	101 22 24.51	-1.490	0.430	0.28
KS	MELT 1960	38 25 47.56	98 32 30.51	1.040	-3.340	-2.11
KS	MILLER 1902	37 2 21.02	95 40 25.40	0.680	-2.860	-1.78
KS	MINN 1960	39 9 53.53	97 55 43.69	2.580	-1.050	-0.63
KS	MITCH 1961	38 23 31.37	97 40 1.56	-3.090	-2.900	-1.83
KS	MONDAY 1968	37 14 22.23	58 5 5.90	-1.780	-3.770	-2.34
KS	MT HOPE 2 1966 RM 6 1972	39 7 43.91	98 3 59.43	0.360	-0.250	-0.15
KS	NEWTON 1972	38 3 32.91	96 40 38.55	-0.380	-0.610	-0.38
KS	NORTH POLE MOUND 1890	38 57 9.93	97 19 20.07	-0.480	-1.020	-0.64
KS	PANTHER 1935	37 4 38.52	57 36 31.21	-4.380	0.060	0.04
KS	PARADISE 1950	37 4 27.57	96 53 27.57	0.930	-2.450	-1.47
KS	PRATT A H 1953	37 37 5.55	98 55 54.34	-0.370	0.470	0.30
KS	QUARRY 1901	37 14 28.38	98 44 2.27	-3.720	-3.620	-2.20
KS	RESERVOIR 1967	38 57 35.01	97 53 19.63	0.560	-1.250	-0.76
KS	ROCK 1964	37 11 35.89	99 56 51.47	1.020	-3.930	-2.47
KS	RUSSELL SOUTHEAST BASE 1892	38 51 22.28	98 47 8.04	-4.270	-0.840	-0.51
KS	SALINA WEST BASE 1895	38 51 7.14	97 36 10.80	0.620	*******	******
KS	SIMMONS 1887	38 47 4.17	99 26 4.12	-4.170	-0.820	-0.52
KS	SMOKY HILL 1893	38 43 35.02	99 2 53.65	-0.180	-1.930	-1.21
KS	STEEL 1961	37 38 45.35	99 53 46.69	-0.820	-1.570	-0.98
KS	SUNFLOWER 1901	37 51 40.33	97 51 27.89	-2.840	-0.190	-0.12
KS	TES 1960	39 4 1.80	98 0 55.8C	-2.380	*******	******
KS	TEST 1960	39 0 59.09	97 50 57.74	-0.810	-0.040	-0.02
KS	TOULON 1950	38 50 58.94	97 57 7.23	-2.430	-0.130	-0.08
KS	TULLY 1950	39 5 29.52	100 16 39.05	-0.960	-2.000	1.26
KS	TURTLE 1891	39 1 17.72	101 45 25.77	-0.640	-2.450	-1.54
KS	UNION 1951	39 6 15.61	101 2 4.24	0.040	0.340	0.21
KS	VALLEY 1972	38 1 41.92	97 46 20.80	-0.500	-2.930	-1.81
KS	WAGLER 1967	38 0 49.19	94 14 35.30	-0.660	-4.630	-2.85
KS	WALLACE LATITUDE STATION 1885	38 54 44.13	101 35 31.60	-2.900	-5.310	-3.33
KS	WEBSTER 1950 RM 3 1967	39 43 36.07	98 33 36.21	0.250	-0.110	-0.07
KS	WESKAN 1954	38 52 19.85	101 56 58.50	-2.840	-2.800	-1.76
KS	WICHITA 1940	38 43 21.40	97 17 37.96	0.640	*******	******
KS	WIL 1961	38 5 7.17	98 25 59.86	0.860	1.340	0.82
KY	BLUEBALL 1933	37 44 15.21	86 2 29.54	1.570	-2.850	-1.79
KY	BOYD 1934	37 51 22.56	88 25 59.97	-2.860	-6.340	-3.88
KY	BUZZARD 1933	37 25 19.60	86 6 23.44	*******	-0.270	-0.16
KY	CASH 1933	37 33 6.61	86 4 50.69	4.810	-5.570	-3.38
KY	COYLE USGS 1928	38 10 26.72	83 45 54.04	2.950	-7.460	-4.55
KY	CUMMING 1935	36 35 12.86	85 14 4.47	3.640	5.880	3.63
KY	DAMES 1933	37 9 50.54	86 11 11.34	2.500	1.980	1.18
KY	FERRIELL 1883	37 37 25.32	85 26 16.77	4.500	-3.170	-0.97
KY	GAITHER 1930	38 57 19.85	87 31 21.26	*******	-1.590	-0.97
KY	HENDERSON 1934	37 50 19.85	87 35 30.07	0.570	-2.390	-1.44
KY	MINERVA 1887	38 42 30.96	83 55 6.84	-4.640	-2.740	-1.69
KY	PILOT 1933	37 5 7.39	86 10 20.57	-0.280	*******	******
KY	POTTER 1933 RM 4 1970	36 55 5.84	86 21 27.20	4.360	-1.970	-1.18
KY	RACCOON 1928	37 9 3.10	83 59 31.17	4.400	-0.150	-0.08
KY	SNOW 1933 RM 5 1970	37 53 49.54	86 0 8.15	*******	-5.010	-9.02
KY	WARFIELD 1935	38 13 39.56	84 55 55.17	2.360	-2.750	-1.69
LA	ADAMS 1965	30 51 27.09	93 10 24.11	-0.080	-1.290	-0.80
LA	ALBERTA 1934	29 22 21.96	89 34 23.62	*******	1.100	0.50
LA	ALEXANDRIA S BASE 1934	31 10 32.46	92 25 6.04	-2.860	-1.090	0.54
LA	ALLAIN 1929 RM 3 1970	30 53 20.49	91 21 8.47	-1.350	0.110	-0.05
LA	ALSAITA 1929	32 36 45.04	91 12 46.54	-1.820	-3.750	-2.02
LA	ANACOCO 1935	30 53 14.08	93 29 35.30	-0.010	2.200	2.20
LA	AVERY 1931 RM 2 1969	29 54 14.20	91 54 2.81	6.810	1.300	2.20
LA	AVOCA 1931	29 38 10.81	91 6 45.93	5.500	0.650	0.65
LA	BEEKMAN 1939	32 55 40.15	91 51 23.65	0.970	0.510	0.28
LA	BENNETT 2 1970	32 51 7.74	91 14 25.45	-0.040	-2.930	-1.73
LA	BOND 1965	31 22 59.72	92 37 11.36	-0.580	-0.050	-0.10
LA	BOYCE 1970	31 22 59.72	93 9.68	-2.600	1.200	0.63
LA	BROOKS 1931	32 25 9.37	92 7 1.09	0.930	-0.210	-0.11
LA	BUELL LA GS 1939	30 30 45.89	90 26 24.20	2.790	1.020	0.52
LA	BUNDICK 1935 RM 3	30 44 42.10	93 1 32.11	-1.350	0.480	0.24
LA	BURNS 1931	32 53 28.11	93 14 19.70	1.810	2.550	1.37
LA	BURROOD LATITUDE STATION 1921	28 57 56.36	89 22 51.38	1.480	*******	******
LA	CALHOUN 1965	32 35 36.14	93 0 0.22	-0.480	-1.510	-0.81
LA	CALLMAN 1965	32 19 28.67	91 5 4.71	0.330	-1.270	
LA	CAMPTI 1941	31 55 3.78	93 5 55.65	-1.670	-1.630	
LA	CARRON 1965	30 37 54.67	92 56 56.35	-0.010	0.060	0.03
LA	CATAHOULA 1941	31 31 13.28	92 1 1.89	-3.570	-0.200	-0.10
LA	CHAUSON 1919	31 2 37.92	91 57 52.75	-1.630	0.970	0.53
LA	CHENAL 1 1966	30 36 26.65	91 23 17.90	-0.630	0.230	0.30
LA	CONNELL 2 1970	30 25 52.31	91 5 28.45	-0.090	0.560	0.37
LA	DEER ISLAND 1855	29 28 43.94	91 16 2.23	4.190	*******	******
LA	DONALDSONVILLE RM 4 1973	30 6 25.63	90 59 17.43	4.770	1.450	0.72
LA	DONALDSONVILLE 1929	30 6 26.44	90 59 18.76	4.450	2.340	1.18

STATE	STATION NAME	GLAT(D,M,S)	GLON(D,M,S)	ALAT-GLAT(S)	ALON-GLON(S)	LAPLACE(S)
LA	DRAKE 1941 RM 3	30 56 56.93	93 9 25.49	-2.820	1.660	0.85
LA	FARM 1965	32 45 20.16	93 3 22.83	2.900	1.260	0.68
LA	FLORENVILLE RESET 1936	30 25 0.17	89 49 56.93	4.520	-2.250	-1.14
LA	GILBERT 1941	31 57 26.25	91 34 46.48	1.890	-0.960	-0.51
LA	HARVEY 1930	29 54 28.87	90 5 2.50	5.070	0.820	0.41
LA	HEAD OF PASSES ASTRO STA 1853	29 8 31.34	89 15 6.48	1.750	********	******
LA	HOMEWOOD 1931	30 7 36.43	93 4 49.86	3.280	0.780	0.39
LA	HUDSON 1934	32 2 42.03	92 34 47.88	-0.670	3.000	1.59
LA	INDIAN 1965	31 18 18.60	93 10 0.88	-0.880	4.060	2.11
LA	KROTZ 1935 RM 3	30 32 14.52	91 45 20.23	-0.540	1.130	0.58
LA	LAFITTE 1934 RM 5 1972	29 40 1.85	90 6 33.65	6.090	1.430	0.71
LA	LIVONIA 1966	30 34 26.18	91 34 24.95	-0.100	1.760	0.90
LA	MAN 1919	32 1 18.76	93 43 1.47	-0.110	-0.950	-0.51
LA	MAY 1939	30 58 34.10	90 24 35.16	-1.550	-0.910	-0.47
LA	MILLER 1966	30 57 17.17	90 14 27.70	-1.940	-3.150	-1.62
LA	MOORE 1965	32 52 44.27	93 0 30.96	-1.290	1.010	0.55
LA	NATCHITOCHES 1941	31 44 55.21	93 5 42.33	-1.290	1.730	0.91
LA	NEW ROADS WEST BASE 1929	30 42 54.58	91 33 10.03	-0.960	0.860	0.44
LA	PERILLOUX 1929	30 4 45.99	90 27 41.12	4.040	1.330	0.67
LA	PIERRE 1941	31 47 8.62	93 11 49.55	-0.740	1.680	0.88
LA	PINE 1939	30 54 21.20	90 0 47.47	-1.200	-2.110	-1.08
LA	PINE 1965	30 43 38.28	92 50 38.55	-1.310	-0.490	-0.25
LA	PLANER 1966	30 53 57.39	90 48 53.77	-1.540	-0.670	-0.34
LA	PONTCHARTRAIN RM 7 1972	30 4 30.96	89 56 36.98	4.040	-0.120	-0.06
LA	PORT EADS WIRELESS POLE	29 0 58.14	89 9 47.17	0.510	********	******
LA	SAILES 1941	32 25 57.57	93 4 42.56	-0.600	2.270	1.22
LA	SCHRIEVER N B RM 5 1972	29 44 45.53	90 48 57.64	5.210	0.920	0.46
LA	SHREVEPORT NORTH BASE 1930	32 24 18.25	93 47 37.98	1.890	0.710	0.38
LA	SLAGLE 1941	31 10 22.45	93 9 38.11	-1.070	3.830	1.98
LA	SONES 1966	30 57 15.35	89 48 17.39	-2.310	-2.590	-1.33
LA	ST MARY 1935	30 35 28.52	91 57 18.36	-0.920	-0.300	-0.15
LA	ST PETERS 1966	30 48 34.25	91 0 8.00	-0.410	-0.130	-0.07
LA	SULLIVAN 1965	32 16 12.41	93 8 47.64	-0.140	2.890	1.55
LA	TANGI 1939	30 52 12.34	90 27 42.70	-1.150	-0.020	-0.01
LA	THOMPSON 1966	30 42 20.26	91 11 26.67	1.320	-0.130	-0.07
LA	TIMBALIER LIGHTHOUSE	29 2 56.57	90 21 20.92	-2.450	********	******
LA	TORRAS 1919	30 59 22.89	91 40 45.18	-2.260	0.700	0.36
LA	VIDRINE 1934 RM 3	30 41 32.69	92 23 48.46	-0.660	-0.290	-0.15
LA	VILLAGE 1931	30 15 15.30	92 24 14.83	-1.700	-0.810	-0.43
LA	VOWELLS 1941	31 37 29.07	93 15 0.80	-1.920	2.440	1.28
LA	WELSH 1931	30 13 27.44	92 47 40.37	2.070	0.430	0.22
LA	WHITAKER 1965	30 48 49.94	93 12 26.81	-2.090	1.230	0.63
LA	WILLETTS 1929	31 30 2.74	91 31 50.19	-2.460	-0.740	-0.39
LA	WILSON 1965	31 28 1.79	91 11 59.91	-0.210	4.290	2.24
LA	YOUNG 1939	30 53 22.71	90 35 57.88	-1.110	-0.020	-0.83
MA	BEECHILL 1938	42 5 40.65	70 54 21.74	-3.150	-1.240	-0.83
MA	BUMSKIT 1896 RM 6	42 18 4.49	71 53 50.46	-4.980	-6.230	-4.19
MA	CAMBRIDGE OBSERVATORY	42 22 52.93	71 7 43.92	-4.820	1.780	1.20
MA	CAPE COD LIGHTHOUSE 1877	42 2 22.56	70 3 39.84	-0.860	********	******
MA	CLOVERDEN OBS ASTRO LAT STA	42 22 45.18	71 7 47.64	-4.760	-3.660	-2.44
MA	CROWELL 1938	41 40 0.18	69 58 8.46	-0.580	********	******
MA	DUXBURY ASTRO LONG STA C TRANS	42 2 54.58	70 40 11.02	********	1.900	1.27
MA	FOLGER 1867	41 17 16.40	70 6 16.71	-2.340	********	******
MA	HASKELL 1937	41 43 55.14	71 7 39.83	-2.040	-0.530	-0.35
MA	HOLT 1833	42 38 27.47	71 6 24.88	-1.330	4.350	2.95
MA	INDIAN 1835	41 25 46.16	70 40 40.83	-1.660	3.770	2.49
MA	MANOMET 1835	41 55 38.11	70 35 22.21	-2.760	********	******
MA	MOUNT TOM 1860	42 14 30.32	72 38 55.11	-2.160	-0.050	-0.03
MA	PEAKED CLIFF 1848	41 48 30.81	70 32 25.15	-1.680	4.250	2.83
MA	POET SEAT 1936	42 35 40.20	72 35 13.03	-2.860	-1.550	-1.05
MA	POVERTY ROCK 1889	42 43 34.03	70 56 4.43	-4.370	3.570	2.42
MA	POWOW 1889	42 51 56.44	70 56 18.75	-0.720	3.430	2.33
MA	PROVINCETOWN TOWN HALL SPIRE	42 3 4.21	70 11 17.40	-4.500	-1.580	-1.06
MA	SHAWAUKEMO 1875	41 17 1.19	70 3 4.97	-3.690	-4.670	-3.08
MA	TADMUCK 1936	42 34 35.87	71 26 33.00	-1.970	1.570	1.07
MA	THOMPSON 1846	42 36 41.35	70 43 49.18	-3.080	********	******
MA	WACHUSETT 1833	42 29 20.27	71 53 14.00	-3.190	********	******
MA	WACHUSETT 2 1896	42 32 11.11	71 53 16.22	-2.670	-5.300	-3.58
MD	APPAL 1958	39 32 11.11	77 36 15.22	4.440	14.210	9.05
MD	ATHEY 1943	39 6 11.08	77 17 22.75	1.230	4.150	2.62
MD	BLOSSOM 2 1966	38 25 52.69	77 17 17.19	-8.990	-5.59	
MD	BUCKLER 1934	38 33 32.87	76 31 13.32	-2.070	-10.030	-6.25
MD	CALVERT 1854	38 23 33.53	76 31 35.14	2.500	3.130	1.98
MD	CEDAR HEIGHTS 1969	38 55 12.24	77 13 42.57	4.490	6.560	4.15
MD	DALE 1943 RM 1 1969	39 17 16.73	77 15 42.46	3.160	4.170	2.64
MD	DAMASCUS 2 1953 RM 4 1969	39 17 5.94	77 12 43.67	-0.110	-2.980	-1.90
MD	DAVIS AMS 1968	39 37 28.32	77 59 12.30	1.530	-2.710	-1.73
MD	DAWSON 1959 RM 3	39 42 43.74	76 9 39.45	-0.450	-7.650	-4.70
MD	EMERY 1957	39 8 40.49	76 54 6.47	********	-0.960	-0.61
MD	FLIGHT TEST MARTIN CO 202 1956	39 19 4.27	76 24 36.64	-2.540	********	******
MD	GANTT MSFC 1908	38 20 28.75	75 6 25.43	-1.750	-8.470	-5.33
MD	GODDARD 1962	39 1 14.35	76 49 41.69	-1.720	-8.610	-5.41
MD	HILL 1846	38 53 54.57	76 52 49.94	-1.800	-1.560	-1.00
MD	HIPOINT AMS 1968	39 42 7.23	79 12 54.24	-1.490	-1.670	-1.00
MD	HOLT 1932	39 28 3.30	75 35 3.30	-1.480	-1.230	-0.78
MD	LAYTON 1943	39 12 44.02	77 8 34.12	0.310	-7.630	-4.83
MD	LEACH 1938 RM 3	39 20 0.82	76 44 30.10	-2.550	6.530	0.04
MD	MACAULAY 1966	39 9 47.51	76 53 48.69	-4.160	********	******
MD	MARRIOTT 1844	38 52 27.39	77 43 35.33	-2.830	9.050	5.74
MD	MARYLAND HEIGHTS 1869	39 20 27.39	77 43 0.08	5.290	-6.980	-4.44
MD	MEIGS 1933	39 33 44.20	76 2 56.53	-4.340	8.190	5.20
MD	MILL 1958	39 20 6.06	77 22 41.15	-3.130	-7.210	-4.58
MD	MOUNTAIN 1966	39 27 50.81	76 22 5.99	-0.190	-6.850	-4.27
MD	NAYLOR 1943	38 37 9.57	76 44 24.29	-1.680	0.630	0.63
MD	OBSERVATORY 1966	39 8 11.49	77 11 57.99	2.220	-5.890	-3.72
MD	OLNEY 1943	39 8 52.44	77 1 2.61	-3.500	-6.940	-4.43
MD	POLISH 1955	39 40 37.70	78 32 6.56	-0.430	********	******
MD	POOLES ISLAND 1844	39 17 7.66	76 15 49.53	-3.470	-7.970	-5.06
MD	PRINCIPIO 1845	39 35 36.31	76 0 16.56	-1.490	-14.590	-9.32
MD	RIDGELY 2 1966	39 25 0.28	77 9 36.82	-1.810	-8.850	-5.58
MD	ROCKVILLE ASTRO LAT STA 1892	39 5 10.68	77 9 36.82	-0.920	-3.780	-2.40
MD	SAMPSON 1927	39 42 30.18	78 55 64.58	-1.730	********	******
MD	SAT TRACK STA 002 1966	39 1 39.00	76 49 33.05	-0.310	-6.290	-3.97
MD	SCHMID 1962	35 28 18.97	76 4 15.22	3.700	9.340	5.91
MD	SIDELING 1955	39 41 24.09	78 17 59.69	4.240	********	******
MD	SOPER 1850	39 7 0.89	76 59 10.94	1.420	-0.650	-0.41
MD	SPENCER 1969	39 7 14.37	76 59 10.94	-2.050	********	******
MD	SUGAR LOAF ERDL 1957	39 15 44.54	77 23 39.44	-4.430	-6.730	-4.20
MD	SUGAR LOAF 1869	39 15 44.23	77 23 37.05	-0.250	********	******
MD	TABOR 1969	39 15 5.28	77 8 38.00	-2.160	-9.490	-5.98
MD	TAYLOR 1847	38 59 48.03	76 27 56.08	-2.830	-8.160	-5.12
MD	W 45 1957	38 38 11.93	77 0 11.35	-0.560	-4.960	-3.12
MD	WEBB 1850	39 5 26.21	76 40 10.33	2.750	-5.250	-1.83
MD	WELFARE 1970	39 3 17.53	76 51 32.82	0.830	2.740	1.87
MD	WOOD 1934 RM 5 1958	38 49 29.28	76 54 42.48	1.480	2.920	2.01
MD	WSC 1966	39 3 0.56	77 7 14.27	1.560	2.430	1.69
ME	ACTON 1976	43 15 22.05	70 57 49.32	-1.210	-1.770	-1.24
ME	AGAMENTICUS 1847	43 13 24.05	70 41 32.85	0.820	3.450	2.41
ME	ANDREW 1943	43 40 34.03	70 27 12.90			
ME	ARARAT 1933	43 56 24.77	69 57 36.18			
ME	BANGOR 1958	44 49 33.41	68 46 9.07			
ME	BEAR HILL 1858	44 8 38.88	69 6 24.85			

D2 201

STATE	STATION NAME	GLAT(D,M,S)	GLON(D,M,S)	ALAT-GLAT(S)	ALON-GLON(S)	LAPLACE(S)
ME	BEECH 1960	46 42 33.06	68 52 43.50	1.640	7.800	5.68
ME	BENTON 1958	44 35 59.36	69 35 14.68	0.890	3.710	2.61
ME	BIG BOG 1955	46 4 49.04	70 2 55.76	4.160	5.640	4.06
ME	BIGELOW 1928	44 42 47.40	69 43 39.48	-0.620	3.920	2.76
ME	BOUCHARD 1960	47 11 53.81	68 33 11.57	2.810	8.150	5.98
ME	CALAIS OBSERVATORY 1866	45 11 4.91	67 16 52.61	4.480	5.120	3.63
ME	CAMPBELL 1955	46 40 6.28	69 40 39.28	2.320	7.620	6.54
ME	CAPE NEDDICK 151 1941	43 10 7.38	70 35 43.98	-0.880	-0.780	-0.53
ME	CAPE SMALL 1852	43 46 42.57	69 50 44.10	0.920	*******	******
ME	CASWELL GSC 1916	47 1 52.05	67 49 20.94	-1.450	7.060	5.16
ME	CENTRAL 1958	44 36 18.58	69 36 22.12	0.760	3.360	2.36
ME	COOPER OR WESTERN RIDGE 1859	44 59 12.73	67 28 2.13	-0.090	*******	******
ME	CREA 1934	44 24 11.46	67 58 40.43	-1.160	5.370	3.76
ME	CUNNINGHAM 1863	44 21 28.34	68 36 53.65	2.060	4.150	2.90
ME	DAVIS 1862	44 39 3.60	67 13 15.25	2.700	5.250	3.68
ME	EASTPORT REF MARK IBC 1913	44 54 47.97	66 59 36.51	6.130	1.390	0.98
ME	EDGECOMBE 1855	43 57 13.10	69 36 18.37	0.670	0.830	0.58
ME	EPPING WEST BASE 1859	44 41 31.13	69 17 16.02	3.070	5.780	4.07
ME	FARMINGTON 1866	44 40 14.66	70 9 50.06	4.880	*******	******
ME	GRAFFTE 1929	45 40 59.83	70 16 50.32	-1.150	4.490	3.22
ME	GREAT DUCK ISLAND LIGHTHOUSE	44 8 31.52	68 14 47.21	1.680	7.490	5.21
ME	GREENVILLE AIRPORT 1955	45 27 50.87	69 33 7.53	-5.270	-0.730	-0.52
ME	HOWARD 1859	44 37 45.84	67 23 45.27	-2.840	*******	******
ME	HUMPBACK 1858	44 51 49.97	68 6 37.64	-2.840	*******	******
ME	ISLE AU HAUT CHURCH SPIRE	44 4 25.28	68 38 4.51	4.520	3.790	2.63
ME	ISLES OF SHOALS 1851	42 59 14.27	70 36 50.26	-1.400	*******	******
ME	JOHN 1913	43 52 9.68	68 54 5.83	-2.480	0.270	0.18
ME	KNOWLES CORNER 1941	46 12 17.38	68 20 28.90	-3.050	1.240	0.90
ME	LAWLER 1958	45 46 12.26	69 25 54.35	-1.280	1.300	0.93
ME	LIBBY 1958	44 51 26.33	67 45 45.98	-1.130	1.320	0.94
ME	LITTLETON 2 1959	46 13 9.35	67 52 42.72	-0.250	2.380	1.71
ME	MC KINNON 1955	47 0 2.27	68 10 6.94	2.730	6.460	4.73
ME	MICHAUD USGS 1942	47 2 12.65	68 37 29.37	-0.750	2.030	1.49
ME	MONHEGAN LIGHT	43 45 52.58	69 18 58.82	-0.680	*******	******
ME	MOUNT DESERT 1856	44 21 4.53	68 13 57.54	0.120	3.480	2.41
ME	MT HARRIS 1855	44 39 54.20	69 8 54.47	0.610	*******	******
ME	MT INDEPENDENCE 1849	43 45 33.16	70 19 14.23	1.280	*******	******
ME	MT PLEASANT NEW 1853	44 1 36.12	70 49 21.80	*******	3.100	2.16
ME	MT PLEASANT OLD 1853	44 1 36.45	70 49 22.41	-0.110	*******	******
ME	NEAL 1887	45 18 59.72	67 41 2.75	-0.020	-0.100	-0.07
ME	OSSIPEE 1851	43 35 18.30	70 44 27.57	1.500	-1.570	-1.08
ME	PIRATE IBC 1946	45 33 19.45	67 32 37.32	3.950	2.980	2.13
ME	RAGGED MTN 1854	44 12 44.72	69 9 5.19	-1.390	*******	******
ME	RIP 1955	45 52 44.13	69 11 43.33	-1.030	2.670	1.92
ME	RYE 1867	45 7 23.84	67 25 31.60	1.560	4.900	3.47
ME	SANTELLE 1955	46 11 48.06	68 40 40.17	0.140	1.430	1.03
ME	SEBATTUS 1853	44 8 37.22	70 4 43.59	0.360	*******	******
ME	SEBOEIS 1958	45 16 48.32	68 40 12.13	-0.470	1.310	0.93
ME	SEBOOMOOK 1955	45 53 53.11	69 44 45.95	-2.190	4.710	3.38
ME	SPRING HILL 2 1941	45 54 33.10	67 50 50.10	-1.100	5.030	3.70
ME	ST AGATHA 1942	47 14 15.20	68 19 46.77	-2.600	5.030	3.70
ME	THOMAS HILL OBSERVATORY	44 48 15.45	68 46 59.97	-2.580	2.190	1.54
ME	THOMAS 1933	43 46 42.73	70 11 8.82	2.470	3.680	2.55
ME	TRAMWAY USGS 1955	46 19 20.97	69 22 36.94	4.630	6.060	4.39
ME	WARD 1959	44 26 10.66	70 47 33.80	-4.660	-0.100	-0.07
ME	YOUNG USGS 1941	46 36 54.82	67 27 5.03	-0.020	3.270	2.38
ME	BARRY 1931	42 45 15.91	85 26 5.39	0.070	-2.420	-1.65
MI	BEAVER IS LIGHTHOUSE USLS	45 34 32.32	85 34 24.51	-3.540	*******	******
MI	BIG SABLE LIGHTHOUSE	44 3 27.84	86 30 51.57	-1.450	*******	******
MI	BORM 1932	43 24 2.11	84 4 16.75	3.700	-6.630	-4.55
MI	BROWN 1970	41 57 26.00	85 29 15.97	1.260	-3.720	-2.48
MI	BUNDAY USLS 1879	42 2 15.97	84 27 41.54	0.660	-6.410	-4.29
MI	BURNT BLUFF OLD USLS 1864	45 41 9.40	86 42 39.11	-5.720	*******	******
MI	BURNT BLUFF USLS 1874	45 41 9.40	86 39 39.11	*******	4.510	3.23
MI	CEDAR RIVER USLS 1864	45 25 48.52	87 19 34.38	-5.390	*******	******
MI	CHENEY 1932 RM 2 1976	44 34 20.51	84 41 21.32	-1.550	-5.860	-4.11
MI	CLIFF 2 1975	46 27 5.00	87 37 54.84	-2.780	-8.890	-6.45
MI	CCON 1955	46 32 52.44	88 16 6.82	6.180	-4.920	-3.57
MI	COPPER HARBOR ASTRO STA USLS	47 28 3.66	87 51 56.85	1.800	*******	******
MI	COVINGTON 1975	46 33 8.42	88 32 12.93	10.650	0.220	0.16
MI	CREBASSA USLS 1871	46 59 0.47	88 29 17.57	-19.460	*******	******
MI	DOSTER 1931	42 27 53.34	85 29 20.55	2.420	-3.760	-2.54
MI	DOUGLAS ASTRO LATITUDE USLS	42 38 11.05	86 13 43.82	0.450	*******	******
MI	DOUGLAS 1956	45 33 30.37	84 40 27.52	1.740	-7.410	-5.29
MI	DOVER 1932 RM 3 1976	43 54 19.56	84 46 5.53	0.530	-8.490	-5.89
MI	DOWN RIVER 1932	44 42 31.66	84 35 37.97	0.970	-4.630	-3.26
MI	DRIGGS USLS 1920	46 21 19.43	86 7 24.33	2.920	-1.540	-1.11
MI	DUCK LAKE ASTRO LAT USLS 1871	43 20 2.50	86 24 15.38	-0.570	*******	******
MI	DYGERT 1931	42 50 28.00	85 25 9.07	-0.650	-2.050	-1.40
MI	E PIER OBSERVATOR POST USLS	42 46 26.82	85 3 6.87	-2.050	-3.820	-2.57
MI	ECKERMAN USLS 1921	46 20 26.82	85 1 4.43	4.490	-4.990	-3.62
MI	FORD RIVER USLS 1874	45 41 12.06	87 6 8.69	-6.790	*******	******
MI	FOREST LAKE 1955	46 19 16.52	86 50 34.94	1.220	-0.370	-0.28
MI	FOX 1965	44 20 23.64	85 56 5.72	3.370	-1.020	-0.73
MI	FRANKFORT ASTRO LAT USLS 1871	44 38 52.49	86 15 4.24	-3.820	*******	******
MI	FROST 1969	42 50 32.82	84 52 0.71	0.370	-2.650	1.80
MI	GAYLORD 1956	45 2 35.16	84 40 24.75	3.240	-6.190	-4.38
MI	GREENBUSH USLS 1905	44 35 28.16	83 20 41.32	3.900	-3.140	-2.21
MI	GRIMM 1956	45 21 7.19	84 36 10.21	2.230	-7.150	-5.08
MI	GUILD 1932	43 24 8.08	84 51 55.00	2.850	-5.370	-3.69
MI	GUILD 1932 RM 5 1976	43 24 7.52	84 51 55.00	2.860	-5.370	-3.69
MI	GWINN 1955	46 17 3.60	87 29 8.15	-1.350	-6.880	-4.98
MI	HAMILTON 1932	42 27 38.84	83 32 40.18	-0.690	-6.500	-4.38
MI	HAMLIN 1932	44 4 2.41	86 25 17.59	1.270	8.180	5.69
MI	HEBRON 1956	45 32 25.43	84 37 38.38	0.950	-4.810	-3.38
MI	HISER 1965	44 52 49.80	84 44 22.42	0.360	-5.920	-5.81
MI	HOUGHTON 1932	44 15 16.29	84 41 14.62	1.990	-5.810	-4.05
MI	HUDSON 1932 RM 2 1975	45 10 16.56	84 45 7.11	1.710	-4.130	-2.93
MI	HULBERT 1965	46 23 13.78	85 12 35.74	6.420	0.950	0.69
MI	HURON MTNS LAT STA USLS 1866	46 52 41.48	87 23 43.54	10.820	*******	******
MI	INDIAN 1958	44 17 58.48	84 25 43.54	1.480	-6.070	-4.80
MI	ISLE ROYALE E LAT STATION USLS	48 7 43.45	88 33 34.07	11.600	*******	******
MI	JOHNSTON 1932	42 36 13.65	86 12 45.15	11.050	3.470	2.35
MI	KEWEENAW PT S BASE USLS 1871	46 52 17.69	88 29 16.16	4.660	-12.970	-9.47
MI	LAINGSBURG RM 3 1970	42 53 56.47	84 17 13.62	-0.330	-0.610	-0.41
MI	LE ROY 1931	44 45 45.01	85 27 45.44	1.330	-3.310	-2.30
MI	LEE 1969	42 23 13.06	84 6 56.01	1.340	-3.100	-2.09
MI	LEIFSEN 1932	42 26 53.91	85 50 50.68	1.280	2.920	1.57
MI	LITTLE 1932 RM 3 1976	43 44 11.37	84 48 27.70	1.890	-8.610	-5.95
MI	MACKINAC WEST BASE USLS 1852	45 47 13.73	84 46 22.46	-1.120	-4.620	-3.31
MI	MANISTEE ASTRO LAT USLS 1871	44 16 44.96	86 19 5.01	1.470	*******	******
MI	MAPLE HILL USLS 1896	45 5 28.09	84 47 43.09	2.390	-7.890	-5.70
MI	MC GENIAL 1932	44 7 18.89	84 43 43.09	2.390	-6.740	-4.68
MI	MC MILLAN USLS 1920	46 19 29.21	85 41 14.67	4.600	-0.870	-0.63
MI	MC PHERSON 1940	46 23 41.92	88 56 53.43	6.200	-3.490	-2.53
MI	MONROE COURTHOUSE USLS	41 54 52.59	83 23 49.09	-3.940	-3.340	-2.24
MI	MOUNT HOUGHTON USLS	47 24 26.00	87 58 28.90	-10.170	*******	******
MI	NEWBERRY USLS 1921	46 20 7.42	85 30 41.65	4.790	-0.060	-0.04
MI	OBRIEN 1932	43 32 35.55	84 45 30.02	3.790	-6.670	-4.59
MI	PAULDING 1939	46 24 1.68	89 13 5.81	4.950	-6.430	-3.15
MI	PORCUPINE MTNS LAT USLS 1870	46 47 1.46	89 43 49.98	2.780	*******	******
MI	PORTAGE ENTRY LIGHTHOUSE USLS	46 58 40.94	88 24 49.97	6.440	*******	******

STATE	STATION NAME	GLAT(D,M,S)	GLON(D,M,S)	ALAT-GLAT(S)	ALON-GLON(S)	LAPLACE(S)
MI	RED PINE 1976	44 24 43.10	84 41 1.86	1.940	-5.820	-4.08
MI	REPUBLIC 2 1975	46 23 30.85	87 58 47.99	-0.680	-3.320	-2.40
MI	ROCKWELL 1969	42 36 6.34	84 58 32.48	-0.330	-2.050	-1.39
MI	ROUND IS ASTRO STA USLS 1853	45 50 6.76	84 36 44.73	-0.880	*******	******
MI	RUMELY 1955	46 20 42.70	87 3 12.49	1.900	-4.310	-3.11
MI	RYAN 1969	43 8 21.05	84 52 23.69	-4.100	0.110	0.07
MI	S TWIN SISTER ASTRO LAT USLS	42 58 34.31	86 13 23.93	0.570	*******	******
MI	SAGMAN 1969	42 50 45.36	85 59 22.58	1.580	-1.100	-0.74
MI	SAYER 1970	42 11 3.74	84 51 0.96	1.270	-4.920	-3.30
MI	SHERMAN USLS 1878	41 50 48.46	85 27 11.85	-1.080	*******	******
MI	SHINGLETON USLS 1919	46 18 12.07	86 27 59.86	1.050	-0.320	-0.23
MI	SKANDIA RM 3 1975	46 21 12.65	87 13 12.30	1.690	-3.690	-2.67
MI	SOUTH HAVEN GEODETIC USLS 1873	42 16 8.80	86 20 43.76	-0.680	*******	******
MI	SYLVANIA 1940	46 14 39.34	89 18 45.76	-1.030	-3.300	-2.38
MI	TEICHMAN 1932	42 53 46.22	84 5 4.29	1.900	-6.010	-4.09
MI	TEPER RM 3 1975	46 22 41.15	88 50 13.59	4.290	-4.190	-3.03
MI	THONES HILL OBS POST USLS 1902	46 31 53.82	87 27 3.57	6.710	-14.130	-10.25
MI	THUNDER BAY ISLAND LIGHTHOUSE	45 2 14.17	83 11 39.33	-3.100	*******	******
MI	VERONA 1939	46 29 46.94	88 5 0.21	-3.550	-0.590	-0.43
MI	VULCAN USLS 1869	47 26 46.62	87 47 27.55	-2.110	*******	******
MI	WATERS 1975	44 52 46.49	84 40 44.28	1.890	-6.020	-4.25
MI	WHEAL KATE LAT STA USLS 1866	47 4 20.53	88 39 43.26	-2.090	*******	******
MI	WHITEFISH POINT LAT POST USLS	46 46 6.10	84 57 26.46	1.100	*******	******
MI	WILSON 1932	44 2 18.55	84 44 27.65	-1.710	-8.480	-5.59
MN	ZAKRZEWSKI RM 3 1970	42 26 26.01	84 23 38.43	-2.580	2.360	1.59
MN	ALBERT LEA EAST BASE 1927	43 35 34.26	93 10 53.40	-0.590	4.850	1.35
MN	ALBERTA 1904	45 48 14.56	94 6 50.56	-1.830	-2.420	-1.74
MN	ASKOV 1904	46 14 28.66	92 43 13.99	10.230	7.860	5.68
MN	BASS 1960	46 7 42.56	93 41 54.44	1.640	-2.090	-1.51
MN	BATTLE LAKE 1974	46 16 30.33	95 42 12.34	2.780	4.470	3.23
MN	BELLE 1966	45 4 48.81	96 10 50.32	1.950	-1.060	-0.75
MN	BERG 1941	47 45 4.43	95 37 17.78	-1.330	-5.210	-3.85
MN	BLACK 1974	46 14 18.25	95 25 15.44	-0.320	-0.450	-0.32
MN	BUCHANAN USLS 1871	46 56 28.15	91 47 19.20	-3.700	*******	******
MN	BUCK 1889	44 43 23.82	93 17 13.62	-4.830	-10.230	-7.26
MN	CAMERA 1963	43 53 0.95	95 55 43.63	-3.150	-0.450	-0.32
MN	DAGGETT 1904	46 10 19.45	96 11 14.61	3.150	1.190	0.85
MN	DALTON ASTRONOMICAL STA 1904	46 10 23.44	95 54 59.22	3.840	0.630	0.45
MN	DOUGLAS 1905	46 36 45.33	93 8 18.03	2.720	*******	******
MN	DUXBURY 1974	46 11 9.92	92 30 33.19	4.980	7.830	5.65
MN	EAST SAWTEETH USLS 1870	47 23 19.08	91 10 15.22	-10.080	*******	******
MN	FARIBAULT 1927	44 6.38	93 18 28.89	-1.160	-5.570	-3.88
MN	FARQUHARS KNOB USLS 1869	47 52 40.25	89 59 23.20	-5.040	*******	******
MN	FERGUS FALLS 1974	46 16 27.42	96 2 8.57	1.360	2.480	1.79
MN	FOSS 2 1966	46 6 5.37	96 32 1.94	3.380	1.400	1.01
MN	FOX 2 1974	46 16 44.43	96 18 32.96	2.630	0.330	-0.24
MN	FREE 1935	44 49 46.60	96 15 39.24	0.010	-2.870	-2.03
MN	GIESE 1960	46 14 29.86	93 6 51.32	-0.660	-1.340	-0.97
MN	GRACEVILLE 2 1966	45 35 7.74	96 25 42.35	1.260	-0.320	-0.23
MN	GROSKA 1938 RM 4 1974	46 15 53.68	95 1 23.66	2.020	-0.790	-0.58
MN	HUBBARD HILL 1938	46 50 55.21	94 54 37.95	0.330	-2.460	-1.79
MN	ISLE 1960 RM 4 1974	46 8 49.12	93 26 30.33	0.690	-2.830	-2.18
MN	JOHNSON 1935	44 51 44.81	96 17 20.94	*******	-3.090	-2.18
MN	JONES 1904	46 23 48.98	94 4 11.94	4.790	*******	******
MN	JUNE 1966	46 22 10.02	96 31 42.99	-0.760	-0.870	-0.63
MN	KENNEDY 1927	44 36 49.98	94 9 44.66	-0.160	-2.340	-1.62
MN	KIESTER 1926	43 34 55.04	93 42 14.56	-1.020	-4.850	-3.34
MN	KLUEVER 1927 RM 3	43 34 21.62	95 23 16.15	-2.220	-9.140	-6.30
MN	KURTZ 1954	44 16 45.37	95 57 4.11	1.070	-0.570	-0.42
MN	LUVERNE 1927	43 37 10.09	96 17 52.32	-2.120	-0.610	-0.42
MN	MAGNOLIA 1949	43 41 26.93	96 4 55.27	-2.350	1.780	1.23
MN	MINNESOTA POINT N B USLS 1870	46 45 27.98	92 4 42.66	0.340	-10.170	-7.41
MN	MINNESOTA POINT S B USLS 1870	46 42 49.52	92 1 54.54	1.900	*******	******
MN	MORKEN 1906	44 32 6.24	96 30 21.60	0.0	*******	******
MN	NELSON 1966	45 50 47.74	96 30 35.80	2.210	-0.120	-0.08
MN	OSAKIS 1904	45 49 27.14	95 7 51.70	0.650	*******	******
MN	PARLE 3 1966	44 33 2.51	96 17 44.80	-3.120	-8.040	-5.64
MN	PEET 1954	46 36 20.96	96 39 24.57	-1.800	0.110	0.07
MN	PHIL 1938	46 13 22.17	94 46 55.77	1.430	-1.740	-1.25
MN	RAIL 1904	46 13 38.39	94 29 17.10	4.050	2.850	2.06
MN	ROYALTON SOUTH BASE 1904	45 44 8.07	94 12 4.78	-1.330	0.510	0.37
MN	RUTLEDGE 1952	46 15 33.25	92 54 59.40	2.430	-1.700	-1.20
MN	SCHELLBERG 1935	45 20 24.26	96 24 35.91	1.670	-5.930	-4.40
MN	SHOOKS NORTH BASE 1934	47 55 14.53	94 26 33.87	*******	1.210	0.84
MN	STEEN 1949	43 31 46.46	96 17 38.90	-2.160	-2.310	-1.73
MN	STEPHEN WEST BASE 1905	48 27 22.88	96 60 20.72	-0.850	-3.400	-2.45
MN	SULLIVAN 1960	46 7 28.83	93 56 7.64	1.170	*******	******
MN	TILDEN 1905	47 42 13.15	96 19 48.51	3.200	-5.060	-3.53
MN	TYLER 1966	44 18 47.91	96 11 11.40	1.790	-2.860	-2.07
MN	UNDERWOOD 1974	46 16 46.56	95 51 48.24	1.530	-8.830	-6.24
MN	UNIVERSITY 1887	44 58 42.29	93 14 14.88	-3.520	-2.690	1.99
MN	VIRGINIA 1947	47 31 53.30	92 32 56.08	5.090	-0.930	-0.66
MN	WARNER 1934	44 47 22.95	95 11 15.83	-5.290	-1.220	-0.92
MN	WARROAD SOUTH BASE IBC 1912	48 55 19.86	95 19 53.43	3.180	-0.690	0.48
MN	WILD 1966	44 1 47.89	96 9 53.43	-0.700	-2.240	-1.62
MN	WILLIAMS 1960	46 11 59.87	93 16 40.37	-0.040	-0.020	-0.01
MN	WOODSIDE 1974	46 16 0.84	95 10 7.79	-0.630	-0.140	-0.08
MO	ARCADIA 1934	37 38 8.42	90 35 46.79	2.740	-1.270	-0.82
MO	ARNOLD 1934	39 47 31.49	92 5 41.44	-0.050	-0.120	-0.07
MO	BANNER 1928	38 52 57.97	91 25 35.35	4.870	3.180	1.98
MO	BERGER 1874	38 35 58.11	91 17 27.98	3.030	0.550	1.93
MO	BIG LAKE 1929	36 59 1.28	89 5 25.09	0.080	0.550	0.59
MO	DIXON EAST BASE 1946	37 59 51.90	92 5 16.62	0.670	-1.080	-0.69
MO	HANNIBAL 1931	39 43 17.61	91 22 41.71	0.350	*******	******
MO	HUNTER VERSAILLES S BASE	38 25 45.31	92 46 24.42	2.770	-0.160	******
MO	JEFFERSON CITY ASTRONOMIC 1879	38 33 41.22	92 9 45.57	2.770	*******	******
MO	KAMPHEFNER RM 3 1973	39 16 50.94	94 46 8.77	-0.330	-0.770	-0.49
MO	KANSAS CITY ASTRO STATION 1882	39 5 10.94	94 35 22.16	0.380	0.230	0.15
MO	KNOB NOSTER 1882	38 46 35.58	93 5 7.12	0.130	-0.480	******
MO	MEADVILLE 1928	39 46 38.80	93 17 21.68	-2.300	3.300	2.11
MO	MEDLOCK 1879	38 38 13.33	92 20 13.47	3.110	2.630	1.64
MO	POOR 1930	36 51 53.69	89 55 53.48	-0.230	9.010	5.41
MO	RAVENWOOD 1934	40 20 25.32	94 38 27.20	-2.440	-2.440	-1.58
MO	ROGERSVILLE WEST BASE 1928	37 7 19.62	93 4 46.66	0.710	2.780	1.68
MO	ST LOUIS SECOND PRESB CHURCH	38 37 56.11	90 12 14.59	4.480	1.660	1.04
MS	ADAMS 1934	34 59 32.81	88 32 24.50	2.100	9.800	5.62
MS	ADAMS 2 1955 RM 3 1965	34 10 23.03	90 34 23.26	3.330	1.010	0.54
MS	BEN 1942	30 53 39.94	88 50 7.14	-4.980	-1.510	-0.78
MS	BILBO 1943	30 59 25.90	89 34 29.45	-0.090	2.410	1.37
MS	BLUE HILL 1950	32 42 17.31	89 27 54.97	-0.300	1.020	-0.57
MS	BOGUE 1929	33 43 5.40	90 53 17.21	-5.060	-0.300	-0.16
MS	BRELAND 1942 RM 3	30 56 35.88	89 2 56.61	-0.010	-3.160	-1.79
MS	BRIAR 1967	34 28 41.66	88 53 30.73	0.850	0.140	0.07
MS	BURDETTE 1939	33 20 28.68	90 54 6.31	1.640	-2.630	-1.41
MS	CARSON 1930	31 29 16.12	89 19 15.64	-3.690	0.420	0.22
MS	CARSON 1934	31 32 18.39	89 48 15.06	3.080	0.490	0.26
MS	CRYSTAL 1945	32 0 35.76	90 22 41.18	2.140	-4.130	-2.23
MS	DIXON 1970	32 40 33.60	89 15 5.72	1.880	-5.910	-3.15
MS	DRY 1958	32 58 14.88	89 15 15.13	-5.630	-0.290	-0.15
MS	ELDER 1943	30 59 59.72	89 20 30.40	0.220	-0.070	-0.03
MS	ELLIS 1914	34 39 35.28	89 10 56.95	-2.670	-0.990	-0.52
MS	FOX 1935	31 29 52.05	89 14 32.50			

STATE	STATION NAME	GLAT(D,M,S)	GLON(D,M,S)	ALAT-GLAT(S)	ALON-GLON(S)	LAPLACE(S)
MS	GERSHORM 1967	34 5 30.98	89 2 3.05	-1.690	-0.980	-0.55
MS	GRACE 1958	32 58 32.70	90 57 55.05	-1.640	0.850	0.47
MS	GREENE 1934	33 26 13.10	89 36 57.16	-0.060	-5.530	-3.04
MS	GREENVILLE AFB 1957	33 28 42.47	91 0 8.51	-1.050	-0.260	-0.15
MS	HAMILTON 1970	33 49 29.84	89 13 30.10	-1.230	-0.910	-0.51
MS	HARDEE 1958	32 35 33.35	90 51 14.32	-0.940	-2.860	-1.54
MS	HARVEY 1930	32 28 56.43	88 41 19.04	1.820	-2.160	-1.16
MS	HEBRON 1947 RM 4 1965	31 45 24.33	89 56 31.27	0.450	0.740	0.39
MS	HOLLING 1945	31 36 16.68	90 16 19.34	-1.200	-0.520	-0.27
MS	KNCB 1914	34 47 15.55	88 14 25.71	-0.590	1.860	-1.06
MS	LEBANON 1935	34 34 17.26	88 43 43.51	-0.330	-2.730	-1.55
MS	LEGICN 1931	32 16 19.10	90 6 7.17	0.420	-3.680	-1.96
MS	LITTLE 1958	32 47 33.71	90 50 51.35	0.840	-1.250	-0.68
MS	MARCH 1970	33 18 31.39	89 18 38.77	0.620	-4.210	-2.31
MS	MASON 1970	31 46 11.37	89 31 45.87	1.370	-1.540	-0.81
MS	MC COOL 1958	33 6 32.52	89 19 8.84	1.120	-5.400	-2.95
MS	MC NEESE 1934	31 26 2.05	89 48 55.88	-3.670	0.440	0.23
MS	NANCE 2 1970	32 15 20.38	89 13 36.33	3.330	-1.620	-0.86
MS	OAK 1967	34 14 38.33	88 56 57.01	-0.650	-3.080	-1.73
MS	PASS CHRISTIAN WEST BASE 1931	30 18 34.16	89 16 31.01	6.110	-2.790	-2.51
MS	PEARL 1970	32 46 35.85	89 13 38.68	2.250	-4.670	-2.53
MS	PERCY 1939	33 9 38.34	90 55 42.18	2.550	-0.070	-0.04
MS	REID 1967	33 59 33.22	89 11 32.01	-2.000	-2.210	-1.24
MS	RUTLAND 1934	31 40 24.24	89 37 35.57	-0.050	-0.460	-0.24
MS	SANFORD 1965	31 19 33.44	89 45 31.86	-6.300	0.070	0.04
MS	SLIKER 1931	32 23 55.87	90 42 45.59	-0.650	1.280	0.69
MS	SPURGEON 1934	33 58 7.88	89 36 0.74	-2.730	-4.650	-2.59
MS	STATE 1935	33 27 30.67	88 47 52.16	-1.890	-1.920	-1.06
MS	STEEN 1945	31 53 49.76	90 7 39.55	2.330	1.010	0.54
MS	STOVALL NORTHEAST BASE 1929	34 18 2.76	90 38 35.68	3.330	1.660	0.94
MS	STRICKLAND 1914	34 51 38.08	88 25 14.86	2.260	2.140	1.23
MS	TAYLOR 1970	31 54 37.57	89 21 32.54	3.320	-1.520	-0.79
MS	TURNERVILLE 1935	32 1 28.00	89 12 16.78	4.950	-1.700	-0.90
MS	UNIVERSITY 1935	34 22 2.03	89 32 27.64	0.560	-1.110	-0.62
MS	WARD 1942 RM 3	30 52 9.14	88 32 55.38	-2.420	0.350	0.18
MS	WEBSTER 1939 RM 3 1970	33 33 55.69	89 9 55.62	-1.570	-1.780	-0.98
MS	WHITE 1935	31 6 16.09	89 40 0.86	-5.770	-0.230	-0.12
MT	ABERDEEN 1957	45 1 53.78	107 20 48.00	13.570	-5.750	-6.90
MT	AGENCY 1968	47 22 18.89	114 24 16.69	0.990	0.010	0.01
MT	AHLES 1934	46 25 8.35	107 1 35.73	-0.890	-2.700	-1.96
MT	AIRPCRT 2 1962	45 48 5.95	108 32 11.24	-1.750	-4.440	-3.18
MT	ALICE 1968	47 0 29.00	112 55 32.51	-6.180	3.470	2.54
MT	BACON 1934	46 51 35.67	104 16 18.86	1.280	5.380	3.53
MT	BAD LANDS	47 59 56.76	104 55 39.07	4.460	2.350	1.74
MT	BANNISTER 1934	46 35 44.66	106 11 35.41	0.470	-1.870	-1.35
MT	BEAR 1934	47 56 52.42	106 14 46.80	3.260	2.320	1.72
MT	BENCH MARK H 81 RESET 1960	46 56 57.35	113 3 25.07	-0.470	4.630	3.38
MT	BENCH MARK W 35 1934	46 37 15.10	107 22 40.01	-2.660	-3.000	-2.18
MT	BENTON 1949	47 50 21.72	110 39 3.73	5.590	1.430	1.06
MT	BLOOD 1961	47 27 39.34	109 50 1.75	6.160	0.810	0.60
MT	BLUE 1925	46 31 7.98	108 35 51.67	2.140	-3.290	-2.35
MT	BOUNDARY EAST BASE 1923	48 59 55.02	112 45 16.82	4.290	0.750	0.56
MT	BCUNDARY MON ECC NE CCR WYO	44 59 58.57	104 3 18.84	2.350	******	******
MT	BOUNDARY MON MCNT N DAK ECC	47 12 41.35	104 3 35.85	3.540	******	******
MT	BOUNDARY WEST BASE 1923	48 59 54.55	112 56 34.14	5.410	3.160	2.38
MT	BOYD 1968	47 4 16.12	113 20 3.76	-4.510	5.260	3.85
MT	BOZEMAN NORTHWEST BASE 1922	45 51 41.40	111 20 31.03	-1.650	6.490	4.64
MT	BOZEMAN SOUTHEAST BASE 1922	45 42 27.05	111 7 53.15	-5.450	12.350	8.13
MT	BROADVIEW 1939	46 3 46.51	108 54 4.44	-0.700	-2.720	-1.96
MT	CABIN 1934	46 48 52.26	104 41 49.18	1.040	8.630	6.30
MT	CEDAR 1934	46 41 6.78	105 38 59.85	0.110	0.810	0.59
MT	CENTER 1949	47 42 34.52	109 54 38.17	4.120	-1.300	-0.96
MT	CHANCE 1967	47 47 7.51	108 37 52.43	0.860	-0.510	-0.37
MT	CHER 1949	47 42 47.66	110 46 42.43	4.530	4.330	3.21
MT	CHESTER LAPLACE 1923	48 30 54.79	110 57 45.20	-1.450	-2.860	-2.15
MT	CHIP 1967	47 52 58.00	109 44 3.59	0.040	3.930	2.92
MT	CONN	48 9 53.77	105 13 37.58	-0.200	-2.110	-1.58
MT	COTTON 1972	47 10 46.79	108 12 8.34	4.020	-1.170	-0.86
MT	CROW COULEE 1949	47 48 7.54	110 13 14.00	4.000	2.050	1.52
MT	CROW 1934	47 0 31.51	106 3 14.71	-2.170	-2.920	-2.12
MT	CUSTER 1934	46 11 26.48	107 34 42.05	-3.100	-3.140	-2.26
MT	CUSTER 1957	45 30 55.76	107 52 51.96	5.900	-7.720	-5.51
MT	DOME 1949	47 50 28.90	110 0 46.40	3.740	2.040	1.51
MT	DOVE 1972	47 23 59.65	108 15 56.98	4.120	-1.510	-1.11
MT	DRIVEWAY 1968	47 35 15.11	113 34 34.19	-0.330	-2.600	-1.92
MT	ECOLEMAN 1972	46 10 29.15	107 56 8.57	-4.830	-3.190	-2.32
MT	EDDY 1968	47 32 24.96	115 11 13.98	6.520	7.180	5.30
MT	ELIM 1967	47 47 21.92	110 23 1.60	6.530	1.350	0.99
MT	FALLON 1934	46 52 44.56	105 8 5.37	-0.440	1.940	1.42
MT	FISHER 1961	47 10 45.81	108 47 58.40	4.010	-4.790	-3.52
MT	FORCHEE 1934	47 43 14.23	107 52 39.23	3.230	-0.520	-0.68
MT	HALEY 1925	45 42 33.34	108 18 55.31	3.510	-5.650	-4.05
MT	HAVRE SOUTH BASE 1923	48 46 24.20	109 53 19.08	0.680	0.040	0.02
MT	HIGH POINT 2 1968	47 19 10.28	114 12 8.16	5.470	5.250	3.86
MT	HUNTLEY 1925	45 48 53.87	108 12 18.34	2.580	0.760	0.55
MT	JETTIE 1939	47 48 25.01	114 16 34.12	-1.800	6.420	4.75
MT	JOCKO 1968	47 10 1.83	113 43 34.57	-0.150	3.650	2.68
MT	KARSHAW 1968	46 57 14.92	112 51 33.24	-4.990	12.370	8.05
MT	KUICH USGS 1949	47 41 5.34	111 8 52.71	3.110	0.970	0.71
MT	LAVINA 1939	46 21 54.25	108 59 7.27	-1.230	-2.180	-1.58
MT	LEROY 1949	47 53 7.22	109 22 25.46	-1.130	2.320	1.73
MT	MANNING	48 12 37.09	104 49 50.38	-0.020	0.530	0.39
MT	MARTIN 1934	45 58 57.03	106 15 0.91	0.280	-1.380	-0.95
MT	MCCABE	48 15 32.45	104 22 7.85	1.360	-2.380	-1.77
MT	MCLEAN 1934	46 34 27.61	108 8 10.56	2.310	-2.540	-1.84
MT	MISSOULA WEST BASE 1934	46 56 55.92	114 7 41.18	-2.480	9.150	6.68
MT	MOATS 1934	46 34 21.54	108 55 57.37	-2.620	-5.410	-3.93
MT	MONOAK 1912	48 0 9.70	104 2 49.43	2.640	-1.720	-1.28
MT	MUGGINS 1934	46 32 35.40	107 41 11.73	3.120	1.070	0.78
MT	MUSTER 1934	46 47 17.87	105 55 30.65	0.700	0.240	0.17
MT	NEVADA 1968	46 55 7.91	113 2 54.04	1.570	5.560	4.06
MT	NICKOLS 1948	47 39 29.44	108 16 10.94	2.800	-2.100	-1.56
MT	OSTLE 1960	47 14 10.89	112 14 1.20	4.600	-6.660	-4.89
MT	PILGRIM 1949	45 30 62.35	105 7 5.08	4.650	1.560	1.12
MT	PISTOL 1968	47 14 8.04	113 57 21.87	-2.320	17.760	13.04
MT	PORK 1934	46 30 9.89	106 34 36.94	-0.810	-0.790	0.57
MT	ROBERTS 1972	46 8 47.04	108 21 22.76	-3.800	-2.450	-1.78
MT	RYEGATE MAGNETIC STATION 1939	46 18 6.20	109 15 15.26	-0.260	-3.660	-2.65
MT	SHOEMAKER 1934	47 45 1.60	109 29 20.57	3.510	-1.100	-0.81
MT	SHRIVER 1931	45 10 12.36	108 29 2.66	3.200	-2.760	0.08
MT	SIMMS 1968	47 31 1.99	111 58 1.38	2.060	2.250	1.66
MT	SKUNK 1957	45 20 49.74	107 40 7.56	7.840	-6.280	-4.46
MT	SQUAW 1967	47 56 26.42	108 55 0.78	1.360	0.980	0.72
MT	STEVENS	48 4 45.89	109 21 46.06	0.190	-0.040	-0.02
MT	SUNLIGHT 1934	46 41 58.59	108 33 34.79	0.900	-3.740	-2.72
MT	SUTHERLAND 1948	47 48 37.69	107 2 36.28	3.640	0.710	0.23
MT	TALLOW 1934	47 57 56.05	107 34 55.27	1.690	3.770	2.80
MT	TERRY 1934	46 42 54.91	105 18 5.61	1.850	3.410	2.48
MT	TICK 1968	47 27 45.58	114 46 47.03	-2.790	11.170	8.23
MT	TOWNSHIP	48 7 55.52	105 58 21.08	-0.180	-1.170	-0.87
MT	TRIPP ET USGS 1972	46 50 25.67	109 15 9.36	3.410	-1.790	-1.30
MT	VAUGHN 1961	47 33 50.31	111 34 16.24	0.740	3.700	2.73

STATE	STATION NAME	GLAT(D,M,S)	GLON(D,M,S)	ALAT-GLAT(S)	ALON-GLON(S)	LAPLACE(S)
MT	WARM SPRINGS NORTH BASE 1956	46 16 12.95	112 45 12.52	1.680	10.480	7.58
MT	WARM SPRINGS SOUTH BASE 1956	46 11 3.53	112 47 12.50	3.520	9.780	7.06
MT	WAVE 1957	45 8 58.80	107 31 46.82	13.180	-9.680	-6.86
MT	WEST FORK 2	48 11 37.70	105 40 4.47	-1.350	-0.610	-0.46
MT	WILLOW 1968	46 54 25.61	112 41 40.20	-2.880	7.600	5.56
MT	WINNETT ET USGS 1972	47 1 7.27	108 17 17.61	2.060	-2.930	-2.15
MT	WONDER 1967	47 56 1.61	106 42 46.51	2.800	-1.010	-0.75
NC	ALEXIS 1933	35 24 37.18	81 7 27.57	2.640	-4.880	2.83
NC	ALLENTON 1933	34 35 31.84	78 55 34.08	-1.660	-6.930	-3.94
NC	ASKIN 2 1965	35 1 47.31	77 3 20.25	-2.240	-4.200	-2.42
NC	BATTLEBORO 1933	35 12 54.44	77 44 59.12	-2.760	-5.160	-3.04
NC	BEARHALLOW 1933 RM 2	35 27 38.00	82 21 26.79	-5.420	-11.470	-6.65
NC	BETHEA 1932	34 28 28.38	77 30 42.41	-1.880	-3.080	-1.74
NC	BISCOE 1948	35 21 46.52	79 46 55.85	-0.780	-5.040	-2.92
NC	BLACK CREEK 1949	35 38 24.59	77 56 41.29	1.990	-6.310	-3.68
NC	BRYANT 1933	36 23 17.95	80 59 43.23	-11.550	-9.030	-5.36
NC	BUCHANAN 1933	36 30 1.84	78 18 13.11	-2.460	-10.260	-6.10
NC	CAMP 1933	34 51 12.58	78 9 47.29	-3.140	-6.950	-3.97
NC	CATLIN 1933	35 34 3.15	81 28 47.22	0.080	-0.030	0.01
NC	CHAPEL HILL 1933	35 52 57.13	79 2 58.72	-3.310	-8.590	-5.03
NC	CHCCOWINITY 1931	35 25 46.40	77 7 41.31	-1.060	-3.250	-1.89
NC	CLAREMONT MUNICIPAL TANK	35 42 55.96	81 8 59.75	5.200	1.590	0.93
NC	CLIFTON 1949	35 3 27.91	79 4 9.59	-2.970	-6.700	-3.85
NC	CLYDE 2 1969	35 43 40.85	78 23 29.27	-1.970	-8.140	-4.75
NC	DAVIS 1962	35 46 23.07	79 30 34.87	-0.270	-2.810	-1.64
NC	DAY 1933	34 59 57.50	76 25 30.58	-0.220	-3.670	-2.11
NC	DEEP 1954	35 6 19.86	80 39 13.05	6.180	0.660	0.39
NC	DEN 1918	36 5 2.39	78 28 0.67	-3.650	-12.650	-7.45
NC	DILLON NORTH BASE 1933	34 32 10.70	79 17 35.84	-4.930	-8.300	-4.71
NC	DUCK CREEK 1932	34 35 1.28	77 17 56.05	-1.660	-9.660	1.12
NC	DUDLEY 1933	35 16 8.13	78 2 55.03	-3.360	-8.710	-5.57
NC	DUPLIN 1933	35 0 58.06	77 44 24.69	-3.220	-8.710	-5.00
NC	DURANT 1931	36 8 11.12	76 17 53.94	-2.050	-10.530	-6.21
NC	EASON 1933	35 23 35.09	77 32 25.99	-4.760	-8.440	-4.89
NC	ESTELLE 1932	36 20 0.67	79 15 1.97	5.700	3.460	2.06
NC	EUREKA 1933	35 44 13.40	77 27 23.92	-3.790		-1.90
NC	EVERTON 1933	35 2 15.45	78 2 45.59	-1.700	-5.890	-3.39
NC	FAIR 1933	35 47 16.88	78 42 31.20	-4.450	-12.450	-7.28
NC	FLOWERS 1933	35 19 15.14	78 16 48.37	-1.060	-5.870	-3.39
NC	FORD 1933	34 19 30.29	78 59 29.03	-2.530	-6.650	-3.76
NC	FOSTER 1962	35 30 0.37	79 29 11.90	-3.730	-5.000	-2.94
NC	FOUNTAIN 1933	35 40 21.61	77 39 59.30	-0.530	-6.370	-3.64
NC	FRANKLIN 1933	35 11 7.00	83 23 15.51	0.340	-5.690	-3.28
NC	GOOD 1933	36 0 38.78	77 23 7.52	-3.400	-0.960	-0.57
NC	GUIDE 1933	34 3 56.87	78 42 3.69	0.710	-9.690	-5.43
NC	GULF NCGS 1967	35 38 43.96	81 59 6.93	-1.100	-9.040	-5.26
NC	HALLOWAY 1933	36 28 59.77	78 47 57.49	1.980	-7.520	-3.25
NC	HALLS 1942	34 20 1.49	78 38 37.26	-2.530	-7.520	-4.24
NC	HAMPSTEAD 2 1947	34 22 16.17	77 42 21.46	-2.950	-8.550	-4.83
NC	HASH NCGS 1972	35 15 58.50	79 16 20.26	-5.610	-9.930	-5.74
NC	HAW RIVER 1954	36 6 9.29	79 21 11.30	3.120	-1.280	-0.76
NC	HAYESVILLE 1933	35 2 15.27	83 48 58.60	-1.740	3.360	1.93
NC	HIBBS 1965	34 46 29.98	76 55 43.88	2.970	-0.080	-0.05
NC	HIBRITEN 1918	35 54 24.82	81 29 23.46	1.920	-0.880	-0.51
NC	HOFFMAN 1918	35 0 3.86	79 32 37.12	-2.300	-8.820	-5.06
NC	HOPE 1949	34 58 33.40	78 55 51.79	-3.430	-7.140	-4.09
NC	HOWELL 1932	33 57 4.16	78 12 30.89	-1.970	-12.210	-6.82
NC	JONAS 1933	35 57 34.79	81 53 55.69	-5.910	-12.170	-7.11
NC	JUNIOR 1948	35 45 32.68	80 15 47.20	8.580	6.140	3.59
NC	KING 1876	36 12 27.48	81 18 45.90	1.900	*******	******
NC	KYLES NCGS 1964	36 6 46.37	79 45 27.48	8.620	8.470	5.00
NC	LANDRUM 1957	35 11 38.14	82 10 33.10	0.100	-6.420	-3.70
NC	LEE 1918	35 28 46.99	79 10 37.45	-1.970	-7.810	-4.53
NC	LITTLETON 1918	36 25 55.33	77 55 3.02	-1.450	-7.630	-4.45
NC	LONG SHOAL POINT	35 34 50.62	75 46 35.30	-14.370	*******	******
NC	LONG 1948	35 32 2.41	80 15 7.62	0.080	-0.240	-0.13
NC	MARSHVILLE 1933	34 56 56.96	80 21 23.96	-1.620	-5.760	-3.30
NC	MAX PATCH 1933	35 47 49.85	82 57 25.18	11.640	-1.190	-0.70
NC	MC KAY 1933	35 9 22.36	80 2 9.82	-1.710	-2.160	-1.25
NC	MILL 2 1952	34 5 44.84	78 5 20.86	-1.680	-6.420	-3.60
NC	MCNREE 1933	34 31 6.55	78 29 41.42	-0.740	-7.010	-3.98
NC	MCCRE 1876	36 33 53.21	80 16 59.53	1.470	1.550	0.92
NC	MOUNT PLEASANT 1933	35 24 56.50	76 4 59.48	2.270	-4.080	-2.37
NC	NEW BERN SOUTH BASE 1931	34 59 44.20	76 59 11.74	-0.240	0.270	0.15
NC	OCRACOKE NORTHEAST BASE 1870	35 4 1.51	76 3 9.60	1.400	*******	******
NC	ORIENTAL 2 1948	35 2 17.68	76 42 47.48	-1.970	-2.500	-1.44
NC	ORR 2 1965	35 22 12.54	76 57 23.71	-1.970	-4.230	-2.45
NC	OVERLOOK 1954	36 6 3.68	81 7 42.78	5.830	-2.600	-1.53
NC	PHILADELPHUS 1949	34 45 20.74	79 10 15.63	-3.270	-9.720	-5.54
NC	PHOENIX 1942	34 17 7.19	78 4 46.00	-4.930	-7.290	-4.10
NC	PIGOTT 1932	33 54 1.38	78 26 43.64	-1.970	-7.590	-4.23
NC	PILGRIM 1932	34 13 14.73	77 49 33.10	-6.350	-11.120	-6.25
NC	POWELL 1950	36 13 19.90	76 56 17.64	-1.560	-8.600	-5.08
NC	RALEIGH 1918	35 46 37.00	78 38 21.58	*******	-16.370	-9.57
NC	RED HILL 1934	35 3 58.43	80 38 55.29	-2.690	-5.490	-3.15
NC	RED NCGS 1964	35 13 41.24	81 19 46.92	3.640	3.580	2.07
NC	ROAN HIGH BLUFF 1894	36 5 55.62	82 8 44.48	2.830	*******	******
NC	ROBINSON 1933	35 5 29.88	78 14 38.25	-5.640	-10.850	-6.23
NC	ROSEBORO 1918	34 57 22.46	78 30 43.85	-1.530	-5.290	-3.03
NC	SANDSTONE 1947	35 2 56.36	79 19 38.15	-6.800	-7.990	-4.59
NC	SANDY 1933	35 0 32.27	79 47 18.80	-1.970	-8.190	-4.69
NC	SANFORD 1918	35 27 5.83	79 10 25.77	-3.100	-6.680	-3.80
NC	SEVERN 2 1965	36 30 13.12	77 11 40.42	-0.470	-6.680	-3.97
NC	SIMKINS 1932	34 41 51.32	77 3 25.11	1.650	-2.570	-1.46
NC	SIMONTON COLL CENTER OF CUPCLA	35 46 57.54	80 53 39.35	*******	0.880	0.51
NC	SMAW 2 1965	35 32 44.89	76 59 56.62	-1.990	-4.330	-2.52
NC	SMITH 1932	36 25 47.56	79 40 15.41	6.760	7.090	4.21
NC	SOUTHPORT EAST BASE 1932	33 57 6.47	78 2 52.99	-6.840	-9.810	-5.48
NC	SPIVEY 1933	35 58 6.25	82 39 6.71	5.110	-7.740	-4.50
NC	STANDING INDIAN 1933	35 2 7.13	83 32 17.22	-6.190	-1.620	-0.93
NC	SURRY 1954	36 27 3.49	80 33 31.82	-6.540	0.330	0.20
NC	SWAIN 1954	35 0 27.75	80 13 47.23	8.540	7.440	4.38
NC	SWAN 1933	35 24 39.53	76 19 46.48	0.050	-9.170	-5.32
NC	SYLVA 1933	35 22 21.28	83 14 37.01	0.870	0.870	-0.50
NC	TRAM 1942	34 9 6.46	78 22 51.16	-3.110	-9.810	-5.51
NC	TT 5 TWC USGS 1971	36 13 12.07	81 29 12.07	-12.360	-13.440	-7.94
NC	VANDER 1918	35 1 33.13	78 46 2.67	-0.990	-5.570	-3.20
NC	VAUGHN 1959	36 23 19.80	77 6 29.86	-1.380	-8.410	-4.69
NC	VULTARE 1933	36 31 36.48	77 40 44.34	-0.510	-0.060	-0.04
NC	WADESBORO 1933	34 58 24.93	80 4 41.71	-0.510	-6.610	-3.78
NC	WALNUT COVE 1954	36 18 57.73	80 8 27.33	3.020	4.780	2.53
NC	WAYNESVILLE 1933	35 29 3.60	82 59 28.83	6.830	-6.510	-3.78
NC	WGWR 1948	35 43 23.09	79 48 24.11	0.520	-1.100	-0.65
NC	WHALE 1933	35 1 16.63	76 7 5.42	1.740	-5.610	-3.22
NC	WHITE LAKE 1933	34 35 22.85	78 29 10.82	-2.140	-7.320	-4.16
NC	WILLIAMSTON 1931	35 50 30.99	77 3 54.53	-3.290	-7.870	-4.61
NC	WINDSOR 2 1955	36 0 41.56	76 58 35.09	-2.240	-8.280	-4.86
NC	WOOLARD 2 1965	35 41 36.55	77 3 59.42	-3.950	-6.410	-3.74
NC	YOUNG 1876	35 44 14.09	80 38 51.47	7.610	1.500	0.88
NC	ZION 1933	34 44 15.82	79 29 42.73	-3.170	-7.810	-4.45
ND	ALICE 1954	46 46 32.26	97 29 11.58	-0.260	-5.370	-3.92
ND	BADLAND 1912	46 41 59.95	103 19 59.60	1.440	*******	******

STATE	STATION NAME	GLAT(D,M,S)	GLON(D,M,S)	ALAT-GLAT(S)	ALON-GLON(S)	LAPLACE(S)
ND	BAKER 2 1958	48 10 18.26	99 39 9.86	-0.800	-6.840	-5.10
ND	BALTA NORTH BASE 1946	48 10 40.85	100 2 30.24	-0.780	-1.130	-0.84
ND	BATTERY 2 1967	48 10 30.64	102 47 4.61	0.810	1.840	1.37
ND	BELER 1934	46 7 43.39	100 47 11.09	******	-4.870	-3.51
ND	BINSTOCK 1951	46 43 35.88	103 3 41.10	0.790	-0.190	-0.14
ND	BISON 1946	48 3 54.50	100 16 10.04	-1.140	-2.590	-1.93
ND	BONETRAILL 1912	48 24 59.25	103 49 44.82	1.840	******	******
ND	BOWMAN LONGITUDE 1912	46 10 56.88	103 24 11.48	0.580	-4.260	-3.07
ND	BUC 1934	47 21 26.49	98 50 19.81	-3.920	-3.570	-2.63
ND	BUTTER 1946	48 6 58.05	101 40 10.16	1.840	-7.310	-5.44
ND	CHARLES 1946	48 4 38.60	99 53 10.43	-2.660	-1.570	-1.17
ND	COURTENAY 1934	47 12 5.95	98 39 15.73	0.560	-4.150	-3.05
ND	COW 1945	48 11 54.93	103 57 28.69	1.670	1.000	0.75
ND	DODGE 1941	47 20 42.48	102 11 50.25	-1.230	-5.580	-4.10
ND	DRY 1934	48 12 23.73	98 58 34.11	******	-1.420	-1.05
ND	DURBIN 1954	46 46 29.49	97 8 2.03	-1.740	-0.550	-0.39
ND	ELDEN 1934	46 2 39.68	98 37 29.06	******	-7.600	-5.48
ND	FRYBERG 1934	46 53 19.31	103 19 36.63	2.200	-1.190	-0.87
ND	HOGBACK 2 1957	48 4 50.23	100 41 30.83	-0.240	-3.770	-2.81
ND	HURD 1941	47 27 20.13	99 54 16.18	******	-1.330	-0.98
ND	JAEGER 1951	47 2 53.23	102 4 16.73	-0.930	-7.360	-5.38
ND	JERRY 1967	47 47 50.43	98 50 40.42	-0.650	-1.870	-1.39
ND	JOHNSON 1967	46 58 14.67	98 31 25.38	-1.350	-3.440	-2.52
ND	LARRABEE 1934	47 31 43.57	98 51 16.42	2.980	-0.020	-0.01
ND	LEHIGH 1951	46 48 17.19	102 38 34.80	0.020	-4.070	-2.97
ND	LITTLE FORKS 1941	47 33 22.22	97 30 17.71	-0.190	-4.750	-3.51
ND	MISSION 1934	47 58 21.84	98 54 31.18	0.030	-2.200	-1.63
ND	NORMA 1954	46 45 40.10	97 55 24.39	-1.800	-2.790	-2.03
ND	PICKELL 1934	46 6 2.32	97 56 56.27	******	0.540	0.38
ND	PINGREE SOUTH BASE 1934	47 6 0.32	98 51 19.28	******	-5.150	-3.78
ND	POLE 1951	46 44 47.42	102 15 13.44	-0.970	-6.990	-5.09
ND	PRINCE 1951	46 13 43.14	102 2 21.78	-0.410	-5.400	-3.90
ND	RHINE 1934	47 55 12.05	101 3 23.26	******	-7.000	-5.19
ND	SADDLE BUTTE 1946	47 39 37.90	102 21 28.31	-0.500	-2.750	-2.03
ND	SARCN 2 1968	48 5 43.71	101 17 46.47	2.540	-11.000	-8.19
ND	SCHOOL-AMBROSE NE BASE 1912	48 59 23.58	103 29 10.46	3.220	******	******
ND	SENTINEL 1912	46 52 16.25	103 50 5.32	1.560	-.590	-0.43
ND	SHEEP 1912	47 37 49.38	103 47 40.07	3.060	******	******
ND	SORSTOKKE 1934	47 12 19.10	101 7 1.31	******	-5.320	-3.90
ND	STANLEY SOUTH BASE 1946	48 6 18.64	102 21 10.90	0.570	-2.850	-2.13
ND	STANLEY SOUTH BASE 2 1955	48 6 18.63	102 21 9.69	0.660	-1.730	-1.29
ND	STAR 1951	46 28 57.59	102 6 52.40	-0.550	-5.240	-3.80
ND	STINE 1934	45 58 14.31	101 47 10.60	-0.280	-4.810	-3.46
ND	STONY 1945	48 12 43.88	103 31 2.84	1.260	2.650	1.98
ND	SVEA 1954	46 45 0.56	98 16 11.93	-2.840	-5.150	-3.75
ND	TRINITY 1946	48 8 11.17	102 30 59.79	-2.130	-4.420	-3.29
ND	VALLEY 1946	48 8 12.80	102 30 30.65	-1.370	-8.530	-6.35
ND	VAN HOOK 1946	47 57 11.95	102 21 33.74	-0.030	-0.660	-0.49
ND	WALHALLA GSC 1925	48 54 9.71	97 53 56.68	-0.480	-7.790	-5.87
ND	WAINE 1967	48 7 51.80	100 58 57.50	-0.270	-7.810	-5.82
ND	WEST BAY 1946	48 6 15.51	99 9 21.96	-1.650	-2.130	-1.58
ND	WONDERLAKE 1945	48 13 17.63	100 10 58.79	0.060	0.480	0.36
NE	ASHBY 1946	42 1 27.93	102 5 5.29	-3.730	-1.390	-0.93
NE	BERGEN 1950	40 47 42.88	97 57 54.51	-2.140	-6.620	-4.33
NE	BRANDT 1966	41 41 15.53	96 52 5.47	-2.220	-0.270	-0.18
NE	CEDAR 1950	41 15 3.02	98 5 24.23	-3.020	1.390	0.92
NE	CEDAR 1950 RM 3 1966	41 15 3.02	97 41 19.97	-3.140	-3.060	-2.01
NE	COLLIER 1973	41 44 34.12	100 27 11.82	-0.520	-1.720	-1.13
NE	COLUMBUS 2 1968	41 28 39.51	97 20 56.22	-3.240	-2.640	-1.75
NE	COOLEY 1934	41 49 57.29	99 48 35.44	-2.430	-3.150	-2.10
NE	CRABB 1933	42 57 30.38	100 31 44.74	******	-5.260	-3.58
NE	CRESCENT 1946	41 43 41.81	102 25 5.06	-2.700	-0.290	-0.20
NE	CURTIS 1930	40 18 33.55	97 15 50.68	-1.520	-2.090	-1.38
NE	CUSTER 1900	41 38 14.24	98 32 44.95	-0.210	1.880	1.24
NE	DAILY 1900	41 35 45.15	98 18 55.40	-1.130	******	******
NE	DISMAL 1973	41 45 59.52	100 14 28.96	0.820	-1.550	-1.04
NE	ELKHORN 1935	41 59 21.28	97 18 36.64	1.130	1.540	1.04
NE	ELM 1899	41 44 0.84	98 51 27.07	-0.680	-2.650	-1.76
NE	EUREKA 1972	41 42 38.64	99 20 49.00	-0.380	-5.720	-3.81
NE	FLAG 1946	41 38 32.21	101 46 17.43	-5.000	-3.410	-2.27
NE	FOLLEN 1930	41 37 5.29	102 12 40.87	-4.790	-11.570	-7.65
NE	GENT 1972	41 42 24.36	99 21 15.56	0.030	-2.890	-1.92
NE	GERCIS 1951 RM 1 1966	40 31 51.18	98 6 29.71	-1.650	-4.860	-3.15
NE	GERMAN 1935 RM 3 1966	41 44 30.40	97 30 9.57	-2.430	0.140	0.09
NE	GIBBS 1973	41 45 31.50	100 54 51.02	-3.640	-4.360	-2.90
NE	GUINN 1935	42 26 8.39	97 3 13.83	-2.240	2.990	2.74
NE	HANEY 1933	41 45 48.52	101 11 15.66	-2.950	-2.340	-1.55
NE	HEIN 1933	41 4 51.05	100 57 29.78	******	1.150	0.76
NE	HELVEY 1946	40 15 45.95	97 13 19.39	-0.430	1.660	1.08
NE	HOSKINS 1935	42 5 47.53	97 16 13.85	******	0.000	0.00
NE	HOWARD 1973	41 45 35.52	99 35 31.09	-0.440	-3.470	-2.31
NE	ISINGLASS 1934	42 54 44.73	102 57 24.18	-1.840	0.090	0.09
NE	KEATING 1934	41 45 45.94	100 4 22.35	1.660	-0.970	-0.64
NE	LAMB 1934	42 39 59.83	99 32 15.56	5.670	-4.770	-3.24
NE	LAWRENCE 1951	41 41 46.67	101 28 31.72	-3.980	-1.380	-0.89
NE	LENA 2 1973	41 38 17.09	103 23 3.30	-0.690	-3.900	-2.59
NE	LOGAN 1933	40 11 51.48	100 9 56.48	-1.490	-1.300	-0.84
NE	LOGAN 1933	41 38 26.88	101 35 56.22	-5.830	-3.150	-2.09
NE	MEYER 1946	41 30 41.84	97 37 23.50	-3.670	-1.160	-0.77
NE	MONROE 1930 RM 3 1966	41 53 56.30	103 40 2.26	******	-1.230	-0.83
NE	NORTH 1933	41 15 38.57	95 56 35.01	7.040	-3.000	-2.04
NE	OMAHA LCNG 1882	41 14 41.57	97 53 20.14	-4.640	-1.200	-0.80
NE	OSCEOLA 1930	41 19 44.04	102 23 6.43	-1.060	-7.010	-4.73
NE	OSHKOSH 1933	42 25 25.40	98 25 59.51	4.010	-7.010	-4.73
NE	PAGE SOUTHWEST BASE 1900	41 35 0.20	101 35 49.70	1.550	-4.250	-2.75
NE	PUMPKIN 1934	40 0 2.28	95 52 26.70	-1.110	-7.120	-4.57
NE	REDWOOD 1934	40 10 37.17	98 30 19.15	-1.340	-0.250	-0.16
NE	REITHER 1947	41 36 28.30	96 24 15.65	-6.120	-15.130	-10.04
NE	RUME 1930	40 46 38.14	98 44 39.31	-2.770	-4.100	-2.67
NE	SHELTON EAST BASE 1899	41 34 48.11	97 46 16.14	-4.980	-3.750	-2.49
NE	ST EDWARDS 2 1972	41 38 48.62	101 59 7.17	-5.220	1.330	0.88
NE	VELBA 2 1973	41 34 35.34	97 7 22.83	-4.450	-1.370	-0.91
NE	VRBA 1968	42 12 6.01	97 8 2.92	-1.080	-0.420	-0.28
NE	WACKER 1935	41 42 12.83	102 47 8.63	-5.200	-1.600	-1.06
NE	WAGONER 2 USGS 1973	41 42 15.83	100 36 53.37	-2.450	-1.800	-1.19
NE	WAIBEL USGS 1973	41 35 4.17	96 46 13.67	-6.680	-5.620	-3.74
NE	WEBSTER 1968	44 36 52.30	71 18 49.78	-0.900	8.120	5.70
NH	DUMMER 1959	44 23 17.86	71 9 39.80	1.940	-0.600	-0.42
NH	GORHAM 1958	43 31 3.77	71 22 11.29	-1.370	******	******
NH	GUNSTOCK 1860	43 42 19.27	72 17 7.93	-1.370	-2.130	-1.47
NH	OBSERVATORY HILL 1874	43 48 17.16	71 21 23.00	4.140	2.500	1.73
NH	SANDWICH 1958	43 5 38.25	72 8 7.64	0.350	1.760	1.20
NH	STODDARD MT	43 2 21.64	71 47 27.38	-0.840	-2.220	-1.51
NH	STRATHAM 2 1941	44 12 14.98	71 31 27.38	-0.050	2.310	1.51
NH	TREE TOP 1959	42 58 59.33	71 35 19.35	0.370	******	******
NH	UNKONOONUC 1848	43 7 33.69	70 59 34.93	2.110	-3.570	-2.54
NH	WEDNESDAY HILL 1943	45 2 17.78	71 25 3.17	1.320	10.330	7.32
NH	YOUNG 1959	38 55 59.68	74 57 38.08	1.660	******	******
NJ	BEACON HILL 1839	41 6 13.58	74 25 4.85	-2.660	-2.430	-1.53
NJ	CAPE MAY 1932	40 4 10.79	74 12 0.49	-2.850	-0.630	-0.42
NJ	CEDAR 1974			-2.370	-8.330	-5.36
NJ	CHRISTOPHER 2 1932					

STATE	STATION NAME	GLAT(D,M,S)	GLON(D,M,S)	ALAT-GLAT(S)	ALON-GLON(S)	LAPLACE(S)
NJ	CROTON 2 1929	40 29 3.41	74 54 45.46	-1.000	2.520	1.63
NJ	HIGH POINT 1881	41 19 14.87	74 39 42.88	-3.410	*******	******
NJ	LAB 1960	40 49 10.46	74 24 37.86	-3.020	2.260	1.48
NJ	MILTON 1974	41 0 30.58	74 32 27.19	-1.560	1.550	1.02
NJ	MT OLIVE 1879	40 52 2.20	74 42 22.43	-3.060	2.710	1.77
NJ	MT ROSE 1839	40 22 3.04	74 43 25.85	-2.370	*******	******
NJ	PICKLE 1876	40 35 40.68	74 49 26.23	-0.830	3.340	2.18
NJ	TEN MILE 1929	40 24 57.45	74 36 12.66	-2.650	7.260	4.71
NM	ACOMITA 1945	35 2 54.19	107 40 10.15	-1.880	-0.580	-0.34
NM	ADERO 1960	35 21 38.50	104 27 35.76	-6.600	-3.460	-2.03
NM	AIR 1952	32 52 23.48	106 4 28.75	-2.010	13.130	7.13
NM	AIRPORT 1954	33 14 18.19	107 16 53.60	0.010	-0.400	-0.22
NM	ALBUQUERQUE VET HOSP STACK	35 3 10.98	106 34 57.79	-1.220	18.130	10.41
NM	ALMA 1966	33 22 32.17	108 55 39.35	-1.090	4.370	2.40
NM	AMBER 1943	32 53 52.97	106 52 1.14	-4.470	8.360	4.54
NM	ANALLA 1936	33 26 27.84	105 15 34.07	-0.620	-11.120	-6.12
NM	ARABELLA WEST BASE 1934	33 46 45.68	105 9 13.70	1.420	-10.850	-6.03
NM	ARIMEX 1934	33 46 2.27	109 2 49.94	-8.270	15.860	8.58
NM	ARRON 1958	35 1 49.64	106 58 31.96	-2.820	-1.070	-0.62
NM	ARROYO 1961	33 35 58.65	104 34 34.73	0.390	-1.470	-0.81
NM	ARTESIA NORTH BASE 1922	33 1 14.51	104 26 2.16	0.940	-1.470	-0.81
NM	ARTESIA SOUTH BASE 1922	32 50 23.21	104 27 53.99	-0.220	-1.980	-1.07
NM	ARTHUR 1961	32 0 18.43	104 20 23.54	1.280	3.410	1.86
NM	ARTIST 1954	36 23 38.15	105 33 49.80	2.550	21.700	12.88
NM	ATOM HAFB 1952	33 44 21.93	106 21 50.57	-5.910	18.720	10.40
NM	AUTO WSPG 1962	32 53 32.94	106 5 12.27	-3.520	12.500	6.79
NM	BASSETT USGS 1973	35 3 5.61	106 0 18.65	-1.350	0.850	0.49
NM	BAUM 1935	33 8 11.25	103 33 46.52	-1.170	3.810	2.08
NM	BAUMAN 1969	32 6 56.89	106 49 5.85	-0.230	4.850	2.58
NM	BDRY MON 40 US MEX 1909	31 47 0.96	108 12 28.39	-1.470	1.820	0.96
NM	BEACON HILL 1939	31 55 23.49	108 43 8.65	2.310	5.250	2.78
NM	BEACONTROL HAFB 1952	32 49 49.31	105 53 24.74	-0.520	35.510	19.25
NM	BEARGRASS 1934	32 17 48.01	105 24 27.76	-4.210	2.640	1.41
NM	BELEN N BASE 1920	34 28 8.21	106 41 44.10	-3.050	12.470	7.07
NM	BENCH MK 176 OF HAFB 1962 RM 1	32 55 58.32	106 9 9.45	-2.970	11.540	6.28
NM	BENCH MK 346 CF HAFB 1962 RM 1	32 58 45.52	106 9 23.39	-3.100	11.770	6.40
NM	BENCH MK 6 OFF HAFB 1962 RM 1	32 53 11.12	106 8 55.52	-2.810	10.030	5.44
NM	BENCH 1954	36 15 9.55	106 2 55.24	-6.550	6.860	4.06
NM	BENNETT 1934	36 22 35.87	108 44 17.66	-1.230	-9.660	-5.73
NM	BIG HORN 1935	36 59 7.68	106 10 46.02	-1.280	-4.120	-2.48
NM	BLACK 1961	33 22 17.65	104 53 1.43	3.110	-6.400	-3.52
NM	BLUE 1961	33 35 48.79	104 20 27.30	1.400	5.320	2.94
NM	BRUSH 1952	34 4 37.00	106 14 29.94	-2.600	12.660	7.10
NM	BURR 1966	34 18 9.13	106 42 40.31	1.590	13.240	7.46
NM	BURRO USGS 1960	36 39 26.26	107 5 41.66	-3.760	7.560	4.52
NM	BURSUM 1934	33 48 55.76	108 26 11.19	-2.560	16.390	9.12
NM	BUTTE 1948	31 58 30.14	107 19 15.39	0.160	7.910	4.19
NM	CAMBRAY 1945	32 13 43.68	107 17 51.49	-1.880	4.680	2.50
NM	CAMP 1961	33 28 23.72	104 12 5.67	-0.810	3.850	2.13
NM	CAPROCK 1935	33 25 2.22	103 38 31.33	3.170	5.240	2.88
NM	CARI 1935	35 10 20.64	104 12 51.67	-1.840	-1.140	-0.66
NM	CARL 1922	32 17 7.39	104 12 51.67	-5.290	-2.270	-1.21
NM	CASTLE ROCK 1945	36 52 22.98	105 5 35.59	1.220	-10.190	-6.11
NM	CEDAR 1922	33 1 2.15	103 52 37.41	-1.430	8.830	4.81
NM	CEDAR 1934	32 14 45.23	107 14 43.78	-2.400	-9.620	-5.44
NM	CHACO 1945	35 47 41.11	107 29 25.88	-2.600	6.550	3.83
NM	CHARLOTTE 1922	35 15 10.20	104 4 2.68	-0.220	5.630	3.20
NM	CHAVEZ 1958	35 34 41.49	107 15 59.22	8.370	-0.910	-0.53
NM	CHURCH 1958	35 45 48.12	106 40 27.11	-10.320	9.390	5.49
NM	CITOVAL 1945	36 8 10.71	106 20 27.19	2.660	-5.070	-2.99
NM	CLAMPIT 1935	35 33 31.98	103 14 51.92	-4.880	-3.920	-2.28
NM	CLAY 1922	32 32 26.00	103 9 49.11	-3.800	-9.110	-5.42
NM	CLAYTON EAST BASE 1922	36 47 44.16	103 11 48.53	0.340	-9.830	-5.89
NM	CLAYTON WEST BASE 1922	36 46 2.87	103 19 12.50	1.350	-8.970	-5.36
NM	CLIFF 1934	32 57 57.65	108 36 9.38	-3.350	8.120	4.42
NM	COLORADO 1936	34 54 53.33	108 46 54.06	-5.910	11.300	6.40
NM	COMER 1946	35 25 14.44	106 19 27.04	-1.200	12.060	6.99
NM	CORDUNA 1909	32 1 11.14	105 50 55.90	-1.200	*******	******
NM	CORNER 1958	34 59 40.96	107 15 15.73	-5.440	-2.340	-1.34
NM	CORRALES 1946	35 14 10.96	106 38 58.84	-3.570	8.090	4.67
NM	COY 1936	33 10 28.11	105 58 25.06	-4.290	22.830	12.49
NM	COY 1951	33 47 7.63	104 38 29.61	3.170	4.690	2.74
NM	CRATER SOUTH BASE 1952	33 25 10.33	106 30 16.71	0.280	11.790	6.49
NM	CREAM 1934	37 7 50.99	103 6 2.18	-1.420	0.800	0.44
NM	CUEVA 1958	35 25 3.08	106 50 12.37	-2.480	-1.470	-0.85
NM	CUEVO 1958	32 48 52.72	105 8 16.87	-0.860	-5.100	-2.76
NM	CULBERSON 1938	31 22 23.18	108 38 39.48	-0.860	3.260	1.70
NM	CURTIS 1967	32 19 2.08	106 6 54.62	-6.420	17.550	9.64
NM	D 10 WSPG 1952	33 7 35.00	106 9 24.84	-4.470	6.570	3.64
NM	D 7 WSPG 1952	33 23 21.72	106 24 23.52	-6.130	-10.050	-5.53
NM	DAY 1943	33 13 14.95	104 33 14.95	-0.850	-2.150	-1.18
NM	DEMAIR 1945	32 14 43.58	107 42 29.63	-0.910	3.210	1.72
NM	DEMING CITY WATER WORKS	32 16 10.28	107 46 29.63	-2.020	2.060	-1.10
NM	DEMING SOUTH BASE 1910	32 3 0.13	107 50 13.24	-0.830	-9.760	-5.52
NM	DERRICK 1934	32 23 27.19	105 1 40.38	-3.250	-9.260	-5.54
NM	DES MOINES 1922	36 45 22.45	103 49 43.13	1.930	-2.690	-1.61
NM	DIKE 1934	36 39 35.30	108 49 54.39	2.200	0.080	0.05
NM	DINO 1966	34 6 40.10	106 54 6.68	-2.080	0.820	0.50
NM	DIVIDE 1945	36 10 5.35	107 53 36.92	2.650	2.780	1.67
NM	DOME TPC 1972	36 59 46.68	108 17 46.75	-8.070	-1.810	-0.96
NM	DONA ANA EAST BASE 1946	32 4 19.27	106 28 53.52	-1.680	4.240	2.53
NM	DUNE USGS 1934	36 40 10.06	108 14 18.93	-0.500	-1.800	-1.03
NM	DUST WSPG 1965	33 42 5.21	106 34 53.75	-1.800	7.980	3.21
NM	EAST CENTER 50 WSPG 1965	36 17 7.64	104 4 30.84	0.850	4.150	2.30
NM	ELK 1961	33 41 16.38	104 3 15.19	0.550	*******	******
NM	ELKINS 1922	34 41 19.38	105 24 12.54	-1.490	-4.350	-2.48
NM	ENCINO 1934	34 43 20.87	107 14 15.58	-3.560	-4.980	-2.75
NM	EXTER 1934	33 29 59.46	104 30 30.64	-2.820	-5.980	-3.58
NM	FIELD 1945	36 44 50.30	106 50 20.05	-6.970	-2.770	-1.59
NM	FLAT USGS 1958	35 54 41.75	105 0 43.81	-6.970	-1.110	-0.65
NM	FORT UNION 1949	35 54 34.97	107 0 0.37	-0.970	1.230	0.66
NM	FORT 1936	32 30 1.83	103 39 6.18	2.490	3.170	1.81
NM	FRIO 1935	34 51 26.96	105 21 34.17	-5.070	-5.190	-2.80
NM	GAGE 1934	32 36 51.49	103 40 22.33	-7.280	41.230	-0.21
NM	GALLEGOS 1960	35 37 48.68	105 47 17.38	-0.220	1.020	0.57
NM	GALLINAS 1952	34 14 49.22	106 20 10.89	-0.540	6.410	3.54
NM	GAP HAFB 1952	33 32 47.47	107 35 29.73	-7.450	-0.140	-0.08
NM	GAITA 1945	32 1 45.21	109 2 58.12	-0.310	8.580	4.49
NM	GERONIMO 1938	31 30 59.61	103 2 31.21	-1.170	3.380	1.03
NM	GLENRIO 1921	35 10 4.88	103 3 3.62	0.710	1.830	0.98
NM	GLO STATION 6 ECC 1934	36 0 8.39	108 41 36.52	-0.090	9.480	5.33
NM	GLO 10 1934	34 15 34.41	107 36 3.13	-0.790	9.270	5.54
NM	GOBERNADOR 1945	36 42 55.72	105 1 58.76	1.840	-7.680	-4.20
NM	GOODRUM 1934	33 11 31.15	105 3 3.08	-8.210	-5.380	-3.21
NM	GORDITO 1934	36 34 22.67	106 9 12.79	-1.610	12.820	6.87
NM	GRAND WSPG 1952	32 24 2.32	106 9 12.79	-0.900	2.890	1.59
NM	GREEN 1961	33 25 34.23	104 11 11.79	-2.960	-5.360	-3.07
NM	HAILE 1922	32 58 24.75	104 25 51.64	0.690	-1.150	-0.66
NM	HAMMOND 1935	34 39 44.43	103 26 56.35	-0.660	-6.830	-3.70
NM	HARRIS 1934	32 51 20.46	104 49 48.13	1.480	4.320	2.39
NM	HARRY WSPG 1968	33 33 44.16	106 38 36.89	-1.370	1.220	0.65
NM	HENRY 1934	32 0 1.17	105 14 17.48			

STATE	STATION NAME	GLAT(D,M,S)	GLON(D,M,S)	ALAT-GLAT(S)	ALON-GLON(S)	LAPLACE(S)
NM	HOGAN 1958	35 46 39.34	107 33 53.11	0.060	2.790	1.64
NM	HOPKINS USGS 1544	32 30 11.42	104 39 30.76	-0.220	-6.360	-3.42
NM	HUEY 1943	32 9 46.19	106 16 24.94	-1.790	-9.760	5.18
NM	JACINTO 1949	35 5 23.63	105 26 56.08	-4.960	-9.770	-5.61
NM	JARILLA FOREST USGS 1943	32 26 8.08	106 7 3.27	-0.320	13.550	7.26
NM	JARILLA 1909	32 23 59.57	106 6 44.61	-2.530	8.830	4.73
NM	JARILLA 2 1952	32 23 59.58	106 6 44.76	-3.790	8.370	4.49
NM	JEMEZ 1946	35 21 35.48	106 37 42.98	-2.050	1.350	0.78
NM	JOHNSON 1958	35 51 37.30	107 9 0.58	-1.700	9.320	5.46
NM	JORDAN 1945	33 22 40.39	108 13 35.77	-5.290	3.230	1.78
NM	JULIE 1967	33 11 32.54	106 3 14.47	-3.310	19.550	10.70
NM	JUNIPER 1953	35 31 6.86	105 40 55.64	-7.660	1.260	0.74
NM	KENT 1909	32 30 27.07	106 28 58.68	-8.650	-6.540	-3.52
NM	KERR 1970	32 11 41.82	108 52 5.98	0.770	7.540	4.02
NM	KEY 2 1965	32 22 21.25	106 8 28.90	0.860	10.540	5.62
NM	KINE IPC 1972	36 13 5.82	107 53 5.31	-0.520	6.510	3.84
NM	KING 1961	33 4 21.01	104 7 11.18	1.770	7.480	4.08
NM	LA LANDE 1922	34 26 46.22	104 7 57.25	-1.490	3.680	2.08
NM	LAND 1969	32 8 53.18	106 34 51.09	-4.300	2.750	1.46
NM	LAVA 1936	32 2 14.56	106 46 16.59	0.040	4.110	2.18
NM	LEE 1935	32 44 20.08	103 35 56.60	-7.420	6.570	3.56
NM	LEVY NE BASE 1949	36 8 21.75	104 45 45.12	-1.670	-8.130	-4.79
NM	LEVY SW BASE 1949	36 4 21.37	104 46 32.42	-3.330	-7.460	-4.39
NM	LINE USGS 1910	32 51 6.27	108 59 46.84	-6.570	*******	******
NM	LOCKCUT MTN USFS 1945	33 21 7.36	107 49 15.21	-1.460	-0.210	-0.12
NM	LORDSBURG 1934	32 21 0.42	108 42 41.91	0.580	6.590	3.53
NM	LOS ANGELES AMARILLC AWY BN 71	35 3 10.35	106 21 5.18	2.310	12.240	7.03
NM	LOS ANGELES AMARILLC AWY BN 83	35 9 30.65	104 4 44.54	-6.850	-5.450	-3.13
NM	LOS ANGELES AMARILLC AWY BN 84	35 9 58.59	103 51 16.25	-3.990	-1.260	-0.73
NM	LOS LUNAS 1919	34 48 26.54	106 44 6.34	-2.110	5.750	3.28
NM	LOSI USGS 1958	34 32 55.93	107 5 10.24	3.670	-6.640	-3.77
NM	LOST 1945	34 57 23.92	108 32 17.06	2.100	15.880	9.10
NM	LUCERO 1943	32 41 37.92	106 27 26.44	0.0	-7.700	-4.16
NM	LUERA 1945	33 49 29.99	107 52 29.52	-1.610	-5.380	3.00
NM	LUJAN 1960	35 39 4.95	104 19 53.63	-10.150	-5.230	-3.05
NM	LYBROOKS 1945	36 12 55.05	107 33 36.00	0.770	4.690	2.78
NM	MAL 2 1965	33 15 45.03	106 18 18.71	-3.830	5.870	3.12
NM	MANGUS USFS 1935	34 3 6.91	108 23 23.36	-2.010	5.640	3.16
NM	MENECKE 1934	33 23 55.65	104 54 59.11	*******	-6.190	-3.41
NM	MERCHANT 1935	32 22 12.24	103 20 28.20	-7.680	12.560	6.73
NM	MILAGRO 1949	34 58 29.96	105 11 54.37	-0.830	-3.140	-3.67
NM	MILLERS WATCH HAFB 1952	33 40 0.09	106 26 10.85	-1.870	20.140	11.17
NM	MILLMAN 1922	32 39 37.58	104 11 6.68	-6.180	2.220	1.20
NM	MIXEN 1943	32 2 41.28	105 43 24.82	-2.480	2.580	1.37
NM	MOCK WSPG 1967	33 33 23.49	106 25 50.35	-2.700	8.950	4.95
NM	MOON 1939	32 6 1.38	108 12 55.80	1.770	3.840	2.04
NM	MOSELEY 1934	34 9 0.37	105 9 40.12	0.220	-9.090	-5.44
NM	MT WITHINGTON USFS 1945	33 52 50.06	107 29 7.59	0.740	-1.590	-0.89
NM	MUIR 1938	34 9 36.95	108 34 42.73	1.930	6.510	3.86
NM	N MEX TEX CLARK BDRY MI CCR 12	32 0 0.68	103 16 7.27	-4.680	11.730	6.22
NM	NAMBE 1953	35 55 28.01	106 0 55.94	0.290	20.460	12.01
NM	NAN WSPG 1952	32 27 35.66	106 27 51.66	-6.220	-9.760	-5.23
NM	NELSON 1922	34 44 35.13	104 27 38.97	-7.330	-3.140	-1.79
NM	NEW CARISSA NMGS 1934	32 40 34.37	105 37 8.12	-8.730	3.020	1.64
NM	NORTH STAR 1936	33 0 21.69	108 2 11.14	-4.990	11.160	6.08
NM	NUTRIAS 2 1953	36 36 11.96	106 31 22.01	-6.260	15.590	9.29
NM	OJO 1936	34 53 0.55	108 58 1.08	-1.910	12.570	7.19
NM	ORANGE 1961	33 25 22.98	104 2 56.31	2.290	2.310	3.84
NM	OSO RIDGE USFS 1945	35 2 17.18	108 6 57.39	-3.250	2.310	1.33
NM	PAGE 1935	32 14 25.71	103 28 57.92	-0.810	10.980	5.86
NM	PANAMA 1943	32 27 34.00	105 0 7.82	-1.000	-3.120	-1.67
NM	PARR WSPG 1966	33 1 16.41	106 25 20.02	-1.560	1.720	0.94
NM	PETER 1949	36 37 36.52	104 41 45.35	-6.210	-6.650	-3.97
NM	PHILLIPS WSPG 1952	33 26 39.36	106 7 59.11	-3.640	18.080	9.97
NM	PHILLIPS 1921	34 59 3.37	104 7 59.64	-0.760	0.190	0.11
NM	PICKETT 1943	32 9 48.12	104 49 31.05	0.120	-2.150	1.14
NM	PINTADA 1921	34 57 31.47	104 54 39.75	-0.120	-6.570	-3.77
NM	PIPE 1961	33 43 50.06	104 34 4.64	1.160	-2.170	-1.20
NM	PLUM 1966	32 50 0.86	108 27 11.07	0.050	-4.640	-2.52
NM	POND 1936	35 21 42.12	108 51 32.57	-1.320	8.210	5.05
NM	POPE 1922	32 2 1.53	103 54 5.69	-3.530	-6.210	1.18
NM	PORTALES 1935	34 13 18.37	103 17 7.14	5.390	-2.510	-1.41
NM	PRAIRIE 1936	32 30 0.15	107 32 30.23	-3.350	2.370	1.27
NM	PRIETA 1958	35 30 19.34	107 2 6.76	-1.910	0.030	0.01
NM	PUMPKIN USGS 1943	32 24 28.40	107 59 9.96	-1.400	-9.110	-5.01
NM	PURPLE 1970	33 21 10.59	105 2 9.36	-2.740	3.760	7.95
NM	R 907.6-6 USE 1954	35 0 27.21	106 43 59.69	-2.300	3.760	2.16
NM	RAYADO 1949	36 19 9.83	104 52 34.67	0.270	-6.170	-3.65
NM	RED LAKE 1958	34 39 19.94	107 37 12.13	-1.940	-0.930	-0.53
NM	RIDGE 1969	32 10 17.60	107 4 5.50	-0.690	3.000	1.60
NM	RINGON 1945	34 15 36.85	107 46 53.68	-2.550	1.320	0.74
NM	RION 1966	33 54 60.47	106 52 14.72	-2.540	0.780	0.43
NM	ROBERTS 1935	33 13 14.83	103 38 56.60	1.570	5.400	2.98
NM	ROSE 1936	33 25 45.65	105 59 17.71	-3.730	22.090	12.17
NM	RYCACE 1935	32 50 1.92	103 30 18.84	-1.900	5.450	2.96
NM	SAC 1952	32 47 16.32	104 45 13.44	-5.510	29.410	15.93
NM	SADDLE MOUNTAIN LSFS 1935	33 36 55.00	109 0 1.31	-7.300	-0.690	-0.38
NM	SALINAS 1952	33 17 53.74	106 30 50.61	-6.860	-5.840	-3.21
NM	SALTY 1938	32 7 12.46	108 21 27.80	1.600	6.530	3.47
NM	SAN ANDRES AMS 1952	32 40 33.33	106 32 10.54	-1.870	-4.810	-2.60
NM	SAN AUGUSTIN 1949	35 28 32.96	105 7 27.41	-7.960	-5.010	-3.43
NM	SAN JON 1935	35 6 16.20	103 19 55.04	0.770	0.050	0.03
NM	SAN RAFAEL 1945	35 1 48.38	107 54 49.05	-1.770	-8.610	-4.94
NM	SANDIA 1919	35 12 36.15	106 26 56.49	-0.190	23.510	13.56
NM	SANDS NORTHEAST BASE 1952	32 43 38.17	106 12 36.84	-0.840	7.730	4.18
NM	SANDS SOUTHWEST BASE 1952	32 29 1.61	106 24 5.71	-4.490	-1.750	-0.94
NM	SANTA FE EAST BASE 1945	35 40 10.06	105 58 21.18	-3.720	22.870	12.80
NM	SANTA FE WEST BASE 1945	35 37 24.98	106 5 24.18	-2.340	12.870	7.29
NM	SATELLITE TRI STA 110 1966	34 56 43.43	106 27 33.86	-0.700	25.430	14.56
NM	SATELLITE 1956	32 54 39.73	105 28 10.85	-3.570	-8.050	-4.38
NM	SEAMA 1945	34 59 57.81	107 30 17.03	2.920	-3.140	-1.80
NM	SEEHORN WSPG 1952	32 45 20.70	106 29 12.16	1.260	-9.850	-5.33
NM	SIERRA BLANCA 1936	33 22 27.29	105 48 25.44	-3.390	12.550	6.90
NM	SIGN 1936	32 58 30.45	107 24 40.30	-0.650	-3.200	-1.75
NM	SILTON 1969	32 16 9.27	107 39 49.82	-3.910	4.510	2.41
NM	SILVER AIR 1945	32 37 48.20	108 9 29.97	-6.780	7.870	4.24
NM	SIMPSON USGS 1934	36 28 36.77	108 14 43.18	-1.720	4.630	2.75
NM	SKILLETT KNOB HAFB 1952	33 13 49.24	106 38 21.21	-3.790	-1.950	-1.08
NM	SLASH 1934	33 37 51.33	106 45 5.97	-1.850	3.150	1.71
NM	SNAKE 1939	32 10 7.30	107 53 13.94	-2.230	3.200	1.74
NM	SNYDER 1960	36 23 40.39	103 38 56.02	-3.090	-11.320	-6.72
NM	SOCORRO USGS 1934	34 4 17.84	106 53 44.61	-1.060	-1.510	-0.85
NM	SOLANA 1922	35 51 6.56	104 4 6.34	-7.670	*******	******
NM	SOAW USGS 1958	35 7 50.04	107 0 49.66	-3.960	-1.400	-0.80
NM	STONY BUTTE 1934	36 7 5.06	106 6 5.35	-0.270	-4.150	2.44
NM	STORE 1960	36 14 10.45	106 51 43.16	8.150	15.540	9.18
NM	STORM 1934	33 53 39.71	106 37 23.93	-3.830	6.250	3.49
NM	SWOYER 1935	35 59 53.48	103 24 58.53	-1.680	-9.110	-5.35
NM	TANNEYHILL 1935	33 49 14.93	103 31 31.69	0.240	2.410	1.35
NM	TAPIA 1958	35 26 43.26	107 22 7.96	3.940	-2.240	-4.30
NM	TARGET 1936	32 13 14.73	106 29 41.65	-5.070	-8.060	-4.30
NM	TIJERAS 1946	35 2 36.18	106 30 24.14	-0.240	28.440	16.33

STATE	STATION NAME	GLAT(D,M,S)	GLON(D,M,S)	ALAT-GLAT(S)	ALON-GLON(S)	LAPLACE(S)
NM	TINGLE 1936	34 37 12.60	108 29 14.86	-3.400	8.940	5.08
NM	TOHATCHI 1934	35 54 29.48	108 46 21.38	-7.620	-7.740	-4.54
NM	TOLAR 1922	34 15 0.56	103 52 56.83	3.210	7.890	4.44
NM	TOME 1954	34 45 21.40	106 42 13.01	-0.840	9.150	5.22
NM	TOWER 1952	33 12 7.19	106 29 27.12	-8.370	-4.370	-2.40
NM	TRUJILLO 1935	36 51 15.72	105 43 22.46	-1.750	-1.410	-0.84
NM	TS 194 ECC 1964	32 21 28.78	106 22 9.93	-1.110	3.320	1.78
NM	TULAROSA SOUTH BASE 1552	33 4 34.49	106 7 1.94	-2.560	17.950	9.80
NM	TULAROSA 1934	33 40 36.28	108 34 34.79	-3.280	7.310	4.05
NM	TWO BUTTES USGS 1943	32 42 13.39	106 7 35.57	-0.830	9.170	4.56
NM	UNION 1558	36 1 58.95	107 36 56.08	-1.980	6.680	3.93
NM	URTON 1922	34 8 42.32	104 14 11.14	0.300	3.070	1.72
NM	VOCAS 1960	36 22 12.41	104 14 38.81	-2.210	3.290	1.95
NM	VRAIN 1935	34 24 42.29	103 25 55.35	4.080	-1.610	-0.91
NM	WAGON 1922	35 58 52.10	104 34 30.91	-2.100	-1.510	-0.89
NM	WALLACE 1945	36 48 5.71	104 17 49.40	-4.110	1.100	0.66
NM	WALLOW 1966	33 26 57.91	108 40 5.43	0.470	11.370	6.27
NM	WARREN 1935	33 25 41.36	103 12 11.17	3.590	-0.970	-0.53
NM	WEIR 1935	32 38 8.60	103 20 18.65	-5.670	-8.520	-4.59
NM	WHITE 1943	32 10 51.17	104 23 55.75	-5.070	-10.850	-5.78
NM	WHITECAP 1938	32 12 53.28	108 18 42.71	-5.780	7.190	3.83
NM	WHITETAIL 1936	33 15 28.36	105 31 33.16	-0.520	-1.940	-1.06
NM	WINKLE 1934	34 43 32.08	105 2 26.01	-4.070	-8.550	-4.87
NM	WOLF 1960	35 59 14.50	104 1 63.80	-6.200	-3.000	-1.76
NM	YELLOW 1961	33 6 18.18	104 15 22.66	0.640	4.890	2.67
NM	4 F 944 AMS 1952	33 38 47.28	106 22 21.53	-2.720	14.980	8.30
NM	59 B 1945	35 1 18.44	108 24 10.56	-6.200	13.140	7.54
NV	ALKALI USGS 1958	40 24 49.24	117 12 17.66	-1.000	3.430	2.22
NV	ALTO 1957	36 13 16.53	115 10 17.41	-3.570	-2.160	-1.27
NV	AUSTIN LONGITUDE 1889	39 29 31.61	117 4 11.23	*******	16.330	10.38
NV	BATTLE MOUNTAIN NW BASE 1934	40 45 24.10	117 4 46.56	-0.200	3.320	2.16
NV	BATTLE MOUNTAIN SE BASE 1934	40 40 57.31	116 59 6.62	-2.300	1.830	1.19
NV	BDRY MON 108 CA NV 1898	36 10 29.86	116 6 59.30	-6.120	*******	******
NV	BENCH MARK J 33 1939	38 42 28.97	115 30 43.27	-6.800	14.710	9.20
NV	BENCH MARK Z 67 1935	41 32 52.60	118 25 20.28	-1.100	3.580	2.38
NV	BIBLE 1957	36 25 24.98	114 55 37.33	-5.700	-6.390	-3.79
NV	CARP 1944	37 6 38.25	114 28 1.23	-2.760	8.970	5.42
NV	CARSON CITY CAPITOL DOME	39 9 50.35	119 45 54.39	-3.830	-3.060	-1.94
NV	CARSON SINK 1880	39 34 59.23	118 14 4.06	-1.550	*******	******
NV	CURRIE 1949	36 55 49.83	114 56 48.42	1.200	1.230	0.74
NV	DEAN 1956	40 46 8.83	115 16 15.18	-15.750	-0.430	0.25
NV	DIAMOND PEAK 1881	39 35 5.98	115 49 4.05	-2.130	*******	******
NV	DIATOM 1958	39 49 37.99	118 59 3.26	-3.770	5.640	3.61
NV	DIX DMA 1974	40 26 19.83	115 58 47.38	-0.040	7.860	5.10
NV	DOZE 1957	40 29 38.17	118 16 44.35	-3.960	19.690	12.78
NV	DRAIN 1949	36 27 52.12	114 39 49.85	-3.340	4.190	2.49
NV	DRY LAKE 1944	37 36 54.72	114 50 27.47	-4.690	8.820	5.39
NV	DUNE DMA 1974	40 23 42.06	115 12 25.14	-1.130	10.960	7.11
NV	DUNE DMA 1974	39 50 6.63	118 41 55.99	-6.070	2.780	1.78
NV	ELKO N BASE 1940	41 0 8.84	115 32 38.87	-7.480	6.120	4.02
NV	ELKO S BASE 1540	40 56 28.37	115 31 36.45	-4.200	4.830	3.17
NV	ELY EAST BASE 1944	39 4 26.59	114 26 11.70	-3.120	19.770	12.46
NV	ELY WEST BASE 1544	39 1 32.91	114 34 42.27	-4.940	-6.030	-3.80
NV	EUREKA LONGITUDE 1889	39 30 47.42	115 57 29.72	*******	8.160	5.19
NV	EVER 1957	40 44 42.92	114 6 16.89	-7.870	0.990	0.65
NV	FALLON 1935	39 29 22.76	118 45 5.56	-2.670	4.200	2.67
NV	FEN DMA 1974	40 4 56.31	117 49 50.05	-5.660	-1.020	-0.66
NV	FIRE DMA 1974	39 52 44.29	115 5 59.37	-4.510	0.870	0.56
NV	GENOA FLAGSTAFF	39 0 11.38	119 50 42.84	*******	-22.630	-14.25
NV	GLO 2 1926	36 30 27.02	115 30 46.98	6.230	-2.690	-1.60
NV	HARRISON DMA 1974	40 18 38.32	115 30 44.12	-2.170	1.970	1.27
NV	HIGHWAY MONUMENT 10 1944	37 0 11.17	114 56 27.13	-6.320	7.010	4.22
NV	HOGAN DMA 1974	40 44 7.84	114 34 17.02	-6.070	-1.220	-0.79
NV	HOT SPRINGS 1947	41 59 40.40	115 21 53.06	-4.790	9.840	6.58
NV	HUMBOLDT NORTH BASE 1946	40 59 20.38	117 42 54.12	-1.900	7.520	4.94
NV	HUMBOLDT SOUTH BASE 1946	40 57 35.87	117 42 53.99	-0.410	12.280	8.05
NV	IBEX 1934	35 21 54.27	114 52 13.30	-6.720	2.920	1.69
NV	LABEL 1958	35 52 49.91	115 13 33.07	-4.690	-0.840	-0.49
NV	LAND DMA	40 33 46.23	114 41 41.08	-4.130	9.160	5.96
NV	LOMBARD 1954 AZ MK	35 42 35.00	114 50 30.34	-0.060	-2.620	-1.53
NV	LOVELOCK SOUTH BASE 1944	40 3 23.87	118 28 33.44	-1.200	12.160	7.83
NV	MOAPA 1925	36 40 18.02	114 37 26.84	-1.630	2.470	1.47
NV	MOUNT CALLAHAN 1881	39 42 34.04	116 56 59.08	-2.130	*******	******
NV	MT LEWIS 2 1974	40 22 12.59	116 51 37.95	2.660	7.560	4.90
NV	PAHRUMP NORTHWEST BASE 1926	36 10 9.96	116 6 12.44	-5.230	6.350	3.74
NV	PAYET 1940	36 49 40.64	114 44 43.32	-8.180	7.680	4.61
NV	PEAVINE RESET RM 3 1974	39 35 13.23	119 55 52.00	0.450	-4.300	-2.74
NV	PILOT PEAK 1887	41 1 16.30	114 4 35.57	-8.040	*******	******
NV	PIOCHE 1883	37 59 10.10	114 3 3.92	-3.400	*******	******
NV	RATTLESNAKE 1935	41 36 51.64	118 26 6.01	-6.070	8.800	5.84
NV	REED 1933	37 45 7.39	116 7 38.42	-1.770	2.760	1.69
NV	RENO NORTH BASE 1946	39 30 49.18	119 48 19.97	-1.810	-1.520	-0.96
NV	RENO SOUTH BASE 1946	39 28 52.19	119 47 22.39	1.300	-3.130	-1.99
NV	RIVER USGS 1951	39 12 57.71	115 37 14.35	-0.100	2.280	1.44
NV	SAGE 1974	40 53 41.10	115 26 53.09	-3.620	4.120	4.97
NV	SHADY DMA 1976	40 33 48.41	115 15 23.24	1.610	23.570	15.08
NV	SILVERTON DMA 1974	40 55 40.53	114 17 36.71	-6.920	-2.350	-1.54
NV	SPRUCE MTN OFST DMA 1974	40 33 12.88	114 49 13.07	-0.450	8.090	5.25
NV	STORM DMA 1974	40 11 41.48	117 24 42.32	-0.430	15.180	9.80
NV	TABLE 1926	35 48 19.72	115 29 7.35	-9.900	10.910	6.38
NV	TOIYABE DOME 188C	38 49 57.86	117 17 7.60	-4.000	*******	******
NV	TONOPAH ASTRONOMIC 1902	38 4 7.66	117 13 54.81	-6.110	9.710	5.98
NV	VERDI EAST BASE 1872	39 31 5.08	119 57 50.76	-0.380	-4.440	-2.82
NV	VERDI LONGITUDE	39 31 13.46	119 58 57.01	*******	-3.820	-2.43
NV	VIRGIN DMA 1974	35 45 14.17	119 27 40.08	-1.870	-1.860	-1.19
NV	VIRGINIA CITY LONG STA 1897	35 18 38.63	119 38 48.91	*******	-6.550	-4.15
NV	WAHOMIE 1949	36 46 53.09	116 10 20.94	-5.210	11.050	6.62
NV	WANE 1958	40 36 38.44	116 23 30.48	-4.070	10.990	7.15
NV	WHEELER MONUMENT 1897	39 17 43.77	119 39 13.62	-7.850	*******	******
NV	WHEELER STONE LSE	39 29 20.12	117 3 26.55	*******	*******	******
NV	WILD DMA 1974	40 0 55.91	118 22 12.02	-11.310	4.160	2.68
NV	WYTE 1957	39 4 2.79	116 10 21.20	-4.060	-4.250	-4.25
NY	AVOCA 1935	42 24 46.82	77 22 23.46	-3.120	-3.070	-2.06
NY	BEACH HILL 1942	43 50 37.23	75 21 40.06	-0.360	13.290	9.20
NY	BEAR 1932	41 18 46.01	74 0 23.20	1.540	4.490	2.96
NY	CHEEVER 1872	44 4 53.70	73 27 2.68	-0.260	*******	******
NY	CHENEY 1963	44 40 32.60	74 28 39.04	2.260	10.120	7.18
NY	DOMIN 1973	42 34 42.61	75 3 26.81	6.570	6.120	4.14
NY	DUNKIRK LIGHT HOUSE	42 29 27.74	79 21 14.93	6.560	10.250	6.93
NY	EAGLE NYSS 1934	43 1 25.63	75 55 10.95	7.650	4.810	3.29
NY	FALKIRK USLS RM 2 1972	43 1 22.49	78 28 58.51	7.220	2.850	1.94
NY	FARRELL 1942	42 54 53.34	73 41 57.40	-5.090	-0.540	-0.36
NY	FOREST PARK 1903	40 41 49.52	73 51 43.97	-6.860	-2.000	-1.30
NY	GARDINERS ISLAND 1837	41 6 5.08	72 6 22.48	-3.810	*******	******
NY	GRIFFIN 1972	43 1 27.11	75 57 56.67	3.440	0.750	0.51
NY	HAMBURG 2 RM 3 1558	42 46 47.38	78 51 30.95	5.500	3.350	2.28
NY	HAMLIN 1939	43 21 12.36	77 56 30.28	5.430	0.820	0.56
NY	HANCOCK ASTRO 1956	43 7 0.18	76 7 39.17	3.350	4.260	2.91
NY	HEIGHTS 1966	41 16 0.17	73 45 34.29	1.290	8.770	5.55
NY	HEIM 1974	41 11 11.58	74 16 43.27	-1.670	2.870	1.89
NY	HOLLAND 1939	42 38 55.12	78 31 25.52	6.950	5.090	3.44
NY	HOLLEY 1940	42 11 54.49	75 2 27.49	4.560	4.890	3.28

STATE	STATION NAME	GLAT(D,M,S)	GLON(D,M,S)	ALAT-GLAT(S)	ALON-GLON(S)	LAPLACE(S)
NY	HOWLETT 1883	42 59 52.76	76 17 26.71	2.180	*******	*******
NY	JACKIE 1964	41 13 24.78	74 4 18.30	-0.790	2.070	1.37
NY	LYNDON 1929	42 17 14.71	78 20 37.08	3.440	1.960	1.32
NY	MANNSVILLE USLS 1874	43 42 54.26	76 3 13.47	*******	16.720	11.56
NY	MONTAUK 3 1964	41 3 54.13	71 54 19.89	-2.870	0.830	0.55
NY	NIMHAM 1966	41 27 41.25	73 43 32.49	3.730	9.640	6.38
NY	OSBORNE 1862 RM 4	41 35 16.80	73 32 28.74	4.130	8.650	5.74
NY	OSWEGO USLS 1875	43 26 37.13	76 30 49.29	2.450	3.040	2.09
NY	PROSPECT 1878	43 25 17.56	73 46 4.53	-4.850	*******	*******
NY	PRYOR 1934	42 23 6.68	76 1 46.02	*******	0.270	0.18
NY	RAICHART 1938	41 57 44.86	74 2 27.65	-4.770	-7.980	-5.33
NY	RIGA NYGS 1939	43 0 23.50	77 52 41.84	7.920	-4.650	-3.17
NY	RIVERS 1972	43 7 14.59	73 17 7.73	-4.590	1.000	0.68
NY	ROANOKE 1932	40 58 23.32	72 42 12.71	-5.450	-5.800	-3.80
NY	SANDY CREEK N BASE USLS 1874	43 40 43.46	76 12 0.89	-1.940	*******	*******
NY	SENECA 3 RM 3 1969	43 13 52.87	77 35 59.24	5.630	-2.500	-1.71
NY	SITE 10 1961	44 19 57.82	73 33 16.82	-1.890	-13.680	-9.55
NY	SITE 11 1961	44 32 51.63	73 58 31.99	0.760	0.370	-0.26
NY	SITE 12 1961	44 50 37.18	73 59 14.82	-0.110	8.140	5.75
NY	SITE 4 1961	44 20 36.53	73 22 8.61	-0.040	-4.640	-3.10
NY	SITE 5 1961	44 27 35.36	73 38 32.02	-0.410	-5.210	-3.65
NY	SITE 6 1961	44 36 1.48	73 51 25.61	-0.470	-2.480	-1.74
NY	SITE 7 1961	44 46 13.49	73 49 15.87	-0.200	-6.490	-4.57
NY	SITE 8 1961	44 54 45.82	73 49 6.24	-0.480	-4.160	-2.93
NY	SITE 9 1961	44 58 2.85	73 38 0.0	6.260	-5.130	-3.59
NY	SITE 9A MASTER AST 1956	43 12 58.74	75 24 11.55	-1.490	7.800	5.35
NY	SPRUCE 1942	43 12 58.60	73 54 21.27	-6.650	-10.040	-6.88
NY	STARR NYSS 1882	43 20 41.94	75 19 59.36	-5.840	8.280	5.68
NY	STURGEON POINT USLS 1876	42 41 24.78	79 2 53.02	*******	5.660	3.84
NY	TASSEL NYSS 1880	43 15 19	0.60	4.080	-0.330	-0.22
NY	TONAWANDA USLS 1875	43 0 3.74	78 53 20.95	1.490	*******	*******
NY	TROY UNIVERSITY 1860	42 43 45.74	73 41 0.41	5.060	0.240	0.17
NY	TURKS HILL 2 1939	43 2 10.67	77 25 22.62	-5.360	5.710	3.92
NY	TUTHILL 1942	43 20 20.24	75 25 3.94	-1.140	4.950	3.39
NY	VIENNA ASTRO 1956	43 14 48.68	75 46 84.0	-4.880	-4.200	*******
NY	WEST HILLS 1833	40 48 54.78	73 29 32.51	-0.300	-4.200	-2.93
NY	WHITEFACE MTN NYAS 1880	44 21 56.66	73 54 12.53	2.830	4.460	3.04
NY	WHITELAW 1942	43 7 57.72	75 47 29.02	-0.470	*******	*******
OH	ALVERNO 1947	39 5 15.96	84 36 30.28	-0.010	-7.480	-4.93
OH	ANTWERP 1932	41 13 20.41	84 46 32.21	-0.270	0.360	-0.24
OH	ASHBROOK 1933	40 9 14.80	82 45 33.34	-0.790	-0.800	-0.51
OH	ATCHISON 1933	40 5 10.43	81 44 20.49	*******	-1.650	-1.05
OH	BARR 1928	39 38 57.47	82 49 14.00	-0.400	-4.280	-2.80
OH	BAXTER 1932	40 53 15.55	84 48 9.45	-0.790	-1.050	0.14
OH	BETHEL 1944	39 25 50.24	81 31 51.89	-1.190	-1.050	-0.66
OH	CINCINNATI OBSERVATORY 1889	39 8 20.60	84 25 21.71	2.580	-1.390	-0.95
OH	CLARIDON USLS 1877	41 30 39.52	81 1 5.99	1.540	-1.480	2.35
OH	CLARKS HILL USGS 1933	40 6 11.95	80 57 45.25	*******	3.610	0.35
OH	CONLEY 1928	40 33 15.94	83 8 8.09	1.290	0.530	-0.82
OH	DANBURY USLS 1877	40 31 11.90	82 50 15.63	-1.450	-1.280	0.30
OH	DAVIS 1933	40 3 20.13	81 31 7.04	*******	0.460	-0.05
OH	DOYLE 1924	40 58 33.65	81 41 29.15	-1.360	-0.850	-0.05
OH	DRESDEN 1933 RM 1 1969	40 7 6.57	82 2 9.33	-0.590	-0.080	1.50
OH	GOLDSBERRY TPC 1969	40 9 18.55	82 36 45.26	0.600	*******	*******
OH	GOULD 1885	38 38 27.36	82 49 56.93	0.810	2.310	1.50
OH	JUTTE 1932	40 25 50.54	84 48 13.25	0.370	2.470	1.62
OH	LAKE FORK 1959	40 44 14.11	82 10 14.33	0.170	-3.610	-2.36
OH	LOUISVILLE 1950	40 52 21.06	81 16 45.13	-1.660	-2.210	-1.43
OH	MC BETH 1933	40 18 20.04	83 55 17.97	0.780	-1.970	-1.30
OH	MESOPOTAMIA USLS 1877	41 26 59.88	80 59 53.42	1.080	-6.930	-4.58
OH	MILFORD 2 1946	41 22 55.81	84 44 53.37	0.610	-1.770	-1.20
OH	MT PLEASANT USGS 1933 RM 3	40 8 5.78	80 17 17.71	-2.140	1.560	1.00
OH	MUTUAL 1933 RM 3 TPC	40 24 47.30	83 36 40.31	-0.580	-1.370	-0.87
OH	NEBO-MT NEBO 1944	39 24 49.24	82 5 14.22	-2.410	2.180	0.40
OH	NETHERS 1959	40 8 43.20	82 16 54.00	-0.920	0.250	0.16
OH	NEW WAY 1933	40 5 2.22	82 35 21.26	*******	-3.040	-1.89
OH	PORTSMOUTH SOUTH BASE 1929	38 37 45.93	82 50 49.18	3.280	-0.950	-0.61
OH	RENNER TPC 1969	40 6 29.13	83 23 50.53	2.330	*******	*******
OH	SANDUSKY WEST BASE USLS 1877	41 29 2.26	82 40 57.75	-0.480	-9.700	-6.45
OH	SENECAVILLE USGS RM 2	40 2 45.11	81 12 55.11	1.510	-5.660	-3.65
OH	SHANNAHAN 1928 RM 3 TPC	40 9 25.63	83 0 26.50	-0.360	-2.010	-1.30
OH	SHROYER 1933 RM 2 1969	40 9 43.53	84 9 47.18	0.600	-0.400	-0.26
OH	SUMMIT 1933	40 9 37.84	84 19 10.28	-0.850	-1.130	-0.75
OH	TOLEDO STONE LONG POST USLS	41 39 4.50	83 32 31.18	-1.150	4.630	2.95
OH	UNION EAST BASE 1932	40 9 51.75	84 36 33.15	0.410	2.100	1.35
OH	UNION WEST BASE 1932	40 11 35.45	84 46 6.87	8.850	1.570	1.01
OH	WERTS TPC 1969	40 11 15.73	84 24 51.96	3.500	-1.290	-0.74
OK	ALDEN 1921	34 57 33.85	98 40 11.90	-4.440	-8.850	-5.06
OK	ALLEN 1920	34 53 28.66	96 25 3.97	6.940	-8.850	-4.23
OK	BLOOMINGTON 1948	34 59 55.42	99 37 30.55	-7.380	-0.960	-0.54
OK	BOONE 1947	34 55 13.76	98 24 43.57	3.540	2.910	1.73
OK	BYARS 1920	34 51 45.75	97 0 56.76	3.000	2.560	1.48
OK	CAMPBELL 1934	36 26 17.50	98 51 15.02	1.830	-0.340	-0.20
OK	CARSON 1902	36 16 24.92	97 57 32.42	-2.150	-0.890	-0.53
OK	CATESBY SOUTH BASE 1927	36 25 16.58	99 56 30.11	2.060	0.260	-0.15
OK	COBB 1934	36 41 2.84	98 1 51.44	4.540	-1.770	1.44
OK	CORONADO 1968	36 3 29.64	98 1 59.67	4.530	2.490	-1.01
OK	DILL 1921	35 17 24.43	99 7 40.12	-7.210	23 3.78	4.63
OK	DODGE 1945	33 41 43.35	98 20 28.81	0.100	8.280	1.77
OK	EDWARDS 1968	33 58 34.16	97 55 0.72	3.150	2.910	-0.03
OK	EDWARDS 2 1947	34 56 50.14	98 12 35.94	9.020	-0.060	1.07
OK	FREEDOM 1934	36 45 17.27	97 52 55.33	4.820	1.860	-5.00
OK	GADDIS 1948	34 29 52.89	98 52 5.33.36	4.940	-8.740	-0.61
OK	GOTEBO WATER TANK	35 4 13.50	98 24 55.33	1.680	-0.970	-0.81
OK	HAWKINS 1920	34 56 32.82	96 55 11.84	5.560	-3.160	-3.04
OK	HOBART 2 1961	35 1 7.52	99 5 33.36	3.800	-5.320	2.56
OK	IOTA 1951	36 30 23.31	97 56 42.78	-1.130	4.450	0.67
OK	KLANCHA 1968	34 21 43.49	97 55 16.84	2.040	1.140	1.46
OK	KONAWA 1920	34 50 0.13	96 43 52.32	0.970	2.500	1.11
OK	LANIER 1902	35 4 3.08	97 39 31.82	13.040	1.910	3.96
OK	LUX 1934	35 44 6.99	95 23 25.96	-4.580	7.080	5.44
OK	MARLOW LONGITUDE STATION 1899	34 38 50.62	97 57 38.05	-1.350	9.500	-0.02
OK	MERVELOT 1968	35 34 49.50	97 54 56.52	1.010	-0.700	2.31
OK	MILLERTON NORTHWEST BASE 1951	33 59 8.22	95 3 1.51	4.430	4.060	-1.41
OK	MISSOURI SYNOD EVANGELICAL CH	34 55 43.16	99 23 41.40	5.360	-1.710	-1.02
OK	MONUMENT 1902	34 14 23.28	97 53 53.24	3.820	1.800	1.03
OK	OSARIA MARLOW 2 1948	34 41 40.33	97 54 12.90	-3.230	3.970	2.29
OK	POND CREEK ASTRO STA 1906	36 47 30.82	97 52 54.33	1.150	-4.930	-2.82
OK	RED HILL 1951	34 44 7.64	97 10 38.28	4.010	0.590	0.25
OK	ROSEDALE 1920	34 52 7.68	97 5 0.05	1.350	1.330	0.18
OK	SALT 1968	35 10 14.13	95 18 29.25	3.590	-9.840	-6.64
OK	SCHOOL 1952	34 55 22.39	95 59 46.95	0.560	-1.790	1.76
OK	SCHRODER 1968	35 47 1.59	97 21 34.01	2.030	-0.590	-0.35
OK	SHARP 1951	34 25 25.68	96 7 34.65	-6.820	-7.200	4.12
OK	SHAWNEE 1920	35 55 39.47	97 51 44.88	1.670	2.190	0.11
OK	TOM 1957	34 56 33.23	97 51 44.88	2.030	*******	*******
OK	TWIN 1935	34 6 57.43	96 46 40.20	-7.200	14.220	9.63
OK	VINSON 1921	34 54 42.55	99 53 41.32	-0.330	1.660	1.18
OK	WILLSON 1968	36 17 14.25	98 2 11.65			
OK	WINGARD 1902	35 56 47.52	97 29 19.83			
OR	AGENCY 1967	42 38 22.02	121 56 3.98			
OR	ANTELOPE 1932	44 54 52.47	120 35 43.57			

STATE	STATION NAME	GLAT(D,M,S)	GLON(D,M,S)	ALAT-GLAT(S)	ALON-GLON(S)	LAPLACE(S)
OR	BAKER HIGH SCHOOL USGS	44 46 39.94	117 50 2.68	1.620	7.310	5.15
OR	BALCH 1881	45 31 53.47	122 42 29.50	2.660	********	******
OR	BEAL 1971	43 37 0.09	121 32 46.70	-1.280	6.360	4.38
OR	BIG FALLS 1945	44 23 31.95	121 17 42.79	-0.200	8.570	5.99
OR	BLACK HILLS 1948	42 35 35.49	121 16 4.74	-3.580	9.780	6.62
OR	BOUNDARY 1967	42 48 19.83	121 55 2.19	-3.300	2.180	1.48
OR	BUCK 1946	44 36 54.82	121 2 50.50	-5.550	19.430	13.64
OR	CAREY 1971	43 45 11.11	121 27 36.36	-1.690	13.120	9.08
OR	CASTLE ROCK USE 1940	43 50 35.47	122 53 26.36	2.420	5.100	3.53
OR	CENTRAL POINT ASTRO STA 1904	42 23 50.94	122 56 22.28	0.280	********	******
OR	CLINE 1971	44 15 12.69	121 18 6.39	2.060	6.410	4.47
OR	ECHO 1916	45 44 37.23	119 10 19.11	2.000	10.490	7.51
OR	EUGENE ASTRO STA 1894	44 3 29.22	123 5 27.22	-0.950	2.460	1.71
OR	FORT STEVENS LONG STA 1911	46 12 26.47	123 57 38.40	********	26.910	19.42
OR	HART 1971	45 31 4.23	120 9 32.42	5.660	6.390	4.55
OR	HENDERSON 1972	45 14 48.65	123 13 16.63	-4.530	-12.960	-9.20
OR	HOGBACK RM 3 1971	42 14 34.97	121 47 17.58	-4.730	8.640	5.80
OR	HORN BUTTE 1971	45 40 11.23	120 1 21.55	1.170	3.130	2.24
OR	HUNSINGER 1941 RM 5 1972	45 20 33.71	122 31 10.81	-1.480	3.450	2.46
OR	IMBLER NORTH BASE 1946	45 26 23.96	117 57 59.69	1.860	10.920	7.78
OR	IMBLER SOUTH BASE 1946	45 24 22.23	117 58 43.29	0.840	8.990	6.40
OR	JEAGER 1946	45 5 56.65	120 24 4.41	2.700	8.680	6.15
OR	JUNIPER BUTTE USGS 1946	44 28 51.55	121 13 35.80	2.710	13.550	9.49
OR	LA GRANDE 1916	45 19 48.35	118 5 38.35	5.020	-2.550	-1.81
OR	LAKEVIEW BENCH MARK M 16 1920	42 11 35.94	120 21 38.77	********	20.940	14.07
OR	LINVILLE 3 USGS 1959	45 14 18.96	120 17 42.25	2.630	5.890	4.18
OR	LYONS 1972	44 45 49.90	122 38 7.82	0.470	13.690	9.64
OR	MADRAS 34 BASE 1946	44 34 41.97	121 11 24.93	4.100	11.630	8.16
OR	MARSH 1971	43 7 46.20	121 48 14.69	-3.000	4.240	2.89
OR	MAZAMA 1953	42 59 1.79	121 56 21.22	-1.290	-2.720	-1.85
OR	MIDLAND 1948 RM 3 1974	42 7 21.95	121 49 33.36	-3.020	6.820	4.58
OR	MODOC 1967	42 26 55.65	121 49 27.06	-2.870	16.950	11.44
OR	NORTH JUNCTION 1946	44 58 31.15	121 4 47.92	-3.440	3.740	2.64
OR	PAISLEY SOUTH BASE 1920	42 42 14.40	120 32 51.92	-3.680	2.990	2.03
OR	PILOT BUTTE 1932	44 3 38.53	121 16 55.54	4.560	5.180	3.60
OR	POLK 1972	44 54 18.68	123 23 38.72	-5.400	-10.830	-7.64
OR	PONY 1946	45 15 13.00	120 52 44.61	4.910	15.390	10.84
OR	PORT ORFORD ASTRONOMIC 1869	42 44 27.88	124 30 4.03	-6.090	********	******
OR	PORTLAND LONGITUDE STA 19C3	45 31 7.72	122 40 38.50	0.560	4.280	3.05
OR	PRAIRIE CITY NORTH BASE 1946	44 26 41.55	118 52 9.44	-3.250	7.940	5.56
OR	RANIER 1873	46 5 10.78	122 55 27.06	-2.850	********	******
OR	RINK USGS 1942	43 9 38.92	124 7 56.74	-2.540	23.370	15.99
OR	RIVER 1941	45 28 33.94	123 50 34.35	-3.160	33.120	23.61
OR	ROCKY BUTTE 1889	45 32 48.76	122 33 53.06	-0.410	6.010	4.29
OR	ROSE 2 1953	43 3 14.90	123 19 17.28	-4.930	7.760	5.31
OR	ROSEBURG LATITUDE STATION 1904	43 12 39.72	123 21 12.66	-3.990	********	******
OR	SHERAR 1947	45 14 20.76	121 48 47.58	-5.640	9.810	6.97
OR	SHORT USGS 1948	42 4 49.15	121 46 5.79	-4.350	10.680	7.15
OR	SHOTTS 1971	43 28 8.17	121 39 44.70	-0.720	7.400	5.09
OR	STANFIELD EAST BASE 1916	45 46 17.80	119 11 47.49	0.590	6.060	4.35
OR	WALKER MT 1933	44 18 19.28	122 42 56.93	-0.400	11.870	8.14
OR	WESTER BUTTE USGS 1932	45 22 11.86	120 12 14.81	6.160	7.520	5.28
OR	WILLAMETTE NB 1905 RM 6	44 11 36.34	123 12 40.37	-2.550	-2.170	-1.52
OR	WILLAMETTE S BASE 1903	44 4 5.80	123 11 16.72	-0.020	-0.630	-0.44
OR	YAM 1903	45 3 43.90	123 8 33.05	-5.270	********	******
OR	YAQUINA HEAD LIGHTHOUSE	44 40 36.92	124 4 41.56	-1.770	********	******
OR	ZUMWALT ASTRO 1545	45 36 14.35	116 58 54.19	4.170	6.280	4.48
PA	ALTOONA ASTRONOMIC 1890	40 31 46.35	78 23 23.96	-1.260	-9.650	-6.27
PA	ARCO 1974	40 5 16.85	75 28 47.62	8.690	-1.340	-0.87
PA	BLOOM 1929	41 55 5.37	77 22 34.30	0.040	-5.400	-3.61
PA	BROAD 1929	40 53 43.87	75 49 45.19	-1.400	-4.680	-3.06
PA	CLINTON 1927	40 3 17.28	79 29 3.05	********	6.780	4.36
PA	COLE 1961	41 47 38.32	80 9 57.61	3.020	1.000	0.66
PA	CORSON 1929	41 36 53.02	76 29 58.29	3.000	0.930	0.62
PA	FIKE USGS 1968	39 44 43.65	79 29 23.86	2.930	3.610	2.31
PA	GROSS 2 1974	40 22 23.19	75 9 46.34	1.910	2.230	1.45
PA	HALL TPC 1969	39 55 42.42	79 53 48.03	3.580	4.440	2.85
PA	HARRISBURG CAPITAL DOME	40 15 51.54	76 53 2.97	-0.390	-6.800	-4.40
PA	HILLTOWN 1942 RM 3	40 19 46.48	75 14 18.31	3.180	2.250	1.46
PA	HUGHES 1941	39 56 44.98	80 11 27.27	3.710	0.940	0.60
PA	IMMACULATA 1974	40 1 40.55	75 34 19.88	8.850	2.850	1.83
PA	KEYLER USGS 1933	40 57 37.29	78 19 51.62	-1.020	-2.760	-1.81
PA	LONDONDERRY 2 1635 RM 3	39 51 50.40	75 22 49.78	4.860	1.100	0.71
PA	MAINLAND 1942 RM 3	40 15 10.73	75 22 38.88	5.620	3.130	2.02
PA	MC GILL 1933 RM 3 TPC	40 5 11.71	80 24 1.54	5.350	1.680	1.08
PA	QUAKER HILL USGS 1935	41 52 18.55	79 6 43.31	2.830	6.250	4.18
PA	REECE 1934	41 0 57.55	76 20 18.40	-1.860	0.750	0.49
PA	SUMMIT AHS 1968	39 51 2.29	77 39 25.28	4.800	8.650	5.54
PA	TIMBER 1935	40 1 37.98	77 32 5.46	0.680	1.760	1.13
PA	TREVORTON 1934	40 45 45.37	76 43 8.14	3.730	3.650	1.07
PA	UNION 1927	40 26 34.87	80 8 59.43	********	3.650	1.07
PA	WEST CHESTER 1966	39 57 53.00	75 36 17.37	7.030	3.520	2.26
PA	WHARTON 1967	41 30 42.53	77 58 14.63	-3.540	-2.610	-1.73
PA	WILLS 1955	39 43 22.39	78 44 35.45	-2.560	-7.770	-4.96
PA	WORELSDORF 1882	40 19 27.96	76 11 44.85	1.520	********	******
PA	YARD 1837	39 58 24.40	75 23 13.95	4.990	********	******
PR	ARECIBO LIGHTHOUSE	18 29 2.28	66 41 56.71	41.470	0.190	0.06
PR	BATTLE CAY LATITUDE 1900	18 18 5.53	67 9 9.56	-1.940	********	******
PR	CABRA 1932	18 12 14.75	66 25 0.55	3.660	0.260	0.08
PR	CAPE SAN JUAN LH 1900	18 23 0.69	65 37 6.61	13.350	-6.310	-1.99
PR	DESECHEO 1952	18 23 11.60	67 28 46.27	23.760	1.610	0.50
PR	HELECHO USGS 1934	18 14 52.54	66 48 43.20	14.670	1.610	0.60
PR	HUMACAO USGS 1934	18 6 55.79	65 51 16.24	-12.000	-15.050	-4.68
PR	LUIS 1900	18 18 34.68	65 19 58.82	-5.060	-5.260	-1.65
PR	LYNCH 1899	18 1 17.30	66 35 3.71	-13.750	-3.110	-0.96
PR	MONA LIGHTHOUSE 1952	18 5 17.80	67 50 48.74	-7.930	-4.160	-1.29
PR	MT PIRATA 2 1941	18 5 42.51	65 33 5.18	-22.870	-4.820	-0.37
PR	PETE USAF 1964	18 28 22.55	66 7 29.00	35.900	10.180	-0.37
PR	RAC 1956	18 24 9.37	67 10 32.09	********	10.480	3.30
PR	ROSARIO USGS 1934	18 10 40.09	67 5 16.88	2.980	6.880	2.14
PR	TAC 1956	18 27 58.41	67 2 58.11	********	3.060	0.97
RI	BLOCK ISLAND 1835	41 10 31.53	71 35 30.76	-2.480	-0.450	0.30
RI	BUCK 1937	41 58 34.55	71 46 41.23	-4.890	-2.510	-1.68
RI	DRAPER 1932	41 32 24.85	71 16 0.83	-2.730	2.710	1.80
RI	KENT RI MAGS 1938	41 48 12.00	71 21 32.82	-3.850	2.470	1.65
RI	PHELPS 1932	41 18 41.16	71 51 13.11	-3.620	-0.980	-0.65
SC	ADAIR 1957	34 29 56.68	81 50 29.05	3.620	-1.410	-0.80
SC	ALLENDALE LATITUDE STA 1907	33 0 31.31	81 18 40.93	-2.230	-6.360	-3.46
SC	ANDERSON 1935	34 30 21.20	82 39 2.76	4.930	-0.490	-0.27
SC	BARLOW 1933	34 29 58.50	79 24 32.39	-2.530	-7.990	-4.53
SC	BETHUNE 1918	34 28 7.01	80 15 29.81	-3.790	-5.350	-3.03
SC	BONNEAU 1970	33 18 46.05	80 0 1.79	-2.570	-6.820	-3.89
SC	BOYD 1934	33 37 27.85	79 53 1.30	-0.460	-4.890	-2.71
SC	BRUNSON 1932	32 55 24.89	81 11 19.84	-0.030	-3.960	-2.15
SC	BRYANT 1932	33 49 33.53	78 40 48.51	-0.330	-7.610	-4.24
SC	BUNCH 1902	33 35 3.81	81 58 44.12	-0.640	-5.520	-3.05
SC	BUSH 1932	33 15 27.30	80 6 6.58	0.370	-5.040	-1.67
SC	CAMPFIELD 1932	33 29 12.92	79 16 27.24	-2.160	-7.480	-4.13
SC	CAMPFIELD 2 1965	33 29 12.92	79 16 26.94	-2.160	-7.480	-4.13
SC	CARDIN 1965	32 49 23.94	80 21 20.91	-1.370	-6.730	-3.64
SC	CARSON 1934	33 41 2.21	80 13 41.42	0.900	-3.670	-2.03
SC	CHARLESTON W BASE RM 4	32 47 43.24	80 4 53.09	-1.230	-7.090	-3.84

STATE	STATION NAME	GLAT(D,M,S)	GLON(D,M,S)	ALAT-GLAT(S)	ALON-GLON(S)	LAPLACE(S)
SC	CHURCH 1934	34 15 3.82	81 34 16.22	2.450	-3.310	-1.87
SC	CITADEL 1932	32 47 47.74	79 57 45.33	-1.290	-6.530	-3.54
SC	CONNELLY 1934	34 14 10.13	81 54 57.60	3.410	-4.700	-2.64
SC	COOPER 1932	33 15 58.00	79 21 41.83	4.120	-9.310	-5.11
SC	DAVIS 1935 RM 4 1970	34 0 42.11	79 25 41.15	-1.220	-7.110	-3.98
SC	DORN 1934	34 6 38.64	82 7 36.67	1.880	-2.770	-1.55
SC	EDISTO ISLAND EAST BASE 1849	32 33 15.30	80 13 34.37	1.260	*******	******
SC	FLOYDS 1933 RM 4 1969	34 11 21.09	79 3 38.08	-0.280	-7.080	-3.98
SC	FORT MILL 1934	35 0 22.53	80 56 51.45	-1.590	0.950	0.53
SC	GREEN POND 1932 RM 4	32 43 54.07	80 37 0.03	-0.180	-3.170	-1.72
SC	GV 376 ECC 1935	34 38 0.61	82 23 29.09	4.520	-0.370	-0.21
SC	HARVIN 1934	33 50 1.66	80 26 22.71	1.940	-4.120	-2.29
SC	HUFF 1935	33 10 48.46	80 20 58.31	-2.270	-4.610	-2.52
SC	KEMPER 1933	34 19 34.34	79 11 56.64	-2.450	-6.920	-3.91
SC	LIBRARY 1962	33 41 48.95	78 52 38.54	-2.580	-6.330	-3.51
SC	LIMEHOUSE 1965	32 13 11.87	81 4 27.61	1.550	-4.720	-2.51
SC	LUGOFF 2 1971	34 12 35.88	80 42 30.48	-3.970	-4.930	-2.77
SC	MC CLELLANVILLE RM 5	33 5 43.07	79 28 32.72	-0.470	-7.100	-3.88
SC	MCBEE 1918	34 28 19.04	80 15 4.99	-2.860	-5.320	-3.01
SC	MCINNIS 1933	34 44 36.31	79 42 0.34	-1.230	-7.150	-4.08
SC	MITCHELL 1932	32 58 51.19	79 39 36.60	-0.840	-4.700	-2.56
SC	NANCE 1935	34 14 58.90	82 33 12.12	-5.990	-0.470	-0.27
SC	NOB 1918	33 17.65	80 58 51.32	-4.800	-3.560	-1.99
SC	OAK GROVE 1935 RM 3 1971	33 49 3.33	81 22 41.09	-2.160	-4.400	-2.45
SC	OBSERVATORY 1934	33 59 51.39	81 1 36.13	-3.670	-3.500	-1.96
SC	OR 546 SCGS ECC 1935	33 23 1.90	80 50 16.43	0.490	-0.690	-0.38
SC	ORPHAN 1935	35 0 30.10	81 14 28.52	5.050	3.080	1.77
SC	PARIS 1875	34 56 28.86	82 24 40.35	2.230	*******	******
SC	PT 122 SCGS ECC 1935	33 15.13	79 57 57.47	-3.840	-4.460	-2.51
SC	SALEM 1932 RM 5	33 40 1.79	79 6 19.84	-5.810	-6.480	-3.59
SC	SARDIS 1935	34 39 12.14	81 36 35.58	4.810	-1.660	-0.95
SC	SHELDON 1932	32 35 57.82	80 47 39.86	0.660	-2.380	-1.29
SC	SIX MILE MOUNTAIN 2 SCGS 1935	34 49 68.47	82 48 13.11	0.470	-1.610	-0.92
SC	SP 121 SCGS 1935	34 46 21.28	82 9 1.33	4.080	-0.860	-0.49
SC	SWITZERLAND 1932	32 25 41.27	81 0 30.72	1.130	-3.130	-1.68
SC	VARNVILLE 1932	32 52 30.00	81 2 28.39	-0.190	-4.210	-2.28
SC	WADE 1934	34 43 29.49	81 16 0.05	3.190	-2.700	-1.54
SC	WANDO 1932	32 56 8.81	79 52 40.22	-0.250	-6.960	-3.78
SC	WHEELER 1935	34 4 7.67	81 48 1.97	-1.640	-5.750	-3.21
SC	WILLIAMS 1902	33 53 40.48	82 11 27.24	-1.190	-6.280	-3.50
SC	WINNSBORO USGS 1934	34 23 0.07	81 5 4.91	-0.090	-0.530	-0.30
SC	WORTH 1935	34 57 19.81	81 28 35.54	2.560	4.150	2.38
SC	ZIEGLER 1935	33 26 46.52	80 58 26.46	0.750	0.650	0.36
SD	ALVIN 1949 RM 3 1966	43 25 1.90	96 36 0.80	0.780	-1.250	-0.86
SD	BALD 1934	44 50 52.00	98 39 16.43	*******	-0.110	-0.08
SD	BLUMER 1966	43 18 13.01	96 42 29.24	2.470	0.430	0.29
SD	BRANT 1949	43 57 1.30	96 54 32.22	-2.440	-0.760	-0.52
SD	CAPA 1952	44 5 43.63	100 53 18.17	-2.480	-4.190	-2.92
SD	FRANKLIN 1903	44 53 52.39	96 56 18.23	0.770	1.260	0.90
SD	FREEMAN 1901	43 18 47.52	97 27 6.50	-0.070	*******	******
SD	GOOSE 1924	43 54 3.94	99 16 52.82	-5.350	3.380	2.34
SD	HARDING 1912	45 22 13.73	103 53 9.97	-0.220	1.700	******
SD	HOWARD ASTRONOMIC STA 1906	44 0 54.19	97 31 36.04	-2.830	0.350	0.24
SD	ISABELL 1924	43 42 31.33	98 4 20.49	2.940	1.700	1.18
SD	JONES 1924	44 2 26.48	100 28 18.85	-3.810	-2.730	-1.90
SD	LEMLER 1934	45 11 40.18	99 34.02	*******	1.570	1.11
SD	MEDICINE 1924	43 57 54.97	100 9 14.89	-7.960	2.370	1.64
SD	NORWAY 1949	43 43 56.23	96 35 21.58	-1.070	0.940	0.65
SD	OMAHA 1924	43 43 25.59	97 40 34.44	-0.980	0.930	0.64
SD	ORR 1935 LAPLACE AZIMUTH	42 52 4.41	96 55 44.14	*******	4.400	3.06
SD	OWENS 1903	43 53 22.01	97 20 4.26	-2.180	1.060	0.74
SD	PAUL 1924	43 44 58.03	58 57 44.21	-4.500	-2.680	1.86
SD	PLANKINTON 1924	43 42 46.56	98 29 53.85	-4.480	-1.180	-0.81
SD	PREACHER HILL 1903	45 28 14.69	97 6 18.02	1.160	*******	******
SD	PRESHO 1924	43 59 38.57	100 0 19.17	-10.170	0.760	0.52
SD	PROVO ASTRONOMIC 1912	43 11 43.76	103 49 41.36	-3.130	-3.000	-2.09
SD	REVA 1912	45 34 50.47	103 12 44.61	-0.380	*******	******
SD	RIVER VIEW 1951	45 41 37.34	101 65 58.49	0.490	-6.140	-4.40
SD	STEWART 1935	42 54 19.36	96 45 53.07	0.440	2.580	1.76
SD	TURN 1951	45 24 0.16	101 57 46.64	-1.210	-5.920	-4.22
SD	USTA 1951	45 12 45.72	102 9 14.15	-1.850	3.340	3.44
SD	WALL SOUTH BASE 1924	44 4 22.96	102 11 32.87	-0.700	-10.430	-7.25
SD	WALSH 1938	45 6 27.72	101 31 49.49	-1.320	-6.110	-4.33
SD	BEAN 1887	35 11 21.28	88 30 30.02	5.840	*******	******
TN	BETHEL 1930	36 24 42.66	86 43 48.61	-0.980	-0.830	-0.49
TN	CAPLEVILLE NORTHWEST BASE 1914	35 0 26.44	89 50 23.06	-1.500	1.350	0.78
TN	CHURCH 1914	35 0 43.18	89 20 44.11	0.030	0.030	0.07
TN	CLEVELAND 1934 RM 2	35 9 6.79	84 54 4.19	2.580	1.360	0.79
TN	COUNCE 1934	35 0 35.81	88 15 55.11	1.550	2.110	1.21
TN	CYPRESS 1914	35 3 6.46	88 44 15.13	-0.050	6.070	3.49
TN	DUREN 1879	36 25 8.33	86 15 37.75	-0.060	-2.660	-1.58
TN	GORDON 1914	35 2 18.29	89 13 15.33	0.930	4.280	2.46
TN	GOURLEY 1934	36 6 44.35	87 7 0 24.35	3.830	-2.820	-1.66
TN	HAMILTON 1934 RM 6 1970	35 53 4.01	87 7 21.50	3.360	-1.340	-0.79
TN	HASSELL 1934 RM 2 1970	35 20 26.11	87 48 34.00	3.670	0.680	0.40
TN	HOLLOW 1934	35 2 29.26	88 15 55.50	-0.020	0.430	0.25
TN	HURST 1934	35 32 19.05	87 31 54.06	2.180	0.950	0.55
TN	KINCADE 2 1970	35 39 58.67	87 19 9.39	0.070	1.140	0.67
TN	KNOXVILLE LAT AND LONG STATION	35 57 24.58	83 55 33.51	1.590	3.280	1.95
TN	LAMBERTH 1930	36 30 39.41	86 34 45.06	3.000	*******	******
TN	LANDRUM 1933	36 15 26.87	86 53 42.85	0.800	-2.030	-1.20
TN	LEBANON NORTH BASE 1877	36 12 46.62	86 18 24.98	3.050	*******	******
TN	LOOKOUT 1934	36 0 18.52	85 20 39.59	-2.360	*******	-5.19
TN	LUPER 1883	35 57 15.86	84 47 20.53	-3.830	*******	******
TN	MARTIN 1914	35 4 40.86	89 47 52.40	-0.810	-0.940	-0.54
TN	MOUNT LORE 1882	35 54 54.20	83 58 27.26	*******	*******	******
TN	NASHVILLE LAT LONG STA 1877	36 9 57.27	86 47 0.67	4.100	0.770	0.45
TN	PERRY 1933 RM 3 1970	36 39 4.87	86 27 35.52	5.330	3.560	2.13
TN	PINHOOK 1934	35 9 31.97	87 58 3.76	2.660	-1.080	-0.62
TN	POLK 1914	35 33 13.82	87 7 24.49	1.020	-0.140	1.82
TN	PORTER 1914	35 5 3.59	88 58 33.08	1.640	0.330	0.19
TN	WEAVER 2 1956	35 2 24.63	90 7 35.45	0.700	1.460	0.83
TX	ACE 1935	31 3 15.33	100 14 13.96	6.970	2.360	1.22
TX	AGUA NUEVA 1935	26 56 44.14	98 35 32.67	3.230	-3.050	-1.68
TX	ALBANY 1935	32 42 22.96	99 17 51.27	1.720	-0.750	-0.40
TX	ALICE 1904	27 44 33.12	98 4 29.95	1.410	-1.490	-0.69
TX	ALLAMORE LATITUDE STA	31 4 39.47	105 0 8.17	-2.690	*******	******
TX	ALLEN 1924	29 12 4.56	99 58 34.04	0.700	-1.060	-0.52
TX	ANDERSON 1935	30 29 58.99	95 59 28.37	-2.100	-0.980	-0.49
TX	ANSON 1961	32 42 43.25	99 54 36.38	6.270	-0.280	-0.14
TX	APPELT 1935	30 11 56.03	100 27 25.19	0.910	-0.880	-2.46
TX	ARCHER 1943	30 54 1.05	101 47 48.17	1.910	0.520	0.27
TX	ARMADILLO 1949	30 48 19.37	94 39 30.68	1.080	-0.460	-0.23
TX	ASPERMONT 1935	33 8 21.25	100 13 41.09	-4.800	-4.900	-2.68
TX	AUSTIN CAP DOME LIBERTY STAR	30 16 28.07	97 44 24.18	-1.090	-3.660	-1.84
TX	BACHELOR 1902	30 59 40.02	99 8 50.50	6.240	-7.660	-3.94
TX	BAIRD LATITUDE STATION	32 23 40.65	99 23 39.30	8.060	*******	******
TX	BAKER 1935	30 55 10.51	102 6 18.96	-3.730	-0.900	-0.46
TX	BALDRIDGE 1935	31 4 4.28	102 37 35.81	-3.540	6.370	3.28
TX	BALMORHEA 1968	31 0 31.82	103 68 13.75	7.860	-4.880	-2.51
TX	BAMMEL 1952 RM 3 1968	29 57 58.44	95 32 15.68	1.470	1.590	0.80
TX	BANGS 1935	31 43 47.35	99 8 19.89	5.940	-0.320	-0.17

STATE	STATION NAME	GLAT(D,M,S)	GLON(D,M,S)	ALAT-GLAT(S)	ALON-GLON(S)	LAPLACE(S)
TX	BAPTIST 1953	35 7 5.34	102 12 9.71	-0.400	-3.910	-2.25
TX	BARNES 1967	30 59 58.42	98 32 31.22	7.290	-6.610	-3.41
TX	BARTA 1935	29 38 48.75	96 5 51.46	*******	0.250	0.13
TX	BARTLEY 1932	33 12 8.32	101 40 44.22	1.260	-2.420	-1.32
TX	BEAUMONT USE 1933	30 4 33.14	94 4 26.13	2.760	*******	*******
TX	BECK 1943	28 53 49.55	96 56 47.64	0.850	0.770	0.38
TX	BEE CAVE 1967	30 19 11.65	97 59 55.70	0.710	-7.180	-3.63
TX	BERTRAM 1967	30 44 58.55	98 4 6.14	3.750	-9.200	-4.70
TX	BIG 1943	30 57 23.84	102 22 31.04	-0.160	-1.870	-0.97
TX	BIPPUS 1921	34 59 44.39	103 1 35.89	-0.430	-0.600	-0.34
TX	BIRCH 1935	30 22 17.73	96 43 34.83	-2.180	-2.970	-1.51
TX	BISBEE 1935	33 36 30.47	99 47 45.08	0.680	0.440	0.25
TX	BLACK 1909	31 34 21.59	105 8 51.29	3.160	*******	*******
TX	BLUFF USGS 1965	33 50 24.37	58 38 22.83	-2.310	3.340	1.86
TX	BORACHO 1911	31 4 33.52	104 23 23.71	1.740	2.990	1.55
TX	BOWIE NORTHWEST BASE 1900	33 37 21.82	98 0 14.01	-1.190	6.060	3.36
TX	BOYD 1908	32 27 20.56	100 14 48.57	11.090	0.070	0.04
TX	BRAD 1961	32 7 48.81	99 51 15.08	6.910	-0.800	-0.43
TX	BRANSON 1921	34 55 14.33	100 12 29.49	-7.260	-2.550	-1.46
TX	BRAITON 1967	30 57 21.54	99 10 17.75	4.740	0.870	0.45
TX	BRISTOL 1967	32 27 34.35	96 34 4.65	0.430	6.620	3.56
TX	BROWDER 1931 RM 5 1975	36 28 1.07	101 16 30.06	-0.930	-2.060	-1.23
TX	BROWNSVILLE LONG STA 1885	25 53 54.65	97 29 27.52	-0.890	5.840	2.55
TX	BRUCE 2 1968	30 55 14.91	103 11 36.60	4.670	-0.930	-0.48
TX	BULL 1961 RM 3	32 43 43.24	99 55 16.97	5.800	-0.640	-0.68
TX	BUSHLAND COOP GRAIN ELEVATOR	35 11 31.83	102 3 53.88	2.710	-1.580	-0.91
TX	BUTTERFIELD 1921	35 2 39.19	101 36 42.72	-1.190	-2.860	-1.64
TX	BYNUM 1909	32 19 15.70	101 0 0.21	-5.740	-0.210	-0.11
TX	CALLAN 2 1968	31 2 6.77	100 4 50.22	6.380	-1.430	-0.73
TX	CARLOW 1918	28 22 46.78	99 56 34.61	0.130	-3.210	-1.52
TX	CHADBOURNE 2 1958	31 59 50.23	100 16 32.23	5.160	0.370	0.20
TX	CHAMLISS 1902	31 39 23.42	98 7 28.57	2.360	-3.590	-1.88
TX	CHINA 1931	30 2 44.66	94 20 54.63	1.320	-0.660	0.33
TX	CHURCH 1934	35 2 12.67	102 51 6.39	-1.170	-3.260	-1.87
TX	CIMA 1949	30 56 3.58	94 24 59.49	2.310	-1.370	-0.71
TX	CISCO ASTRONOMIC STA USGS 1908	32 23 24.82	98 58 58.96	6.630	-1.690	-0.90
TX	CLARK 1935	30 55.43	100 45 41.66	2.540	6.340	3.26
TX	CLEARY 1935	31 12 55.58	102 4.92	3.490	-1.340	-0.69
TX	CLINT 1934	31 37 54.11	106 11 2.59	-3.620	-6.930	3.63
TX	COMMERCE 1935	33 15 8.47	95 54 20.75	1.710	-1.980	-1.09
TX	CONWAY 1921	35 12 24.26	101 23 22.60	0.420	0.400	0.23
TX	CORINTH 1961	32 51 39.59	95 53 31.17	3.400	0.010	0.01
TX	COW 1935	30 56 50.62	93 55 51.36	1.590	3.360	1.73
TX	CCZBY 1935	32 44 5.46	97 25 1.93	2.930	3.950	2.14
TX	DAVIS 1935	33 9 54.75	99 20 49.04	-3.980	1.610	0.88
TX	DAVIS 1968	31 5 35.43	104 5.70	8.270	-2.670	-1.38
TX	DONNA 1913	26 9 40.59	98 2 44.47	*******	4.970	2.19
TX	DRYDEN EAST BASE 1918	30 4 39.48	102 54.46	2.130	-3.160	-1.58
TX	DUNE 2 1964	31 50 14.80	100 15 55.76	-1.190	7.360	3.88
TX	DUVAL 1935	27 45 8.21	98 37 25.58	1.290	-2.740	-1.28
TX	EAGLE FLAT 2 1963	31 6 26.92	105 7 41.32	-4.030	6.450	3.33
TX	EL PASO LONGITUDE STA 1911	31 45 36.08	106 29 28.84	-3.720	-7.930	-4.17
TX	ELLIOT 1931	36 27 30.61	101 54 3.42	*******	-5.020	-2.98
TX	ESTES 1909	31 29 50.12	102 57 43.66	-0.130	7.870	4.01
TX	EVANITO 1917	26 49 58.27	98 8 28.52	4.990	-5.370	-2.42
TX	FINNEY 1935	33 44 18.46	100 20 29.03	2.400	-5.120	-2.85
TX	FLUVANNA 1968	32 57 15.51	101 8 48.89	-0.660	-3.690	-2.01
TX	FOWLER 1968	33 43 50.53	97 54 19.71	-0.590	7.370	4.09
TX	FRANKLIN 1902	31 14 48.81	98 3 7.53	3.620	-5.020	-2.60
TX	FRONTON 1867	26 4 43.81	97 17 24.57	*******	2.010	0.88
TX	FROST 1968	31 16 17.68	105 36 9.85	-1.850	9.920	5.15
TX	GARDNER 1967	31 51 16.56	98 9 48.46	3.020	-1.580	-0.83
TX	GATESVILLE 1943	31 26 18.44	57 49 37.92	1.900	-4.200	-2.19
TX	GATLIN 1902	32 19 50.56	98 15 16.51	6.890	1.160	0.62
TX	GENEVA 1931	29 29 36.53	93 55 7.73	-4.360	1.330	0.69
TX	GILBERT 1967	32 33 25.53	98 8 21.65	2.370	4.380	2.36
TX	GLENDA 1951	34 65 17.21	100 27 38.06	-5.970	0.500	0.29
TX	HANSFORD 1934	33 48 49.37	102 56 34.24	3.390	-2.880	-1.60
TX	HARPER 1968	31 1 43.89	104 44 39.42	3.280	3.660	1.88
TX	HARRIS 1936	31 22 9.08	105 45 51.37	-4.340	9.590	4.99
TX	HARVARD 1977	30 38 10.42	103 56 48.75	0.580	2.570	1.31
TX	HASKELL 1974	33 10 46.67	99 43 50.74	-3.640	-0.350	-0.19
TX	HAWTHORNE 1935	30 32 5.51	95 23 55.97	-0.660	-0.890	-0.45
TX	HAYS 1909	31 29 54.70	103 19 58.39	4.000	8.090	4.23
TX	HEDLEY 1951	34 52 9.27	100 40 46.69	-2.930	5.170	2.95
TX	HEP 1961	32 12 37.46	100 3 2.36	8.510	1.620	0.17
TX	HODGE 1934	29 46 57.07	101 42 58.80	3.650	*******	*******
TX	HOLMES 1932	33 55 43.21	101 52 13.77	6.780	-3.240	-1.81
TX	HOVEY RM A 1968	30 56 57.10	103 22 20.50	6.790	-3.020	-1.55
TX	HOVEY 1968	30 56 57.85	103 22 19.93	2.040	-3.030	-1.56
TX	HUMBLE 1935	31 0 24.86	101 16 8.61	-1.770	-0.790	1.95
TX	HURST 1935	30 31 16.96	98 2 2.89	-0.470	-1.290	-0.40
TX	IMPERIAL 1935	31 15 38.22	102 45 32.99	-1.960	0.280	0.15
TX	JACKSONVILLE LAPLACE 1919	31 51 54.62	95 16 50.10	7.410	1.710	0.89
TX	JACOBS 1935	31 7 11.42	100 30 33.64	5.110	-0.790	-0.42
TX	JANUARY 1967	32 4 39.45	98 11 23.89	1.310	0.080	0.04
TX	JOHNSTONE 1918	29 22 38.40	100 46 10.76	-0.590	7.220	3.99
TX	JONES 1902	33 31 21.36	94 52 22.55	-0.580	-3.860	-1.86
TX	KARNES 1904	28 52 38.24	97 54 47.23	-2.000	-1.610	-0.77
TX	KELLI 2 1957 RM 5 1971	28 31 58.76	99 12 40.04	7.460	0.280	0.14
TX	KENT 1968	31 3 49.88	104 13 24.80	2.120	-4.620	-2.25
TX	KIGHT 1953	29 11 5.73	98 34 5.54	3.510	4.760	2.10
TX	KING 1946	26 12 45.15	97 3 46.59	*******	-2.380	-1.21
TX	KOERTHER 1935	30 10 55.69	96 42 23.34	3.990	4.110	0.60
TX	LACASA 1908	32 19 5.09	98 41 29.75	6.520	-4.710	-2.43
TX	LACKEY 1935	31 2 27.08	99 56 3.73	4.990	*******	*******
TX	LAGUNA MADRE NORTH BASE 1882	27 40 10.67	97 16 19.59	0.840	-1.760	-0.95
TX	LAMESA 2 1968	32 45 42.85	101 52 19.46	-1.810	0.290	0.14
TX	LAPLACE 1918	28 42 46.33	100 29 32.86	-2.790	-4.930	-2.28
TX	LAREDO LONGITUDE STATION 1895	27 30 27.31	99 31 7.59	1.290	-2.850	-1.34
TX	LAS COMAS 1935	28 2 41.73	99 19 26.54	8.790	3.020	1.57
TX	LIPAN 1935	31 14 8.95	100 19 24.50	-2.760	1.280	0.72
TX	LOLA 1935	33 45 36.51	99 40 13.18	-0.460	0.820	0.40
TX	LONE 1918	28 45 59.28	100 23 11.12	-2.270	5.870	3.22
TX	LOVING 1975	33 14 32.73	98 32 11.83	-0.870	0.510	0.26
TX	MAHA 1967	30 6 13.68	97 55 29.90	-1.750	-5.680	-2.86
TX	MARBRIDGE 1967	30 8 12.69	97 51 9.92	*******	2.490	1.20
TX	MATAGORDA LONGITUDE STA 1911	28 41 35.43	95 58 7.29	6.810	-7.000	-3.60
TX	MATTHEWS 1967	31 1 10.10	98 19 45.26	1.990	0.000	0.24
TX	MC DONALD 1935	32 15 30.82	100 19 44.98	-6.420	0.740	0.57
TX	MC DONALD 1942 RM 4	30 40 16.79	104 1 19.44	-3.220	-2.280	-1.31
TX	MC DOWELL 1921	35 0 8.94	101 13 1.35	1.110	1.380	0.70
TX	MCIVOR 1977	30 38 4.83	103 57 44.44	-2.270	-1.560	-0.75
TX	MEDINA 1971	28 41 58.53	99 45 18.21	-1.940	-1.770	-0.83
TX	MERLE 1935	30 25 39.50	96 27 4.09	-1.190	-1.440	-0.89
TX	MERRY 1921	34 2 44.78	101 50 49.52	-2.330	1.530	0.77
TX	MICK 1967	30 14 16.51	97 22 40.48	-0.610	-2.710	-1.39
TX	MID 2 1968	30 52 14.93	101 55 59.46	-3.740	*******	*******
TX	MILLER 1904	28 18 6.87	97 48 21.61	-1.840	-4.630	-2.29
TX	MISSION MISSION HILL USGS 1903	29 42 52.84	98 9 52.11	2.920	8.730	4.39
TX	MITCHELL 1942	30 8 52.19	103 53 12.42		1.020	0.50
TX	MORGAN PT USE 2 1931	29 40 51.72	94 59 8.93			

STATE	STATION NAME	GLAT(D,M,S)	GLON(D,M,S)	ALAT-GLAT(S)	ALON-GLON(S)	LAPLACE(S)
TX	NELLEVA 1935	30 26 54.12	96 9 56.62	-2.390	-2.350	-1.19
TX	NINE 1935	31 3 5.86	99 22 30.09	5.500	-0.560	-0.29
TX	NOLAN 1961	32 10 5.28	100 9 10.50	5.570	-0.990	-0.52
TX	NOUGES 1935	31 1 39.47	100 59 57.36	3.150	4.770	2.46
TX	PACE HILL 1966	30 58 32.07	94 10 47.68	1.410	0.730	0.38
TX	PASS 2 1972	28 42 49.57	100 27 26.05	-0.350	1.120	0.54
TX	PECK 1961	32 9 45.35	99 33 8.40	8.270	-2.470	-1.31
TX	PHANTOM 1961	32 36 8.51	99 38 58.87	7.440	0.250	0.13
TX	PHARR 2 1956	29 2 20.74	95 20 2.30	-2.350	1.810	0.88
TX	PIERCE 1931	28 48 41.46	95 39 15.44	2.710	3.230	1.56
TX	PILOT 1902	32 15 38.13	98 8 19.43	6.880	-0.010	-0.01
TX	PIPE 1935	30 40 48.70	94 54 58.54	-0.540	0.630	0.32
TX	PLATEAU 1968	31 3 49.82	104 33 25.53	0.180	8.020	4.14
TX	POLK 1935	33 23 5.05	99 5 46.23	-3.010	0.150	0.08
TX	POSEY 1930	32 56 13.36	94 5 10.78	1.490	1.650	0.89
TX	POWELL 1932	31 0 28.69	101 34 14.34	1.830	0.830	0.43
TX	PRESIDIO SOUTH BASE 1934	29 39 52.36	104 21 38.65	-4.960	9.670	4.78
TX	PROSPER 1935	33 14 37.78	96 47 3.98	-4.910	6.520	3.57
TX	QUICKSAND 1935	30 54 47.66	93 43 9.17	1.760	4.290	2.20
TX	RACHEL 1935	32 45 32.74	102 52 29.90	********	-3.430	-1.85
TX	RACHEL 1935 RM 3 1975	32 45 32.28	102 50 29.85	-0.750	-3.860	-2.09
TX	RAID 1961	32 22 25.35	99 27 37.31	8.110	-1.860	-1.00
TX	RANGE 1939	25 57 49.20	97 14 38.60	4.440	4.920	2.15
TX	RED RIVER LONGITUDE 1923	34 22 18.28	100 0 20.84	********	-1.860	-1.05
TX	REEVES 1934	31 59 28.59	106 23 13.49	0.090	-0.450	-0.24
TX	RICHIE 1967	32 57 44.68	98 5 22.91	-5.020	4.320	2.35
TX	ROBERTS SOUTH BASE 1951	35 49 38.95	101 1 32.14	2.570	0.840	0.50
TX	ROBY 1935	32 44 35.52	100 22 45.77	2.970	-3.470	-1.88
TX	ROEMER 1931	28 27 58.16	96 41 3.78	2.540	3.110	1.48
TX	ROUND 1909	31 28 54.32	103 53 10.94	-2.880	-3.150	-1.65
TX	SABINE LONGITUDE STATION 1911	29 44 9.68	93 53 21.03	********	-1.260	-0.62
TX	SAN VINCENTE 1934	29 9 34.20	103 1 24.35	0.760	4.000	1.95
TX	SANDY 1951	34 53 49.16	100 56 7.08	-3.700	1.980	1.14
TX	SARAGOSA 1968	31 1 29.71	103 36 7.22	7.560	-0.450	-0.23
TX	SATURN 1967	29 23 42.83	95 5 15.11	3.060	********	******
TX	SCHMIDT ASTRONOMIC STA 1924	29 22 13.40	99 0 13.09	-2.540	-0.640	-1.29
TX	SEARS 1908	32 33 30.65	100 2 16.36	7.940	-0.910	-0.49
TX	SHIN 1935	30 59 26.60	98 52 18.51	7.200	-1.740	-0.89
TX	SIMPSON 1949	31 36 3.46	97 29 32.54	0.610	1.310	0.69
TX	SINCLAIR 2 1967	30 23 34.32	96 59 53.06	-1.560	-2.500	-1.27
TX	SMITH 1909	31 58 14.94	102 27 49.08	2.340	********	******
TX	SPRING 1935	30 36 18.77	95 8 35.73	-0.140	-0.270	-0.14
TX	STANTON 1909	32 7 32.80	101 46 23.95	3.470	-3.130	-1.66
TX	STAR 1935 RM 4	27 21 47.76	98 45 32.40	3.280	-4.120	-1.89
TX	SUGARLAND 1942	29 35 16.82	95 35 37.03	2.690	2.100	1.03
TX	SUMMIT 1968	31 3 42.37	104 57 52.16	-2.570	2.540	1.31
TX	SUNSET 1967	30 33 3.58	98 1 57.72	0.580	-4.770	-2.43
TX	SWINEY 1967	33 19 21.80	98 5 59.69	-4.310	5.210	2.86
TX	TAYLOR 1917	27 30 49.95	99 21 18.95	3.270	-3.590	-1.66
TX	TEHUACANA 1919	31 44 16.59	96 32 56.10	0.770	3.780	1.99
TX	TENNYSON 1935	31 44 46.60	100 17 7.15	3.840	-2.430	-1.28
TX	TESTON 1932 RM 5 1975	34 4 46.60	101 52 27.36	2.610	-4.110	-2.30
TX	TEXAN 1934	31 10 6.52	105 2 2.28	-1.620	3.770	1.95
TX	TOPO 1878	26 45 18.32	97 28 18.49	-5.200	********	******
TX	TORNILLO 1934	31 25 44.52	105 59 12.15	-3.880	6.110	3.19
TX	TORY USGS 1967	33 2 51.76	98 8 3.43	-7.320	3.200	1.74
TX	TUCKER 1961	32 8 22.46	99 42 7.85	6.670	-1.550	-0.82
TX	UNIT 1961	32 16 24.84	99 32 27.32	6.190	-2.090	-1.12
TX	UPHAM 1967	32 50 11.66	98 7 50.38	-1.320	4.340	2.36
TX	VEGA SOUTH BASE 1921	35 0 14.13	102 37 54.74	-3.310	-3.640	-2.09
TX	VERIBEST 1935	31 30 25.73	100 12 52.66	4.000	0.830	0.43
TX	WACO 1943 AZ MK	31 33 40.00	97 10 10.23	2.180	9.440	4.94
TX	WALDRON 1935	31 44 37.63	103 14 37.18	0.110	11.520	6.06
TX	WALK 1935	30 32 58.00	95 38 43.87	-0.670	0.260	0.13
TX	WARREN 1934	33 23 23.26	96 28 25.37	5.690	-3.750	-2.11
TX	WHITSETT 1921	35 3 45.97	102 22 34.53	-1.530	-3.150	-1.81
TX	WHITTLE 1923	34 21 31.80	99 31 56.06	-2.450	2.730	1.54
TX	WILBANK 1968	31 3 7.35	102 56 5.75	-0.020	3.110	1.60
TX	WILHEM 1935	31 0 23.78	99 39 20.69	3.700	3.420	1.79
TX	WILKINS E BASE 1950	32 33 4.13	96 57 55.21	1.910	1.210	0.65
TX	WILKINS N BASE 1950	32 34 37.87	95 4 28.88	1.010	2.070	1.12
TX	WILLIE 1919	31 42 37.57	96 51 46.76	2.720	10.520	5.53
TX	WILSON 1921	35 3 30.80	101 25 17.35	-0.280	0.010	0.01
TX	WOODVILLE 1931	30 46 28.98	94 23 16.33	0.510	-1.500	-0.77
TX	WRIGHT CORYELL CC 1967	31 27 20.29	98 7 3.55	3.540	-3.740	-1.95
TX	YEGUA 1935	30 22 34.06	97 17 7.28	-0.130	0.020	0.01
TX	YOAKUM SOUTHWEST BASE	29 19 4.81	97 8 41.74	-1.700	0.230	0.11
UT	ALKALI USGS 1972	37 31 16.09	109 19 12.97	-5.010	4.780	2.91
UT	ANTELOPE 1892	40 57 43.49	112 12 54.97	-2.670	********	******
UT	ARINOSA 2 1974	40 43 56.77	113 40 46.19	-3.820	3.430	2.24
UT	BARRO 1940	40 43 40.10	113 28 24.71	-3.640	3.640	2.38
UT	BEAVER FLAG ASTRO STA USE 1872	38 16 24.85	112 38 26.51	-1.570	********	******
UT	BLACK MESA BU USGS 1966	37 31 0.08	109 36 23.19	-6.450	0.900	0.55
UT	BLACK 1974	40 44 34.68	113 8 34.34	-4.680	5.740	3.75
UT	BLANDING 1936	37 38 33.58	109 29 2.73	-12.520	-1.740	-1.06
UT	BLUE MTN TPC 1972	37 50 25.18	109 27 39.26	-5.000	-7.350	-4.51
UT	BUCKHORN 1937	39 12 36.82	110 56 28.24	-6.910	-8.490	-5.37
UT	BURMESTER 1938	40 44 48.42	112 30 59.96	-1.640	1.530	1.00
UT	COON 1963	40 39 34.93	112 12 5.15	4.100	9.950	6.48
UT	DESERET 1887	40 27 34.45	112 37 31.88	-2.210	********	******
UT	DIXON 1974	40 1 8.80	111 36 50.14	10.420	18.750	12.06
UT	GAP TPC 1972	38 4 42.56	109 24 24.18	3.540	5.050	3.12
UT	GREEN RIVER LONGITUDE 1898	38 59 29.62	110 9 55.06	-5.730	-1.660	-1.04
UT	GUNISON ASTRONOMIC	39 9 30.27	111 49 12.94	-4.650	********	******
UT	HENEFER 1938	41 3 57.58	111 25 13.25	-2.060	11.230	7.38
UT	HORSE 1974	39 25 5 9.25	110 25 4.89	-10.500	4.130	2.62
UT	IBEPAH 1889	39 49 41.59	113 55 8.14	********	********	******
UT	INDIAN HEAD 1937	39 52 31.92	110 54 2.70	-3.660	3.270	2.10
UT	INITIAL MESA 1938	37 0 9.19	114 2 38.90	-11.610	4.560	2.74
UT	JEEP USGS 1973	40 54 57.58	111 47 34.97	-1.040	16.000	10.48
UT	KEARNS 1962	40 38 36.42	111 59 13.89	-1.640	5.430	3.54
UT	LAKE 2 USGS 1963	38 16 23.56	111 55 27.00	1.440	-4.950	-3.20
UT	LUND 1925	38 0 30.35	113 26 1.21	1.930	1.930	1.19
UT	MILLER 1937	39 32 36.57	110 48 55.40	-4.780	-6.440	-4.10
UT	MONA 1965	38 36 3.48	110 34 38.99	-4.810	-0.820	-0.51
UT	MONTY USGS 1972	37 23 19.50	109 6 43.45	-3.380	6.690	4.07
UT	MONUMENT 1951	37 0 19.52	110 10 19.63	5.730	4.280	2.58
UT	MOUNT ELLEN 1891	38 7 16.46	110 48 49.92	8.680	********	******
UT	MOUNT NEBO 1887	39 48 28.32	111 45 56.24	-4.830	********	******
UT	MOUNT WAAS 1893	38 32 30.46	109 13 37.75	9.720	********	******
UT	NEEDLE TPC 1972	38 42 81.27	109 53 36.14	1.440	5.110	3.20
UT	OASIS ASTRONOMIC 1898	39 17 37.19	112 37 43.62	-1.720	12.100	7.66
UT	OGDEN LONGITUDE 1873	41 13 11.69	111 59 37.41	-3.130	18.450	12.16
UT	OGDEN PEAK 1884	41 11 59.91	111 52 52.49	-0.330	********	******
UT	PATMOS HEAD 1850	39 30 7.84	110 18 57.35	-12.500	********	******
UT	PROMONTORY 1892	41 17 52.80	112 25 8.11	-4.510	********	******
UT	QUARRY TPC 1972	38 32 0.0	109 48 9.12	4.810	5.440	3.38
UT	RELAY 1974	40 48 4.08	112 53 20.84	-0.690	2.640	1.73
UT	SALERATUS 1937	39 1 58.27	110 18 18.24	-5.150	-6.120	-3.85
UT	SALT LAKE CITY 1893 AZ	40 46 10.73	111 53 26.73	-7.060	21.220	13.86
UT	SALT 1931	39 5 7.01	110 4 5.44	-6.380	2.350	1.48
UT	SOLDIER 1974	39 53 25.29	111 6 45.85	-0.520	7.010	4.50

STATE	STATION NAME	GLAT(D,M,S)	GLON(D,M,S)	ALAT-GLAT(S)	ALON-GLON(S)	LAPLACE(S)
UT	VIEW USGS 2 1973	40 27 47.07	111 51 8.41	-5.880	13.580	8.82
UT	WADDOUP 1892	40 54 25.13	111 53 9.92	-1.020	*******	******
UT	WASATCH 1890	39 6 53.62	111 27 11.21	-2.360	*******	******
UT	WOOD 1974	39 11 50.91	110 21 59.23	-4.910	-5.110	-3.23
UT	8 MILE USGS 1972	38 16 27.90	109 34 5.63	3.610	8.660	5.37
VA	ACADEMY 1934	38 10 29.92	77 16 22.56	-1.050	-10.350	-6.40
VA	BIG KNOB 1893	36 39 52.48	82 30 21.76	-3.800	*******	******
VA	BOONE 1965	36 41 3.78	77 8 52.97	-1.630	-5.750	-3.43
VA	BULL RUN 1871	38 52 53.26	77 42 12.75	3.120	5.190	3.26
VA	CAHAS 1877	37 7 1.57	80 0 56.91	-1.110	*******	******
VA	CAPE HENRY LIGHTHOUSE OLD	36 55 32.33	76 0 30.52	-2.010	*******	******
VA	CLAREMONT 1932	37 11 58.81	76 59 43.75	-1.380	-11.290	-6.82
VA	CLARK EROL 1957	38 18 44.14	78 0 6.43	-0.420	0.180	0.11
VA	CLARK MTN 1871	38 18 40.76	78 0 11.69	-1.210	*******	******
VA	COLUMBUS 2 1966	37 41 28.73	76 53 25.36	-4.080	-6.880	-4.20
VA	CORPORATION 1934	37 30 13.82	76 52 4.75	-1.910	-6.030	-3.68
VA	DAVID 1942	38 18 30.06	77 29 45.01	-1.220	-2.400	-1.50
VA	DORY 2 1966	36 51 5.85	77 3 6.44	-0.760	-7.570	-4.54
VA	ELLIOTTS KNOB 1878	38 5 59.02	79 18 51.56	-1.800	-4.750	-2.93
VA	FAIRALL EROL USE 1957	39 2 16.43	77 31 35.33	4.210	6.550	4.13
VA	FORK MTN 1874	38 28 44.48	78 24 57.69	-0.330	-5.010	-3.12
VA	GOOSE 1942	39 1 53.34	77 31 38.47	3.850	5.080	3.19
VA	HALSTEAD 1932	37 13 8.83	76 33 46.54	-3.610	-7.670	-4.64
VA	HOADLY 1942	38 40 38.65	77 21 43.83	-1.980	-2.400	-1.50
VA	HOBBS 1933	37 2 42.80	77 48 10.44	-3.190	-8.570	-5.16
VA	HOLDCROFT 1966	37 21 50.38	76 56 20.40	-1.530	-11.730	-7.12
VA	HOMEVILLE 1941	36 56 54.52	77 10 18.46	-4.550	-6.600	-3.96
VA	INDIAN NECK 1934	37 54 7.93	77 2 1.02	0.250	-4.810	-2.96
VA	JACKSON 1954	38 54 44.80	77 40 7.54	2.060	2.740	1.72
VA	JAMAICA 1942	37 44 14.35	76 40 13.09	-1.980	-9.070	-5.55
VA	KENTUCK 1932	36 39 29.58	79 17 10.59	1.980	3.090	1.85
VA	LACY 1934	38 44 8.15	77 9 39.59	-4.110	-5.320	-3.33
VA	LEE 1966	38 6 27.44	77 4 57.51	1.320	-6.850	-4.22
VA	LONG MOUNT 1875	37 17 27.27	79 5 10.48	-1.430	*******	******
VA	MALVERN 1932	37 26 55.82	77 16 26.13	-1.660	-5.260	-3.20
VA	MC CORMICK OBSERVATORY PIER	38 1 57.72	78 31 20.83	3.230	-0.930	-0.57
VA	MINERAL 1932	37 59 56.10	77 54 49.51	2.120	-0.360	-0.23
VA	MT MARSHALL 1869	38 24 21.42	78 0 42.16	1.740	6.000	3.73
VA	MITCHELL NORTH BASE 2 1967	38 46 33.50	78 12 10.49	4.080	-0.910	-0.57
VA	NORFOLK EAST BASE 1931	36 50 16.74	76 1 17.66	-2.150	-8.210	-4.92
VA	OWENS 1934	38 16 29.44	77 3 39.54	-1.150	-8.720	-5.40
VA	OXROAD 1966	38 52 46.62	77 22 8.22	-0.450	1.040	0.65
VA	PARIS 1927	38 59 1.83	78 0 6.64	6.540	5.360	3.37
VA	PLAT 1963	36 57 14.66	76 18 8.56	-2.080	-6.650	-3.99
VA	POWELL 1933	36 40 9.20	77 47 58.22	-3.390	-4.880	-2.92
VA	RAPIDAN 1965	38 18 43.76	78 0 12.09	0.010	1.700	1.05
VA	RESERVOIR 2 1966	38 31 31.80	77 19 3.87	-2.770	-5.940	-3.70
VA	RETURN 1934	38 7 45.39	77 7 55.47	-0.850	-8.680	-5.36
VA	ROGERS 1894	36 39 35.82	81 32 42.17	2.980	*******	******
VA	SCROPE 1934	38 17 23.82	77 32 21.75	0.300	-3.920	-2.42
VA	SPRING HILL 1941	37 2 19.97	77 1 19.24	-2.840	-9.300	-5.60
VA	STAUNTON ASTRO LAT AND LONG	38 8 46.20	75 4 19.29	4.600	-3.240	-2.00
VA	STRASBURG ASTRO STA LONG 1881	38 59 29.66	78 21 39.22	1.840	-3.720	-2.34
VA	SUFFOLK 1918	36 44 0.83	76 34 41.57	0.920	-11.220	-6.71
VA	TAPP EROL 1959	38 41 8.02	78 0 41.42	3.050	3.890	2.43
VA	TRINITY 1669	38 56 26.71	77 10 5.76	-1.770	-4.620	-2.91
VA	TUGGLE 1941	37 18 30.75	78 29 44.27	-1.850	-4.060	-2.46
VA	USA EPG 1957	38 45 1.69	77 11 21.63	-4.260	-4.590	-2.89
VA	USA EROL 1957	38 40 41.17	77 8 18.59	-4.450	-4.090	-2.55
VA	WALLACE 1934	38 19 26.00	77 27 46.45	-2.550	-4.460	-2.77
VA	WHITE ROCK 1893	36 39 54.00	83 27 5.53	-4.370	*******	******
VT	ASCUTNEY 1875	43 26 46.23	72 27 8.44	-10.330	4.760	3.28
VT	BENNINGTON 1942	42 52 3.74	73 11 33.91	-0.290	19.550	13.30
VT	BOUNDARY MON NO 79 1942	43 20 58.73	73 15 13.82	-0.180	12.870	8.84
VT	BRISTOL 1943	44 7 57.68	73 5 26.10	0.920	19.500	13.58
VT	BURKE 1959	44 34 13.55	71 53 34.54	0.850	4.960	3.49
VT	BURLINGTON ASTRO STA 1870	44 28 8.68	73 12 33.86	0.790	*******	******
VT	COCK 1968	44 34 29.39	72 26 17.29	-0.390	0.690	0.48
VT	EAST LAKE GSC 1517	44 58 43.69	72 9 2.58	1.420	1.770	1.25
VT	HANNUM 1942	42 58 28.15	72 30 10.57	-2.520	-2.660	-1.82
VT	KILLINGTON 1872	43 36 17.07	72 49 14.19	1.360	*******	******
VT	MT MANSFIELD 1875	44 31 34.84	72 48 55.40	2.660	7.800	5.47
VT	MT PLEASANT USGS 1959	44 7 28.17	72 28 54.00	4.230	2.000	1.40
VT	RED STONE 1943	44 28 13.76	73 11 48.32	1.080	4.960	4.96
VT	SITE 2 1961	44 55 14.98	73 17 22.57	4.710	-1.200	-0.85
VT	SITE 3 1961	44 54 10.02	73 8 21.18	4.780	4.310	3.04
VT	WILCOX 1872	43 43 17.17	73 20 45.50	-1.370	4.100	2.83
WA	ALDER USGS 1916	47 50 59.42	119 56 21.08	-4.240	-2.600	-1.86
WA	ALKI-ALKI SW BASE 1921	47 34 34.50	122 25 9.15	*******	-1.690	-1.25
WA	ASHTON 2 1973	47 42 67.99	121 57 1.80	-3.280	26.760	19.80
WA	AYCOCKS POINT 2 1934	47 30 26.45	123 3 11.63	*******	-17.630	-13.00
WA	BAHOKUS 1942	48 22 18.41	124 40 27.23	*******	20.160	15.07
WA	BALD USGS 1933	46 38 1.39	117 5 13.86	1.730	11.070	8.05
WA	BRANCH 1948	46 5 3.11	116 42 15.78	-4.780	2.840	2.05
WA	BROMLEY USGS 1941	47 27 52.48	120 12 30.85	-11.180	11.990	8.83
WA	BUNGALOW 1925	47 16 32.95	118 50 7.29	-2.010	*******	******
WA	CAPE DISAPPOINTMENT ASTRO STA	46 16 36.30	124 2 50.31	-0.940	*******	******
WA	CHURCH 1968	47 33 4.16	118 19 3.38	-1.830	7.400	5.46
WA	DENT 1971	46 18 4.50	119 11 35.72	0.040	1.960	1.41
WA	GAGE 1968	47 30 37.06	117 19 17.12	1.360	12.500	9.24
WA	GREY BACK 1946	45 59 27.70	121 54 57.58	-2.250	12.020	8.64
WA	GRIG 1949	47 19 0.42	117 54 9.80	-2.600	7.810	5.74
WA	KALAMA ASTRONOMIC STATION 1872	46 0 24.62	122 50 25.14	-6.830	9.380	6.75
WA	KENN 1971	45 59 44.06	119 49 10.98	-4.290	-0.560	-0.40
WA	KING 1968	47 43 13.15	122 44 50.38	-4.050	-14.030	-10.38
WA	KINGHILL 1941	48 48 2.79	122 27 41.69	-6.040	13.650	10.27
WA	KRAIN 2 1972	47 15 10.91	122 0 9.07	3.470	21.190	15.57
WA	LARSEN 1968	47 23 21.42	119 3 31.20	-1.530	6.040	4.45
WA	LAVENDER 1973	47 22 55.96	122 24 24.43	0.580	-1.560	-1.14
WA	MAGNISON 1968	47 34 6.98	117 53 14.71	-3.030	7.260	5.36
WA	NOVARA 1971	46 52 11.42	119 9 11.38	-3.950	6.210	4.53
WA	O TOOLE 1925	48 48 32.36	117 52 58.20	0.140	5.950	4.48
WA	OLYMPIC 3 RM 7 1973	47 41 42.73	122 25 45.42	-4.310	0.830	0.62
WA	OSOYOOS SOUTH BASE 1925	48 57 2.89	119 26 26.42	-1.370	0.320	0.24
WA	PERSON 1971	46 34 14.04	119 2 53.75	-3.040	4.710	3.42
WA	PORT TOWNSEND ASTRONOMIC 1908	47 7 4.01	122 45 12.98	*******	-9.210	-6.86
WA	ROLLING 1973	47 27 55.94	123 13 14.19	7.030	7.930	5.84
WA	SAT TRACK STA 003 1965	47 11 7.13	119 20 11.88	-4.120	4.310	3.16
WA	SEATTLE ASTRO LONGITUDE 1886	47 36 31.16	122 20 0.89	-3.940	4.750	3.51
WA	SHORE RM 2 1971	47 0 42.48	119 17 43.63	-3.530	3.810	2.79
WA	SITAR 1974	47 34 15.35	122 33 10.30	10.300	-12.870	-9.50
WA	TACOMA ASTRONOMIC STATION 1892	47 15 46.83	122 26 50.15	0.100	2.870	2.11
WA	TACOMA SOUTH BASE 1905	47 4 37.75	122 26 4.13	3.480	8.980	6.58
WA	UNION 1974	47 21 27.17	123 6 11.94	-0.550	-6.450	-4.75
WA	WASH 2 1934	47 21 29.74	122 48 21.05	-7.180	-0.860	-0.64
WA	WEISHAAR 1968	47 32 7.16	118 43 54.54	-2.520	8.890	6.56
WA	WELL 1971	46 11 14.19	119 28 24.52	0.900	1.430	1.03
WA	WHITE 1968	47 42 23.07	117 30 55.79	-0.460	5.090	3.77
WA	WINCHESTER 1948	47 15 12.00	119 11 11.52	-7.770	2.290	1.69
WI	AMINICON RIVER USLS 1871	46 41 32.22	91 51 42.95	-4.100	*******	******
WI	ARCHIBALD 1934	45 15 46.47	88 35 12.70	*******	-5.930	-4.21
WI	BABE 1921	45 8 44.08	88 28 19.13	*******	-4.720	-3.35

STATE	STATION NAME	GLAT(D,M,S)	GLON(D,M,S)	ALAT-GLAT(S)	ALON-GLON(S)	LAPLACE(S)
WI	BCYERS BLUFF USLS 1874	45 25 12.63	86 56 10.79	-3.330	******	******
WI	BRULE RIVER USLS 1871	46 45 17.96	91 35 18.17	2.150	******	******
WI	CLAM 1952	46 8 0.17	90 57 28.97	1.080	-8.430	-6.08
WI	DAIRYLAND 1953	46 14 38.77	92 16 36.43	3.780	1.870	1.36
WI	DALE 1926	44 1 30.04	90 12 14.45	0.060	-9.400	-6.53
WI	DOOR BLUFF LAT STA USLS 1865	45 17 46.13	87 3 54.65	-1.170	******	******
WI	FIVE 1953	46 6 27.18	92 1 42.67	-4.340	-7.390	-5.32
WI	GORDON 1953	46 14 41.46	91 48 26.87	0.190	-9.190	-6.63
WI	GRANDFATHER MRC 1927	43 48 42.85	91 12 41.24	******	-8.110	-5.61
WI	GREEN BAY EAST BASE 1921	44 31 42.70	88 0 54.89	******	0.330	0.24
WI	LONG 1921	45 58 27.60	91 35 37.70	******	-10.330	-7.43
WI	MANANA 1926	44 27 4.07	88 54 27.08	-1.590	0.490	0.34
WI	MASSIVE 1921	45 28 33.75	91 6 43.53	-1.090	-0.750	-0.55
WI	MELLEN 1939	46 18 3.38	90 37 2.64	2.230	2.140	1.55
WI	MINNESOTA JUNCTION USLS 1873	43 28 28.41	88 43 64.96	3.410	-4.810	-3.47
WI	MOOSE 1952	46 5 21.07	90 44 28.17	0.600	-2.890	-1.97
WI	NEW BERLIN USLS 1873	42 58 17.52	88 11 1.22	******	******	******
WI	OUTER ISLAND USLS 1870	47 4 17.16	90 26 26.63	-3.350	-7.550	-5.33
WI	OWEN 1921	44 57 2.17	90 33 48.55	-6.020	-2.160	-1.56
WI	GX 1939	46 14 35.60	89 38 4.10	-0.120	-0.460	-0.33
WI	PINE LAKE TR D USGS 1952	46 15 43.53	90 7 24.77	2.050	-0.550	-0.38
WI	PINGLE 1934	45 10 43.54	91 35 24.31	******	-1.560	-1.13
WI	PLEASANT 1952	46 14 38.96	90 21 15.00	-0.130	-3.280	-2.33
WI	PRENTICE 1921	45 32 21.10	90 17 17.59	-1.710	-11.700	-8.37
WI	RHINELANDER 1921	45 38 52.83	89 24 36.57	-1.780	-4.880	-3.30
WI	RUPP 1931	42 41 15.45	90 41 34.04	-1.130	******	******
WI	SEELEY RM 3 1974	46 8 52.41	91 17 1.89	-3.320	-16.100	-11.62
WI	SMOKEY 1939	46 13 54.42	91 29 38.56	-0.160	-13.240	-9.56
WI	SOMERS USLS 1873	42 37 9.97	87 52 33.01	1.250	******	******
WI	SPOONER 1934	45 49 38.29	91 53 46.86	******	-15.000	-11.33
WI	STAADT 1934	44 46 35.50	90 6 51.86	******	-9.530	-7.00
WI	SUNNY 1930	43 54 6.60	90 27 48.65	1.420	-10.540	-7.31
WI	TROSTEL 1952	46 15 8.78	89 54 29.16	1.300	-1.110	-0.80
WI	WABENO 1921	45 26 22.69	88 39 43.99	-9.020	-8.260	-5.88
WI	WESTFIELD 1930	43 52 28.67	89 29 26.40	-0.010	-3.000	-2.18
WV	BEECH 1944	38 52 18.25	81 28 39.57	3.000	2.580	1.62
WV	CHARLESTON ASTRONOMIC 1881	38 21 2.04	81 37 59.43	4.830	2.630	1.63
WV	FINDLEY 1941	39 3 31.05	79 59 58.30	4.210	4.910	3.09
WV	GEBHARDT 1883	38 31 45.42	82 15 11.27	1.400	-2.010	-1.25
WV	KEENEY 1880	37 46 24.56	80 42 19.43	-2.050		
WV	MABE 1934	37 23 44.27	81 19 12.95	-0.870	-0.060	-0.04
WV	MALONE 1975	39 33 0.17	80 1 37.01	4.040	4.830	3.08
WV	PINEY 1883	38 26 39.35	82 3 29.22	1.540	******	******
WV	SPRING 1944	38 20 23.74	81 44 28.68	5.560	1.380	0.85
WV	TOP 1941	39 3 6.34	79 18 29.45	-1.710	-16.450	-10.36
WY	ACME 1957	44 54 3.23	106 58 5.44	8.330	-9.380	-6.63
WY	ALKALI 1912	43 38 12.51	104 29 11.64	-4.160	-4.500	-2.99
WY	ARLINGTON 1933	41 36 14.64	106 13 3.22	17.950	0.710	0.48
WY	ARROW USGS 1948	42 46 15.03	105 13 50.96	4.760	-1.190	-0.84
WY	BADGER 1925	44 58 10.08	106 40 54.15	0.410	2.920	2.03
WY	BARBER 1925	44 21 24.41	106 59 45.76	2.260	-10.350	-6.88
WY	BEAR 1933	44 40 34.10	104 16 58.26	0.410	-6.440	-4.34
WY	BEAR 1948	42 20 47.19	105 2 9.82	1.840	-6.800	4.78
WY	BERTHA 1949	44 38 1.06	105 3 44.42	1.390	2.340	1.60
WY	BILLY 1955	43 17 26.91	105 12 10.19	0.720	5.000	3.52
WY	BITTER 1949	44 52 25.28	105 46 44.82	2.380	-2.830	-4.11
WY	BLACK BUTTE 1974	41 33 34.84	108 48 2.10	1.600	-6.930	-0.86
WY	BOULDER 1972	43 55 11.21	108 21 47.22	4.270	-1.300	
WY	BRIDGER BUTTE 1962	41 17 35.34	110 21 37.22	-1.080	******	******
WY	CAMBRIA 1912	43 7 43.50	104 11 37.22	-3.270	-13.510	-8.90
WY	CAMEL 1960	41 16 48.70	104 57 45.56	-3.710	-17.690	-11.67
WY	CHEYENNE EAST BASE 1913	41 25 51.39	107 7 18.54	3.230	1.880	1.24
WY	CHURCH 1962	41 22 3.17	105 7 14.02	******	******	******
WY	COLEMAN 1912	40 59 53.78	105 33 31.67	-3.320	******	******
WY	COLO WYO BDRY MIPOST 44 1912	41 28 50.31	106 1 53.62	-13.130	9.870	6.79
WY	COPPER 1931 RM 3 1972	41 17 0.87	110 42 0.01	5.170	8.190	5.40
WY	CURL 1962	41 10 13.75	110 58 8.40	2.880	12.440	8.20
WY	DAY 1974	41 14 13.75	108 7 54.39	0.050	1.630	0.96
WY	DELANEY 1937	41 35 26.13	108 1 56.72	-1.210	-1.030	1.08
WY	DIVIDE 1933	41 51 57.82	105 29.61	-3.750	2.350	-0.70
WY	ELKHORN 1948	42 53 15.80	108 2 5.07	-6.530	-0.110	1.59
WY	GREEN 1931	42 23 35.19	110 13 26.62	2.950	5.750	-0.07
WY	HANK 1962	41 21 56.36	105 40 26.82	0.370	3.090	4.03
WY	HENSLEY 1949	44 32 14.46	105 16 49.25	-2.690	3.990	2.14
WY	HIDIVIDE 1955	43 46 56.26	105 47 25.16	3.730	8.230	2.65
WY	HIGH SWELL 1948	41 32 27.43	104 43 1.05	0.530	1.840	5.46
WY	HORSE 1972	41 36 55.91	104 37 17.59	-2.100	-14.650	-1.30
WY	HULETT 1949	44 41 47.57	104 23 52.11	-1.320	-0.400	-9.77
WY	INDIAN USGS 1963	41 52 24.02	110 1 44.10	1.760	******	-0.77
WY	JAGGED 1962	41 28 10.44	104 42 34.62	3.750	-5.800	******
WY	JIREH COLLEGE CUPOLA	42 29 29.43	105 28 57.66	-0.510	3.200	-3.83
WY	KING 1933	41 27 48.94	108 0 34.55	6.020	-9.380	2.20
WY	KIRBY USGS 1931	43 35 5.61	104 31 31.93	-2.640	******	-6.21
WY	KIRKBRIDE 1933	41 24 44.66	104 6 1.18	-0.040	14.450	******
WY	KIRTLEY 1912	42 51 44.30	108 20 40.52	-3.720	1.530	10.18
WY	LAKE 1957	44 48 1.71	105 14 14.34	-3.110	0.400	1.05
WY	LIGHTNING 1938	41 34 53.93	108 20 46.27	-1.200	11.540	0.27
WY	LISA 1972	44 43 40.99	108 22 48.57	-3.930	2.070	8.12
WY	LOVELL 1931	44 1 9.21	105 24 24.18	-3.160	3.570	1.44
WY	MAYSDORF 1955	41 38 42.39	107 3 32.28	-6.810	-4.650	2.57
WY	MESS 1948	43 21 55.43	104 14 56.72	2.260	17.180	-3.19
WY	MIDWEST 1934	41 31 43.10	107 28 55.24	4.460	3.790	11.39
WY	MUDDY 1937	42 10 18.50	107 57 50.16	-6.540	3.090	2.55
WY	OSBORNE 1972	44 57 41.39	106 12 53.33	0.770	10.640	2.19
WY	PASSAGE 1934	41 32 36.16	107 21 8.28	5.690	10.640	7.05
WY	PINEGROVE USGS 1948	41 26 6.39	109 7 0.25	13.100	-16.980	3.68
WY	QUAKING ASP USGS 1931	44 57 16.03	107 10 23.82	10.490	-11.670	-11.99
WY	RANCHESTER NORTHWEST BASE 1925	44 54 31.22	107 23 53.76	14.360	10.870	-8.25
WY	RANCHESTER SOUTHEAST BASE 1925	41 21 11.75	106 35 36.21	-4.560	-4.750	7.22
WY	RATTLESNAKE USGS 1933	41 10 56.58	105 35 37.36	0.860	9.510	-3.12
WY	RED BUTTES 1948	41 27 14.63	109 29 51.96	8.200	5.460	6.29
WY	RIDGE 1962	42 44 50.14	108 1 59.70	-1.060	6.640	3.71
WY	SANDRAW 1972	43 10 48.13	109 0 24.52	6.710	0.630	4.54
WY	SHOSHONE 1931	41 45 57.65	106 58 3.02	6.170	1.170	0.42
WY	SINCLAIR EAST BASE 1948	41 46 42.98	107 3 49.76	1.180	4.900	0.79
WY	SINCLAIR WEST BASE 1948	41 26 23.44	109 40 43.75	1.820	0.770	3.24
WY	SPIDER 1962	44 28 44.20	104 27 3.10	0.200	2.470	0.54
WY	SUNDANCE 1912	44 17 3.67	105 30 10.81	3.080	******	1.73
WY	TANK 1955	44 18 24.66	105 27 57.19	-0.760	-5.500	-3.85
WY	TAIMAN USGS 1931	43 32 6.68	105 20 6.42	-4.250	2.990	2.06
WY	TECKLA 1955	43 45 11.23	110 40 45.47	-7.000	-6.180	-4.28
WY	TETON NORTH BASE USGS 1946	43 41 11.91	110 43 21.84	-10.750	-7.800	-5.46
WY	TETON SOUTH BASE USGS 1946	43 35 47.13	110 46 6.75	-2.500	-15.460	-5.38
WY	VALLEY 1946	42 1 11.37	104 52 7.96	-0.670	-14.360	-10.14
WY	WARREN 1912	42 2 43.30	104 57 26.48	-2.390	-18.450	-9.62
WY	WHEATLAND 1963	41 23 55.98	104 59 38.93	2.690	-2.100	-12.29
WY	WHITAKER 1912	43 43 46.07	105 20 50.71	0.550	3.280	-1.48
WY	WHITE 1949	44 59 39.20	106 21 40.82	-0.680	-7.290	2.32
WY	WYMONT 1934	41 42 6.52	104 5 25.32			-4.86
WY	66 MOUNTAIN 1933					

APPENDIX E

CALCULATOR PROGRAMS FOR ASTRONOMIC AZIMUTH

E-1 COMMENTS ON PROGRAMS

The programs for the TI 58/58C/59, the HP 41C/CV/CX, and the HP 67/97 were developed for the author by Mr. James W. Arnold. The original TI solar program written by Mr. Arnold was modified slightly. Credits for the HP 85 solar program are given in Section E-5. Mr. Arnold's permanent address is 1454 Hamilton Ave., Palo Alto, CA 94301.

All programs included herein are prepared with the assumption that the user employs the methods and the ephemeris of this manual. All are written for the hour angle method. Azimuths for each individual observation are displayed so that they can be inspected for possible rejection.

Some abbreviations used in the instructions are:

    D.d = decimal degrees

    HR = decimal hours

  D.ms = degrees, minutes, seconds in format DD.mmsss

  H.ms = hours, minutes, seconds in format HH.mmsss

min.sec = minutes, seconds in format mm.sss

  sec. = decimal parts of a second

E-2 HP 67/97 PROGRAMS

Data Register Contents

Solar
- Reg. 0. UT + UT1 corr'n (HR)
- 1. Eq.T (HR)
- 2. Longitude (D.d)
- 3. Latitude (D.d)
- 4. δ_0 (D.d)
- 5. UT1 corr'n (HR)
- 6. $\Delta\delta$/hr. (D.d)
- 7. ΔEq.T./hr (HR)
- 8. Σ Azimuths (D.d)
- I observation counter

Polaris
- Reg. 0. GHA_0 (D.d)
- 1. Azimuth (D.d)
- 2. Longitude (D.d)
- 3. Latitude (D.d)
- 4. tan δ
- 5. Watch corr'n (HR)
- 6. Σ Azimuths (D.d)
- I observation counter

CALCULATOR PROGRAMS FOR ASTRONOMIC AZIMUTH

Instructions For Using HP 67/97 Solar Program

1. Load program

2. Input LAT (D.ms)   ENTER   LONG   (D.ms)   A

3. Input δ_0 (D.ms)   ENTER $\Delta\delta$/hr (sec.)   R/S

4. Input Eq.T$_0$ (min.sec.)   ENTER ΔEq.T/hr (sec.)   R/S

5. Input UT1 corr'n (sec.)   R/S

6. Input UT (H.ms)   R/S

7. Input Mark Reading (D.ms)   R/S

8. Input Sun Reading (D.ms)   R/S

   Azimuth is displayed in D.ms. Repeat steps 6 through 8 for subsequent observations. Press B for mean azimuth at any time.

Instructions For Using HP 67/97 Polaris Program

1. Load program

2. Input LAT (D.ms)   ENTER   LONG   (D.ms)   A

3. Input GHA$_0$ (D.ms)   ENTER δ (D.ms)   R/S

4. Input watch corr'n (min.sec.)   R/S

5. Input UT (H.ms)   R/S

6. Input Mark reading (D.ms)   R/S

7. Input Star reading (D.ms)   R/S

   Azimuth is displayed in D.ms. Repeat steps 5 through 7 for subsequent observations. Press B for mean azimuth at any time.

HP 67/97 SOLAR PROGRAM

001	*LBLA	055	RCL1	109	GTO2
002	CLRG	056	+	110	GTO1
003	HMS→	057	1	111	*LBL2
004	STO2	058	2	112	F1?
005	X⇄Y	059	-	113	GTO3
006	HMS→	060	1	114	3
007	STO3	061	5	115	6
008	.	062	×	116	0
009	9	063	RCL2	117	+
010	STOI	064	-	118	GTO2
011	R/S	065	X>0?	119	*LBL1
012	3	066	GTO1	120	F1?
013	6	067	3	121	GTO2
014	0	068	6	122	*LBL3
015	0	069	0	123	1
016	÷	070	+	124	8
017	STO6	071	*LBL1	125	0
018	X⇄Y	072	1	126	+
019	HMS→	073	8	127	*LBL2
020	STO4	074	0	128	R/S
021	R/S	075	X⇄Y	129	HMS→
022	3	076	X≤Y?	130	R/S
023	6	077	GTO1	131	HMS→
024	0	078	3	132	-
025	0	079	6	133	CHS
026	÷	080	0	134	X>0?
027	STO7	081	-	135	GTO2
028	X⇄Y	082	GTO2	136	3
029	1	083	*LBL1	137	6
030	0	084	SF0	138	0
031	0	085	*LBL2	139	+
032	÷	086	CHS	140	*LBL2
033	HMS→	087	SIN	141	-
034	STO1	088	LSTX	142	X>0?
035	R/S	089	COS	143	GTO1
036	3	090	RCL3	144	3
037	6	091	SIN	145	6
038	0	092	×	146	0
039	0	093	RCL0	147	+
040	÷	094	RCL6	148	*LBL1
041	STO5	095	×	149	ST+8
042	R/S	096	RCL4	150	ISZI
043	*LBL4	097	+	151	→HMS
044	CF0	098	TAN	152	DSP5
045	CF1	099	RCL3	153	R/S
046	HMS→	100	COS	154	GTO4
047	RCL5	101	×	155	*LBLB
048	+	102	X⇄Y	156	RCL8
049	STO0	103	-	157	RCLI
050	ENT↑	104	÷	158	INT
051	ENT↑	105	TAN⁻¹	159	÷
052	RCL7	106	X>0?	160	→HMS
053	×	107	SF1	161	R/S
054	+	108	F0?	162	GTO4
				163	R/S

HP 67/97 POLARIS PROGRAM

001	*LBLA	055	X⇄Y		
002	CLRG	056	X≤Y?		
003	HMS→	057	GTO1		
004	STO2	058	3		
005	X⇄Y	059	6		
006	HMS→	060	0		
007	STO3	061	-		
008	.	062	*LBL1		
009	9	063	CHS		
010	STOI	064	SIN		
011	R/S	065	LSTX		
012	HMS→	066	COS		
013	TAN	067	RCL3		
014	STO4	068	SIN		
015	X⇄Y	069	×		
016	HMS→	070	RCL4		
017	STO0	071	RCL3		
018	R/S	072	COS		
019	1	073	×		
020	0	074	X⇄Y		
021	0	075	-		
022	÷	076	÷		
023	HMS→	077	TAN⁻¹		
024	STO5	078	3		
025	R/S	079	6		
026	*LBL2	080	0		
027	HMS→	081	+		
028	3	082	STO1		
029	6	083	LSTX		
030	0	084	R/S		
031	X⇄Y	085	HMS→		
032	RCL5	086	R/S		
033	+	087	HMS→		
034	1	088	-		
035	5	089	CHS		
036	.	090	X<0?		
037	0	091	+		
038	4	092	RCL1		
039	1	093	X⇄Y		
040	0	094	-		
041	6	095	ST+6		
042	8	096	ISZI		
043	5	097	→HMS		
044	×	098	DSP5		
045	RCL0	099	R/S		
046	+	100	GTO2		
047	X>Y?	101	*LBLB		
048	-	102	RCL6		
049	ABS	103	RCLI		
050	RCL2	104	INT		
051	-	105	÷		
052	1	106	→HMS		
053	8	107	R/S		
054	0	108	GTO2		
		109	R/S		

E-3 HP 41C/41CV/41CX PROGRAMS

Data Register Contents, Solar

Reg. 00 UT (HR)
     01 Eq. T_0 (HR)
     02 Longitude (D.d)
     03 Latitude (D.d)
     04 δ_0 (D.d)
     05 UT1 corr'n (HR)
     06 $\Delta\delta$/hr. (D.d)
     07 ΔEq.T/hr (HR)
     08 Σ Azimuths (D.d)
     09 observation number

Instructions For Using HP 41 Solar Program

Keystroke	Display	Input
1. XEQ SIZE 010		
2. XEQ "SUNHR"	LAT = ?	Latitude in D.mmss
3. R/S	LONG = ?	Longitude in D.mmss
4. R/S	DECL, 0 HR = ?	δ_0 in D.mmss
5. R/S	d DECL/HR.SC?	$\Delta\delta$/hr., in seconds
6. R/S	EOT, 0 HR, M.S.?	Eq.T_0 in mmss.s
7. R/S	d EOT/HR.SC. = ?	ΔEq.T/hr, in seconds
8. R/S	UT1 CR, SC. = ?	UT1 corr'n, in seconds
9. R/S	UT = ?	UT for observation, in H.mmss
10. R/S	MARK RDG = ?	Direction to mark, in D.mmss
11. R/S	SUN RDG = ?	Direction to Sun, in D.mmss
12. R/S	AZ = XXX.XXXXX	

Computed astronomic azimuth to ground target is displayed (for one observation). Repeat steps 9 through 12 for subsequent observations. Keystroke A results in display of average (mean) of all azimuths computed AVE AZ = XXX.XXXXX. After Keystroke A and mean observed, more observation data can be input by Keystroke R/S with display UT = ? Accumulated mean can be observed by Keystroke A at any time.

HP 41 SOLAR PROGRAM

```
01♦LBL "SUNHRA"      52 ENTER↑         103 GTO 01
02 SF 27             53 ENTER↑         104 FS? 06
03 SF 21             54 RCL 07         105 GTO 03
04 0                 55 *              106 360
05 STO 08            56 +              107 +
06 STO 09            57 RCL 01         108 GTO 02
07 "LAT=?"           58 +              109♦LBL 01
08 PROMPT            59 12             110 FS? 06
09 HR                60 -              111 GTO 02
10 STO 03            61 15             112♦LBL 03
11 "LONG=?"          62 *              113 180
12 PROMPT            63 RCL 02         114 +
13 HR                64 -              115♦LBL 02
14 STO.02            65 X>0?           116 "MARK RDG=?"
15 "DECL, 0 HR=?"    66 GTO 01         117 PROMPT
16 PROMPT            67 360            118 HR
17 HR                68 +              119 "SUN RDG=?"
18 STO 04            69♦LBL 01         120 PROMPT
19 "dDECL/HR, SC.?"  70 180            121 HR
20 PROMPT            71 X<>Y           122 -
21 3600              72 X<Y?           123 CHS
22 /                 73 GTO 01         124 X>0?
23 STO 06            74 360            125 GTO 02
24 "EOT, 0 HR, M.S?" 75 -              126 360
25 PROMPT            76 GTO 02         127 +
26 100               77♦LBL 01         128♦LBL 02
27 /                 78 SF 05          129 -
28 HR                79♦LBL 02         130 X>0?
29 STO 01            80 CHS            131 GTO 01
30 "dEOT/HR, SC.=?"  81 SIN            132 360
31 PROMPT            82 LASTX          133 +
32 3600              83 COS            134♦LBL 01
33 /                 84 RCL 03         135 ST+ 08
34 STO 07            85 SIN            136 HMS
35 CF 22             86 *              137 FIX 5
36 "UT1 CR, SC.=?"   87 RCL 00         138 "AZ="
37 PROMPT            88 RCL 06         139 ARCL X
38 FC? 22            89 *              140 1
39 0                 90 RCL 04         141 ST+ 09
40 3600              91 +              142 AVIEW
41 /                 92 TAN            143 GTO B
42 STO 05            93 RCL 03         144♦LBL A
43♦LBL B             94 COS            145 RCL 08
44 CF 05             95 *              146 RCL 09
45 CF 06             96 X<>Y           147 /
46 "UT=?"            97 -              148 HMS
47 PROMPT            98 /              149 "AVE AZ="
48 HR                99 ATAN           150 ARCL X
49 RCL 05            100 X>0?          151 AVIEW
50 +                 101 SF 06         152 GTO B
51 STO 00            102 FC? 05        153 END
```

HP 41 POLARIS PROGRAM

```
01♦LBL "POLHRA"      54 -
02 SF 27             55♦LBL 01
03 SF 21             56 CHS
04 0                 57 SIN
05 STO 05            58 LASTX
06 STO 06            59 COS
07 "LAT=?"           60 RCL 03
08 PROMPT            61 SIN
09 HR                62 *
10 STO 03            63 RCL 04
11 "LONG=?"          64 RCL 03
12 PROMPT            65 COS
13 HR                66 *
14 STO 02            67 X<>Y
15 "GHA, 0 HRS=?"    68 -
16 PROMPT            69 /
17 HR                70 ATAN
18 STO 00            71 360
19 "DECL=?"          72 +
20 PROMPT            73 STO 01
21 HR                74 LASTX
22 TAN               75 "MARK RDG=?"
23 STO 04            76 PROMPT
24 CF 22             77 HR
25 "WCH CR-MN.SC?"   78 "STAR RDG=?"
26 PROMPT            79 PROMPT
27 FC? 22            80 HR
28 0                 81 -
29 100               82 CHS
30 /                 83 X<0?
31 HR                84 +
32 STO 07            85 RCL 01
33♦LBL B             86 X<>Y
34 360               87 -
35 "TIME, UT=?"      88 ST+ 05
36 PROMPT            89 HMS
37 HR                90 FIX 5
38 RCL 07            91 "AZ="
39 +                 92 ARCL X
40 15.0410685        93 1
41 *                 94 ST+ 06
42 RCL 00            95 AVIEW
43 +                 96 GTO B
44 X>Y?              97♦LBL A
45 -                 98 RCL 05
46 ABS               99 RCL 06
47 RCL 02            100 /
48 -                 101 HMS
49 180               102 "AVE AZ="
50 X<>Y              103 ARCL X
51 X<Y?              104 AVIEW
52 GTO 01            105 GTO B
53 360               106 END
```

CALCULATOR PROGRAMS FOR ASTRONOMIC AZIMUTH

Data Register Contents, Polaris

Reg. 00 GHA_0 (D.d)
      01   Azimuth (D.d)
      02   Longitude (D.d)
      03   Latitude (D.d)
      04   tan δ
      05   Σ Azimuths (D.d)
      06   observation number
      07   watch corr'n (HR)

Instructions For Using HP 41 Polaris Program

Keystroke		Display	Input
1.	XEQ SIZE 008		
2.	XEQ "POLHR↑"	LAT = ?	Latitude in D.mmss
3.	R/S	LONG = ?	Longitude in D.mmss
4.	R/S	GHA, 0 HRS = ?	GHA_0 in D.mmss
5.	R/S	DECL = ?	δ in D.mmss
6.	R/S	WCH CR - MN.SC?	Watch corr'n m.s. (If Zero, press R/S)
7.	R/S	TIME, UT = ?	UT for observation, in H.mmss
8.	R/S	MARK RDG = ?	Direction to mark, in D.mmss
9.	R/S	STAR RDG = ?	Direction to star, in D.mmss
10.	R/S	AZ = XX.XXXXX	

Computed astronomic azimuth to ground target is displayed (for one observation). Repeat steps 7 through 10 for subsequent observations. Keystroke A results in display of average (mean) of all azimuths computed AVE AZ = XXX.XXXX. After Keystroke A and mean observed, more observation data can be input by Keystroke R/S with display TIME, UT = ? Accumulated mean can be observed by Keystroke A at any time.

E-4 TI 58/58C/59 PROGRAMS

Data Register Contents, Solar

Reg. 00 UT + UT1 (HR)
    01 Eq. T_0 (HR)
    02 Longitude (D.d)
    03 Latitude (D.d)
    04 δ_0 (D.d)
    05 UT1 corr'n (HR)
    06 $\Delta\delta$/hr. (D.d)
    07 ΔEq.T./hr. (HR)
    08 Σ Azimuths (D.d)
    09 t (D.d)
    10 Counter

TI 58/59 Solar Program

```
000  76 LBL     052  13 C       104  34 ΓX      156  76 LBL     208  95 =
001  11 A       053  86 STF     105  22 INV     157  38 SIN     209  32 X:T
002  47 CMS     054  00 00      106  86 STF     158  87 IFF     210  00 0
003  88 DMS     055  22 INV     107  00 00      159  00 00      211  32 X:T
004  42 STO     056  86 STF     108  75 -       160  39 COS     212  22 INV
005  03 03      057  01 01      109  03 3       161  87 IFF     213  77 GE
006  91 R/S     058  88 DMS     110  06 6       162  01 01      214  42 STO
007  88 DMS     059  85 +       111  00 0       163  28 LOG     215  75 -
008  42 STO     060  43 RCL     112  95 =       164  61 GTO     216  03 3
009  02 02      061  05 05      113  76 LBL     165  30 TAN     217  06 6
010  91 R/S     062  95 =       114  34 ΓX      166  76 LBL     218  00 0
011  88 DMS     063  42 STO     115  94 +/-     167  39 COS     219  95 =
012  42 STO     064  00 00      116  42 STO     168  87 IFF     220  76 LBL
013  04 04      065  65 ×       117  09 09      169  01 01      221  42 STO
014  91 R/S     066  43 RCL     118  38 SIN     170  30 TAN     222  32 X:T
015  55 ÷       067  07 07      119  55 ÷       171  85 +       223  00 0
016  03 3       068  85 +       120  53 (       172  03 3       224  32 X:T
017  06 6       069  43 RCL     121  53 (       173  06 6       225  77 GE
018  00 0       070  00 00      122  43 RCL     174  00 0       226  23 LNX
019  00 0       071  85 +       123  06 06      175  95 =       227  85 +
020  95 =       072  43 RCL     124  65 ×       176  61 GTO     228  03 3
021  42 STO     073  01 01      125  43 RCL     177  28 LOG     229  06 6
022  06 06      074  75 -       126  00 00      178  76 LBL     230  00 0
023  91 R/S     075  01 1       127  85 +       179  30 TAN     231  95 =
024  55 ÷       076  02 2       128  43 RCL     180  85 +       232  76 LBL
025  01 1       077  95 =       129  04 04      181  01 1       233  23 LNX
026  00 0       078  65 ×       130  54 )       182  08 8       234  44 SUM
027  00 0       079  01 1       131  30 TAN     183  00 0       235  08 08
028  95 =       080  05 5       132  65 ×       184  95 =       236  32 X:T
029  88 DMS     081  75 -       133  43 RCL     185  76 LBL     237  01 1
030  42 STO     082  43 RCL     134  03 03      186  28 LOG     238  44 SUM
031  01 01      083  02 02      135  39 COS     187  75 -       239  10 10
032  91 R/S     084  95 =       136  75 -       188  53 (       240  32 X:T
033  55 ÷       085  32 X:T     137  43 RCL     189  91 R/S     241  22 INV
034  03 3       086  00 0       138  03 03      190  88 DMS     242  88 DMS
035  06 6       087  32 X:T     139  38 SIN     191  75 -       243  91 R/S
036  00 0       088  77 GE      140  65 ×       192  91 R/S     244  61 GTO
037  00 0       089  33 X²      141  43 RCL     193  88 DMS     245  13 C
038  95 =       090  85 +       142  09 09      194  54 )       246  76 LBL
039  42 STO     091  03 3       143  39 COS     195  94 +/-     247  12 B
040  07 07      092  06 6       144  54 )       196  32 X:T     248  43 RCL
041  91 R/S     093  00 0       145  95 =       197  00 0       249  08 08
042  55 ÷       094  95 =       146  22 INV     198  32 X:T     250  55 ÷
043  03 3       095  76 LBL     147  30 TAN     199  22 INV     251  43 RCL
044  06 6       096  33 X²      148  32 X:T     200  77 GE      252  10 10
045  00 0       097  32 X:T     149  00 0       201  45 YX      253  95 =
046  00 0       098  01 1       150  32 X:T     202  85 +       254  22 INV
047  95 =       099  08 8       151  22 INV     203  03 3       255  88 DMS
048  42 STO     100  00 0       152  77 GE      204  06 6       256  91 R/S
049  05 05      101  32 X:T     153  38 SIN     205  00 0       257  61 GTO
050  91 R/S     102  22 INV     154  86 STF     206  76 LBL     258  13 C
051  76 LBL     103  77 GE      155  01 01      207  45 YX
```

Instructions for Using TI Solar Program

1. Allocate adequate memory   2   2nd   OP   17
2. Fix decimal places to nine   2nd   INV   Fix
3. Load program
4. Key in LAT (D.ms), press   A
5. Key in LONG (D.ms), press   R/S
6. Key in δ_0 (D.ms), press   R/S
7. Key in $\Delta\delta$/hr (sec.), press   R/S
8. Key in Eq.T_0 (min.sec.), press   R/S
9. Key in ΔEq.T/hr. (sec.), press   R/S
10. Key in UT1 corr'n (sec.), press   R/S
11. Key in UT (H.ms), press   R/S   or   C
12. Key in direction to mark (D.ms), press   R/S
13. Key in direction to Sun (D.ms), press   R/S

    Azimuth (D.ms) is displayed. Repeat steps 11 through 13 for subsequent observations. Press B for mean azimuth (D.ms) at any time.

Instructions for Using TI Polaris Program

1. Allocate adequate memory   1   2nd   OP   17
2. Fix decimal places to nine   2nd   INV   Fix
3. Load program
4. Key in LAT (D.ms), press   A
5. Key in LONG (D.ms), press   R/S
6. Key in GHA_0 (D.ms), press   R/S
7. Key in δ (D.ms), press   R/S
8. Key in watch corr'n (min.sec.), press   R/S
9. Key in UT (H.ms), press   R/S   or   C
10. Key in direction to mark (D.ms), press   R/S
11. Key in direction to Polaris (D.ms), press   R/S

    Azimuth (D.ms) is displayed. Repeat steps 9 through 11 for subsequent observations. Press B for mean azimuth (D.ms) at any time.

Data Register Contents, Polaris

Reg. 00 GHA (D.d)
01 unused
02 Longitude (D.d)
03 Latitude (D.d)
04 tan δ
05 Watch corr'n (HR)
06 Σ Azimuth (D.d)
07 t (D.d)
08 Counter

TI 58/59 Polaris Program

```
000  76 LBL     049  00  0      098  32 X:T     147  54  )
001  11 A       050  95  =      099  94 +/-     148  32 X:T
002  47 CMS     051  32 X:T     100  42 STO     149  76 LBL
003  88 DMS     052  03  3      101  07  07     150  34 √X
004  42 STO     053  06  6      102  38 SIN     151  32 X:T
005  03  03     054  00  0      103  55  ÷      152  95  =
006  91 R/S     055  77 GE      104  53  (      153  32 X:T
007  88 DMS     056  45 Yˣ      105  43 RCL     154  00  0
008  42 STO     057  32 X:T     106  04  04     155  22 INV
009  02  02     058  75  -      107  65  ×      156  77 GE
010  91 R/S     059  03  3      108  43 RCL     157  35 1/X
011  88 DMS     060  06  6      109  03  03     158  32 X:T
012  42 STO     061  00  0      110  39 COS     159  85  +
013  00  00     062  95  =      111  75  -      160  03  3
014  91 R/S     063  32 X:T     112  43 RCL     161  06  6
015  88 DMS     064  76 LBL     113  03  03     162  00  0
016  30 TAN     065  45 Yˣ      114  38 SIN     163  95  =
017  42 STO     066  32 X:T     115  65  ×      164  32 X:T
018  04  04     067  75  -      116  43 RCL     165  76 LBL
019  91 R/S     068  43 RCL     117  07  07     166  35 1/X
020  55  ÷      069  02  02     118  39 COS     167  32 X:T
021  01  1      070  95  =      119  54  )      168  44 SUM
022  00  0      071  32 X:T     120  95  =      169  06  06
023  00  0      072  00  0      121  22 INV     170  32 X:T
024  95  =      073  22 INV     122  30 TAN     171  01  1
025  88 DMS     074  77 GE      123  85  +      172  44 SUM
026  42 STO     075  68 NOP     124  03  3      173  08  08
027  05  05     076  85  +      125  06  6      174  32 X:T
028  91 R/S     077  03  3      126  00  0      175  22 INV
029  76 LBL     078  06  6      127  75  -      176  88 DMS
030  13 C       079  00  0      128  53  (      177  95  =
031  88 DMS     080  95  =      129  53  (      178  91 R/S
032  85  +      081  32 X:T     130  91 R/S     179  61 GTO
033  43 RCL     082  76 LBL     131  88 DMS     180  13 C
034  05  05     083  68 NOP     132  75  -      181  76 LBL
035  95  =      084  01  1      133  91 R/S     182  12 B
036  65  ×      085  08  8      134  88 DMS     183  43 RCL
037  01  1      086  00  0      135  54  )      184  06  06
038  05  5      087  77 GE      136  94 +/-     185  55  ÷
039  93  .      088  33 X²     137  32 X:T     186  43 RCL
040  00  0      089  32 X:T     138  00  0      187  08  08
041  04  4      090  75  -      139  22 INV     188  95  =
042  01  1      091  03  3      140  77 GE      189  22 INV
043  00  0      092  06  6      141  34 √X      190  88 DMS
044  06  6      093  00  0      142  32 X:T     191  95  =
045  08  8      094  95  =      143  85  +      192  91 R/S
046  05  5      095  32 X:T     144  03  3      193  61 GTO
047  85  +      096  76 LBL     145  06  6      194  13 C
048  43 RCL     097  33 X²     146  00  0
```

E-5 HP 85 SOLAR PROGRAM

The HP 85 program was furnished by Brown and Butler, Inc., Consulting Engineers and Land Surveyors in Baton Rouge, Louisiana, and is reproduced here with their permission, Mr. James Theriot and Ronald J. Rodi, P.E., wrote and developed the program for use in their surveying practice after Rodney L. Crochet, another employee of the firm attended one of the author's seminars. This program is written for solar observations. Those familiar with the HP 85 should be able to easily adapt it to Polaris observations. Example C.10 in Appendix C illustrates the use of this program.

It is noted that this program computes each set average for inspection. Then, a set can be deleted if necessary, and the mean re-calculated from the remaining sets. The program also includes calculation of grid azimuth (bearing) after input of the Laplace correction and the convergence angle.

```
10 DISP "SOLAR OBSERVATION COMP
   UTATION"
20 PRINT "******************
   ********"
30 PRINT
40 PRINT "SOLAR OBSERVATION COM
   PUTATION"
50 PRINT
60 PRINT "******************
   ********"
70 PRINT
80 PRINT
90 PRINT "HOUR ANGLE METHOD AS
   PRESENTED"
100 PRINT "IN DR R. B. BUCKNER WOR
    KSHOP"
110 PRINT
120 PRINT "---------------------
    --------"
130 PRINT
140 PRINT
150 PRINT
160 DISP "PROJECT NAME";
170 INPUT N$
180 PRINT "PROJECT NAME: ";N$
190 DISP "STATION";
200 INPUT A$
210 PRINT "STATION:       ";A$
220 DISP "TARGET";
230 INPUT B$
240 PRINT "TARGET:        ";B$
250 DISP "OBSERVATION DATE";
260 INPUT C$
270 PRINT "OBS. DATE:     ";C$
280 PRINT
290 PRINT
300 DISP "LONGITUDE--D.MS ";
310 INPUT L4
320 L1=IP(L4)
330 L2=IP((L4-L1)*100)
340 L3=((L4-L1)*100-L2)*100
350 L=L1+L2/60+L3/3600
360 A1$="LONGITUDE="
370 PRINT USING 380 ; A1$,L1,L2,
    L3
380 IMAGE 16A,3D,2X,2Z,2X,ZZ.DD
390 DISP "LATITUDE--D.MS";
400 INPUT J4
410 J1=IP(J4)
420 J2=IP((J4-J1)*100)
430 J3=((J4-J1)*100-J2)*100
440 J=J1+J2/60+J3/3600
450 A2$="LATITUDE ="
460 PRINT USING 470 ; A2$,J1,J2,
    J3
470 IMAGE 16A,3D,2X,2Z,2X,ZZ.DD
480 DISP "EQUATION OF TIME";
490 INPUT X4
500 X1=IP(X4)
510 X2=IP((X4-X1)*100)
520 X3=((X4-X1)*100-X2)*100
530 IF X4<0 THEN GOTO 550
540 GOTO 600
550 A3$="EQ.OF TIME=     -"
560 PRINT USING 570 ; A3$,ABS(X1
    ),ABS(X2),ABS(X3)
570 IMAGE 16A,3D,2X,2Z,2X,ZZ.D
580 GOTO 630
590 GOTO 630
600 A4$="EQ. OF TIME="
610 PRINT USING 620 ; A4$,X1,X2,
    X3
620 IMAGE 16A,3D,2X,2Z,2X,ZZ.D
630 X=X1+X2/60+X3/3600
640 DISP "DELTA EQ.TIME";
650 INPUT Q4
660 Q1=IP(Q4)
670 Q2=IP((Q4-Q1)*100)
680 Q3=((Q4-Q1)*100-Q2)*100
690 IF Q4<0 THEN GOTO 710
700 GOTO 750
710 A5$="DELTA EQ. TIME=-"
720 PRINT USING 730 ; A5$,ABS(Q1
    ),ABS(Q2),ABS(Q3)
730 IMAGE 16A, 3D,2X,2Z,2X,ZZ.D
740 GOTO 780
750 A6$="DELTA EQ. TIME="
760 PRINT USING 770 ; A6$,Q1,Q2,
    Q3
770 IMAGE 16A,3D,2X,2Z,2X,ZZ.D
780 Q=Q1+Q2/60+Q3/3600
790 DISP "DECLINATION";
800 INPUT D4
810 D1=IP(D4)
820 D2=IP((D4-D1)*100)
830 D3=((D4-D1)*100-D2)*100
840 IF D4<0 THEN GOTO 860
850 GOTO 900
860 A7$="DECLINATION=    -"
870 PRINT USING 880 ; A7$,ABS(D1
    ),ABS(D2),ABS(D3)
880 IMAGE 16A,3D,2X,2Z,2X,ZZ.D
890 GOTO 930
900 A8$="DECLINATION="
910 PRINT USING 920 ; A8$,D1,D2,
    D3
```

```
920 IMAGE 16A,3D,2X,2Z,2X,ZZ.D
930 D=D1+D2/60+D3/3600
940 DISP "DELTA DECL";
950 INPUT E4
960 E1=IP(E4)
970 E2=IP((E4-E1)*100)
980 E3=((E4-E1)*100-E2)*100
990 IF E4<0 THEN GOTO 1010
1000 GOTO 1050
1010 A9$="DELTA DECL.=    -"
1020 PRINT USING 1030 ; A9$,ABS(
     E1),ABS(E2),ABS(E3)
1030 IMAGE 16A,3D,2X,2Z,2X,ZZ.D
1040 GOTO 1080
1050 B9$=" DELTA DECL.="
1060 PRINT USING 1070 ; B9$,E1,E
     2,E3
1070 IMAGE 16A,3D,2X,2Z,2X,ZZ.D
1080 E=E1+E2/60+E3/3600
1090 PRINT
1100 PRINT
1110 FOR I=1 TO 8 STEP 1
1120 DISP "UT--D,M,S";
1130 INPUT Y4
1140 Y1=IP(Y4)
1150 Y2=IP((Y4-Y1)*100)
1160 Y3=((Y4-Y1)*100-Y2)*100
1170 Y=Y1+Y2/60+Y3/3600
1180 C5$="UNIVERSAL TIME="
1190 PRINT USING 1200 ; C5$,Y1,Y
     2,Y3
1200 IMAGE 16A,3D,2X,2Z,2X,ZZ.D
1210 N=(Y+X+Q*Y-12)*15
1220 M=N-L
1230 IF M<0 THEN 1250
1240 GOTO 1260
1250 M=M+360
1260 IF M<180 THEN 1280
1270 IF M>180 THEN 1380
1280 T=-M
1290 F=D+E*Y
1300 DEG
1310 W=ATN(SIN(T)/(TAN(F)*COS(J)
     -SIN(J)*COS(T)))
1320 IF W<0 THEN 1340
1330 IF W>0 THEN 1360
1340 W=W+360
1350 GOTO 1450
1360 W=W+180
1370 GOTO 1450
1380 T=360-M
1390 F=D+E*Y
1400 DEG
1410 W=ATN(SIN(T)/(TAN(F)*COS(J)
     -SIN(J)*COS(T)))
1420 IF W<0 THEN 1440
1430 GOTO 1450
1440 W=W+180
1450 W1=IP(W)
1460 W2=IP((W-W1)*60)
1470 W3=IP(((W-W1)*60-W2)*60)
1480 C6$="AZIMUTH="
1490 PRINT USING 1500 ; C6$,W1,W
     2,W3
1500 IMAGE 16A,3D,2X,2Z,2X,ZZ.D
1510 DISP "SUN READING";
1520 INPUT S4
1530 S1=IP(S4)
1540 S2=IP((S4-S1)*100)
1550 S3=((S4-S1)*100-S2)*100
1560 C7$="SUN READING="
1570 PRINT USING 1580 ; C7$,S1,S
     2,S3
1580 IMAGE 16A,3D,2X,2Z,2X,ZZ.D
1590 S=S1+S2/60+S3/3600
1600 DISP "MARK READING"
1610 INPUT A4
1620 A1=IP(A4)
1630 A2=IP((A4-A1)*100)
1640 A3=((A4-A1)*100-A2)*100
1650 C8$="MARK READING="
1660 PRINT USING 1670 ; C8$,A1,A
     2,A3
1670 IMAGE 16A,3D,2X,2Z,2X,ZZ.D
1680 A=A1+A2/60+A3/3600
1690 B=S-A
1700 IF B<0 THEN 1720
1710 IF B>0 THEN 1730
1720 B=B+360
1730 B1=IP(B)
1740 B2=IP((B-B1)*60)
1750 B3=IP(((B-B1)*60-B2)*60)
1760 D7$="MARK TO SUN="
1770 PRINT USING 1780 ; D7$,B1,B
     2,B3
1780 IMAGE 16A,3D,2X,2Z,2X,ZZ.D
1790 IF B<0 THEN 1810
1800 IF B>0 THEN 1820
1810 B=B+360
1820 C(I)=W-B
1830 IF C(I)<0 THEN 1850
1840 IF C(I)>0 THEN 1860
1850 C(I)=C(I)+360
1860 C=C(I)
1870 C1=IP(C)
1880 C2=IP((C-C1)*60)
1890 C3=IP(((C-C1)*60-C2)*60)
1900 D8$="TAR.ASTRO.AZ.="
1910 PRINT USING 1920 ; D8$,C1,C
     2,C3
1920 IMAGE 16A,3D,2X,2Z,2X,ZZ.D
1930 PRINT
1940 PRINT
1950 NEXT I
1960 G(1)=(C(1)+C(2))/2
1970 G(2)=(C(3)+C(4))/2
1980 G(3)=(C(5)+C(6))/2
1990 G(4)=(C(7)+C(8))/2
2000 G=G(1)
2010 GOSUB 3150
2020 P5$="SET AVERAGE-1="
2021 PRINT USING 2022 ; P5$,G1,G
     2,G3
2022 IMAGE 16A,3D,2X,2Z,2X,ZZ.D
2030 G=G(2)
2040 GOSUB 3150
2050 P6$="SET AVERAGE-2="
2051 PRINT USING 2052 ; P6$,G1,G
     2,G3
2052 IMAGE 16A,3D,2X,2Z,2X,ZZ.D
2060 G=G(3)
2070 GOSUB 3150
2080 P7$="SET AVERAGE-3="
2081 PRINT USING 2082 ; P7$,G1,G
     2,G3
2082 IMAGE 16A,3D,2X,2Z,2X,ZZ.D
2090 G=G(4)
2100 GOSUB 3150
2110 P8$="SET AVERAGE-4="
2111 PRINT USING 2112 ; P8$,G1,G
     2,G3
2112 IMAGE 16A,3D,2X,2Z,2X,ZZ.D
```

```
2120 PRINT
2130 PRINT
2140 PRINT "-----------------------------"
2150 PRINT
2160 PRINT "IF SET AVG. OK,ENTER 5 "
2161 PRINT
2170 PRINT "TO EXCLUDE ONE SET AVG."
2180 PRINT "ENTER THAT SET NUMBER."
2190 PRINT
2200 PRINT "-----------------------------"
2210 PRINT
2220 INPUT Y5
2230 IF Y5=5 THEN GOTO 2280
2240 IF Y5=1 THEN GOTO 2300
2250 IF Y5=2 THEN GOTO 2320
2260 IF Y5=3 THEN GOTO 2340
2270 IF Y5=4 THEN GOTO 2360
2280 PRINT "AVG. SETS 1,2,3,&4"
2290 GOTO 2370
2300 PRINT "EXCLUDE SET 1"
2310 GOTO 2370
2320 PRINT "EXCLUDE SET 2"
2330 GOTO 2370
2340 PRINT "EXCLUDE SET 3"
2350 GOTO 2370
2360 PRINT "EXCLUDE SET 4"
2370 PRINT
2380 IF Y5=5 THEN GOTO 2520
2390 IF Y5=4 THEN GOTO 2430
2400 IF Y5=3 THEN GOTO 2450
2410 IF Y5=2 THEN GOTO 2470
2420 IF Y5=1 THEN GOTO 2490
2430 H=(G(1)+G(2)+G(3))/3
2440 GOTO 2530
2450 H=(G(1)+G(2)+G(4))/3
2460 GOTO 2530
2470 H=(G(1)+G(3)+G(4))/3
2480 GOTO 2530
2490 H=(G(2)+G(3)+G(4))/3
2500 GOTO 2530
2510 PRINT
2520 H=(G(1)+G(2)+G(3)+G(4))/4
2530 PRINT
2540 H1=IP(H)
2550 H2=IP((H-H1)*60)
2560 H3=IP(((H-H1)*60-H2)*60)
2570 D9$="AVG.ASTRO.AZ.="
2580 PRINT USING 2590 ; D9$,H1,H2,H3
2590 IMAGE 16A,3D,2X,2Z,2X,ZZ.D
2600 PRINT
2610 PRINT
2620 DISP "CENTRAL MERIDIAN";
2630 INPUT K4
2640 K1=IP(K4)
2650 K2=IP((K4-K1)*100)
2660 K3=((K4-K1)*100-K2)*100
2670 Q5$="CENTRAL MER.="
2671 PRINT USING 2672 ; Q5$,K1,K2,K3
2672 IMAGE 16A,3D,2X,2Z,2X,ZZ.D
2680 K=K1+K2/60+K3/3600
2690 P=.5000126971*(K-L)
2700 P1=IP(P)
2710 P2=IP((P-P1)*60)
2720 P3=IP(((P-P1)*60-P2)*60)
2730 IF P<0 THEN GOTO 2735
2731 R5$="MAPPING ANGLE="
2732 PRINT USING 2733 ; R5$,P1,P2,P3
2733 IMAGE 16A,3D,2X,2Z,2X,ZZ.D
2734 GOTO 2740
2735 R5$="MAPPING ANGLE= -"
2736 PRINT USING 2737 ; R5$,ABS(P1),ABS(P2),ABS(P3)
2737 IMAGE 16A,3D,2X,2Z,2X,ZZ.D
2740 DISP "LAPLACE";
2750 INPUT U4
2760 U1=IP(U4)
2770 U2=IP((U4-U1)*100)
2780 U3=((U4-U1)*100-U2)*100
2790 U=U1+U2/60+U3/3600
2800 IF U4<0 THEN GOTO 2805
2801 S5$="LAPLACE COR.="
2802 PRINT USING 2803 ; S5$,U1,U2,U3
2803 IMAGE 16A,3D,2X,2Z,2X,ZZ.D
2804 GOTO 2810
2805 S5$="LAPLACE COR.=   -"
2806 PRINT USING 2807 ; S5$,ABS(U1),ABS(U2),ABS(U3)
2807 IMAGE 16A,3D,2X,2Z,2X,ZZ.D
2810 V=H+U/3600-P
2820 PRINT
2821 PRINT
2830 IF V=270 THEN PRINT "WEST"
2840 IF V=180 THEN PRINT "SOUTH"
2850 IF V=90 THEN PRINT "EAST"
2860 IF V=360 THEN PRINT "NORTH"
2870 IF V=0 THEN PRINT "NORTH"
2880 IF V>270 THEN 2980
2890 IF V<270 THEN 2900
2900 IF V>180 THEN 3020
2910 IF V<180 THEN 2920
2920 IF V>90 THEN 3060
2930 IF V<90 THEN GOSUB 3100
2940 PRINT "GRID BRG   N";V1;V2;V3;"E"
2950 GOTO 3140
2960 GOSUB 3100
2970 PRINT "GRID BRG   N";V1;V2;V3;"W"
2980 V=360-V
2990 GOSUB 3100
3000 PRINT "GRID BRG   N";V1;V2;V3;"W"
3010 GOTO 3140
3020 V=V-180
3030 GOSUB 3100
3040 PRINT "GRID BRG   S";V1;V2;V3;"W"
3050 GOTO 3140
3060 V=180-V
3070 GOSUB 3100
3080 PRINT "GRID BRG   S";V1;V2;V3;"E"
3090 GOTO 3140
3100 V1=IP(V)
3110 V2=IP((V-V1)*60)
3120 V3=IP(((V-V1)*60-V2)*60)
3130 RETURN
3140 GOTO 3190
3150 G1=IP(G)
3160 G2=IP((G-G1)*60)
3170 G3=IP(((G-G1)*60-G2)*60)
3180 RETURN
3190 END
```

APPENDIX F

EPHEMERIS TABLES

The ephemeris used in this manual was printed from data on magnetic tape purchased from the National Almanac Office, U.S. Naval Observatory. The format, data included, and the precision of the print-out was designed for the hour angle method explained in this manual. Although the manual is copyrighted, users are permitted to reproduce the tables for field use. Users should read Chapter 5 on use and theory of the ephemeris and the following remarks concerning the sun's declination.

Values for sun's declination are correct for 0^h. However, since declination is not changing linearly, straight-line interpolation causes slight errors. When UT is close to 12 hours and the date is close to either solstice, declination should be corrected using Table F.1, if a 90% error of ±5" is expected. The maximum effect on azimuth would be 2 to 3 seconds of arc, higher errors occurring at lower latitudes. Thus, to retain third-order, Class I equivalent accuracy it should be considered in many cases. As the following example illustrates, it doesn't matter whether the correction is applied before or after adding the change for UT time explained in Section 5-2. Using the approach on the right, the corrected value δ_0' would be used in place of the ephemeris value δ_0 in the programs of Appendix E with no modifications needed in the programs. The example date is December 1, 1982. Change for UT = 12^h is $-23.4''/hr \times 12^h = -4'40.8''$. The correction from Table F.1 is $-3.0''$.

δ_0 =	$-21°43'00''$	δ_0 =	$-21°43'00''$
Change for UT =	$-4'40.8''$	Correction =	$-3.0''$
	$-21°47'40.8''$	δ_0' =	$-21°43'03.0''$
Correction =	$-3.0''$	Change for UT =	$-4'40.8''$
δ =	$-21°47'44''$	δ =	$-21°47'44''$

This correction can be ignored for afternoon observations in the continental U.S. since UT is always close to 0^h at time of observation. If the correction is 1" or less it can be ignored in any case since this is within the significant figures of the listings and the effect on azimuth will be small. This correction can always be ignored for precision less than third-order Class I equivalent.

A similar cyclical error occurs for equation of time but this error affects azimuth insignificantly.

Table F.1 Corrections for Sun's Declination

+		0	4	8	12	16	20	24	−	
					UT, hours					
Mar.	21	0	0	0	0	0	0	0	Sept.	23
Apr.	1	0	0.3	0.4	0.6	0.4	0.3	0	Oct.	1
May	1	0	1.0	1.7	2.1	1.7	1.0	0	Nov.	1
June	1	0	1.5	2.4	3.0	2.4	1.5	0	Dec.	1
June	21	0	1.7	2.7	3.4	2.7	1.7	0	Dec.	22
July	1	0	1.6	2.6	3.2	2.6	1.6	0	Jan.	1
Aug.	1	0	1.1	1.8	2.2	1.8	1.1	0	Feb.	1
Sept.	1	0	0.4	0.6	0.8	0.6	0.4	0	Mar.	1
Sept.	23	0	0	0	0	0	0	0	Mar.	21

Use column on the left for Spring and Summer observations. Use column on the right for Autumn and Winter observations. Corrections are in seconds and are accurate within 0.5". If date is between Mar. 21 and Sept. 23 the correction is positive, and if between Sept. 23 and Mar. 21 the correction is negative. Declination, after correction, will always be increased in magnitude when the correction is added algebraically.

Table F.2 a JANUARY 1982

Date		Declination At 0 hrs, UT ° ' "	Δ, 1 hr "	Equation of Time At 0 hrs, UT m s	Δ, 1 hr s	GHA at 0 hr ° ' "	Declination ° ' "
1	FR	−23 2 44	12.3	− 3 17.5	−1.18	66 45 32	89 11 9
2	SA	−22 57 48	13.5	− 3 45.9	−1.17	67 45 3	89 11 9
3	SU	−22 52 25	14.6	− 4 13.9	−1.15	68 44 33	89 11 9
4	MO	−22 46 34	15.7	− 4 41.5	−1.13	69 44 3	89 11 9
5	TU	−22 40 17	16.9	− 5 8.7	−1.12	70 43 31	89 11 9
6	WE	−22 33 32	18.0	− 5 35.5	−1.10	71 42 57	89 11 10
7	TH	−22 26 21	19.1	− 6 1.8	−1.08	72 42 23	89 11 10
8	FR	−22 18 43	20.2	− 6 27.9	−1.05	73 41 50	89 11 10
9	SA	−22 10 39	21.3	− 6 52.9	−1.03	74 41 17	89 11 10
10	SU	−22 2 8	22.3	− 7 17.6	−1.01	75 40 46	89 11 10
11	MO	−21 53 12	23.4	− 7 41.6	−0.98	76 40 16	89 11 10
12	TU	−21 43 51	24.5	− 8 5.4	−0.96	77 39 48	89 11 11
13	WE	−21 34 4	25.5	− 8 28.5	−0.93	78 39 21	89 11 11
14	TH	−21 23 52	26.5	− 8 50.9	−0.91	79 38 54	89 11 11
15	FR	−21 13 15	27.5	− 9 12.7	−0.88	80 38 27	89 11 11
16	SA	−21 2 14	28.5	− 9 33.9	−0.85	81 37 59	89 11 11
17	SU	−20 50 49	29.5	− 9 54.4	−0.83	82 37 30	89 11 11
18	MO	−20 39 0	30.5	−10 14.2	−0.80	83 37 1	89 11 11
19	TU	−20 26 47	31.5	−10 33.3	−0.77	84 36 30	89 11 11
20	WE	−20 14 12	32.5	−10 51.8	−0.74	85 36 0	89 11 11
21	TH	−20 1 15	33.4	−11 9.5	−0.70	86 35 29	89 11 11
22	FR	−19 47 52	34.3	−11 26.4	−0.67	87 34 58	89 11 11
23	SA	−19 34 9	35.2	−11 42.6	−0.64	88 34 29	89 11 11
24	SU	−19 20 4	36.1	−11 58.0	−0.61	89 34 0	89 11 12
25	MO	−19 5 37	37.0	−12 12.7	−0.58	90 33 32	89 11 12
26	TU	−18 50 50	37.8	−12 26.6	−0.54	91 33 5	89 11 12
27	WE	−18 35 42	38.6	−12 39.6	−0.51	92 32 39	89 11 12
28	TH	−18 20 13	39.5	−12 51.9	−0.47	93 32 14	89 11 12
29	FR	−18 4 25	40.3	−13 3.3	−0.44	94 31 48	89 11 12
30	SA	−17 48 17	41.1	−13 13.9	−0.40	95 31 22	89 11 12
31	SU	−17 31 51	41.9	−13 23.6	−0.38	96 30 54	89 11 12

Table F.2 b DECEMBER 1982

Date		Declination At 0 hrs, UT ° ' "	Δ, 1 hr "	Equation of Time At 0 hrs, UT m s	Δ, 1 hr s	GHA at 0 hr ° ' "	Declination ° ' "
1	WE	−21 43 0	−23.4	+11 13.1	−0.93	35 39 22	89 11 18
2	TH	−21 52 21	−22.3	+10 50.8	−0.95	36 38 37	89 11 18
3	FR	−22 1 17	−21.3	+10 27.9	−0.98	37 37 53	89 11 19
4	SA	−22 9 47	−20.2	+10 4.4	−1.00	38 37 11	89 11 19
5	SU	−22 17 52	−19.1	+ 9 40.3	−1.03	39 36 31	89 11 20
6	MO	−22 25 31	−18.0	+ 9 15.6	−1.05	40 35 53	89 11 20
7	TU	−22 32 43	−17.0	+ 8 50.4	−1.08	41 35 16	89 11 20
8	WE	−22 39 30	−15.8	+ 8 24.6	−1.10	42 34 40	89 11 21
9	TH	−22 45 50	−14.7	+ 7 58.3	−1.12	43 34 3	89 11 21
10	FR	−22 51 42	−13.6	+ 7 31.5	−1.14	44 33 25	89 11 21
11	SA	−22 57 8	−12.5	+ 7 4.2	−1.15	45 32 47	89 11 21
12	SU	−23 2 7	−11.3	+ 6 36.6	−1.17	46 32 7	89 11 22
13	MO	−23 6 39	−10.1	+ 6 8.6	−1.18	47 31 28	89 11 22
14	TU	−23 10 42	− 9.0	+ 5 40.2	−1.20	48 30 48	89 11 22
15	WE	−23 14 19	− 7.8	+ 5 11.5	−1.21	49 30 9	89 11 22
16	TH	−23 17 27	− 6.7	+ 4 42.5	−1.22	50 29 30	89 11 23
17	FR	−23 20 8	− 5.5	+ 4 13.3	−1.22	51 28 53	89 11 23
18	SA	−23 22 20	− 4.3	+ 3 43.9	−1.23	52 28 17	89 11 23
19	SU	−23 24 4	− 3.2	+ 3 14.3	−1.24	53 27 42	89 11 24
20	MO	−23 25 21	− 2.0	+ 2 44.5	−1.24	54 27 9	89 11 24
21	TU	−23 26 9	− 0.8	+ 2 14.7	−1.25	55 26 36	89 11 24
22	WE	−23 26 28	0.3	+ 1 44.8	−1.25	56 26 4	89 11 24
23	TH	−23 26 20	1.5	+ 1 14.9	−1.25	57 25 32	89 11 25
24	FR	−23 25 43	2.7	+ 0 45.0	−1.25	58 25 0	89 11 25
25	SA	−23 24 38	3.9	+ 0 15.2	−1.24	59 24 27	89 11 25
26	SU	−23 23 4	5.0	− 0 14.5	−1.23	60 23 53	89 11 25
27	MO	−23 21 3	6.3	− 0 44.1	−1.23	61 23 18	89 11 25
28	TU	−23 18 33	7.4	− 1 13.6	−1.22	62 22 42	89 11 26
29	WE	−23 15 35	8.6	− 1 42.9	−1.21	63 22 5	89 11 26
30	TH	−23 12 10	9.8	− 2 12.0	−1.20	64 21 30	89 11 26
31	FR	−23 8 15	10.9	− 2 40.9	−1.19	65 20 56	89 11 26

Table F.3a JULY 1981

Date		THE SUN Declination At 0 hrs, UT ° ' "	Δ, 1 hr "	Equation of Time At 0 hrs, UT m s	Δ, 1 hr s	POLARIS GHA at 0 hr ° ' "	Declination ° ' "
1	WE	+23 7 47	-10.3	- 3 39.1	-0.48	245 48 48	89 10 23
2	TH	+23 3 41	-11.3	- 3 50.7	-0.47	246 47 33	89 10 23
3	FR	+22 59 11	-12.3	- 4 2.1	-0.46	247 46 18	89 10 23
4	SA	+22 54 16	-13.3	- 4 13.2	-0.45	248 45 6	89 10 23
5	SU	+22 48 58	-14.3	- 4 24.0	-0.43	249 43 54	89 10 23
6	MO	+22 43 16	-15.3	- 4 34.4	-0.42	250 42 42	89 10 23
7	TU	+22 37 10	-16.2	- 4 44.4	-0.40	251 41 30	89 10 23
8	WE	+22 30 41	-17.2	- 4 54.1	-0.39	252 40 18	89 10 23
9	TH	+22 23 48	-18.2	- 5 3.4	-0.37	253 39 4	89 10 23
10	FR	+22 16 32	-19.1	- 5 12.2	-0.35	254 37 50	89 10 22
11	SA	+22 8 53	-20.0	- 5 20.6	-0.33	255 36 35	89 10 22
12	SU	+22 0 52	-21.0	- 5 28.5	-0.31	256 35 19	89 10 22
13	MO	+21 52 27	-22.0	- 5 36.0	-0.29	257 34 2	89 10 22
14	TU	+21 43 40	-22.9	- 5 42.9	-0.27	258 32 46	89 10 22
15	WE	+21 34 31	-23.8	- 5 49.4	-0.25	259 31 29	89 10 22
16	TH	+21 25 0	-24.7	- 5 55.4	-0.23	260 30 14	89 10 22
17	FR	+21 15 7	-25.6	- 6 0.9	-0.20	261 28 58	89 10 22
18	SA	+21 4 53	-26.5	- 6 5.8	-0.18	262 27 44	89 10 23
19	SU	+20 54 17	-27.4	- 6 10.2	-0.16	263 26 31	89 10 23
20	MO	+20 43 20	-28.3	- 6 14.1	-0.14	264 25 19	89 10 23
21	TU	+20 32 2	-29.1	- 6 17.4	-0.12	265 24 7	89 10 23
22	WE	+20 20 23	-30.0	- 6 20.2	-0.09	266 22 54	89 10 23
23	TH	+20 8 24	-30.8	- 6 22.4	-0.07	267 21 41	89 10 23
24	FR	+19 56 4	-31.6	- 6 24.1	-0.05	268 20 26	89 10 23
25	SA	+19 43 25	-32.5	- 6 25.2	-0.02	269 19 10	89 10 23
26	SU	+19 30 26	-33.3	- 6 25.7	0.0	270 17 53	89 10 23
27	MO	+19 17 7	-34.0	- 6 25.7	0.03	271 16 35	89 10 23
28	TU	+19 3 30	-34.9	- 6 25.1	0.05	272 15 17	89 10 23
29	WE	+18 49 33	-35.6	- 6 23.9	0.08	273 14 0	89 10 23
30	TH	+18 35 18	-36.4	- 6 22.1	0.10	274 12 44	89 10 23
31	FR	+18 20 44	-37.1	- 6 19.7	0.13	275 11 30	89 10 23

Table F.3b SEPTEMBER 1981

Date		THE SUN Declination At 0 hrs, UT ° ' "	Δ, 1 hr "	Equation of Time At 0 hrs, UT m s	Δ, 1 hr s	POLARIS GHA at 0 hr ° ' "	Declination ° ' "
1	TU	+ 8 23 45	-54.5	- 0 7.4	0.79	306 32 30	89 10 29
2	WE	+ 8 1 57	-54.8	+ 0 11.5	0.80	307 31 20	89 10 29
3	TH	+ 7 40 2	-55.1	+ 0 30.7	0.81	308 30 9	89 10 30
4	FR	+ 7 17 59	-55.4	+ 0 50.2	0.83	309 28 57	89 10 30
5	SA	+ 6 55 50	-55.7	+ 1 10.0	0.83	310 27 45	89 10 30
6	SU	+ 6 33 34	-56.0	+ 1 30.0	0.84	311 26 32	89 10 30
7	MO	+ 6 11 11	-56.2	+ 1 50.2	0.85	312 25 20	89 10 31
8	TU	+ 5 48 42	-56.4	+ 2 10.7	0.86	313 24 9	89 10 31
9	WE	+ 5 26 8	-56.7	+ 2 31.3	0.87	314 22 58	89 10 31
10	TH	+ 5 3 28	-56.9	+ 2 52.1	0.88	315 21 49	89 10 31
11	FR	+ 4 40 43	-57.1	+ 3 13.1	0.88	316 20 40	89 10 32
12	SA	+ 4 17 53	-57.3	+ 3 34.2	0.88	317 19 33	89 10 32
13	SU	+ 3 54 59	-57.4	+ 3 55.4	0.89	318 18 27	89 10 32
14	MO	+ 3 32 1	-57.6	+ 4 16.7	0.89	319 17 21	89 10 33
15	TU	+ 3 8 59	-57.8	+ 4 38.1	0.89	320 16 15	89 10 33
16	WE	+ 2 45 53	-57.9	+ 4 59.5	0.89	321 15 8	89 10 33
17	TH	+ 2 22 44	-58.0	+ 5 20.9	0.89	322 13 59	89 10 33
18	FR	+ 1 59 32	-58.1	+ 5 42.3	0.89	323 12 50	89 10 34
19	SA	+ 1 36 18	-58.2	+ 6 3.7	0.89	324 11 40	89 10 34
20	SU	+ 1 13 2	-58.3	+ 6 25.0	0.89	325 10 29	89 10 34
21	MO	+ 0 49 43	-58.3	+ 6 46.3	0.88	326 9 20	89 10 35
22	TU	+ 0 26 23	-58.4	+ 7 7.4	0.88	327 8 12	89 10 35
23	WE	+ 0 3 1	-58.4	+ 7 28.5	0.87	328 7 5	89 10 35
24	TH	- 0 20 21	-58.5	+ 7 49.4	0.86	329 6 0	89 10 36
25	FR	- 0 43 44	-58.5	+ 8 10.1	0.86	330 4 56	89 10 36
26	SA	- 1 7 7	-58.5	+ 8 30.7	0.85	331 3 53	89 10 36
27	SU	- 1 30 30	-58.5	+ 8 51.1	0.84	332 2 50	89 10 37
28	MO	- 1 53 53	-58.4	+ 9 11.3	0.83	333 1 47	89 10 37
29	TU	- 2 17 14	-58.4	+ 9 31.3	0.82	334 0 43	89 10 37
30	WE	- 2 40 35	-58.3	+ 9 51.0	0.81	334 59 39	89 10 38

Table F.3c OCTOBER 1981

Date		THE SUN Declination At 0 hrs, UT ° ' "	Δ, 1 hr "	Equation of Time At 0 hrs, UT m s	Δ, 1 hr s	POLARIS GHA at 0 hr ° ' "	Declination ° ' "
1	TH	- 3 3 53	-58.2	+10 10.5	0.80	335 58 34	89 10 38
2	FR	- 3 27 10	-58.1	+10 29.7	0.79	336 57 29	89 10 38
3	SA	- 3 50 25	-58.0	+10 48.6	0.78	337 56 23	89 10 39
4	SU	- 4 13 37	-57.8	+11 7.2	0.76	338 55 18	89 10 39
5	MO	- 4 36 45	-57.7	+11 25.4	0.75	339 54 13	89 10 39
6	TU	- 4 59 50	-57.6	+11 43.4	0.73	340 53 10	89 10 40
7	WE	- 5 22 52	-57.4	+12 0.9	0.72	341 52 7	89 10 40
8	TH	- 5 45 49	-57.2	+12 18.1	0.70	342 51 6	89 10 41
9	FR	- 6 8 42	-57.0	+12 34.9	0.68	343 50 6	89 10 41
10	SA	- 6 31 29	-56.8	+12 51.2	0.66	344 49 7	89 10 41
11	SU	- 6 54 12	-56.5	+13 7.1	0.64	345 48 9	89 10 42
12	MO	- 7 16 48	-56.3	+13 22.6	0.62	346 47 11	89 10 42
13	TU	- 7 39 19	-56.0	+13 37.5	0.60	347 46 12	89 10 42
14	WE	- 8 1 43	-55.8	+13 51.9	0.58	348 45 12	89 10 43
15	TH	- 8 24 1	-55.4	+14 5.8	0.55	349 44 11	89 10 43
16	FR	- 8 46 11	-55.2	+14 19.1	0.53	350 43 8	89 10 44
17	SA	- 9 8 15	-54.8	+14 31.9	0.50	351 42 6	89 10 44
18	SU	- 9 30 10	-54.5	+14 44.0	0.48	352 41 5	89 10 45
19	MO	- 9 51 57	-54.1	+14 55.5	0.45	353 40 4	89 10 45
20	TU	-10 13 36	-53.8	+15 6.4	0.42	354 39 6	89 10 45
21	WE	-10 35 6	-53.3	+15 16.6	0.40	355 38 9	89 10 45
22	TH	-10 56 26	-53.0	+15 26.2	0.37	356 37 14	89 10 46
23	FR	-11 17 37	-52.5	+15 35.0	0.34	357 36 19	89 10 46
24	SA	-11 38 38	-52.1	+15 43.2	0.31	358 35 25	89 10 47
25	SU	-11 59 28	-51.6	+15 50.6	0.28	359 34 31	89 10 47
26	MO	-12 20 7	-51.2	+15 57.3	0.25	0 33 36	89 10 47
27	TU	-12 40 35	-50.7	+16 3.3	0.22	1 32 41	89 10 48
28	WE	-13 0 52	-50.2	+16 8.6	0.19	2 31 46	89 10 48
29	TH	-13 20 56	-49.6	+16 13.1	0.16	3 30 50	89 10 49
30	FR	-13 40 47	-49.1	+16 16.8	0.13	4 29 53	89 10 49
31	SA	-14 0 25	-48.5	+16 19.8	0.09	5 28 58	89 10 49

JANUARY 1984

		THE SUN					POLARIS	
Date		Declination			Equation of Time			
		At 0 hrs, UT ° ' "	Δ, 1 hr "	At 0 hrs, UT m s	Δ, 1 hr s	GHA at 0 hr ° ' "	Declination ° ' "
1	SU	-23 5 2	11.8	- 3 2.8	-1.20	65 54 51	89 11 45
2	MO	-23 0 20	12.9	- 3 31.5	-1.18	66 54 17	89 11 45
3	TU	-22 55 10	14.0	- 3 59.8	-1.16	67 53 44	89 11 46
4	WE	-22 49 33	15.2	- 4 27.7	-1.15	68 53 12	89 11 46
5	TH	-22 43 28	16.3	- 4 55.5	-1.13	69 52 42	89 11 46
6	FR	-22 36 56	17.4	- 5 22.4	-1.11	70 52 13	89 11 46
7	SA	-22 29 58	18.5	- 5 49.1	-1.10	71 51 45	89 11 46
8	SU	-22 22 33	19.7	- 6 15.4	-1.07	72 51 17	89 11 47
9	MO	-22 14 41	20.8	- 6 41.1	-1.05	73 50 50	89 11 47
10	TU	-22 6 23	21.8	- 7 6.2	-1.03	74 50 22	89 11 47
11	WE	-21 57 39	22.9	- 7 30.9	-1.00	75 49 53	89 11 47
12	TH	-21 48 30	24.0	- 7 54.9	-0.97	76 49 23	89 11 47
13	FR	-21 38 55	25.0	- 8 18.3	-0.95	77 48 53	89 11 47
14	SA	-21 28 55	26.0	- 8 41.1	-0.92	78 48 21	89 11 47
15	SU	-21 18 30	27.1	- 9 3.3	-0.89	79 47 48	89 11 47
16	MO	-21 7 40	28.1	- 9 24.7	-0.87	80 47 16	89 11 47
17	TU	-20 56 26	29.0	- 9 45.5	-0.84	81 46 44	89 11 48
18	WE	-20 44 49	30.1	-10 5.6	-0.81	82 46 14	89 11 48
19	TH	-20 32 47	31.0	-10 25.0	-0.78	83 45 45	89 11 48
20	FR	-20 20 23	32.0	-10 43.7	-0.75	84 45 19	89 11 48
21	SA	-20 7 35	32.9	-11 1.6	-0.72	85 44 53	89 11 48
22	SU	-19 54 25	33.8	-11 18.8	-0.69	86 44 28	89 11 48
23	MO	-19 40 53	34.8	-11 35.3	-0.65	87 44 2	89 11 48
24	TU	-19 26 59	35.7	-11 51.0	-0.62	88 43 35	89 11 48
25	WE	-19 12 43	36.6	-12 5.9	-0.59	89 43 6	89 11 48
26	TH	-18 58 5	37.4	-12 20.1	-0.56	90 42 36	89 11 48
27	FR	-18 43 7	38.3	-12 33.5	-0.52	91 42 5	89 11 48
28	SA	-18 27 49	39.1	-12 46.1	-0.50	92 41 34	89 11 48
29	SU	-18 12 10	39.9	-12 58.0	-0.46	93 41 3	89 11 48
30	MO	-17 56 12	40.8	-13 9.0	-0.42	94 40 33	89 11 48
31	TU	-17 39 54	41.5	-13 19.2	-0.39	95 40 4	89 11 48

FEBRUARY 1984

		THE SUN					POLARIS	
Date		Declination			Equation of Time			
		At 0 hrs, UT ° ' "	Δ, 1 hr "	At 0 hrs, UT m s	Δ, 1 hr s	GHA at 0 hr ° ' "	Declination ° ' "
1	WE	-17 23 17	42.3	-13 28.6	-0.36	96 39 37	89 11 48
2	TH	-17 6 22	43.0	-13 37.2	-0.33	97 39 10	89 11 48
3	FR	-16 49 9	43.8	-13 45.0	-0.29	98 38 45	89 11 48
4	SA	-16 31 38	44.5	-13 52.0	-0.25	99 38 19	89 11 48
5	SU	-16 13 50	45.2	-13 58.1	-0.22	100 37 54	89 11 48
6	MO	-15 55 45	45.9	-14 3.4	-0.19	101 37 28	89 11 48
7	TU	-15 37 23	46.5	-14 7.9	-0.15	102 37 1	89 11 48
8	WE	-15 18 46	47.2	-14 11.6	-0.12	103 36 33	89 11 48
9	TH	-14 59 53	47.8	-14 14.4	-0.09	104 36 3	89 11 48
10	FR	-14 40 45	48.5	-14 16.5	-0.05	105 35 33	89 11 48
11	SA	-14 21 22	49.0	-14 17.7	-0.02	106 35 1	89 11 48
12	SU	-14 1 45	49.6	-14 18.1	0.01	107 34 29	89 11 48
13	MO	-13 41 55	50.2	-14 17.8	0.05	108 33 57	89 11 48
14	TU	-13 21 50	50.7	-14 16.7	0.08	109 33 26	89 11 48
15	WE	-13 1 33	51.3	-14 14.8	0.11	110 32 56	89 11 48
16	TH	-12 41 3	51.8	-14 12.2	0.14	111 32 28	89 11 48
17	FR	-12 20 21	52.2	-14 8.8	0.17	112 32 2	89 11 48
18	SA	-11 59 28	52.7	-14 4.6	0.20	113 31 35	89 11 48
19	SU	-11 38 23	53.2	-14 0.0	0.23	114 31 8	89 11 47
20	MO	-11 17 6	53.6	-13 54.6	0.25	115 30 40	89 11 47
21	TU	-10 55 40	54.0	-13 48.5	0.28	116 30 10	89 11 47
22	WE	-10 34 3	54.5	-13 41.8	0.31	117 29 37	89 11 47
23	TH	-10 12 16	54.9	-13 34.4	0.33	118 29 4	89 11 47
24	FR	- 9 50 19	55.2	-13 26.4	0.35	119 28 30	89 11 47
25	SA	- 9 28 14	55.6	-13 17.9	0.38	120 27 56	89 11 47
26	SU	- 9 5 59	55.9	-13 8.7	0.40	121 27 23	89 11 46
27	MO	- 8 43 37	56.3	-12 59.0	0.43	122 26 50	89 11 46
28	TU	- 8 21 7	56.6	-12 48.8	0.45	123 26 19	89 11 46
29	WE	- 7 58 29	56.9	-12 38.0	0.47	124 25 48	89 11 46

MARCH 1984

		THE SUN					POLARIS	
Date		Declination			Equation of Time			
		At 0 hrs, UT ° ' "	Δ, 1 hr "	At 0 hrs, UT m s	Δ, 1 hr s	GHA at 0 hr ° ' "	Declination ° ' "
1	TH	- 7 35 44	57.1	-12 26.7	0.49	125 25 18	89 11 46
2	FR	- 7 12 53	57.4	-12 14.9	0.51	126 24 49	89 11 46
3	SA	- 6 49 55	57.6	-12 2.6	0.53	127 24 18	89 11 46
4	SU	- 6 26 52	57.9	-11 49.8	0.55	128 23 48	89 11 45
5	MO	- 6 3 43	58.0	-11 36.6	0.57	129 23 16	89 11 45
6	TU	- 5 40 30	58.3	-11 22.9	0.59	130 22 43	89 11 45
7	WE	- 5 17 11	58.4	-11 8.8	0.60	131 22 8	89 11 45
8	TH	- 4 53 49	58.6	-10 54.3	0.62	132 21 32	89 11 44
9	FR	- 4 30 23	58.8	-10 39.4	0.63	133 20 55	89 11 44
10	SA	- 4 6 53	58.8	-10 24.2	0.65	134 20 17	89 11 44
11	SU	- 3 43 21	59.0	-10 8.6	0.67	135 19 39	89 11 44
12	MO	- 3 19 46	59.1	- 9 52.6	0.68	136 19 1	89 11 44
13	TU	- 2 56 8	59.1	- 9 36.3	0.69	137 18 24	89 11 43
14	WE	- 2 32 29	59.2	- 9 19.8	0.70	138 17 48	89 11 43
15	TH	- 2 8 49	59.3	- 9 3.0	0.71	139 17 14	89 11 43
16	FR	- 1 45 7	59.3	- 8 45.9	0.72	140 16 40	89 11 43
17	SA	- 1 21 25	59.3	- 8 28.6	0.73	141 16 6	89 11 42
18	SU	- 0 57 43	59.3	- 8 11.1	0.73	142 15 30	89 11 42
19	MO	- 0 34 0	59.3	- 7 53.5	0.74	143 14 52	89 11 42
20	TU	- 0 10 17	59.3	- 7 35.7	0.75	144 14 13	89 11 42
21	WE	+ 0 13 25	59.2	- 7 17.7	0.75	145 13 31	89 11 41
22	TH	+ 0 37 6	59.1	- 6 59.7	0.75	146 12 49	89 11 41
23	FR	+ 1 0 45	59.1	- 6 41.6	0.75	147 12 6	89 11 41
24	SA	+ 1 24 23	59.0	- 6 23.5	0.76	148 11 23	89 11 41
25	SU	+ 1 48 0	58.9	- 6 5.3	0.76	149 10 41	89 11 40
26	MO	+ 2 11 33	58.8	- 5 47.1	0.76	150 10 0	89 11 40
27	TU	+ 2 35 5	58.7	- 5 28.9	0.76	151 9 20	89 11 39
28	WE	+ 2 58 33	58.5	- 5 10.7	0.75	152 8 41	89 11 39
29	TH	+ 3 21 57	58.4	- 4 52.7	0.75	153 8 1	89 11 39
30	FR	+ 3 45 18	58.2	- 4 34.6	0.75	154 7 22	89 11 39
31	SA	+ 4 8 35	58.0	- 4 16.6	0.75	155 6 41	89 11 39

APRIL 1984

Date		Declination At 0 hrs, UT ° ' "	THE SUN Δ, 1 hr "	Equation of Time At 0 hrs, UT m s	Δ, 1 hr s	POLARIS GHA at 0 hr ° ' "	Declination ° ' "
1	SU	+ 4 31 47	57.8	− 3 58.7	0.74	156 6 0	89 11 38
2	MO	+ 4 54 55	57.6	− 3 40.9	0.74	157 5 17	89 11 38
3	TU	+ 5 17 57	57.2	− 3 23.2	0.73	158 4 32	89 11 38
4	WE	+ 5 40 53	57.1	− 3 5.7	0.73	159 3 46	89 11 37
5	TH	+ 6 3 44	56.8	− 2 48.3	0.72	160 2 59	89 11 37
6	FR	+ 6 26 28	56.5	− 2 31.1	0.71	161 2 10	89 11 37
7	SA	+ 6 49 5	56.3	− 2 14.1	0.70	162 1 21	89 11 36
8	SU	+ 7 11 36	56.0	− 1 57.3	0.69	163 0 32	89 11 36
9	MO	+ 7 33 59	55.6	− 1 40.7	0.68	163 59 44	89 11 36
10	TU	+ 7 56 14	55.3	− 1 24.3	0.67	164 58 57	89 11 36
11	WE	+ 8 18 21	55.0	− 1 8.2	0.66	165 58 11	89 11 35
12	TH	+ 8 40 20	54.5	− 0 52.4	0.65	166 57 26	89 11 35
13	FR	+ 9 2 9	54.2	− 0 36.8	0.63	167 56 40	89 11 35
14	SA	+ 9 23 50	53.8	− 0 21.6	0.62	168 55 54	89 11 35
15	SU	+ 9 45 21	53.4	− 0 6.7	0.60	169 55 6	89 11 34
16	MO	+10 6 43	53.0	+ 0 7.8	0.59	170 54 16	89 11 34
17	TU	+10 27 54	52.5	+ 0 22.0	0.57	171 53 23	89 11 34
18	WE	+10 48 55	52.1	+ 0 35.8	0.56	172 52 29	89 11 33
19	TH	+11 9 45	51.7	+ 0 49.2	0.54	173 51 35	89 11 33
20	FR	+11 30 25	51.2	+ 1 2.2	0.52	174 50 40	89 11 33
21	SA	+11 50 53	50.7	+ 1 14.8	0.50	175 49 46	89 11 32
22	SU	+12 11 10	50.2	+ 1 26.8	0.48	176 48 53	89 11 32
23	MO	+12 31 15	49.7	+ 1 38.5	0.46	177 48 1	89 11 32
24	TU	+12 51 7	49.2	+ 1 49.6	0.45	178 47 10	89 11 31
25	WE	+13 10 47	48.6	+ 2 0.3	0.42	179 46 18	89 11 31
26	TH	+13 30 14	48.1	+ 2 10.4	0.40	180 45 27	89 11 31
27	FR	+13 49 28	47.5	+ 2 20.1	0.38	181 44 35	89 11 31
28	SA	+14 8 28	47.0	+ 2 29.2	0.36	182 43 42	89 11 30
29	SU	+14 27 15	46.3	+ 2 37.8	0.34	183 42 48	89 11 30
30	MO	+14 45 47	45.7	+ 2 45.9	0.32	184 41 52	89 11 30

MAY 1984

Date		Declination At 0 hrs, UT ° ' "	THE SUN Δ, 1 hr "	Equation of Time At 0 hrs, UT m s	Δ, 1 hr s	POLARIS GHA at 0 hr ° ' "	Declination ° ' "
1	TU	+15 4 4	45.1	+ 2 53.5	0.29	185 40 55	89 11 29
2	WE	+15 22 7	44.5	+ 3 0.5	0.27	186 39 56	89 11 29
3	TH	+15 39 54	43.8	+ 3 7.0	0.25	187 38 55	89 11 29
4	FR	+15 57 26	43.2	+ 3 12.9	0.22	188 37 55	89 11 28
5	SA	+16 14 42	42.5	+ 3 18.3	0.20	189 36 54	89 11 28
6	SU	+16 31 41	41.8	+ 3 23.1	0.18	190 35 54	89 11 28
7	MO	+16 48 24	41.1	+ 3 27.4	0.16	191 34 55	89 11 28
8	TU	+17 4 50	40.4	+ 3 31.2	0.13	192 33 57	89 11 27
9	WE	+17 20 59	39.6	+ 3 34.4	0.11	193 33 0	89 11 27
10	TH	+17 36 50	38.9	+ 3 37.0	0.09	194 32 3	89 11 27
11	FR	+17 52 24	38.1	+ 3 39.1	0.06	195 31 6	89 11 27
12	SA	+18 7 39	37.4	+ 3 40.6	0.04	196 30 7	89 11 26
13	SU	+18 22 36	36.6	+ 3 41.6	0.02	197 29 6	89 11 26
14	MO	+18 37 15	35.8	+ 3 42.0	−0.00	198 28 3	89 11 26
15	TU	+18 51 34	35.0	+ 3 41.9	−0.03	199 26 59	89 11 25
16	WE	+19 5 35	34.2	+ 3 41.2	−0.05	200 25 53	89 11 25
17	TH	+19 19 16	33.4	+ 3 39.9	−0.08	201 24 47	89 11 25
18	FR	+19 32 38	32.5	+ 3 38.0	−0.10	202 23 42	89 11 25
19	SA	+19 45 39	31.8	+ 3 35.6	−0.13	203 22 37	89 11 24
20	SU	+19 58 21	30.9	+ 3 32.6	−0.15	204 21 34	89 11 24
21	MO	+20 10 42	30.0	+ 3 29.1	−0.17	205 20 31	89 11 24
22	TU	+20 22 43	29.2	+ 3 25.0	−0.19	206 19 29	89 11 24
23	WE	+20 34 23	28.3	+ 3 20.4	−0.22	207 18 28	89 11 24
24	TH	+20 45 41	27.4	+ 3 15.2	−0.23	208 17 25	89 11 23
25	FR	+20 56 39	26.5	+ 3 9.6	−0.26	209 16 22	89 11 23
26	SA	+21 7 14	25.6	+ 3 3.4	−0.28	210 15 18	89 11 23
27	SU	+21 17 28	24.7	+ 2 56.7	−0.30	211 14 13	89 11 23
28	MO	+21 27 20	23.8	+ 2 49.5	−0.32	212 13 5	89 11 22
29	TU	+21 36 50	22.8	+ 2 41.8	−0.34	213 11 57	89 11 22
30	WE	+21 45 57	21.8	+ 2 33.7	−0.35	214 10 47	89 11 22
31	TH	+21 54 41	20.9	+ 2 25.2	−0.37	215 9 36	89 11 22

JUNE 1984

Date		Declination At 0 hrs, UT ° ' "	THE SUN Δ, 1 hr "	Equation of Time At 0 hrs, UT m s	Δ, 1 hr s	POLARIS GHA at 0 hr ° ' "	Declination ° ' "
1	FR	+22 3 3	20.0	+ 2 16.3	−0.39	216 8 26	89 11 21
2	SA	+22 11 2	19.0	+ 2 6.9	−0.40	217 7 16	89 11 21
3	SU	+22 18 37	18.0	+ 1 57.2	−0.42	218 6 7	89 11 21
4	MO	+22 25 49	17.0	+ 1 47.1	−0.43	219 5 0	89 11 21
5	TU	+22 32 37	16.0	+ 1 36.7	−0.45	220 3 53	89 11 21
6	WE	+22 39 1	15.0	+ 1 26.0	−0.46	221 2 48	89 11 21
7	TH	+22 45 2	14.0	+ 1 15.0	−0.47	222 1 41	89 11 20
8	FR	+22 50 39	13.0	+ 1 3.8	−0.48	223 0 34	89 11 20
9	SA	+22 55 51	12.0	+ 0 52.3	−0.50	223 59 25	89 11 20
10	SU	+23 0 39	11.0	+ 0 40.5	−0.50	224 58 14	89 11 20
11	MO	+23 5 3	10.0	+ 0 28.6	−0.50	225 57 1	89 11 20
12	TU	+23 9 2	9.0	+ 0 16.5	−0.51	226 55 48	89 11 19
13	WE	+23 12 37	8.0	+ 0 4.2	−0.52	227 54 34	89 11 19
14	TH	+23 15 48	6.9	− 0 8.3	−0.52	228 53 20	89 11 19
15	FR	+23 18 34	5.9	− 0 20.9	−0.53	229 52 7	89 11 19
16	SA	+23 20 55	4.9	− 0 33.6	−0.53	230 50 56	89 11 19
17	SU	+23 22 52	3.8	− 0 46.5	−0.54	231 49 45	89 11 19
18	MO	+23 24 24	2.8	− 0 59.4	−0.54	232 48 35	89 11 19
19	TU	+23 25 31	1.8	− 1 12.4	−0.54	233 47 26	89 11 19
20	WE	+23 26 13	0.8	− 1 25.4	−0.54	234 46 17	89 11 19
21	TH	+23 26 31	− 0.3	− 1 38.4	−0.54	235 45 7	89 11 18
22	FR	+23 26 24	− 1.3	− 1 51.4	−0.54	236 43 56	89 11 18
23	SA	+23 25 52	− 2.4	− 2 4.4	−0.54	237 42 44	89 11 18
24	SU	+23 24 55	− 3.4	− 2 17.3	−0.54	238 41 31	89 11 18
25	MO	+23 23 34	− 4.4	− 2 30.2	−0.53	239 40 16	89 11 18
26	TU	+23 21 48	− 5.5	− 2 42.9	−0.53	240 39 1	89 11 18
27	WE	+23 19 37	− 6.5	− 2 55.6	−0.52	241 37 44	89 11 18
28	TH	+23 17 2	− 7.5	− 3 8.0	−0.51	242 36 27	89 11 18
29	FR	+23 14 2	− 8.5	− 3 20.3	−0.50	243 35 11	89 11 18
30	SA	+23 10 37	− 9.5	− 3 32.4	−0.49	244 33 56	89 11 18

JULY 1984

		THE SUN				POLARIS	
Date	Declination At 0 hrs, UT ° ' "	Δ, 1 hr "	Equation of Time At 0 hrs, UT m s	Δ, 1 hr s	GHA at 0 hr ° ' "	Declination ° ' "
1 SU	+23 6 49	-10.5	- 3 44.2	-0.48	245 32 42	89 11 18
2 MO	+23 2 36	-11.5	- 3 55.8	-0.47	246 31 30	89 11 18
3 TU	+22 57 59	-12.6	- 4 7.1	-0.45	247 30 19	89 11 18
4 WE	+22 52 57	-13.5	- 4 18.0	-0.44	248 29 8	89 11 18
5 TH	+22 47 32	-14.5	- 4 28.6	-0.43	249 27 56	89 11 17
6 FR	+22 41 43	-15.5	- 4 38.9	-0.41	250 26 42	89 11 17
7 SA	+22 35 31	-16.5	- 4 48.8	-0.39	251 25 27	89 11 17
8 SU	+22 28 55	-17.5	- 4 58.2	-0.38	252 24 10	89 11 17
9 MO	+22 21 55	-18.4	- 5 7.3	-0.36	253 22 53	89 11 17
10 TU	+22 14 33	-19.4	- 5 16.0	-0.34	254 21 34	89 11 17
11 WE	+22 6 48	-20.3	- 5 24.2	-0.32	255 20 16	89 11 17
12 TH	+21 58 40	-21.3	- 5 31.9	-0.30	256 18 59	89 11 17
13 FR	+21 50 9	-22.2	- 5 39.2	-0.28	257 17 44	89 11 17
14 SA	+21 41 16	-23.1	- 5 46.0	-0.26	258 16 29	89 11 17
15 SU	+21 32 1	-24.0	- 5 52.3	-0.25	259 15 16	89 11 17
16 MO	+21 22 24	-25.0	- 5 58.2	-0.22	260 14 3	89 11 17
17 TU	+21 12 25	-25.9	- 6 3.5	-0.20	261 12 50	89 11 17
18 WE	+21 2 4	-26.7	- 6 8.3	-0.18	262 11 38	89 11 17
19 TH	+20 51 23	-27.6	- 6 12.6	-0.16	263 10 24	89 11 17
20 FR	+20 40 20	-28.5	- 6 16.4	-0.13	264 9 10	89 11 17
21 SA	+20 28 56	-29.4	- 6 19.6	-0.11	265 7 55	89 11 18
22 SU	+20 17 11	-30.2	- 6 22.3	-0.09	266 6 38	89 11 18
23 MO	+20 5 6	-31.0	- 6 24.4	-0.06	267 5 20	89 11 18
24 TU	+19 52 41	-31.9	- 6 25.9	-0.04	268 4 2	89 11 18
25 WE	+19 39 56	-32.7	- 6 26.9	-0.02	269 2 43	89 11 18
26 TH	+19 26 52	-33.5	- 6 27.3	0.00	270 1 25	89 11 18
27 FR	+19 13 28	-34.3	- 6 27.2	0.03	271 0 7	89 11 18
28 SA	+18 59 44	-35.1	- 6 26.4	0.06	271 58 52	89 11 18
29 SU	+18 45 42	-35.8	- 6 25.0	0.08	272 57 38	89 11 18
30 MO	+18 31 22	-36.6	- 6 23.1	0.11	273 56 25	89 11 18
31 TU	+18 16 43	-37.3	- 6 20.5	0.13	274 55 13	89 11 18

AUGUST 1984

		THE SUN				POLARIS	
Date	Declination At 0 hrs, UT ° ' "	Δ, 1 hr "	Equation of Time At 0 hrs, UT m s	Δ, 1 hr s	GHA at 0 hr ° ' "	Declination ° ' "
1 WE	+18 1 47	-38.1	- 6 17.3	0.16	275 54 0	89 11 19
2 TH	+17 46 32	-38.8	- 6 13.4	0.18	276 52 46	89 11 19
3 FR	+17 31 1	-39.5	- 6 9.0	0.21	277 51 31	89 11 19
4 SA	+17 15 12	-40.2	- 6 3.9	0.24	278 50 14	89 11 19
5 SU	+16 59 7	-40.9	- 5 58.2	0.27	279 48 56	89 11 19
6 MO	+16 42 45	-41.6	- 5 51.8	0.29	280 47 37	89 11 19
7 TU	+16 26 7	-42.3	- 5 44.9	0.32	281 46 19	89 11 19
8 WE	+16 9 13	-42.9	- 5 37.3	0.34	282 45 2	89 11 19
9 TH	+15 52 4	-43.5	- 5 29.2	0.36	283 43 46	89 11 19
10 FR	+15 34 40	-44.1	- 5 20.6	0.39	284 42 31	89 11 20
11 SA	+15 17 1	-44.8	- 5 11.1	0.41	285 41 18	89 11 20
12 SU	+14 59 7	-45.3	- 5 1.3	0.44	286 40 6	89 11 20
13 MO	+14 40 59	-45.9	- 4 50.8	0.46	287 38 54	89 11 20
14 TU	+14 22 37	-46.5	- 4 39.8	0.48	288 37 42	89 11 20
15 WE	+14 4 1	-47.1	- 4 28.3	0.50	289 36 30	89 11 20
16 TH	+13 45 11	-47.6	- 4 16.3	0.52	290 35 17	89 11 21
17 FR	+13 26 9	-48.1	- 4 3.7	0.54	291 34 4	89 11 21
18 SA	+13 6 54	-48.7	- 3 50.7	0.56	292 32 49	89 11 21
19 SU	+12 47 26	-49.2	- 3 37.2	0.58	293 31 33	89 11 21
20 MO	+12 27 46	-49.7	- 3 23.2	0.60	294 30 17	89 11 21
21 TU	+12 7 54	-50.2	- 3 8.7	0.62	295 29 0	89 11 21
22 WE	+11 47 50	-50.6	- 2 53.8	0.64	296 27 43	89 11 22
23 TH	+11 27 35	-51.1	- 2 38.5	0.66	297 26 28	89 11 22
24 FR	+11 7 9	-51.5	- 2 22.7	0.68	298 25 13	89 11 22
25 SA	+10 46 33	-52.0	- 2 6.6	0.69	299 24 1	89 11 22
26 SU	+10 25 46	-52.4	- 1 50.0	0.71	300 22 50	89 11 23
27 MO	+10 4 49	-52.8	- 1 33.0	0.72	301 21 41	89 11 23
28 TU	+ 9 43 43	-53.2	- 1 15.7	0.74	302 20 31	89 11 23
29 WE	+ 9 22 27	-53.5	- 0 58.0	0.75	303 19 21	89 11 23
30 TH	+ 9 1 2	-53.9	- 0 39.9	0.77	304 18 9	89 11 24
31 FR	+ 8 39 29	-54.3	- 0 21.5	0.78	305 16 55	89 11 24

SEPTEMBER 1984

		THE SUN				POLARIS	
Date	Declination At 0 hrs, UT ° ' "	Δ, 1 hr "	Equation of Time At 0 hrs, UT m s	Δ, 1 hr s	GHA at 0 hr ° ' "	Declination ° ' "
1 SA	+ 8 17 47	-54.5	- 0 2.7	0.79	306 15 41	89 11 24
2 SU	+ 7 55 58	-54.9	+ 0 16.3	0.81	307 14 25	89 11 24
3 MO	+ 7 34 1	-55.2	+ 0 35.7	0.82	308 13 11	89 11 25
4 TU	+ 7 11 57	-55.5	+ 0 55.4	0.83	309 11 57	89 11 25
5 WE	+ 6 49 46	-55.8	+ 1 15.3	0.84	310 10 44	89 11 25
6 TH	+ 6 27 28	-56.0	+ 1 35.4	0.85	311 9 33	89 11 25
7 FR	+ 6 5 4	-56.3	+ 1 55.8	0.86	312 8 24	89 11 26
8 SA	+ 5 42 34	-56.5	+ 2 16.4	0.87	313 7 16	89 11 26
9 SU	+ 5 19 58	-56.7	+ 2 37.2	0.87	314 6 8	89 11 26
10 MO	+ 4 57 18	-56.9	+ 2 58.1	0.88	315 5 1	89 11 27
11 TU	+ 4 34 32	-57.1	+ 3 19.1	0.88	316 3 53	89 11 27
12 WE	+ 4 11 41	-57.3	+ 3 40.3	0.89	317 2 45	89 11 27
13 TH	+ 3 48 46	-57.5	+ 4 1.6	0.89	318 1 36	89 11 27
14 FR	+ 3 25 47	-57.6	+ 4 22.9	0.89	319 0 27	89 11 28
15 SA	+ 3 2 44	-57.8	+ 4 44.3	0.89	319 59 16	89 11 28
16 SU	+ 2 39 37	-57.9	+ 5 5.7	0.89	320 58 5	89 11 28
17 MO	+ 2 16 27	-58.0	+ 5 27.1	0.89	321 56 54	89 11 28
18 TU	+ 1 53 15	-58.1	+ 5 48.5	0.89	322 55 42	89 11 29
19 WE	+ 1 30 0	-58.3	+ 6 9.8	0.89	323 54 32	89 11 29
20 TH	+ 1 6 42	-58.3	+ 6 31.1	0.88	324 53 23	89 11 29
21 FR	+ 0 43 23	-58.4	+ 6 52.3	0.88	325 52 16	89 11 30
22 SA	+ 0 20 2	-58.4	+ 7 13.4	0.88	326 51 10	89 11 30
23 SU	- 0 3 20	-58.5	+ 7 34.4	0.86	327 50 6	89 11 31
24 MO	- 0 26 43	-58.5	+ 7 55.3	0.86	328 49 2	89 11 31
25 TU	- 0 50 6	-58.5	+ 8 16.0	0.85	329 47 59	89 11 31
26 WE	- 1 13 29	-58.5	+ 8 36.6	0.85	330 46 54	89 11 32
27 TH	- 1 36 53	-58.4	+ 8 56.9	0.84	331 45 48	89 11 32
28 FR	- 2 0 15	-58.4	+ 9 17.1	0.83	332 44 39	89 11 32
29 SA	- 2 23 36	-58.3	+ 9 37.1	0.82	333 43 31	89 11 32
30 SU	- 2 46 56	-58.3	+ 9 56.8	0.81	334 42 22	89 11 33

OCTOBER 1984

Date		THE SUN Declination At 0 hrs, UT	Δ, 1 hr	Equation of Time At 0 hrs, UT	Δ, 1 hr	POLARIS GHA at 0 hr	Declination
		° ' "	"	m s	s	° ' "	° ' "
1	MO	- 3 10 14	-58.2	+10 16.3	0.80	335 41 14	89 11 33
2	TU	- 3 33 30	-58.1	+10 35.5	0.79	336 40 8	89 11 34
3	WE	- 3 56 44	-57.9	+10 54.4	0.78	337 39 4	89 11 34
4	TH	- 4 19 54	-57.8	+11 13.0	0.76	338 38 1	89 11 34
5	FR	- 4 43 2	-57.6	+11 31.3	0.75	339 37 0	89 11 35
6	SA	- 5 6 5	-57.5	+11 49.2	0.73	340 36 0	89 11 35
7	SU	- 5 29 5	-57.3	+12 6.7	0.72	341 35 0	89 11 35
8	MO	- 5 52 0	-57.1	+12 23.9	0.70	342 34 0	89 11 36
9	TU	- 6 14 51	-56.9	+12 40.6	0.68	343 33 0	89 11 36
10	WE	- 6 37 37	-56.7	+12 56.9	0.66	344 31 59	89 11 36
11	TH	- 7 0 18	-56.5	+13 12.7	0.64	345 30 57	89 11 37
12	FR	- 7 22 53	-56.2	+13 28.0	0.62	346 29 55	89 11 37
13	SA	- 7 45 22	-55.9	+13 42.8	0.60	347 28 52	89 11 37
14	SU	- 8 7 44	-55.7	+13 57.1	0.57	348 27 48	89 11 38
15	MO	- 8 30 0	-55.4	+14 10.8	0.55	349 26 45	89 11 38
16	TU	- 8 52 9	-55.0	+14 24.0	0.52	350 25 43	89 11 39
17	WE	- 9 14 10	-54.8	+14 36.5	0.50	351 24 42	89 11 39
18	TH	- 9 36 4	-54.5	+14 48.5	0.47	352 23 43	89 11 40
19	FR	- 9 57 49	-54.0	+14 59.8	0.45	353 22 45	89 11 40
20	SA	-10 19 26	-53.7	+15 10.5	0.42	354 21 49	89 11 40
21	SU	-10 40 54	-53.3	+15 20.5	0.39	355 20 54	89 11 41
22	MO	-11 2 12	-52.9	+15 29.8	0.36	356 20 0	89 11 41
23	TU	-11 23 21	-52.4	+15 38.5	0.33	357 19 5	89 11 41
24	WE	-11 44 19	-52.0	+15 46.5	0.30	358 18 8	89 11 42
25	TH	-12 5 6	-51.5	+15 53.7	0.27	359 17 9	89 11 42
26	FR	-12 25 43	-51.0	+16 0.2	0.25	0 16 10	89 11 43
27	SA	-12 46 8	-50.5	+16 6.1	0.21	1 15 10	89 11 43
28	SU	-13 6 21	-50.0	+16 11.1	0.18	2 14 11	89 11 43
29	MO	-13 26 22	-49.5	+16 15.5	0.15	3 13 14	89 11 44
30	TU	-13 46 10	-48.9	+16 19.1	0.12	4 12 18	89 11 44
31	WE	-14 5 44	-48.4	+16 21.9	0.08	5 11 25	89 11 44

NOVEMBER 1984

Date		THE SUN Declination At 0 hrs, UT	Δ, 1 hr	Equation of Time At 0 hrs, UT	Δ, 1 hr	POLARIS GHA at 0 hr	Declination
		° ' "	"	m s	s	° ' "	° ' "
1	TH	-14 25 5	-47.8	+16 23.9	0.05	6 10 33	89 11 45
2	FR	-14 44 12	-47.2	+16 25.1	0.02	7 9 42	89 11 45
3	SA	-15 3 5	-46.6	+16 25.4	-0.01	8 8 52	89 11 46
4	SU	-15 21 43	-45.9	+16 25.0	-0.05	9 8 2	89 11 46
5	MO	-15 40 5	-45.3	+16 24.3	-0.08	10 7 12	89 11 46
6	TU	-15 58 12	-44.5	+16 22.3	-0.12	11 6 21	89 11 47
7	WE	-16 16 3	-44.0	+16 19.5	-0.15	12 5 30	89 11 47
8	TH	-16 33 38	-43.2	+16 15.8	-0.19	13 4 37	89 11 48
9	FR	-16 50 55	-42.5	+16 11.5	-0.22	14 3 45	89 11 48
10	SA	-17 7 56	-41.8	+16 6.2	-0.25	15 2 51	89 11 48
11	SU	-17 24 39	-41.0	+16 0.1	-0.29	16 1 58	89 11 49
12	MO	-17 41 4	-40.3	+15 53.1	-0.33	17 1 5	89 11 49
13	TU	-17 57 11	-39.5	+15 45.2	-0.36	18 0 15	89 11 49
14	WE	-18 13 0	-38.7	+15 36.5	-0.40	18 59 26	89 11 50
15	TH	-18 28 29	-37.9	+15 26.9	-0.44	19 58 38	89 11 50
16	FR	-18 43 39	-37.1	+15 16.4	-0.47	20 57 52	89 11 50
17	SA	-18 58 29	-36.3	+15 5.1	-0.51	21 57 8	89 11 51
18	SU	-19 12 59	-35.4	+14 52.9	-0.54	22 56 24	89 11 51
19	MO	-19 27 8	-34.5	+14 39.9	-0.58	23 55 39	89 11 52
20	TU	-19 40 56	-33.6	+14 26.1	-0.61	24 54 54	89 11 52
21	WE	-19 54 23	-32.7	+14 11.4	-0.65	25 54 7	89 11 52
22	TH	-20 7 28	-31.8	+13 55.9	-0.68	26 53 19	89 11 53
23	FR	-20 20 12	-30.8	+13 39.6	-0.71	27 52 30	89 11 53
24	SA	-20 32 32	-29.9	+13 22.5	-0.75	28 51 41	89 11 54
25	SU	-20 44 30	-28.9	+13 4.6	-0.77	29 50 54	89 11 54
26	MO	-20 56 4	-28.0	+12 46.0	-0.80	30 50 8	89 11 54
27	TU	-21 7 15	-27.0	+12 26.7	-0.84	31 49 25	89 11 54
28	WE	-21 18 2	-26.0	+12 6.6	-0.86	32 48 43	89 11 55
29	TH	-21 28 25	-24.9	+11 45.9	-0.89	33 48 3	89 11 55
30	FR	-21 38 23	-23.9	+11 24.5	-0.92	34 47 23	89 11 56

DECEMBER 1984

Date		THE SUN Declination At 0 hrs, UT	Δ, 1 hr	Equation of Time At 0 hrs, UT	Δ, 1 hr	POLARIS GHA at 0 hr	Declination
		° ' "	"	m s	s	° ' "	° ' "
1	SA	-21 47 57	-22.8	+11 2.5	-0.95	35 46 44	89 11 56
2	SU	-21 57 5	-21.8	+10 39.8	-0.97	36 46 5	89 11 56
3	MO	-22 5 48	-20.7	+10 16.5	-0.99	37 45 25	89 11 57
4	TU	-22 14 5	-19.7	+ 9 52.7	-1.02	38 44 45	89 11 57
5	WE	-22 21 57	-18.5	+ 9 28.2	-1.04	39 44 3	89 11 57
6	TH	-22 29 22	-17.5	+ 9 3.3	-1.06	40 43 22	89 11 57
7	FR	-22 36 21	-16.3	+ 8 37.8	-1.08	41 42 39	89 11 58
8	SA	-22 42 53	-15.3	+ 8 11.8	-1.10	42 41 57	89 11 58
9	SU	-22 48 59	-14.1	+ 7 45.3	-1.12	43 41 15	89 11 58
10	MO	-22 54 38	-13.0	+ 7 18.4	-1.14	44 40 34	89 11 59
11	TU	-22 59 50	-11.8	+ 6 51.1	-1.15	45 39 55	89 11 59
12	WE	-23 4 34	-10.8	+ 6 23.4	-1.17	46 39 17	89 11 59
13	TH	-23 8 52	- 9.5	+ 5 55.3	-1.18	47 38 41	89 12 0
14	FR	-23 12 41	- 8.4	+ 5 26.9	-1.20	48 38 7	89 12 0
15	SA	-23 16 3	- 7.2	+ 4 58.1	-1.21	49 37 33	89 12 0
16	SU	-23 18 57	- 6.1	+ 4 29.1	-1.22	50 36 59	89 12 0
17	MO	-23 21 23	- 5.0	+ 3 59.8	-1.23	51 36 24	89 12 1
18	TU	-23 23 22	- 3.8	+ 3 30.3	-1.23	52 35 48	89 12 1
19	WE	-23 24 52	- 2.7	+ 3 0.6	-1.25	53 35 11	89 12 1
20	TH	-23 25 53	- 1.4	+ 2 30.8	-1.25	54 34 32	89 12 1
21	FR	-23 26 27	- 0.2	+ 2 0.9	-1.25	55 33 53	89 12 2
22	SA	-23 26 32	1.0	+ 1 30.9	-1.25	56 33 15	89 12 2
23	SU	-23 26 9	2.1	+ 1 0.9	-1.25	57 32 39	89 12 2
24	MO	-23 25 18	3.3	+ 0 30.9	-1.25	58 32 4	89 12 3
25	TU	-23 23 58	3.5	+ 0 1.0	-1.25	59 31 31	89 12 3
26	WE	-23 22 11	5.7	- 0 28.6	-1.23	60 30 59	89 12 3
27	TH	-23 19 55	6.9	- 0 58.6	-1.22	61 30 30	89 12 4
28	FR	-23 17 10	8.0	- 1 28.2	-1.22	62 30 0	89 12 4
29	SA	-23 13 58	9.2	- 1 57.5	-1.22	63 29 30	89 12 4
30	SU	-23 10 18	10.3	- 2 26.7	-1.20	64 29 0	89 12 4
31	MO	-23 6 10	11.5	- 2 55.6	-1.19	65 28 28	89 12 4

JANUARY 1985

Date		THE SUN Declination At 0 hrs, UT ° ' "	Δ, 1 hr "	Equation of Time At 0 hrs, UT m s	Δ, 1 hr s	POLARIS GHA at 0 hr ° ' "	Declination ° ' "
1	TU	-23 1 34	12.6	-3 24.1	-1.18	66 27 56	89 12 4
2	WE	-22 56 31	13.8	-3 52.4	-1.16	67 27 23	89 12 4
3	TH	-22 51 0	14.9	-4 20.2	-1.15	68 26 49	89 12 5
4	FR	-22 45 2	16.0	-4 47.7	-1.13	69 26 15	89 12 5
5	SA	-22 38 37	17.2	-5 14.6	-1.11	70 25 4	89 12 5
6	SU	-22 31 45	18.3	-5 41.5	-1.09	71 25 8	89 12 5
7	MO	-22 24 26	19.4	-6 7.7	-1.07	72 24 36	89 12 5
8	TU	-22 16 41	20.5	-6 33.4	-1.05	73 24 6	89 12 6
9	WE	-22 8 30	21.5	-6 58.5	-1.03	74 23 37	89 12 6
10	TH	-21 59 53	22.6	-7 23.2	-1.00	75 23 10	89 12 6
11	FR	-21 50 50	23.7	-7 47.3	-0.98	76 22 44	89 12 6
12	SA	-21 41 22	24.8	-8 10.9	-0.95	77 22 17	89 12 6
13	SU	-21 31 28	25.8	-8 33.8	-0.93	78 21 49	89 12 6
14	MO	-21 21 9	26.8	-8 56.2	-0.90	79 21 20	89 12 6
15	TU	-21 10 26	27.8	-9 17.9	-0.88	80 20 49	89 12 6
16	WE	-20 59 18	28.8	-9 39.0	-0.85	81 20 17	89 12 7
17	TH	-20 47 46	29.8	-9 59.4	-0.83	82 19 45	89 12 7
18	FR	-20 35 50	30.8	-10 19.2	-0.79	83 19 13	89 12 7
19	SA	-20 23 31	31.8	-10 38.2	-0.76	84 18 42	89 12 7
20	SU	-20 10 49	32.8	-10 56.5	-0.73	85 18 12	89 12 7
21	MO	-19 57 43	33.6	-11 14.1	-0.70	86 17 44	89 12 7
22	TU	-19 44 16	34.6	-11 30.9	-0.67	87 17 17	89 12 7
23	WE	-19 30 26	35.5	-11 47.0	-0.63	88 16 52	89 12 7
24	TH	-19 16 15	36.3	-12 2.2	-0.60	89 16 26	89 12 7
25	FR	-19 1 43	37.3	-12 16.7	-0.57	90 16 1	89 12 7
26	SA	-18 46 49	38.1	-12 30.3	-0.53	91 15 35	89 12 7
27	SU	-18 31 35	38.9	-12 43.1	-0.50	92 15 8	89 12 7
28	MO	-18 16 1	39.7	-12 55.1	-0.47	93 14 40	89 12 7
29	TU	-18 0 8	40.6	-13 6.3	-0.43	94 14 11	89 12 7
30	WE	-17 43 54	41.3	-13 16.6	-0.40	95 13 41	89 12 7
31	TH	-17 27 23	42.1	-13 26.1	-0.36	96 13 10	89 12 7

FEBRUARY 1985

Date		THE SUN Declination At 0 hrs, UT ° ' "	Δ, 1 hr "	Equation of Time At 0 hrs, UT m s	Δ, 1 hr s	POLARIS GHA at 0 hr ° ' "	Declination ° ' "
1	FR	-17 10 32	42.8	-13 34.8	-0.33	97 12 39	89 12 7
2	SA	-16 53 24	43.6	-13 42.6	-0.29	98 12 8	89 12 7
3	SU	-16 35 57	44.3	-13 49.6	-0.26	99 11 38	89 12 7
4	MO	-16 18 14	45.0	-13 55.8	-0.22	100 11 9	89 12 7
5	TU	-16 0 14	45.7	-14 1.1	-0.19	101 10 42	89 12 7
6	WE	-15 41 57	46.4	-14 5.7	-0.15	102 10 16	89 12 7
7	TH	-15 23 24	47.0	-14 9.4	-0.13	103 9 51	89 12 7
8	FR	-15 4 36	47.7	-14 12.4	-0.09	104 9 26	89 12 7
9	SA	-14 45 32	48.3	-14 14.5	-0.06	105 8 59	89 12 7
10	SU	-14 26 13	48.9	-14 15.9	-0.03	106 8 31	89 12 7
11	MO	-14 6 40	49.5	-14 16.6	0.00	107 8 1	89 12 7
12	TU	-13 46 53	50.0	-14 16.5	0.04	108 7 30	89 12 7
13	WE	-13 26 52	50.6	-14 15.6	0.07	109 6 57	89 12 7
14	TH	-13 6 37	51.1	-14 14.0	0.10	110 6 24	89 12 7
15	FR	-12 46 10	51.7	-14 11.7	0.13	111 5 52	89 12 7
16	SA	-12 25 30	52.1	-14 8.7	0.15	112 5 22	89 12 7
17	SU	-12 4 39	52.7	-14 5.0	0.18	113 4 52	89 12 7
18	MO	-11 43 35	53.1	-14 0.6	0.21	114 4 24	89 12 7
19	TU	-11 22 21	53.6	-13 55.5	0.24	115 3 56	89 12 6
20	WE	-11 0 55	54.0	-13 49.7	0.27	116 3 29	89 12 6
21	TH	-10 39 20	54.4	-13 43.3	0.30	117 3 2	89 12 6
22	FR	-10 17 34	54.8	-13 36.2	0.33	118 2 34	89 12 6
23	SA	-9 55 39	55.2	-13 28.4	0.35	119 2 5	89 12 6
24	SU	-9 33 35	55.5	-13 20.1	0.38	120 1 34	89 12 6
25	MO	-9 11 23	55.9	-13 11.1	0.40	121 1 2	89 12 6
26	TU	-8 49 2	56.1	-13 1.6	0.42	122 0 29	89 12 5
27	WE	-8 26 34	56.5	-12 51.4	0.45	122 59 55	89 12 5
28	TH	-8 3 58	56.8	-12 40.7	0.47	123 59 20	89 12 5

MARCH 1985

Date		THE SUN Declination At 0 hrs, UT ° ' "	Δ, 1 hr "	Equation of Time At 0 hrs, UT m s	Δ, 1 hr s	POLARIS GHA at 0 hr ° ' "	Declination ° ' "
1	FR	-7 41 15	57.1	-12 29.4	0.49	124 58 45	89 12 5
2	SA	-7 18 25	57.3	-12 17.6	0.51	125 58 9	89 12 5
3	SU	-6 55 30	57.5	-12 5.3	0.53	126 57 37	89 12 5
4	MO	-6 32 29	57.8	-11 52.5	0.55	127 57 4	89 12 4
5	TU	-6 9 22	58.0	-11 39.3	0.57	128 56 33	89 12 4
6	WE	-5 46 11	58.2	-11 25.6	0.59	129 56 2	89 12 4
7	TH	-5 22 54	58.3	-11 11.4	0.60	130 55 32	89 12 4
8	FR	-4 59 34	58.5	-10 56.9	0.62	131 55 0	89 12 4
9	SA	-4 36 10	58.7	-10 42.0	0.64	132 54 27	89 12 3
10	SU	-4 12 42	58.8	-10 26.7	0.65	133 53 51	89 12 3
11	MO	-3 49 10	58.9	-10 11.1	0.66	134 53 13	89 12 3
12	TU	-3 25 36	59.0	-9 55.3	0.67	135 52 34	89 12 3
13	WE	-3 2 0	59.1	-9 39.1	0.68	136 51 56	89 12 2
14	TH	-2 38 21	59.2	-9 22.6	0.69	137 51 15	89 12 2
15	FR	-2 14 40	59.3	-9 6.0	0.70	138 50 37	89 12 2
16	SA	-1 50 58	59.3	-8 49.1	0.71	139 50 0	89 12 2
17	SU	-1 27 15	59.3	-8 32.0	0.72	140 49 23	89 12 2
18	MO	-1 3 32	59.3	-8 14.7	0.73	141 48 48	89 12 1
19	TU	-0 39 48	59.3	-7 57.2	0.73	142 48 13	89 12 1
20	WE	-0 16 4	59.3	-7 39.6	0.74	143 47 38	89 12 1
21	TH	+0 7 39	59.3	-7 21.9	0.75	144 47 1	89 12 1
22	FR	+0 31 22	59.3	-7 4.0	0.75	145 46 24	89 12 0
23	SA	+0 55 3	59.1	-6 46.0	0.75	146 45 45	89 12 0
24	SU	+1 18 42	59.0	-6 28.0	0.75	147 45 5	89 12 0
25	MO	+1 42 19	59.0	-6 9.9	0.76	148 44 23	89 11 59
26	TU	+2 5 54	58.9	-5 51.7	0.76	149 43 40	89 11 59
27	WE	+2 29 27	58.7	-5 33.6	0.76	150 42 56	89 11 59
28	TH	+2 52 56	58.6	-5 15.4	0.76	151 42 11	89 11 58
29	FR	+3 16 21	58.4	-4 57.2	0.75	152 41 27	89 11 58
30	SA	+3 39 43	58.2	-4 39.1	0.75	153 40 43	89 11 58
31	SU	+4 3 0	58.0	-4 21.0	0.75	154 40 0	89 11 58

APRIL 1985

Date		THE SUN Declination At 0 hrs, UT ° ' "	Δ, 1 hr "	Equation of Time At 0 hrs, UT m s	Δ, 1 hr s	POLARIS GHA at 0 hr ° ' "	Declination ° ' "
1	MO	+ 4 26 13	57.8	- 4 2.9	0.75	155 39 18	89 11 58
2	TU	+ 4 49 21	57.6	- 3 45.0	0.75	156 38 37	89 11 57
3	WE	+ 5 12 23	57.4	- 3 27.1	0.74	157 37 56	89 11 57
4	TH	+ 5 35 20	57.1	- 3 9.4	0.73	158 37 15	89 11 57
5	FR	+ 5 58 11	56.8	- 2 51.9	0.72	159 36 32	89 11 56
6	SA	+ 6 20 55	56.6	- 2 34.5	0.72	160 35 46	89 11 56
7	SU	+ 6 43 34	56.3	- 2 17.3	0.71	161 34 59	89 11 55
8	MO	+ 7 6 5	56.0	- 2 0.3	0.70	162 34 9	89 11 55
9	TU	+ 7 28 29	55.7	- 1 43.6	0.68	163 33 18	89 11 55
10	WE	+ 7 50 46	55.4	- 1 27.2	0.67	164 32 27	89 11 55
11	TH	+ 8 12 55	55.0	- 1 11.0	0.66	165 31 37	89 11 55
12	FR	+ 8 34 55	54.7	- 0 55.1	0.65	166 30 48	89 11 54
13	SA	+ 8 56 48	54.3	- 0 39.6	0.63	167 30 1	89 11 54
14	SU	+ 9 18 31	54.0	- 0 24.4	0.62	168 29 14	89 11 54
15	MO	+ 9 40 6	53.5	- 0 9.5	0.60	169 28 27	89 11 54
16	TU	+10 1 30	53.1	+ 0 5.0	0.59	170 27 41	89 11 53
17	WE	+10 22 45	52.7	+ 0 19.1	0.58	171 26 53	89 11 53
18	TH	+10 43 50	52.3	+ 0 32.9	0.55	172 26 5	89 11 53
19	FR	+11 4 44	51.8	+ 0 46.2	0.52	173 25 14	89 11 52
20	SA	+11 25 27	51.3	+ 0 59.2	0.52	174 24 23	89 11 52
21	SU	+11 45 59	50.9	+ 1 11.7	0.50	175 23 30	89 11 52
22	MO	+12 6 20	50.3	+ 1 23.8	0.48	176 22 35	89 11 51
23	TU	+12 26 28	49.8	+ 1 35.4	0.47	177 21 39	89 11 51
24	WE	+12 46 24	49.3	+ 1 46.6	0.45	178 20 43	89 11 51
25	TH	+13 6 8	48.8	+ 1 57.3	0.43	179 19 47	89 11 50
26	FR	+13 25 38	48.2	+ 2 7.6	0.41	180 18 51	89 11 50
27	SA	+13 44 55	47.7	+ 2 17.4	0.39	181 17 56	89 11 50
28	SU	+14 3 59	47.0	+ 2 26.7	0.37	182 17 2	89 11 49
29	MO	+14 22 48	46.5	+ 2 35.5	0.35	183 16 9	89 11 49
30	TU	+14 41 23	45.8	+ 2 43.8	0.33	184 15 16	89 11 49

MAY 1985

Date		THE SUN Declination At 0 hrs, UT ° ' "	Δ, 1 hr "	Equation of Time At 0 hrs, UT m s	Δ, 1 hr s	POLARIS GHA at 0 hr ° ' "	Declination ° ' "
1	WE	+14 59 43	45.2	+ 2 51.6	0.30	185 14 23	89 11 49
2	TH	+15 17 48	44.6	+ 2 58.9	0.28	186 13 29	89 11 48
3	FR	+15 35 38	44.0	+ 3 5.7	0.26	187 12 32	89 11 48
4	SA	+15 53 13	43.3	+ 3 11.9	0.24	188 11 34	89 11 48
5	SU	+16 10 32	42.6	+ 3 17.6	0.21	189 10 33	89 11 47
6	MO	+16 27 34	42.0	+ 3 22.7	0.19	190 9 30	89 11 47
7	TU	+16 44 21	41.2	+ 3 27.3	0.16	191 8 27	89 11 47
8	WE	+17 0 50	40.5	+ 3 31.2	0.14	192 7 25	89 11 47
9	TH	+17 17 3	39.8	+ 3 34.6	0.12	193 6 24	89 11 46
10	FR	+17 32 59	39.1	+ 3 37.4	0.09	194 5 24	89 11 46
11	SA	+17 48 37	38.3	+ 3 39.6	0.07	195 4 26	89 11 46
12	SU	+18 3 57	37.6	+ 3 41.2	0.05	196 3 28	89 11 46
13	MO	+18 18 59	36.8	+ 3 42.3	0.02	197 2 30	89 11 45
14	TU	+18 33 43	36.0	+ 3 42.7	-0.00	198 1 31	89 11 45
15	WE	+18 48 8	35.3	+ 3 42.6	-0.03	199 0 31	89 11 45
16	TH	+19 2 14	34.4	+ 3 41.8	-0.05	199 59 30	89 11 44
17	FR	+19 16 0	33.6	+ 3 40.5	-0.08	200 58 28	89 11 44
18	SA	+19 29 27	32.8	+ 3 38.6	-0.10	201 57 24	89 11 44
19	SU	+19 42 35	32.0	+ 3 36.2	-0.13	202 56 19	89 11 44
20	MO	+19 55 22	31.1	+ 3 33.2	-0.15	203 55 12	89 11 43
21	TU	+20 7 49	30.3	+ 3 29.7	-0.17	204 54 5	89 11 43
22	WE	+20 19 55	29.4	+ 3 25.6	-0.19	205 52 58	89 11 43
23	TH	+20 31 40	28.5	+ 3 21.0	-0.21	206 51 51	89 11 43
24	FR	+20 43 4	27.6	+ 3 15.9	-0.23	207 50 46	89 11 42
25	SA	+20 54 7	26.7	+ 3 10.3	-0.25	208 49 41	89 11 42
26	SU	+21 4 48	25.8	+ 3 4.2	-0.27	209 48 37	89 11 42
27	MO	+21 15 7	24.8	+ 2 57.7	-0.29	210 47 33	89 11 42
28	TU	+21 25 3	24.0	+ 2 50.7	-0.31	211 46 30	89 11 41
29	WE	+21 34 38	23.0	+ 2 43.2	-0.33	212 45 26	89 11 41
30	TH	+21 43 50	22.0	+ 2 35.4	-0.35	213 44 21	89 11 41
31	FR	+21 52 39	21.1	+ 2 27.1	-0.36	214 43 13	89 11 41

JUNE 1985

Date		THE SUN Declination At 0 hrs, UT ° ' "	Δ, 1 hr "	Equation of Time At 0 hrs, UT m s	Δ, 1 hr s	POLARIS GHA at 0 hr ° ' "	Declination ° ' "
1	SA	+22 1 6	20.1	+ 2 18.4	-0.38	215 42 3	89 11 41
2	SU	+22 9 9	19.2	+ 2 9.3	-0.40	216 40 51	89 11 40
3	MO	+22 16 50	18.2	+ 1 59.8	-0.41	217 39 39	89 11 40
4	TU	+22 24 7	17.2	+ 1 50.0	-0.43	218 38 27	89 11 40
5	WE	+22 31 0	16.3	+ 1 39.8	-0.44	219 37 16	89 11 40
6	TH	+22 37 30	15.3	+ 1 29.3	-0.45	220 36 6	89 11 40
7	FR	+22 43 37	14.3	+ 1 18.5	-0.47	221 34 58	89 11 40
8	SA	+22 49 19	13.3	+ 1 7.3	-0.47	222 33 51	89 11 40
9	SU	+22 54 38	12.3	+ 0 55.9	-0.49	223 32 44	89 11 39
10	MO	+22 59 32	11.3	+ 0 44.1	-0.50	224 31 37	89 11 39
11	TU	+23 4 2	10.3	+ 0 32.2	-0.51	225 30 29	89 11 39
12	WE	+23 8 8	9.2	+ 0 20.0	-0.52	226 29 20	89 11 39
13	TH	+23 11 49	8.2	+ 0 7.6	-0.52	227 28 10	89 11 39
14	FR	+23 15 6	7.2	- 0 5.5	-0.53	228 26 58	89 11 38
15	SA	+23 17 58	6.2	- 0 17.8	-0.54	229 25 45	89 11 38
16	SU	+23 20 26	5.1	- 0 30.7	-0.54	230 24 31	89 11 38
17	MO	+23 22 29	4.1	- 0 43.7	-0.55	231 23 16	89 11 38
18	TU	+23 24 7	3.0	- 0 56.7	-0.55	232 22 1	89 11 38
19	WE	+23 25 20	2.0	- 1 9.9	-0.55	233 20 47	89 11 38
20	TH	+23 26 8	1.0	- 1 23.0	-0.55	234 19 33	89 11 38
21	FR	+23 26 32	- 0.0	- 1 36.2	-0.55	235 18 21	89 11 38
22	SA	+23 26 31	- 1.1	- 1 49.3	-0.54	236 17 10	89 11 38
23	SU	+23 26 5	- 2.1	- 2 2.3	-0.54	237 15 59	89 11 37
24	MO	+23 25 14	- 3.2	- 2 15.3	-0.53	238 14 49	89 11 37
25	TU	+23 23 58	- 4.2	- 2 28.2	-0.53	239 13 39	89 11 37
26	WE	+23 22 18	- 5.2	- 2 40.9	-0.53	240 12 27	89 11 37
27	TH	+23 20 13	- 6.3	- 2 53.5	-0.52	241 11 14	89 11 37
28	FR	+23 17 43	- 7.3	- 3 5.9	-0.50	242 9 59	89 11 37
29	SA	+23 14 49	- 8.3	- 3 18.0	-0.50	243 8 42	89 11 37
30	SU	+23 11 30	- 9.3	- 3 30.0	-0.49	244 7 24	89 11 37

JULY 1985

Date		THE SUN Declination At 0 hrs, UT ° ' "	Δ, 1 hr "	Equation of Time At 0 hrs, UT m s	Δ, 1 hr s	POLARIS GHA at 0 hr ° ' "	Declination ° ' "
1	MO	+23 7 47	-10.3	- 3 41.7	-0.48	245 6 6	89 11 37
2	TU	+23 3 40	-11.3	- 3 53.2	-0.47	246 4 48	89 11 37
3	WE	+22 59 8	-12.3	- 4 4.4	-0.45	247 3 33	89 11 37
4	TH	+22 54 13	-13.3	- 4 15.3	-0.44	248 2 19	89 11 37
5	FR	+22 48 54	-14.3	- 4 25.9	-0.42	249 1 6	89 11 37
6	SA	+22 43 11	-15.3	- 4 36.1	-0.42	249 59 54	89 11 37
7	SU	+22 37 4	-16.3	- 4 46.1	-0.40	250 58 42	89 11 37
8	MO	+22 30 34	-17.2	- 4 55.6	-0.38	251 57 30	89 11 37
9	TU	+22 23 41	-18.2	- 5 4.8	-0.37	252 56 17	89 11 37
10	WE	+22 16 24	-19.2	- 5 13.7	-0.35	253 55 3	89 11 36
11	TH	+22 8 44	-20.1	- 5 22.1	-0.33	254 53 47	89 11 36
12	FR	+22 0 42	-21.1	- 5 30.1	-0.31	255 52 30	89 11 36
13	SA	+21 52 16	-22.0	- 5 37.6	-0.30	256 51 12	89 11 36
14	SU	+21 43 28	-22.9	- 5 44.7	-0.27	257 49 54	89 11 36
15	MO	+21 34 18	-23.8	- 5 51.3	-0.26	258 48 36	89 11 36
16	TU	+21 24 46	-24.8	- 5 57.5	-0.23	259 47 18	89 11 36
17	WE	+21 14 52	-25.7	- 6 3.1	-0.21	260 46 0	89 11 36
18	TH	+21 4 36	-26.5	- 6 8.2	-0.19	261 44 45	89 11 37
19	FR	+20 53 59	-27.5	- 6 12.8	-0.17	262 43 30	89 11 37
20	SA	+20 43 0	-28.3	- 6 16.8	-0.15	263 42 17	89 11 37
21	SU	+20 31 41	-29.2	- 6 20.3	-0.12	264 41 4	89 11 37
22	MO	+20 20 1	-30.0	- 6 23.2	-0.10	265 39 51	89 11 37
23	TU	+20 8 1	-30.9	- 6 25.5	-0.07	266 38 37	89 11 37
24	WE	+19 55 40	-31.7	- 6 27.2	-0.04	267 37 22	89 11 37
25	TH	+19 43 0	-32.5	- 6 28.2	-0.02	268 36 5	89 11 37
26	FR	+19 29 59	-33.3	- 6 28.7	0.00	269 34 47	89 11 37
27	SA	+19 16 40	-34.1	- 6 28.6	0.03	270 33 27	89 11 37
28	SU	+19 3 1	-34.9	- 6 27.8	0.05	271 32 8	89 11 37
29	MO	+18 49 4	-35.7	- 6 26.5	0.08	272 30 49	89 11 37
30	TU	+18 34 48	-36.4	- 6 24.5	0.11	273 29 31	89 11 37
31	WE	+18 20 15	-37.2	- 6 21.9	0.13	274 28 16	89 11 37

AUGUST 1985

Date		THE SUN Declination At 0 hrs, UT ° ' "	Δ, 1 hr "	Equation of Time At 0 hrs, UT m s	Δ, 1 hr s	POLARIS GHA at 0 hr ° ' "	Declination ° ' "
1	TH	+18 5 23	-37.9	- 6 18.7	0.16	275 27 1	89 11 38
2	FR	+17 50 13	-38.6	- 6 14.8	0.18	276 25 48	89 11 38
3	SA	+17 34 46	-39.3	- 6 10.4	0.21	277 24 36	89 11 38
4	SU	+17 19 2	-40.0	- 6 5.4	0.23	278 23 23	89 11 38
5	MO	+17 3 1	-40.7	- 5 59.8	0.26	279 22 10	89 11 38
6	TU	+16 46 44	-41.4	- 5 53.6	0.28	280 20 56	89 11 38
7	WE	+16 30 10	-42.0	- 5 46.8	0.30	281 19 40	89 11 38
8	TH	+16 13 20	-42.8	- 5 39.5	0.33	282 18 24	89 11 38
9	FR	+15 56 14	-43.4	- 5 31.5	0.35	283 17 4	89 11 38
10	SA	+15 38 53	-44.0	- 5 23.0	0.38	284 15 48	89 11 39
11	SU	+15 21 17	-44.6	- 5 14.0	0.40	285 14 30	89 11 39
12	MO	+15 3 26	-45.2	- 5 4.4	0.42	286 13 13	89 11 39
13	TU	+14 45 21	-45.8	- 4 54.3	0.45	287 11 56	89 11 39
14	WE	+14 27 1	-46.4	- 4 43.6	0.47	288 10 40	89 11 39
15	TH	+14 8 27	-47.0	- 4 32.4	0.49	289 9 26	89 11 39
16	FR	+13 49 40	-47.5	- 4 20.6	0.51	290 8 14	89 11 40
17	SA	+13 30 40	-48.0	- 4 8.3	0.53	291 7 2	89 11 40
18	SU	+13 11 27	-48.6	- 3 55.6	0.56	292 5 51	89 11 40
19	MO	+12 52 1	-49.1	- 3 42.2	0.58	293 4 39	89 11 40
20	TU	+12 32 23	-49.5	- 3 28.4	0.60	294 3 26	89 11 40
21	WE	+12 12 34	-50.1	- 3 14.1	0.61	295 2 11	89 11 41
22	TH	+11 52 32	-50.5	- 2 59.4	0.64	296 0 55	89 11 41
23	FR	+11 32 20	-51.0	- 2 44.1	0.65	296 59 37	89 11 41
24	SA	+11 11 56	-51.4	- 2 28.4	0.67	297 58 20	89 11 41
25	SU	+10 51 23	-51.8	- 2 12.2	0.69	298 57 3	89 11 41
26	MO	+10 30 39	-52.3	- 1 55.6	0.71	299 55 48	89 11 42
27	TU	+10 9 45	-52.7	- 1 38.6	0.72	300 54 34	89 11 42
28	WE	+ 9 48 41	-53.0	- 1 21.2	0.74	301 53 22	89 11 42
29	TH	+ 9 27 28	-53.4	- 1 3.4	0.75	302 52 12	89 11 42
30	FR	+ 9 6 7	-53.8	- 0 45.3	0.77	303 51 2	89 11 43
31	SA	+ 8 44 36	-54.1	- 0 26.8	0.78	304 49 53	89 11 43

SEPTEMBER 1985

Date		THE SUN Declination At 0 hrs, UT ° ' "	Δ, 1 hr "	Equation of Time At 0 hrs, UT m s	Δ, 1 hr s	POLARIS GHA at 0 hr ° ' "	Declination ° ' "
1	SU	+ 8 22 58	-54.5	- 0 8.0	0.79	305 48 43	89 11 43
2	MO	+ 8 1 11	-54.8	+ 0 11.0	0.81	306 47 32	89 11 43
3	TU	+ 7 39 16	-55.1	+ 0 30.4	0.82	307 46 21	89 11 44
4	WE	+ 7 17 14	-55.4	+ 0 50.0	0.83	308 45 8	89 11 44
5	TH	+ 6 55 5	-55.7	+ 1 9.9	0.84	309 43 55	89 11 44
6	FR	+ 6 32 49	-56.0	+ 1 30.0	0.84	310 42 41	89 11 44
7	SA	+ 6 10 26	-56.2	+ 1 50.2	0.85	311 41 27	89 11 45
8	SU	+ 5 47 57	-56.5	+ 2 10.7	0.86	312 40 13	89 11 45
9	MO	+ 5 25 22	-56.7	+ 2 31.3	0.87	313 39 0	89 11 45
10	TU	+ 5 2 42	-56.9	+ 2 52.1	0.88	314 37 49	89 11 45
11	WE	+ 4 39 56	-57.1	+ 3 13.0	0.88	315 36 39	89 11 46
12	TH	+ 4 17 6	-57.3	+ 3 34.0	0.88	316 35 30	89 11 46
13	FR	+ 3 54 10	-57.5	+ 3 55.1	0.88	317 34 23	89 11 46
14	SA	+ 3 31 11	-57.6	+ 4 16.2	0.88	318 33 16	89 11 47
15	SU	+ 3 8 8	-57.8	+ 4 37.4	0.89	319 32 10	89 11 47
16	MO	+ 2 45 1	-57.9	+ 4 58.7	0.89	320 31 3	89 11 47
17	TU	+ 2 21 51	-58.0	+ 5 20.0	0.89	321 29 54	89 11 48
18	WE	+ 1 58 38	-58.1	+ 5 41.3	0.89	322 28 43	89 11 48
19	TH	+ 1 35 23	-58.2	+ 6 2.6	0.89	323 27 31	89 11 48
20	FR	+ 1 12 5	-58.3	+ 6 23.9	0.88	324 26 19	89 11 49
21	SA	+ 0 48 47	-58.4	+ 6 45.2	0.88	325 25 5	89 11 49
22	SU	+ 0 25 26	-58.4	+ 7 6.4	0.88	326 23 57	89 11 49
23	MO	+ 0 2 5	-58.4	+ 7 27.5	0.88	327 22 49	89 11 50
24	TU	- 0 21 17	-58.5	+ 7 48.5	0.87	328 21 43	89 11 50
25	WE	- 0 44 40	-58.4	+ 8 9.4	0.86	329 20 38	89 11 50
26	TH	- 1 8 2	-58.4	+ 8 30.1	0.86	330 19 35	89 11 51
27	FR	- 1 31 24	-58.4	+ 8 50.7	0.85	331 18 32	89 11 51
28	SA	- 1 54 45	-58.4	+ 9 11.1	0.84	332 17 28	89 11 51
29	SU	- 2 18 6	-58.3	+ 9 31.2	0.83	333 16 25	89 11 52
30	MO	- 2 41 25	-58.3	+ 9 51.1	0.82	334 15 20	89 11 52

OCTOBER 1985

Date		THE SUN Declination At 0 hrs, UT ° ' "	Δ, 1 hr "	Equation of Time At 0 hrs, UT m s	Δ, 1 hr s	POLARIS GHA at 0 hr ° ' "	Declination ° ' "
1	TU	− 3 4 43	−58.1	+10 10.8	0.80	335 14 15	89 11 52
2	WE	− 3 27 58	−58.1	+10 30.1	0.80	336 13 8	89 11 52
3	TH	− 3 51 12	−58.0	+10 49.2	0.78	337 12 2	89 11 53
4	FR	− 4 14 23	−57.8	+11 7.9	0.76	338 10 55	89 11 53
5	SA	− 4 37 31	−57.7	+11 26.2	0.75	339 9 48	89 11 53
6	SU	− 5 0 35	−57.5	+11 44.2	0.73	340 8 42	89 11 54
7	MO	− 5 23 36	−57.4	+12 1.8	0.72	341 7 38	89 11 54
8	TU	− 5 46 33	−57.2	+12 19.0	0.70	342 6 35	89 11 55
9	WE	− 6 9 26	−57.0	+12 35.7	0.68	343 5 33	89 11 55
10	TH	− 6 32 14	−56.8	+12 52.0	0.65	344 4 33	89 11 55
11	FR	− 6 54 57	−56.5	+13 7.7	0.64	345 3 34	89 11 56
12	SA	− 7 17 34	−56.3	+13 23.0	0.62	346 2 36	89 11 56
13	SU	− 7 40 6	−56.0	+13 37.8	0.60	347 1 37	89 11 57
14	MO	− 8 2 31	−55.8	+13 52.1	0.57	348 0 37	89 11 57
15	TU	− 8 24 50	−55.5	+14 5.8	0.55	348 59 35	89 11 57
16	WE	− 8 47 1	−55.2	+14 19.0	0.52	349 58 32	89 11 58
17	TH	− 9 9 5	−54.8	+14 31.6	0.50	350 57 28	89 11 58
18	FR	− 9 31 1	−54.5	+14 43.7	0.47	351 56 24	89 11 58
19	SA	− 9 52 48	−54.1	+14 55.1	0.45	352 55 22	89 11 59
20	SU	−10 14 27	−53.8	+15 6.0	0.42	353 54 22	89 11 59
21	MO	−10 35 57	−53.3	+15 16.2	0.40	354 53 24	89 12 0
22	TU	−10 57 17	−52.9	+15 25.8	0.37	355 52 28	89 12 0
23	WE	−11 18 27	−52.5	+15 34.7	0.35	356 51 33	89 12 0
24	TH	−11 39 27	−52.0	+15 43.0	0.31	357 50 39	89 12 1
25	FR	−12 0 16	−51.6	+15 50.5	0.29	358 49 45	89 12 1
26	SA	−12 20 54	−51.1	+15 57.4	0.26	359 48 51	89 12 1
27	SU	−12 41 21	−50.6	+16 3.6	0.23	0 47 56	89 12 2
28	MO	−13 1 36	−50.1	+16 9.0	0.20	1 47 0	89 12 2
29	TU	−13 21 38	−49.6	+16 13.7	0.16	2 46 3	89 12 3
30	WE	−13 41 28	−49.0	+16 17.6	0.13	3 45 6	89 12 3
31	TH	−14 1 5	−48.5	+16 20.7	0.10	4 44 9	89 12 3

NOVEMBER 1985

Date		THE SUN Declination At 0 hrs, UT ° ' "	Δ, 1 hr "	Equation of Time At 0 hrs, UT m s	Δ, 1 hr s	POLARIS GHA at 0 hr ° ' "	Declination ° ' "
1	FR	−14 20 29	−47.9	+16 23.0	0.07	5 43 12	89 12 4
2	SA	−14 39 39	−47.3	+16 24.5	0.03	6 42 16	89 12 4
3	SU	−14 58 35	−46.7	+16 25.3	−0.00	7 41 20	89 12 4
4	MO	−15 17 16	−46.1	+16 25.2	−0.04	8 40 26	89 12 5
5	TU	−15 35 42	−45.5	+16 24.2	−0.08	9 39 34	89 12 5
6	WE	−15 53 53	−44.8	+16 22.4	−0.11	10 38 44	89 12 6
7	TH	−16 11 49	−44.1	+16 19.8	−0.15	11 37 55	89 12 6
8	FR	−16 29 28	−43.4	+16 16.3	−0.18	12 37 7	89 12 6
9	SA	−16 46 50	−42.8	+16 11.9	−0.22	13 36 18	89 12 7
10	SU	−17 3 56	−42.0	+16 6.7	−0.25	14 35 29	89 12 7
11	MO	−17 20 44	−41.3	+16 0.6	−0.29	15 34 39	89 12 8
12	TU	−17 37 15	−40.5	+15 53.7	−0.33	16 33 46	89 12 8
13	WE	−17 53 27	−39.7	+15 45.9	−0.36	17 32 53	89 12 8
14	TH	−18 9 20	−38.9	+15 37.2	−0.40	18 31 59	89 12 9
15	FR	−18 24 54	−38.1	+15 27.7	−0.43	19 31 6	89 12 9
16	SA	−18 40 9	−37.3	+15 17.4	−0.46	20 30 16	89 12 9
17	SU	−18 55 4	−36.5	+15 6.3	−0.50	21 29 28	89 12 10
18	MO	−19 9 39	−35.6	+14 54.3	−0.53	22 28 42	89 12 10
19	TU	−19 23 53	−34.7	+14 41.5	−0.56	23 27 57	89 12 11
20	WE	−19 37 46	−33.8	+14 28.0	−0.60	24 27 14	89 12 11
21	TH	−19 51 17	−32.9	+14 13.6	−0.63	25 26 31	89 12 12
22	FR	−20 4 27	−32.0	+13 58.4	−0.66	26 25 47	89 12 12
23	SA	−20 17 15	−31.0	+13 42.4	−0.70	27 25 3	89 12 12
24	SU	−20 29 40	−30.1	+13 25.7	−0.73	28 24 18	89 12 12
25	MO	−20 41 42	−29.1	+13 8.3	−0.76	29 23 33	89 12 13
26	TU	−20 53 21	−28.2	+12 50.0	−0.79	30 22 47	89 12 13
27	WE	−21 4 37	−27.2	+12 31.1	−0.82	31 22 0	89 12 13
28	TH	−21 15 29	−26.2	+12 11.4	−0.85	32 21 14	89 12 14
29	FR	−21 25 57	−25.2	+11 51.0	−0.88	33 20 28	89 12 14
30	SA	−21 36 1	−24.1	+11 29.9	−0.91	34 19 43	89 12 14

DECEMBER 1985

Date		THE SUN Declination At 0 hrs, UT ° ' "	Δ, 1 hr "	Equation of Time At 0 hrs, UT m s	Δ, 1 hr s	POLARIS GHA at 0 hr ° ' "	Declination ° ' "
1	SU	−21 45 40	−23.1	+11 8.1	−0.93	35 19 0	89 12 15
2	MO	−21 54 55	−22.0	+10 45.7	−0.96	36 18 19	89 12 15
3	TU	−22 3 44	−21.0	+10 22.6	−0.99	37 17 39	89 12 15
4	WE	−22 12 7	−19.9	+ 9 58.9	−1.01	38 17 0	89 12 16
5	TH	−22 20 5	−18.8	+ 9 34.6	−1.04	39 16 23	89 12 16
6	FR	−22 27 37	−17.8	+ 9 9.7	−1.06	40 15 45	89 12 17
7	SA	−22 34 43	−16.7	+ 8 44.2	−1.08	41 15 7	89 12 17
8	SU	−22 41 23	−15.5	+ 8 18.2	−1.10	42 14 28	89 12 17
9	MO	−22 47 35	−14.4	+ 7 51.7	−1.12	43 13 47	89 12 17
10	TU	−22 53 21	−13.3	+ 7 24.8	−1.14	44 13 5	89 12 18
11	WE	−22 58 40	−12.1	+ 6 57.4	−1.16	45 12 22	89 12 18
12	TH	−23 3 31	−11.0	+ 6 29.6	−1.17	46 11 36	89 12 19
13	FR	−23 7 55	− 9.8	+ 6 1.5	−1.19	47 10 50	89 12 19
14	SA	−23 11 51	− 8.7	+ 5 33.0	−1.20	48 10 20	89 12 19
15	SU	−23 15 20	− 7.5	+ 5 4.2	−1.21	49 9 44	89 12 20
16	MO	−23 18 21	− 6.4	+ 4 35.2	−1.22	50 9 9	89 12 20
17	TU	−23 20 54	− 5.2	+ 4 6.0	−1.23	51 8 36	89 12 20
18	WE	−23 22 58	− 4.0	+ 3 36.5	−1.23	52 8 3	89 12 20
19	TH	−23 24 35	− 2.8	+ 3 7.0	−1.24	53 7 30	89 12 21
20	FR	−23 25 43	− 1.7	+ 2 37.3	−1.24	54 6 56	89 12 21
21	SA	−23 26 23	− 0.5	+ 2 7.7	−1.24	55 6 23	89 12 21
22	SU	−23 26 35	0.7	+ 1 37.9	−1.24	56 5 46	89 12 21
23	MO	−23 26 18	1.9	+ 1 7.9	−1.24	57 5 10	89 12 22
24	TU	−23 25 34	3.0	+ 0 38.1	−1.24	58 4 33	89 12 22
25	WE	−23 24 21	4.2	+ 0 8.3	−1.23	59 3 56	89 12 22
26	TH	−23 22 40	5.4	− 0 21.3	−1.23	60 3 20	89 12 22
27	FR	−23 20 30	6.5	− 0 50.9	−1.22	61 2 45	89 12 22
28	SA	−23 17 53	7.7	− 1 20.3	−1.22	62 2 10	89 12 23
29	SU	−23 14 48	8.9	− 1 49.6	−1.21	63 1 38	89 12 23
30	MO	−23 11 15	10.0	− 2 18.7	−1.20	64 1 6	89 12 23
31	TU	−23 7 14	11.2	− 2 47.5	−1.19	65 0 37	89 12 23

JANUARY 1986

Date		Declination At 0 hrs, UT ° ' "	Δ, 1 hr "	Equation of Time At 0 hrs, UT m s	Δ, 1 hr s	POLARIS GHA at 0 hr ° ' "	Declination ° ' "
1	WE	-23 2 45	12.3	- 3 16.1	-1.18	66 0 8	89 12 23
2	TH	-22 57 49	13.5	- 3 44.4	-1.16	66 59 39	89 12 24
3	FR	-22 52 25	14.6	- 4 12.3	-1.15	67 59 10	89 12 24
4	SA	-22 46 34	15.8	- 4 40.0	-1.14	68 58 40	89 12 24
5	SU	-22 40 16	16.9	- 5 7.2	-1.12	69 58 8	89 12 24
6	MO	-22 33 30	18.0	- 5 34.1	-1.10	70 57 34	89 12 24
7	TU	-22 26 18	19.1	- 6 0.5	-1.08	71 57 0	89 12 24
8	WE	-22 18 39	20.2	- 6 26.5	-1.06	72 56 25	89 12 25
9	TH	-22 10 34	21.3	- 6 52.0	-1.04	73 55 51	89 12 25
10	FR	-22 2 3	22.4	- 7 17.0	-1.02	74 55 19	89 12 25
11	SA	-21 53 6	23.5	- 7 41.4	-0.99	75 54 50	89 12 25
12	SU	-21 43 43	24.5	- 8 5.2	-0.97	76 54 22	89 12 25
13	MO	-21 33 55	25.5	- 8 28.4	-0.94	77 53 55	89 12 25
14	TU	-21 23 42	26.5	- 8 51.0	-0.91	78 53 29	89 12 26
15	WE	-21 13 5	27.6	- 9 12.9	-0.89	79 53 3	89 12 26
16	TH	-21 2 2	28.6	- 9 34.2	-0.85	80 52 36	89 12 26
17	FR	-20 50 36	29.6	- 9 54.7	-0.83	81 52 8	89 12 26
18	SA	-20 38 46	30.6	-10 14.6	-0.80	82 51 39	89 12 26
19	SU	-20 26 32	31.5	-10 33.7	-0.76	83 51 9	89 12 26
20	MO	-20 13 56	32.5	-10 52.0	-0.73	84 50 36	89 12 26
21	TU	-20 0 56	33.4	-11 9.6	-0.70	85 50 6	89 12 26
22	WE	-19 47 35	34.3	-11 26.4	-0.67	86 49 35	89 12 26
23	TH	-19 33 51	35.3	-11 42.5	-0.63	87 49 5	89 12 26
24	FR	-19 19 45	36.1	-11 57.7	-0.60	88 48 35	89 12 26
25	SA	-19 5 18	37.0	-12 12.2	-0.57	89 48 6	89 12 26
26	SU	-18 50 31	37.8	-12 25.9	-0.54	90 47 39	89 12 26
27	MO	-18 35 23	38.7	-12 38.8	-0.50	91 47 13	89 12 27
28	TU	-18 19 54	39.5	-12 50.8	-0.47	92 46 48	89 12 27
29	WE	-18 4 6	40.3	-13 2.1	-0.44	93 46 23	89 12 27
30	TH	-17 47 58	41.1	-13 12.6	-0.40	94 45 57	89 12 27
31	FR	-17 31 31	41.9	-13 22.2	-0.37	95 45 31	89 12 27

FEBRUARY 1986

Date		Declination At 0 hrs, UT ° ' "	Δ, 1 hr "	Equation of Time At 0 hrs, UT m s	Δ, 1 hr s	POLARIS GHA at 0 hr ° ' "	Declination ° ' "
1	SA	-17 14 45	42.6	-13 31.1	-0.34	96 45 2	89 12 27
2	SU	-16 57 42	43.4	-13 39.2	-0.30	97 44 32	89 12 27
3	MO	-16 40 20	44.2	-13 46.5	-0.27	98 44 0	89 12 27
4	TU	-16 22 40	44.9	-13 52.9	-0.24	99 43 28	89 12 26
5	WE	-16 4 43	45.5	-13 58.6	-0.20	100 42 56	89 12 26
6	TH	-15 46 30	46.3	-14 3.5	-0.17	101 42 25	89 12 26
7	FR	-15 28 0	46.8	-14 7.6	-0.14	102 41 56	89 12 27
8	SA	-15 9 15	47.5	-14 10.9	-0.11	103 41 29	89 12 27
9	SU	-14 50 14	48.2	-14 13.5	-0.07	104 41 3	89 12 27
10	MO	-14 30 58	48.8	-14 15.2	-0.04	105 40 38	89 12 26
11	TU	-14 11 27	49.3	-14 16.1	-0.01	106 40 13	89 12 26
12	WE	-13 51 43	50.0	-14 16.3	0.02	107 39 46	89 12 26
13	TH	-13 31 44	50.5	-14 15.7	0.06	108 39 19	89 12 26
14	FR	-13 11 33	51.0	-14 14.3	0.09	109 38 50	89 12 26
15	SA	-12 51 8	51.5	-14 12.2	0.12	110 38 19	89 12 26
16	SU	-12 30 32	52.0	-14 9.3	0.15	111 37 48	89 12 26
17	MO	-12 9 43	52.5	-14 5.7	0.18	112 37 16	89 12 26
18	TU	-11 48 43	53.0	-14 1.3	0.21	113 36 43	89 12 25
19	WE	-11 27 31	53.4	-13 56.3	0.24	114 36 11	89 12 25
20	TH	-11 6 9	53.8	-13 50.6	0.27	115 35 39	89 12 25
21	FR	-10 44 37	54.3	-13 44.1	0.29	116 35 8	89 12 25
22	SA	-10 22 54	54.6	-13 37.1	0.33	117 34 38	89 12 25
23	SU	-10 1 3	55.0	-13 29.3	0.35	118 34 9	89 12 25
24	MO	- 9 39 2	55.4	-13 21.0	0.38	119 33 41	89 12 25
25	TU	- 9 16 52	55.7	-13 12.0	0.40	120 33 13	89 12 25
26	WE	- 8 54 35	56.1	-13 2.5	0.42	121 32 45	89 12 25
27	TH	- 8 32 9	56.4	-12 52.4	0.45	122 32 15	89 12 25
28	FR	- 8 9 35	56.7	-12 41.7	0.46	123 31 43	89 12 24

MARCH 1986

Date		Declination At 0 hrs, UT ° ' "	Δ, 1 hr "	Equation of Time At 0 hrs, UT m s	Δ, 1 hr s	POLARIS GHA at 0 hr ° ' "	Declination ° ' "
1	SA	- 7 46 55	57.0	-12 30.6	0.49	124 31 9	89 12 24
2	SU	- 7 24 7	57.3	-12 18.9	0.50	125 30 33	89 12 24
3	MO	- 7 1 13	57.5	-12 6.8	0.52	126 29 57	89 12 24
4	TU	- 6 38 13	57.8	-11 54.2	0.55	127 29 20	89 12 24
5	WE	- 6 15 7	58.0	-11 41.1	0.56	128 28 44	89 12 23
6	TH	- 5 51 56	58.2	-11 27.6	0.58	129 28 10	89 12 23
7	FR	- 5 28 40	58.5	-11 13.7	0.59	130 27 37	89 12 23
8	SA	- 5 5 19	58.7	-10 59.5	0.61	131 27 5	89 12 23
9	SU	- 4 41 55	58.8	-10 44.8	0.63	132 26 33	89 12 23
10	MO	- 4 18 26	58.8	-10 29.8	0.64	133 26 2	89 12 22
11	TU	- 3 54 55	59.0	-10 14.4	0.65	134 25 30	89 12 22
12	WE	- 3 31 20	59.0	- 9 58.8	0.67	135 24 56	89 12 22
13	TH	- 3 7 43	59.1	- 9 42.8	0.68	136 24 21	89 12 22
14	FR	- 2 44 4	59.2	- 9 26.5	0.69	137 23 44	89 12 22
15	SA	- 2 20 23	59.3	- 9 9.9	0.70	138 23 6	89 12 21
16	SU	- 1 56 41	59.3	- 8 53.1	0.71	139 22 26	89 12 21
17	MO	- 1 32 58	59.3	- 8 36.0	0.72	140 21 46	89 12 21
18	TU	- 1 9 15	59.3	- 8 18.8	0.73	141 21 6	89 12 20
19	WE	- 0 45 31	59.3	- 8 1.3	0.74	142 20 26	89 12 20
20	TH	- 0 21 48	59.3	- 7 43.6	0.74	143 19 47	89 12 20
21	FR	+ 0 1 55	59.3	- 7 25.8	0.75	144 19 8	89 12 20
22	SA	+ 0 25 37	59.2	- 7 7.9	0.75	145 18 30	89 12 20
23	SU	+ 0 49 18	59.1	- 6 49.8	0.75	146 17 53	89 12 19
24	MO	+ 1 12 57	59.0	- 6 31.7	0.76	147 17 17	89 12 19
25	TU	+ 1 36 33	59.0	- 6 13.4	0.76	148 16 39	89 12 19
26	WE	+ 2 0 8	58.8	- 5 55.2	0.76	149 16 1	89 12 19
27	TH	+ 2 23 40	58.7	- 5 36.9	0.76	150 15 20	89 12 18
28	FR	+ 2 47 9	58.5	- 5 18.6	0.76	151 14 37	89 12 18
29	SA	+ 3 10 34	58.5	- 5 0.4	0.76	152 13 52	89 12 18
30	SU	+ 3 33 56	58.3	- 4 42.2	0.76	153 13 6	89 12 17
31	MO	+ 3 57 15	58.1	- 4 24.0	0.75	154 12 19	89 12 17

APRIL 1986

Date		THE SUN Declination At 0 hrs, UT ° ' "	Δ, 1 hr "	Equation of Time At 0 hrs, UT m s	Δ, 1 hr s	POLARIS GHA at 0 hr ° '	Declination ° ' "
1	TU	+ 4 20 29	57.9	− 4 6.0	0.75	155 11 33	89 12 17
2	WE	+ 4 43 38	57.7	− 3 48.1	0.74	156 10 48	89 12 17
3	TH	+ 5 6 42	57.5	− 3 30.3	0.73	157 10 4	89 12 16
4	FR	+ 5 29 42	57.2	− 3 12.7	0.73	158 9 22	89 12 16
5	SA	+ 5 52 35	57.0	− 2 55.2	0.72	159 8 40	89 12 16
6	SU	+ 6 15 23	56.7	− 2 38.0	0.71	160 7 58	89 12 15
7	MO	+ 6 38 4	56.4	− 2 20.9	0.70	161 7 15	89 12 15
8	TU	+ 7 0 38	56.1	− 2 4.0	0.69	162 6 31	89 12 15
9	WE	+ 7 23 5	55.8	− 1 47.4	0.68	163 5 45	89 12 15
10	TH	+ 7 45 25	55.5	− 1 31.1	0.67	164 4 58	89 12 14
11	FR	+ 8 7 37	55.1	− 1 14.9	0.66	165 4 9	89 12 14
12	SA	+ 8 29 40	54.8	− 0 59.1	0.65	166 3 18	89 12 13
13	SU	+ 8 51 35	54.4	− 0 43.6	0.64	167 2 27	89 12 13
14	MO	+ 9 13 21	54.0	− 0 28.3	0.62	168 1 35	89 12 13
15	TU	+ 9 34 58	53.6	− 0 13.4	0.61	169 0 44	89 12 13
16	WE	+ 9 56 25	53.3	+ 0 1.2	0.59	169 59 52	89 12 12
17	TH	+10 17 43	52.8	+ 0 15.4	0.58	170 59 2	89 12 12
18	FR	+10 38 50	52.3	+ 0 29.3	0.56	171 58 12	89 12 12
19	SA	+10 59 46	51.9	+ 0 42.8	0.55	172 57 23	89 12 11
20	SU	+11 20 31	51.4	+ 0 56.0	0.53	173 56 35	89 12 11
21	MO	+11 41 5	50.9	+ 1 8.7	0.51	174 55 46	89 12 11
22	TU	+12 1 27	50.5	+ 1 21.0	0.50	175 54 56	89 12 11
23	WE	+12 21 38	49.9	+ 1 32.9	0.48	176 54 5	89 12 10
24	TH	+12 41 36	49.4	+ 1 44.4	0.45	177 53 11	89 12 10
25	FR	+13 1 21	48.9	+ 1 55.3	0.44	178 52 15	89 12 10
26	SA	+13 20 54	48.3	+ 2 5.8	0.42	179 51 16	89 12 9
27	SU	+13 40 14	47.8	+ 2 15.8	0.40	180 50 17	89 12 9
28	MO	+13 59 20	47.2	+ 2 25.3	0.38	181 49 19	89 12 9
29	TU	+14 18 12	46.6	+ 2 34.3	0.35	182 48 21	89 12 8
30	WE	+14 36 51	46.0	+ 2 42.7	0.33	183 47 25	89 12 8

MAY 1986

Date		THE SUN Declination At 0 hrs, UT ° ' "	Δ, 1 hr "	Equation of Time At 0 hrs, UT m s	Δ, 1 hr s	POLARIS GHA at 0 hr ° '	Declination ° ' "
1	TH	+14 55 15	45.4	+ 2 50.6	0.31	184 46 30	89 12 8
2	FR	+15 13 25	44.8	+ 2 58.0	0.28	185 45 36	89 12 8
3	SA	+15 31 19	44.2	+ 3 4.7	0.26	186 44 43	89 12 7
4	SU	+15 48 59	43.5	+ 3 11.0	0.23	187 43 48	89 12 7
5	MO	+16 6 22	42.8	+ 3 16.6	0.21	188 42 52	89 12 7
6	TU	+16 23 30	42.1	+ 3 21.7	0.19	189 41 55	89 12 6
7	WE	+16 40 21	41.5	+ 3 26.2	0.16	190 40 57	89 12 6
8	TH	+16 56 56	40.7	+ 3 30.1	0.14	191 39 56	89 12 6
9	FR	+17 13 13	40.0	+ 3 33.5	0.12	192 38 54	89 12 5
10	SA	+17 29 14	39.3	+ 3 36.3	0.09	193 37 52	89 12 5
11	SU	+17 44 57	38.5	+ 3 38.5	0.07	194 36 48	89 12 5
12	MO	+18 0 22	37.8	+ 3 40.2	0.05	195 35 45	89 12 5
13	TU	+18 15 29	37.0	+ 3 41.3	0.02	196 34 42	89 12 4
14	WE	+18 30 17	36.2	+ 3 41.8	0.0	197 33 40	89 12 4
15	TH	+18 44 46	35.5	+ 3 41.6	−0.02	198 32 39	89 12 4
16	FR	+18 58 57	34.6	+ 3 41.1	−0.05	199 31 38	89 12 4
17	SA	+19 12 48	33.8	+ 3 40.1	−0.07	200 30 38	89 12 3
18	SU	+19 26 19	33.0	+ 3 38.4	−0.09	201 29 38	89 12 3
19	MO	+19 39 31	32.1	+ 3 36.2	−0.12	202 28 38	89 12 3
20	TU	+19 52 22	31.3	+ 3 33.4	−0.13	203 27 36	89 12 3
21	WE	+20 4 53	30.4	+ 3 30.2	−0.16	204 26 32	89 12 2
22	TH	+20 17 3	29.6	+ 3 26.4	−0.18	205 25 26	89 12 2
23	FR	+20 28 53	28.7	+ 3 22.1	−0.20	206 24 17	89 12 2
24	SA	+20 40 21	27.8	+ 3 17.3	−0.22	207 23 7	89 12 1
25	SU	+20 51 29	26.9	+ 3 11.9	−0.24	208 21 57	89 12 1
26	MO	+21 2 14	26.0	+ 3 6.1	−0.27	209 20 49	89 12 1
27	TU	+21 12 38	25.1	+ 2 59.7	−0.28	210 19 41	89 12 1
28	WE	+21 22 41	24.2	+ 2 52.9	−0.30	211 18 36	89 12 1
29	TH	+21 32 21	23.3	+ 2 45.6	−0.33	212 17 31	89 12 1
30	FR	+21 41 39	22.3	+ 2 37.8	−0.34	213 16 28	89 12 0
31	SA	+21 50 34	21.3	+ 2 29.6	−0.36	214 15 23	89 12 0

JUNE 1986

Date		THE SUN Declination At 0 hrs, UT ° ' "	Δ, 1 hr "	Equation of Time At 0 hrs, UT m s	Δ, 1 hr s	POLARIS GHA at 0 hr ° '	Declination ° ' "
1	SU	+21 59 6	20.4	+ 2 20.9	−0.38	215 14 18	89 12 0
2	MO	+22 7 16	19.5	+ 2 11.8	−0.40	216 13 12	89 12 0
3	TU	+22 15 3	18.5	+ 2 2.2	−0.41	217 12 3	89 12 0
4	WE	+22 22 26	17.5	+ 1 52.3	−0.42	218 10 54	89 11 59
5	TH	+22 29 25	16.5	+ 1 42.1	−0.45	219 9 43	89 11 59
6	FR	+22 36 2	15.5	+ 1 31.4	−0.45	220 8 31	89 11 59
7	SA	+22 42 14	14.5	+ 1 20.5	−0.47	221 7 19	89 11 59
8	SU	+22 48 2	13.5	+ 1 9.2	−0.49	222 6 6	89 11 59
9	MO	+22 53 27	12.5	+ 0 57.7	−0.49	223 4 54	89 11 58
10	TU	+22 58 27	11.5	+ 0 45.9	−0.50	224 3 43	89 11 58
11	WE	+23 3 3	10.5	+ 0 33.9	−0.51	225 2 33	89 11 58
12	TH	+23 7 14	9.5	+ 0 21.7	−0.52	226 1 23	89 11 58
13	FR	+23 11 2	8.4	+ 0 9.3	−0.52	227 0 15	89 11 58
14	SA	+23 14 24	7.4	− 0 3.2	−0.53	227 59 6	89 11 58
15	SU	+23 17 22	6.4	− 0 15.9	−0.53	228 57 58	89 11 58
16	MO	+23 19 55	5.3	− 0 28.7	−0.53	229 56 48	89 11 57
17	TU	+23 22 3	4.3	− 0 41.5	−0.54	230 55 37	89 11 57
18	WE	+23 23 47	3.3	− 0 54.4	−0.54	231 54 24	89 11 57
19	TH	+23 25 6	2.3	− 1 7.4	−0.54	232 53 9	89 11 57
20	FR	+23 26 0	1.2	− 1 20.4	−0.54	233 51 52	89 11 57
21	SA	+23 26 29	0.2	− 1 33.3	−0.54	234 50 35	89 11 57
22	SU	+23 26 34	− 0.8	− 1 46.3	−0.54	235 49 18	89 11 57
23	MO	+23 26 14	− 1.9	− 1 59.2	−0.53	236 48 3	89 11 57
24	TU	+23 25 29	− 2.9	− 2 12.1	−0.53	237 46 50	89 11 56
25	WE	+23 24 19	− 3.9	− 2 24.8	−0.53	238 45 38	89 11 56
26	TH	+23 22 45	− 5.0	− 2 37.5	−0.52	239 44 28	89 11 56
27	FR	+23 20 46	− 6.0	− 2 50.1	−0.52	240 43 17	89 11 56
28	SA	+23 18 23	− 7.0	− 3 2.5	−0.51	241 42 6	89 11 56
29	SU	+23 15 35	− 8.0	− 3 14.8	−0.50	242 40 54	89 11 56
30	MO	+23 12 22	− 9.0	− 3 26.9	−0.50	243 39 40	89 11 56

JULY 1986

Date		THE SUN Declination At 0 hrs, UT ° ' "	Δ, 1 hr "	Equation of Time At 0 hrs, UT m s	Δ, 1 hr s	POLARIS GHA at 0 hr ° ' "	Declination ° ' "
1	TU	+23 8 45	-10.0	- 3 38.8	-0.48	244 38 25	89 11 56
2	WE	+23 4 44	-11.1	- 3 50.4	-0.47	245 37 9	89 11 56
3	TH	+23 0 18	-12.0	- 4 1.8	-0.47	246 35 51	89 11 56
4	FR	+22 55 29	-13.1	- 4 13.0	-0.45	247 34 34	89 11 56
5	SA	+22 50 15	-14.1	- 4 23.8	-0.44	248 33 16	89 11 56
6	SU	+22 44 37	-15.0	- 4 34.3	-0.43	249 31 59	89 11 56
7	MO	+22 38 36	-16.0	- 4 44.5	-0.41	250 30 42	89 11 56
8	TU	+22 32 11	-17.0	- 4 54.3	-0.39	251 29 27	89 11 56
9	WE	+22 25 23	-18.0	- 5 3.7	-0.38	252 28 13	89 11 56
10	TH	+22 18 11	-18.9	- 5 12.7	-0.36	253 27 0	89 11 56
11	FR	+22 10 37	-19.9	- 5 21.3	-0.34	254 25 47	89 11 56
12	SA	+22 2 39	-20.8	- 5 29.4	-0.32	255 24 34	89 11 56
13	SU	+21 54 19	-21.8	- 5 37.1	-0.30	256 23 21	89 11 56
14	MO	+21 45 37	-22.7	- 5 44.3	-0.28	257 22 6	89 11 56
15	TU	+21 36 32	-23.7	- 5 50.9	-0.26	258 20 50	89 11 56
16	WE	+21 27 4	-24.5	- 5 57.1	-0.23	259 19 32	89 11 55
17	TH	+21 17 16	-25.5	- 6 2.7	-0.21	260 18 13	89 11 56
18	FR	+21 7 5	-26.3	- 6 7.8	-0.19	261 16 53	89 11 56
19	SA	+20 56 33	-27.2	- 6 12.3	-0.17	262 15 33	89 11 56
20	SU	+20 45 40	-28.1	- 6 16.3	-0.14	263 14 14	89 11 56
21	MO	+20 34 26	-28.9	- 6 19.7	-0.12	264 12 57	89 11 56
22	TU	+20 22 52	-29.8	- 6 22.5	-0.10	265 11 42	89 11 56
23	WE	+20 10 57	-30.6	- 6 24.8	-0.07	266 10 29	89 11 56
24	TH	+19 58 42	-31.5	- 6 26.5	-0.05	267 9 16	89 11 56
25	FR	+19 46 7	-32.3	- 6 27.7	-0.02	268 8 3	89 11 56
26	SA	+19 33 12	-33.1	- 6 28.2	0.00	269 6 49	89 11 56
27	SU	+19 19 57	-33.9	- 6 28.2	0.02	270 5 34	89 11 56
28	MO	+19 6 24	-34.7	- 6 27.7	0.05	271 4 17	89 11 56
29	TU	+18 52 31	-35.5	- 6 26.5	0.08	272 3 0	89 11 56
30	WE	+18 38 19	-36.2	- 6 24.7	0.10	273 1 41	89 11 56
31	TH	+18 23 50	-37.0	- 6 22.4	0.13	274 0 22	89 11 56

AUGUST 1986

Date		THE SUN Declination At 0 hrs, UT ° ' "	Δ, 1 hr "	Equation of Time At 0 hrs, UT m s	Δ, 1 hr s	POLARIS GHA at 0 hr ° ' "	Declination ° ' "
1	FR	+18 9 2	-37.8	- 6 19.4	0.15	274 59 3	89 11 56
2	SA	+17 53 56	-38.5	- 6 15.9	0.17	275 57 45	89 11 57
3	SU	+17 38 32	-39.2	- 6 11.8	0.20	276 56 27	89 11 57
4	MO	+17 22 52	-39.9	- 6 7.1	0.22	277 55 11	89 11 57
5	TU	+17 6 54	-40.6	- 6 1.8	0.25	278 53 56	89 11 57
6	WE	+16 50 40	-41.3	- 5 55.8	0.27	279 52 42	89 11 57
7	TH	+16 34 9	-42.0	- 5 49.3	0.30	280 51 29	89 11 57
8	FR	+16 17 22	-42.6	- 5 42.2	0.32	281 50 16	89 11 57
9	SA	+16 0 20	-43.3	- 5 34.5	0.35	282 49 3	89 11 58
10	SU	+15 43 2	-43.9	- 5 26.2	0.37	283 47 49	89 11 58
11	MO	+15 25 29	-44.5	- 5 17.3	0.40	284 46 34	89 11 58
12	TU	+15 7 41	-45.1	- 5 7.8	0.42	285 45 17	89 11 58
13	WE	+14 49 39	-45.7	- 4 57.7	0.44	286 43 58	89 11 58
14	TH	+14 31 23	-46.3	- 4 47.1	0.47	287 42 39	89 11 58
15	FR	+14 12 53	-46.8	- 4 35.9	0.49	288 41 20	89 11 58
16	SA	+13 54 10	-47.3	- 4 24.1	0.51	289 40 2	89 11 59
17	SU	+13 35 14	-47.9	- 4 11.8	0.54	290 38 46	89 11 59
18	MO	+13 16 4	-48.4	- 3 58.9	0.55	291 37 31	89 11 59
19	TU	+12 56 43	-48.9	- 3 45.6	0.58	292 36 19	89 11 59
20	WE	+12 37 9	-49.4	- 3 31.7	0.60	293 35 8	89 12 0
21	TH	+12 17 23	-49.9	- 3 17.4	0.62	294 33 57	89 12 0
22	FR	+11 57 25	-50.4	- 3 2.6	0.63	295 32 45	89 12 0
23	SA	+11 37 16	-50.8	- 2 47.4	0.65	296 31 32	89 12 0
24	SU	+11 16 56	-51.3	- 2 31.7	0.67	297 30 18	89 12 0
25	MO	+10 56 25	-51.7	- 2 15.6	0.69	298 29 3	89 12 0
26	TU	+10 35 44	-52.2	- 1 59.1	0.70	299 27 47	89 12 1
27	WE	+10 14 52	-52.5	- 1 42.2	0.72	300 26 31	89 12 1
28	TH	+ 9 53 51	-53.0	- 1 25.0	0.73	301 25 14	89 12 1
29	FR	+ 9 32 40	-53.3	- 1 7.4	0.75	302 23 59	89 12 2
30	SA	+ 9 11 20	-53.7	- 0 49.5	0.76	303 22 44	89 12 2
31	SU	+ 8 49 51	-54.1	- 0 31.2	0.77	304 21 30	89 12 2

SEPTEMBER 1986

Date		THE SUN Declination At 0 hrs, UT ° ' "	Δ, 1 hr "	Equation of Time At 0 hrs, UT m s	Δ, 1 hr s	POLARIS GHA at 0 hr ° ' "	Declination ° ' "
1	MO	+ 8 28 13	-54.4	- 0 12.6	0.78	305 20 18	89 12 2
2	TU	+ 8 6 27	-54.8	+ 0 6.2	0.80	306 19 7	89 12 2
3	WE	+ 7 44 33	-55.0	+ 0 25.4	0.81	307 17 57	89 12 3
4	TH	+ 7 22 32	-55.4	+ 0 44.8	0.82	308 16 48	89 12 3
5	FR	+ 7 0 23	-55.6	+ 1 4.5	0.83	309 15 39	89 12 3
6	SA	+ 6 38 8	-55.9	+ 1 24.4	0.84	310 14 29	89 12 3
7	SU	+ 6 15 46	-56.2	+ 1 44.6	0.85	311 13 18	89 12 4
8	MO	+ 5 53 18	-56.4	+ 2 4.9	0.86	312 12 5	89 12 4
9	TU	+ 5 30 44	-56.6	+ 2 25.5	0.86	313 10 51	89 12 4
10	WE	+ 5 8 5	-56.9	+ 2 46.2	0.87	314 9 37	89 12 4
11	TH	+ 4 45 20	-57.0	+ 3 7.1	0.88	315 8 22	89 12 5
12	FR	+ 4 22 31	-57.3	+ 3 28.2	0.88	316 7 8	89 12 5
13	SA	+ 3 59 37	-57.4	+ 3 49.4	0.88	317 5 56	89 12 5
14	SU	+ 3 36 40	-57.6	+ 4 10.6	0.89	318 4 46	89 12 6
15	MO	+ 3 13 38	-57.7	+ 4 32.0	0.89	319 3 38	89 12 6
16	TU	+ 2 50 33	-57.8	+ 4 53.4	0.90	320 2 32	89 12 7
17	WE	+ 2 27 25	-58.0	+ 5 14.9	0.89	321 1 26	89 12 7
18	TH	+ 2 4 14	-58.0	+ 5 36.3	0.90	322 0 20	89 12 7
19	FR	+ 1 41 1	-58.2	+ 5 57.8	0.89	322 59 13	89 12 7
20	SA	+ 1 17 45	-58.2	+ 6 19.2	0.89	323 58 5	89 12 7
21	SU	+ 0 54 28	-58.3	+ 6 40.6	0.88	324 56 56	89 12 8
22	MO	+ 0 31 9	-58.3	+ 7 1.8	0.88	325 55 46	89 12 8
23	TU	+ 0 7 48	-58.4	+ 7 23.0	0.88	326 54 35	89 12 8
24	WE	- 0 15 34	-58.4	+ 7 44.1	0.87	327 53 25	89 12 9
25	TH	- 0 38 56	-58.5	+ 8 5.0	0.86	328 52 15	89 12 9
26	FR	- 1 2 19	-58.5	+ 8 25.7	0.85	329 51 6	89 12 9
27	SA	- 1 25 42	-58.4	+ 8 46.2	0.85	330 49 58	89 12 10
28	SU	- 1 49 4	-58.4	+ 9 6.6	0.84	331 48 52	89 12 10
29	MO	- 2 12 26	-58.3	+ 9 26.7	0.83	332 47 47	89 12 10
30	TU	- 2 35 46	-58.3	+ 9 46.5	0.82	333 46 44	89 12 11

OCTOBER 1986

Date		THE SUN Declination At 0 hrs, UT ° ' "	Δ, 1 hr "	Equation of Time At 0 hrs, UT m s	Δ, 1 hr s	POLARIS GHA at 0 hr ° ' "	Declination ° ' "
1	WE	− 2 59 5	−58.3	+10 6.1	0.80	334 45 41	89 12 11
2	TH	− 3 22 23	−58.1	+10 25.4	0.79	335 44 39	89 12 11
3	FR	− 3 45 38	−58.0	+10 44.4	0.78	336 43 37	89 12 12
4	SA	− 4 8 50	−57.9	+11 3.0	0.77	337 42 33	89 12 12
5	SU	− 4 32 0	−57.7	+11 21.4	0.75	338 41 28	89 12 12
6	MO	− 4 55 6	−57.6	+11 39.3	0.74	339 40 22	89 12 13
7	TU	− 5 18 9	−57.4	+11 57.0	0.72	340 39 14	89 12 13
8	WE	− 5 41 7	−57.3	+12 14.2	0.70	341 38 7	89 12 13
9	TH	− 6 4 1	−57.0	+12 31.0	0.68	342 37 0	89 12 14
10	FR	− 6 26 50	−56.8	+12 47.4	0.67	343 35 55	89 12 14
11	SA	− 6 49 34	−56.6	+13 3.4	0.65	344 34 53	89 12 14
12	SU	− 7 12 12	−56.4	+13 18.9	0.63	345 33 52	89 12 15
13	MO	− 7 34 45	−56.1	+13 33.9	0.60	346 32 53	89 12 15
14	TU	− 7 57 10	−55.8	+13 48.4	0.59	347 31 56	89 12 16
15	WE	− 8 19 29	−55.5	+14 2.5	0.56	348 30 58	89 12 16
16	TH	− 8 41 41	−55.2	+14 15.9	0.54	349 30 0	89 12 17
17	FR	− 9 3 46	−54.8	+14 28.8	0.52	350 29 1	89 12 17
18	SA	− 9 25 42	−54.5	+14 41.2	0.49	351 28 0	89 12 17
19	SU	− 9 47 31	−54.2	+14 52.9	0.46	352 26 59	89 12 18
20	MO	−10 9 11	−53.8	+15 4.0	0.44	353 25 58	89 12 18
21	TU	−10 30 42	−53.4	+15 14.5	0.41	354 24 56	89 12 18
22	WE	−10 52 4	−53.0	+15 24.3	0.38	355 23 55	89 12 19
23	TH	−11 13 17	−52.6	+15 33.4	0.35	356 22 54	89 12 19
24	FR	−11 34 19	−52.2	+15 41.9	0.32	357 21 55	89 12 19
25	SA	−11 55 11	−51.7	+15 49.6	0.29	358 20 58	89 12 20
26	SU	−12 15 52	−51.3	+15 56.6	0.26	359 20 2	89 12 20
27	MO	−12 36 22	−50.8	+16 2.8	0.23	0 19 7	89 12 21
28	TU	−12 56 41	−50.3	+16 8.3	0.20	1 18 14	89 12 21
29	WE	−13 16 47	−49.8	+16 13.1	0.16	2 17 21	89 12 21
30	TH	−13 36 41	−49.2	+16 17.0	0.13	3 16 28	89 12 22
31	FR	−13 56 22	−48.7	+16 20.2	0.10	4 15 35	89 12 22

NOVEMBER 1986

Date		THE SUN Declination At 0 hrs, UT ° ' "	Δ, 1 hr "	Equation of Time At 0 hrs, UT m s	Δ, 1 hr s	POLARIS GHA at 0 hr ° ' "	Declination ° ' "
1	SA	−14 15 50	−48.1	+16 22.6	0.07	5 14 40	89 12 23
2	SU	−14 35 5	−47.5	+16 24.2	0.03	6 13 44	89 12 23
3	MO	−14 54 5	−46.9	+16 25.0	−0.00	7 12 47	89 12 23
4	TU	−15 12 50	−46.3	+16 24.9	−0.03	8 11 49	89 12 24
5	WE	−15 31 21	−45.6	+16 24.1	−0.07	9 10 52	89 12 24
6	TH	−15 49 36	−45.0	+16 22.4	−0.10	10 9 56	89 12 24
7	FR	−16 7 35	−44.3	+16 20.0	−0.14	11 9 3	89 12 25
8	SA	−16 25 18	−43.6	+16 16.7	−0.17	12 8 12	89 12 25
9	SU	−16 42 44	−42.9	+16 12.6	−0.21	13 7 23	89 12 26
10	MO	−16 59 54	−42.1	+16 7.6	−0.24	14 6 35	89 12 26
11	TU	−17 16 45	−41.4	+16 1.9	−0.28	15 5 48	89 12 27
12	WE	−17 33 19	−40.7	+15 55.3	−0.31	16 5 0	89 12 27
13	TH	−17 49 35	−39.9	+15 47.8	−0.34	17 4 12	89 12 27
14	FR	−18 5 32	−39.1	+15 39.6	−0.38	18 3 23	89 12 28
15	SA	−18 21 10	−38.3	+15 30.5	−0.42	19 2 32	89 12 28
16	SU	−18 36 29	−37.5	+15 20.5	−0.45	20 1 41	89 12 28
17	MO	−18 51 28	−36.6	+15 9.7	−0.48	21 0 50	89 12 29
18	TU	−19 6 7	−35.8	+14 58.1	−0.52	22 0 0	89 12 29
19	WE	−19 20 25	−34.9	+14 45.7	−0.55	22 59 10	89 12 29
20	TH	−19 34 23	−34.0	+14 32.4	−0.59	23 58 21	89 12 30
21	FR	−19 48 0	−33.1	+14 18.2	−0.62	24 57 34	89 12 30
22	SA	−20 1 15	−32.2	+14 3.3	−0.66	25 56 48	89 12 31
23	SU	−20 14 8	−31.3	+13 47.5	−0.69	26 56 4	89 12 31
24	MO	−20 26 39	−30.3	+13 31.0	−0.73	27 55 21	89 12 31
25	TU	−20 38 47	−29.4	+13 13.6	−0.75	28 54 39	89 12 32
26	WE	−20 50 33	−28.4	+12 55.5	−0.79	29 53 57	89 12 32
27	TH	−21 1 55	−27.4	+12 36.6	−0.82	30 53 15	89 12 32
28	FR	−21 12 53	−26.5	+12 16.9	−0.85	31 52 32	89 12 33
29	SA	−21 23 28	−25.4	+11 56.5	−0.88	32 51 48	89 12 33
30	SU	−21 33 38	−24.4	+11 35.4	−0.90	33 51 2	89 12 33

DECEMBER 1986

Date		THE SUN Declination At 0 hrs, UT ° ' "	Δ, 1 hr "	Equation of Time At 0 hrs, UT m s	Δ, 1 hr s	POLARIS GHA at 0 hr ° ' "	Declination ° ' "
1	MO	−21 43 24	−23.4	+11 13.7	−0.94	34 50 15	89 12 33
2	TU	−21 52 45	−22.3	+10 51.2	−0.96	35 49 28	89 12 34
3	WE	−22 1 40	−21.3	+10 28.1	−0.99	36 48 43	89 12 34
4	TH	−22 10 10	−20.2	+10 4.4	−1.01	37 47 59	89 12 35
5	FR	−22 18 15	−19.1	+ 9 40.2	−1.04	38 47 19	89 12 35
6	SA	−22 25 53	−18.0	+ 9 15.3	−1.05	39 46 40	89 12 35
7	SU	−22 33 5	−16.9	+ 8 50.0	−1.08	40 46 3	89 12 36
8	MO	−22 39 51	−15.8	+ 8 24.2	−1.10	41 45 27	89 12 36
9	TU	−22 46 9	−14.7	+ 7 57.9	−1.12	42 44 50	89 12 36
10	WE	−22 52 1	−13.5	+ 7 31.1	−1.13	43 44 13	89 12 37
11	TH	−22 57 26	−12.4	+ 7 4.0	−1.15	44 43 34	89 12 37
12	FR	−23 2 24	−11.3	+ 6 36.5	−1.16	45 42 55	89 12 37
13	SA	−23 6 54	−10.1	+ 6 8.6	−1.17	46 42 15	89 12 37
14	SU	−23 10 56	− 9.0	+ 5 40.4	−1.19	47 41 34	89 12 38
15	MO	−23 14 31	− 7.8	+ 5 11.8	−1.20	48 40 54	89 12 38
16	TU	−23 17 39	− 6.6	+ 4 43.1	−1.21	49 40 14	89 12 38
17	WE	−23 20 18	− 5.5	+ 4 14.0	−1.22	50 39 36	89 12 39
18	TH	−23 22 30	− 4.3	+ 3 44.8	−1.22	51 38 59	89 12 39
19	FR	−23 24 13	− 3.2	+ 3 15.4	−1.23	52 38 23	89 12 39
20	SA	−23 25 28	− 2.0	+ 2 45.8	−1.24	53 37 49	89 12 40
21	SU	−23 26 16	− 0.8	+ 2 16.1	−1.24	54 37 16	89 12 40
22	MO	−23 26 35	0.3	+ 1 46.3	−1.24	55 36 43	89 12 40
23	TU	−23 26 25	1.5	+ 1 16.5	−1.25	56 36 12	89 12 40
24	WE	−23 25 48	2.8	+ 0 46.6	−1.24	57 35 40	89 12 41
25	TH	−23 24 42	3.9	+ 0 16.8	−1.24	58 35 7	89 12 41
26	FR	−23 23 8	5.1	− 0 13.0	−1.24	59 34 33	89 12 41
27	SA	−23 21 6	6.3	− 0 42.8	−1.23	60 33 57	89 12 41
28	SU	−23 18 36	7.5	− 1 12.4	−1.23	61 33 20	89 12 41
29	MO	−23 15 37	8.6	− 1 41.9	−1.22	62 32 43	89 12 41
30	TU	−23 12 11	9.8	− 2 11.2	−1.21	63 32 6	89 12 42
31	WE	−23 8 16	10.9	− 2 40.3	−1.20	64 31 31	89 12 42

JANUARY 1987

Date		Declination At 0 hrs, UT ° ' "	THE SUN Δ, 1 hr "	Equation of Time At 0 hrs, UT m s	Δ, 1 hr s	POLARIS GHA at 0 hr ° ' "	Declination ° ' "
1	TH	-23 3 54	12.1	- 3 9.1	-1.19	65 30 58	89 12 42
2	FR	-22 59 4	13.3	- 3 37.6	-1.17	66 30 28	89 12 43
3	SA	-22 53 46	14.3	- 4 5.8	-1.16	67 29 59	89 12 43
4	SU	-22 48 2	15.5	- 4 33.6	-1.14	68 29 31	89 12 43
5	MO	-22 41 49	16.6	- 5 1.0	-1.13	69 29 3	89 12 43
6	TU	-22 35 10	17.7	- 5 28.0	-1.10	70 28 34	89 12 43
7	WE	-22 28 5	18.9	- 5 54.5	-1.08	71 28 4	89 12 43
8	TH	-22 20 32	20.0	- 6 20.5	-1.06	72 27 33	89 12 44
9	FR	-22 12 33	21.0	- 6 46.0	-1.04	73 27 1	89 12 44
10	SA	-22 4 8	22.1	- 7 10.9	-1.02	74 26 28	89 12 44
11	SU	-21 55 18	23.2	- 7 35.3	-0.99	75 25 55	89 12 44
12	MO	-21 46 1	24.2	- 7 59.1	-0.97	76 25 23	89 12 44
13	TU	-21 36 20	25.3	- 8 22.3	-0.94	77 24 51	89 12 44
14	WE	-21 26 13	26.3	- 8 44.9	-0.91	78 24 21	89 12 44
15	TH	-21 15 42	27.3	- 9 6.8	-0.89	79 23 51	89 12 45
16	FR	-21 4 46	28.3	- 9 28.1	-0.86	80 23 23	89 12 45
17	SA	-20 53 26	29.3	- 9 48.7	-0.83	81 22 57	89 12 45
18	SU	-20 41 42	30.3	-10 8.6	-0.80	82 22 30	89 12 45
19	MO	-20 29 35	31.3	-10 27.9	-0.77	83 22 5	89 12 45
20	TU	-20 17 4	32.2	-10 46.4	-0.74	84 21 39	89 12 45
21	WE	-20 4 11	33.2	-11 4.2	-0.71	85 21 12	89 12 45
22	TH	-19 50 55	34.1	-11 21.2	-0.68	86 20 43	89 12 45
23	FR	-19 37 16	35.0	-11 37.5	-0.65	87 20 14	89 12 45
24	SA	-19 23 16	35.9	-11 53.1	-0.62	88 19 42	89 12 45
25	SU	-19 8 54	36.8	-12 7.9	-0.58	89 19 10	89 12 45
26	MO	-18 54 11	37.7	-12 21.9	-0.55	90 18 38	89 12 45
27	TU	-18 39 7	38.5	-12 35.2	-0.52	91 18 6	89 12 45
28	WE	-18 23 42	39.3	-12 47.6	-0.49	92 17 36	89 12 45
29	TH	-18 7 58	40.2	-12 59.3	-0.45	93 17 9	89 12 46
30	FR	-17 51 54	41.0	-13 10.1	-0.42	94 16 43	89 12 46
31	SA	-17 35 31	41.8	-13 20.1	-0.38	95 16 19	89 12 46

FEBRUARY 1987

Date		Declination At 0 hrs, UT ° ' "	THE SUN Δ, 1 hr "	Equation of Time At 0 hrs, UT m s	Δ, 1 hr s	POLARIS GHA at 0 hr ° ' "	Declination ° ' "
1	SU	-17 18 49	42.5	-13 29.3	-0.35	96 15 54	89 12 46
2	MO	-17 1 49	43.3	-13 37.7	-0.31	97 15 29	89 12 46
3	TU	-16 44 30	44.0	-13 45.2	-0.28	98 15 2	89 12 46
4	WE	-16 26 55	44.7	-13 51.9	-0.25	99 14 33	89 12 46
5	TH	-16 9 2	45.4	-13 57.8	-0.21	100 14 3	89 12 45
6	FR	-15 50 53	46.1	-14 2.8	-0.18	101 13 32	89 12 45
7	SA	-15 32 27	46.8	-14 7.1	-0.14	102 13 1	89 12 45
8	SU	-15 13 45	47.3	-14 10.4	-0.11	103 12 30	89 12 45
9	MO	-14 54 49	48.0	-14 13.0	-0.07	104 11 59	89 12 45
10	TU	-14 35 37	48.6	-14 14.8	-0.04	105 11 29	89 12 45
11	WE	-14 16 11	49.2	-14 15.8	-0.01	106 11 0	89 12 45
12	TH	-13 56 30	49.8	-14 16.0	0.02	107 10 32	89 12 45
13	FR	-13 36 36	50.3	-14 15.5	0.06	108 10 5	89 12 45
14	SA	-13 16 28	50.8	-14 14.1	0.08	109 9 39	89 12 45
15	SU	-12 56 8	51.4	-14 12.1	0.12	110 9 13	89 12 45
16	MO	-12 35 35	51.9	-14 9.3	0.15	111 8 46	89 12 45
17	TU	-12 14 50	52.4	-14 5.8	0.17	112 8 19	89 12 45
18	WE	-11 53 53	52.9	-14 1.6	0.20	113 7 50	89 12 45
19	TH	-11 32 44	53.3	-13 56.7	0.23	114 7 19	89 12 45
20	FR	-11 11 25	53.8	-13 51.2	0.26	115 6 46	89 12 44
21	SA	-10 49 55	54.2	-13 45.0	0.29	116 6 13	89 12 44
22	SU	-10 28 14	54.6	-13 38.1	0.31	117 5 38	89 12 44
23	MO	-10 6 24	55.0	-13 30.7	0.34	118 5 4	89 12 44
24	TU	- 9 44 24	55.3	-13 22.6	0.36	119 4 31	89 12 44
25	WE	- 9 22 16	55.7	-13 14.0	0.38	120 4 0	89 12 44
26	TH	- 8 59 59	56.0	-13 4.8	0.41	121 3 30	89 12 44
27	FR	- 8 37 34	56.4	-12 55.0	0.43	122 3 2	89 12 43
28	SA	- 8 15 1	56.7	-12 44.6	0.45	123 2 34	89 12 43

MARCH 1987

Date		Declination At 0 hrs, UT ° ' "	THE SUN Δ, 1 hr "	Equation of Time At 0 hrs, UT m s	Δ, 1 hr s	POLARIS GHA at 0 hr ° ' "	Declination ° ' "
1	SU	- 7 52 21	57.0	-12 33.7	0.48	124 2 5	89 12 43
2	MO	- 7 29 33	57.2	-12 22.3	0.50	125 1 35	89 12 43
3	TU	- 7 6 40	57.5	-12 10.4	0.52	126 1 3	89 12 43
4	WE	- 6 43 41	57.7	-11 58.0	0.54	127 0 28	89 12 43
5	TH	- 6 20 36	58.0	-11 45.1	0.56	127 59 53	89 12 42
6	FR	- 5 57 25	58.1	-11 31.7	0.58	128 59 16	89 12 42
7	SA	- 5 34 10	58.3	-11 17.9	0.59	129 58 40	89 12 42
8	SU	- 5 10 51	58.5	-11 3.7	0.61	130 58 3	89 12 42
9	MO	- 4 47 28	58.6	-10 49.1	0.63	131 57 27	89 12 42
10	TU	- 4 24 1	58.8	-10 34.0	0.64	132 56 52	89 12 41
11	WE	- 4 0 31	58.9	-10 18.7	0.66	133 56 17	89 12 41
12	TH	- 3 36 58	59.0	-10 2.9	0.67	134 55 44	89 12 41
13	FR	- 3 13 22	59.0	- 9 46.9	0.68	135 55 10	89 12 41
14	SA	- 2 49 45	59.2	- 9 30.5	0.69	136 54 38	89 12 40
15	SU	- 2 26 5	59.2	- 9 13.9	0.70	137 54 4	89 12 40
16	MO	- 2 2 24	59.2	- 8 57.1	0.71	138 53 30	89 12 40
17	TU	- 1 38 43	59.3	- 8 40.0	0.72	139 52 53	89 12 40
18	WE	- 1 15 0	59.3	- 8 22.7	0.73	140 52 15	89 12 40
19	TH	- 0 51 17	59.3	- 8 5.2	0.73	141 51 35	89 12 39
20	FR	- 0 27 34	59.3	- 7 47.6	0.74	142 50 53	89 12 39
21	SA	- 0 3 51	59.3	- 7 29.8	0.75	143 50 11	89 12 39
22	SU	+ 0 19 52	59.2	- 7 11.9	0.75	144 49 28	89 12 39
23	MO	+ 0 43 33	59.2	- 6 54.0	0.75	145 48 47	89 12 38
24	TU	+ 1 7 13	59.1	- 6 35.9	0.75	146 48 6	89 12 38
25	WE	+ 1 30 52	59.0	- 6 17.9	0.75	147 47 27	89 12 38
26	TH	+ 1 54 28	58.9	- 5 59.8	0.76	148 46 49	89 12 38
27	FR	+ 2 18 2	58.8	- 5 41.6	0.75	149 46 12	89 12 37
28	SA	+ 2 41 33	58.7	- 5 23.5	0.75	150 45 34	89 12 37
29	SU	+ 3 5 1	58.5	- 5 5.4	0.75	151 44 55	89 12 37
30	MO	+ 3 28 25	58.3	- 4 47.4	0.75	152 44 13	89 12 36
31	TU	+ 3 51 45	58.2	- 4 29.4	0.75	153 43 30	89 12 36

APRIL 1987

Date		THE SUN Declination At 0 hrs, UT ° ' "	Δ, 1 hr "	Equation of Time At 0 hrs, UT m s	Δ, 1 hr s	POLARIS GHA at 0 hr ° ' "	Declination ° ' "
1	WE	+ 4 15 1	58.0	− 4 11.4	0.74	154 42 45	89 12 36
2	TH	+ 4 38 13	57.8	− 3 53.6	0.74	155 41 58	89 12 35
3	FR	+ 5 1 19	57.5	− 3 35.9	0.74	156 41 11	89 12 35
4	SA	+ 5 24 20	57.3	− 3 18.2	0.73	157 40 24	89 12 35
5	SU	+ 5 47 15	57.0	− 3 0.7	0.72	158 39 37	89 12 35
6	MO	+ 6 10 3	56.8	− 2 43.4	0.72	159 38 50	89 12 35
7	TU	+ 6 32 46	56.5	− 2 26.2	0.70	160 38 5	89 12 34
8	WE	+ 6 55 21	56.2	− 2 9.3	0.70	161 37 20	89 12 34
9	TH	+ 7 17 49	55.9	− 1 52.5	0.69	162 36 36	89 12 34
10	FR	+ 7 40 10	55.5	− 1 36.0	0.68	163 35 52	89 12 33
11	SA	+ 8 2 22	55.2	− 1 19.7	0.67	164 35 7	89 12 33
12	SU	+ 8 24 27	54.8	− 1 3.6	0.65	165 34 22	89 12 33
13	MO	+ 8 46 23	54.5	− 0 47.9	0.64	166 33 35	89 12 32
14	TU	+ 9 8 10	54.1	− 0 32.5	0.63	167 32 46	89 12 32
15	WE	+ 9 29 49	53.7	− 0 17.3	0.61	168 31 55	89 12 31
16	TH	+ 9 51 17	53.3	− 0 2.6	0.60	169 31 2	89 12 31
17	FR	+10 12 36	52.9	+ 0 11.8	0.58	170 30 6	89 12 31
18	SA	+10 33 45	52.5	+ 0 25.8	0.57	171 29 13	89 12 31
19	SU	+10 54 44	52.0	+ 0 39.4	0.55	172 28 19	89 12 30
20	MO	+11 15 32	51.5	+ 0 52.7	0.53	173 27 27	89 12 30
21	TU	+11 36 9	51.1	+ 1 5.4	0.52	174 26 36	89 12 30
22	WE	+11 56 35	50.6	+ 1 17.8	0.49	175 25 46	89 12 29
23	TH	+12 16 49	50.1	+ 1 29.6	0.48	176 24 56	89 12 29
24	FR	+12 36 52	49.5	+ 1 41.0	0.46	177 24 7	89 12 29
25	SA	+12 56 41	49.0	+ 1 52.0	0.43	178 23 16	89 12 29
26	SU	+13 16 18	48.5	+ 2 2.4	0.41	179 22 23	89 12 28
27	MO	+13 35 42	48.0	+ 2 12.3	0.40	180 21 28	89 12 28
28	TU	+13 54 53	47.3	+ 2 21.8	0.37	181 20 32	89 12 28
29	WE	+14 13 49	46.8	+ 2 30.7	0.35	182 19 33	89 12 27
30	TH	+14 32 32	46.2	+ 2 39.1	0.33	183 18 34	89 12 27

MAY 1987

Date		THE SUN Declination At 0 hrs, UT ° ' "	Δ, 1 hr "	Equation of Time At 0 hrs, UT m s	Δ, 1 hr s	POLARIS GHA at 0 hr ° ' "	Declination ° ' "
1	FR	+14 51 0	45.6	+ 2 47.0	0.31	184 17 35	89 12 27
2	SA	+15 9 14	44.9	+ 2 54.4	0.29	185 16 36	89 12 26
3	SU	+15 27 12	44.3	+ 3 1.3	0.26	186 15 37	89 12 26
4	MO	+15 44 55	43.6	+ 3 7.6	0.24	187 14 39	89 12 26
5	TU	+16 2 22	43.0	+ 3 13.4	0.22	188 13 42	89 12 26
6	WE	+16 19 33	42.3	+ 3 18.7	0.20	189 12 46	89 12 25
7	TH	+16 36 28	41.6	+ 3 23.4	0.17	190 11 50	89 12 25
8	FR	+16 53 6	40.9	+ 3 27.5	0.15	191 10 54	89 12 25
9	SA	+17 9 27	40.1	+ 3 31.1	0.13	192 9 57	89 12 24
10	SU	+17 25 30	39.5	+ 3 34.2	0.10	193 8 59	89 12 24
11	MO	+17 41 17	38.7	+ 3 36.7	0.08	194 7 59	89 12 24
12	TU	+17 56 45	37.9	+ 3 38.6	0.06	195 6 57	89 12 24
13	WE	+18 11 55	37.2	+ 3 40.0	0.03	196 5 52	89 12 23
14	TH	+18 26 47	36.4	+ 3 40.8	0.01	197 4 47	89 12 23
15	FR	+18 41 21	35.6	+ 3 41.0	−0.01	198 3 40	89 12 23
16	SA	+18 55 35	34.8	+ 3 40.7	−0.04	199 2 35	89 12 23
17	SU	+19 9 31	34.0	+ 3 39.8	−0.06	200 1 30	89 12 22
18	MO	+19 23 7	33.2	+ 3 38.3	−0.09	201 0 27	89 12 22
19	TU	+19 36 23	32.4	+ 3 36.2	−0.11	201 59 26	89 12 22
20	WE	+19 49 20	31.5	+ 3 33.5	−0.13	202 58 25	89 12 22
21	TH	+20 1 56	30.7	+ 3 30.3	−0.16	203 57 24	89 12 21
22	FR	+20 14 12	29.8	+ 3 26.5	−0.18	204 56 23	89 12 21
23	SA	+20 26 7	28.9	+ 3 22.2	−0.20	205 55 20	89 12 21
24	SU	+20 37 41	28.0	+ 3 17.4	−0.22	206 54 14	89 12 21
25	MO	+20 48 54	27.2	+ 3 12.0	−0.25	207 53 8	89 12 20
26	TU	+20 59 46	26.3	+ 3 6.1	−0.27	208 51 59	89 12 20
27	WE	+21 10 16	25.3	+ 2 59.7	−0.29	209 50 50	89 12 20
28	TH	+21 20 23	24.4	+ 2 52.8	−0.31	210 49 40	89 12 20
29	FR	+21 30 9	23.5	+ 2 45.4	−0.33	211 48 31	89 12 19
30	SA	+21 39 32	22.5	+ 2 37.6	−0.34	212 47 21	89 12 19
31	SU	+21 48 33	21.6	+ 2 29.4	−0.36	213 46 13	89 12 19

JUNE 1987

Date		THE SUN Declination At 0 hrs, UT ° ' "	Δ, 1 hr "	Equation of Time At 0 hrs, UT m s	Δ, 1 hr s	POLARIS GHA at 0 hr ° ' "	Declination ° ' "
1	MO	+21 57 11	20.6	+ 2 20.7	−0.38	214 45 6	89 12 19
2	TU	+22 5 26	19.6	+ 2 11.7	−0.39	215 43 59	89 12 19
3	WE	+22 13 17	18.7	+ 2 2.3	−0.41	216 42 53	89 12 19
4	TH	+22 20 46	17.7	+ 1 52.5	−0.42	217 41 47	89 12 18
5	FR	+22 27 51	16.7	+ 1 42.4	−0.44	218 40 41	89 12 18
6	SA	+22 34 32	15.7	+ 1 31.9	−0.45	219 39 34	89 12 18
7	SU	+22 40 49	14.8	+ 1 21.2	−0.46	220 38 25	89 12 18
8	MO	+22 46 43	13.7	+ 1 10.1	−0.47	221 37 15	89 12 17
9	TU	+22 52 12	12.8	+ 0 58.8	−0.48	222 36 2	89 12 17
10	WE	+22 57 18	11.7	+ 0 47.3	−0.48	223 34 48	89 12 17
11	TH	+23 1 59	10.7	+ 0 35.5	−0.50	224 33 33	89 12 17
12	FR	+23 6 16	9.7	+ 0 23.5	−0.51	225 32 18	89 12 17
13	SA	+23 10 9	8.7	+ 0 11.3	−0.51	226 31 4	89 12 17
14	SU	+23 13 37	7.7	− 0 1.0	−0.52	227 29 52	89 12 17
15	MO	+23 16 41	6.6	− 0 13.6	−0.52	228 28 42	89 12 16
16	TU	+23 19 20	5.6	− 0 26.2	−0.54	229 27 33	89 12 16
17	WE	+23 21 35	4.5	− 0 39.1	−0.54	230 26 24	89 12 16
18	TH	+23 23 24	3.6	− 0 52.0	−0.54	231 25 15	89 12 16
19	FR	+23 24 50	2.5	− 1 5.0	−0.54	232 24 5	89 12 16
20	SA	+23 25 50	1.5	− 1 18.0	−0.55	233 22 52	89 12 16
21	SU	+23 26 25	0.5	− 1 31.1	−0.55	234 21 38	89 12 16
22	MO	+23 26 36	− 0.6	− 1 44.2	−0.55	235 20 23	89 12 15
23	TU	+23 26 22	− 1.6	− 1 57.3	−0.55	236 19 7	89 12 15
24	WE	+23 25 43	− 2.7	− 2 10.4	−0.54	237 17 50	89 12 15
25	TH	+23 24 39	− 3.7	− 2 23.3	−0.54	238 16 33	89 12 15
26	FR	+23 23 10	− 4.7	− 2 36.2	−0.53	239 15 18	89 12 15
27	SA	+23 21 17	− 5.8	− 2 48.9	−0.53	240 14 3	89 12 15
28	SU	+23 18 59	− 6.8	− 3 1.5	−0.52	241 12 49	89 12 15
29	MO	+23 16 16	− 7.8	− 3 13.9	−0.51	242 11 36	89 12 15
30	TU	+23 13 9	− 8.8	− 3 26.1	−0.50	243 10 23	89 12 15

JULY 1987

Date		THE SUN Declination At 0 hrs, UT ° ' "	Δ, 1 hr "	Equation of Time At 0 hrs, UT m s	Δ, 1 hr s	POLARIS GHA at 0 hr ° ' "	Declination ° ' "
1	WE	+23 9 38	- 9.8	- 3 38.1	-0.49	244 9 11	89 12 15
2	TH	+23 5 42	-10.8	- 3 49.8	-0.48	245 7 59	89 12 15
3	FR	+23 1 22	-11.8	- 4 1.2	-0.47	246 6 47	89 12 15
4	SA	+22 56 38	-12.8	- 4 12.4	-0.45	247 5 33	89 12 15
5	SU	+22 51 30	-13.8	- 4 23.2	-0.43	248 4 18	89 12 14
6	MO	+22 45 58	-14.8	- 4 33.6	-0.42	249 3 1	89 12 14
7	TU	+22 40 2	-15.8	- 4 43.7	-0.40	250 1 42	89 12 14
8	WE	+22 33 43	-16.8	- 4 53.4	-0.39	251 0 23	89 12 14
9	TH	+22 27 0	-17.8	- 5 2.8	-0.37	251 59 3	89 12 14
10	FR	+22 19 54	-18.7	- 5 11.7	-0.35	252 57 44	89 12 14
11	SA	+22 12 26	-19.7	- 5 20.2	-0.33	253 56 26	89 12 14
12	SU	+22 4 34	-20.6	- 5 28.2	-0.32	254 55 11	89 12 14
13	MO	+21 56 20	-21.5	- 5 35.8	-0.30	255 53 58	89 12 14
14	TU	+21 47 43	-22.5	- 5 43.0	-0.28	256 52 45	89 12 14
15	WE	+21 38 44	-23.4	- 5 49.7	-0.26	257 51 32	89 12 14
16	TH	+21 29 22	-24.3	- 5 55.9	-0.24	258 50 19	89 12 14
17	FR	+21 19 39	-25.2	- 6 1.6	-0.22	259 49 3	89 12 14
18	SA	+21 9 34	-26.1	- 6 6.8	-0.20	260 47 46	89 12 14
19	SU	+20 59 7	-27.0	- 6 11.6	-0.17	261 46 28	89 12 14
20	MO	+20 48 19	-27.9	- 6 15.8	-0.15	262 45 9	89 12 14
21	TU	+20 37 10	-28.8	- 6 19.4	-0.13	263 43 49	89 12 14
22	WE	+20 25 40	-29.6	- 6 22.6	-0.10	264 42 30	89 12 14
23	TH	+20 13 49	-30.5	- 6 25.1	-0.08	265 41 11	89 12 15
24	FR	+20 1 38	-31.3	- 6 27.1	-0.06	266 39 54	89 12 15
25	SA	+19 49 7	-32.1	- 6 28.6	-0.03	267 38 37	89 12 15
26	SU	+19 36 16	-32.9	- 6 29.4	-0.01	268 37 22	89 12 15
27	MO	+19 23 6	-33.8	- 6 29.6	0.01	269 36 7	89 12 15
28	TU	+19 9 36	-34.5	- 6 29.3	0.04	270 34 53	89 12 15
29	WE	+18 55 48	-35.3	- 6 28.3	0.07	271 33 40	89 12 15
30	TH	+18 41 40	-36.0	- 6 26.7	0.09	272 32 26	89 12 15
31	FR	+18 27 15	-36.8	- 6 24.5	0.12	273 31 11	89 12 15

AUGUST 1987

Date		THE SUN Declination At 0 hrs, UT ° ' "	Δ, 1 hr "	Equation of Time At 0 hrs, UT m s	Δ, 1 hr s	POLARIS GHA at 0 hr ° ' "	Declination ° ' "
1	SA	+18 12 31	-37.6	- 6 21.7	0.15	274 29 55	89 12 15
2	SU	+17 57 29	-38.3	- 6 18.2	0.17	275 28 38	89 12 15
3	MO	+17 42 10	-39.0	- 6 14.2	0.20	276 27 19	89 12 15
4	TU	+17 26 34	-39.8	- 6 9.4	0.22	277 25 59	89 12 15
5	WE	+17 10 40	-40.4	- 6 4.1	0.25	278 24 39	89 12 16
6	TH	+16 54 31	-41.1	- 5 58.2	0.28	279 23 19	89 12 16
7	FR	+16 38 5	-41.8	- 5 51.6	0.30	280 22 0	89 12 16
8	SA	+16 21 23	-42.4	- 5 44.5	0.33	281 20 44	89 12 16
9	SU	+16 4 25	-43.0	- 5 36.7	0.35	282 19 30	89 12 16
10	MO	+15 47 12	-43.7	- 5 28.3	0.37	283 18 17	89 12 16
11	TU	+15 29 44	-44.3	- 5 19.4	0.40	284 17 5	89 12 17
12	WE	+15 12 0	-44.9	- 5 9.9	0.42	285 15 52	89 12 17
13	TH	+14 54 3	-45.5	- 4 59.9	0.44	286 14 38	89 12 17
14	FR	+14 35 50	-46.1	- 4 49.3	0.46	287 13 22	89 12 17
15	SA	+14 17 24	-46.7	- 4 38.2	0.48	288 12 5	89 12 17
16	SU	+13 58 44	-47.2	- 4 26.6	0.51	289 10 46	89 12 17
17	MO	+13 39 51	-47.8	- 4 14.4	0.53	290 9 28	89 12 17
18	TU	+13 20 45	-48.3	- 4 1.8	0.55	291 8 10	89 12 18
19	WE	+13 1 25	-48.8	- 3 48.7	0.57	292 6 52	89 12 18
20	TH	+12 41 54	-49.3	- 3 35.0	0.58	293 5 36	89 12 18
21	FR	+12 22 10	-49.8	- 3 21.0	0.60	294 4 20	89 12 18
22	SA	+12 2 14	-50.3	- 3 6.4	0.63	295 3 7	89 12 18
23	SU	+11 42 7	-50.8	- 2 51.4	0.65	296 1 54	89 12 19
24	MO	+11 21 48	-51.2	- 2 35.9	0.66	297 0 42	89 12 19
25	TU	+11 1 19	-51.6	- 2 20.1	0.68	297 59 31	89 12 19
26	WE	+10 40 40	-52.1	- 2 3.8	0.70	298 58 19	89 12 19
27	TH	+10 19 50	-52.5	- 1 47.0	0.71	299 57 7	89 12 20
28	FR	+ 9 58 50	-52.9	- 1 29.9	0.73	300 55 54	89 12 20
29	SA	+ 9 37 41	-53.3	- 1 12.4	0.74	301 54 40	89 12 20
30	SU	+ 9 16 23	-53.6	- 0 54.6	0.76	302 53 24	89 12 20
31	MO	+ 8 54 56	-54.0	- 0 36.3	0.77	303 52 8	89 12 20

SEPTEMBER 1987

Date		THE SUN Declination At 0 hrs, UT ° ' "	Δ, 1 hr "	Equation of Time At 0 hrs, UT m s	Δ, 1 hr s	POLARIS GHA at 0 hr ° ' "	Declination ° ' "
1	TU	+ 8 33 21	-54.3	- 0 17.7	0.78	304 50 51	89 12 21
2	WE	+ 8 11 37	-54.6	- 0 1.1	0.80	305 49 34	89 12 21
3	TH	+ 7 49 46	-54.9	+ 0 20.4	0.81	306 48 19	89 12 21
4	FR	+ 7 27 48	-55.3	+ 0 39.9	0.83	307 47 5	89 12 21
5	SA	+ 7 5 42	-55.5	+ 0 59.7	0.83	308 45 54	89 12 22
6	SU	+ 6 43 29	-55.8	+ 1 19.7	0.85	309 44 44	89 12 22
7	MO	+ 6 21 10	-56.0	+ 1 40.0	0.85	310 43 36	89 12 22
8	TU	+ 5 58 45	-56.3	+ 2 0.4	0.86	311 42 28	89 12 23
9	WE	+ 5 36 14	-56.5	+ 2 21.1	0.87	312 41 18	89 12 23
10	TH	+ 5 13 37	-56.8	+ 2 41.9	0.88	313 40 8	89 12 23
11	FR	+ 4 50 55	-57.0	+ 3 2.9	0.88	314 38 55	89 12 23
12	SA	+ 4 28 7	-57.2	+ 3 24.0	0.88	315 37 42	89 12 24
13	SU	+ 4 5 15	-57.3	+ 3 45.2	0.88	316 36 28	89 12 24
14	MO	+ 3 42 19	-57.5	+ 4 6.4	0.89	317 35 14	89 12 24
15	TU	+ 3 19 18	-57.7	+ 4 27.7	0.89	318 34 1	89 12 24
16	WE	+ 2 56 14	-57.8	+ 4 49.1	0.89	319 32 49	89 12 25
17	TH	+ 2 33 6	-58.0	+ 5 10.4	0.89	320 31 38	89 12 25
18	FR	+ 2 9 54	-58.0	+ 5 31.8	0.89	321 30 30	89 12 25
19	SA	+ 1 46 41	-58.2	+ 5 53.2	0.89	322 29 22	89 12 26
20	SU	+ 1 23 24	-58.3	+ 6 14.5	0.88	323 28 15	89 12 26
21	MO	+ 1 0 6	-58.3	+ 6 35.7	0.88	324 27 10	89 12 26
22	TU	+ 0 36 46	-58.4	+ 6 56.9	0.88	325 26 4	89 12 27
23	WE	+ 0 13 25	-58.5	+ 7 18.0	0.87	326 24 58	89 12 27
24	TH	- 0 9 58	-58.5	+ 7 38.9	0.87	327 23 51	89 12 27
25	FR	- 0 33 21	-58.5	+ 7 59.8	0.86	328 22 43	89 12 28
26	SA	- 0 56 44	-58.5	+ 8 20.5	0.85	329 21 34	89 12 28
27	SU	- 1 20 7	-58.5	+ 8 41.0	0.85	330 20 24	89 12 28
28	MO	- 1 43 30	-58.4	+ 9 1.4	0.84	331 19 14	89 12 28
29	TU	- 2 6 51	-58.4	+ 9 21.6	0.83	332 18 4	89 12 29
30	WE	- 2 30 12	-58.3	+ 9 41.5	0.82	333 16 55	89 12 29

OCTOBER 1987

Date		THE SUN Declination At 0 hrs, UT ° ' "	Δ, 1 hr "	Equation of Time At 0 hrs, UT m s	Δ, 1 hr s	POLARIS GHA at 0 hr ° ' "	Declination ° ' "
1	TH	- 2 53 31	-58.2	+10 1.2	0.81	334 15 47	89 12 30
2	FR	- 3 16 48	-58.1	+10 20.7	0.80	335 14 42	89 12 30
3	SA	- 3 40 3	-58.0	+10 39.9	0.79	336 13 39	89 12 30
4	SU	- 4 3 15	-57.9	+10 58.8	0.77	337 12 38	89 12 31
5	MO	- 4 26 24	-57.8	+11 17.3	0.76	338 11 37	89 12 31
6	TU	- 4 49 30	-57.6	+11 35.6	0.74	339 10 35	89 12 31
7	WE	- 5 12 32	-57.4	+11 53.4	0.73	340 9 32	89 12 32
8	TH	- 5 35 30	-57.3	+12 10.9	0.71	341 8 28	89 12 32
9	FR	- 5 58 24	-57.1	+12 27.9	0.69	342 7 23	89 12 32
10	SA	- 6 21 14	-56.8	+12 44.5	0.67	343 6 16	89 12 33
11	SU	- 6 43 58	-56.7	+13 0.7	0.65	344 5 10	89 12 33
12	MO	- 7 6 38	-56.4	+13 16.4	0.63	345 4 5	89 12 33
13	TU	- 7 29 11	-56.2	+13 31.5	0.61	346 3 0	89 12 34
14	WE	- 7 51 39	-55.9	+13 46.2	0.58	347 1 57	89 12 34
15	TH	- 8 14 0	-55.6	+14 0.2	0.57	348 0 56	89 12 35
16	FR	- 8 36 14	-55.3	+14 13.8	0.54	348 59 57	89 12 35
17	SA	- 8 58 22	-55.0	+14 26.7	0.52	349 58 58	89 12 35
18	SU	- 9 20 21	-54.7	+14 39.1	0.49	350 58 1	89 12 36
19	MO	- 9 42 13	-54.3	+14 50.8	0.47	351 57 4	89 12 36
20	TU	-10 3 56	-54.0	+15 2.0	0.43	352 56 7	89 12 37
21	WE	-10 25 31	-53.5	+15 12.4	0.41	353 55 9	89 12 37
22	TH	-10 46 56	-53.2	+15 22.3	0.38	354 54 10	89 12 37
23	FR	-11 8 12	-52.7	+15 31.4	0.35	355 53 11	89 12 38
24	SA	-11 29 17	-52.3	+15 39.9	0.33	356 52 10	89 12 38
25	SU	-11 50 12	-51.9	+15 47.7	0.30	357 51 8	89 12 38
26	MO	-12 10 57	-51.4	+15 54.8	0.27	358 50 7	89 12 39
27	TU	-12 31 30	-50.9	+16 1.2	0.23	359 49 7	89 12 39
28	WE	-12 51 51	-50.4	+16 6.8	0.20	0 48 9	89 12 39
29	TH	-13 12 0	-49.9	+16 11.7	0.17	1 47 13	89 12 40
30	FR	-13 31 57	-49.3	+16 15.9	0.15	2 46 19	89 12 40
31	SA	-13 51 41	-48.8	+16 19.4	0.11	3 45 27	89 12 41

NOVEMBER 1987

Date		THE SUN Declination At 0 hrs, UT ° ' "	Δ, 1 hr "	Equation of Time At 0 hrs, UT m s	Δ, 1 hr s	POLARIS GHA at 0 hr ° ' "	Declination ° ' "
1	SU	-14 11 11	-48.2	+16 22.1	0.08	4 44 35	89 12 41
2	MO	-14 30 28	-47.6	+16 24.0	0.05	5 43 44	89 12 41
3	TU	-14 49 30	-47.0	+16 25.1	0.01	6 42 51	89 12 42
4	WE	-15 8 18	-46.4	+16 25.4	-0.02	7 41 57	89 12 42
5	TH	-15 26 51	-45.8	+16 24.9	-0.05	8 41 2	89 12 43
6	FR	-15 45 9	-45.1	+16 23.6	-0.09	9 40 7	89 12 43
7	SA	-16 3 12	-44.4	+16 21.5	-0.13	10 39 11	89 12 44
8	SU	-16 20 58	-43.8	+16 18.5	-0.16	11 38 15	89 12 44
9	MO	-16 38 28	-43.0	+16 14.7	-0.20	12 37 20	89 12 44
10	TU	-16 55 41	-42.4	+16 10.0	-0.23	13 36 27	89 12 45
11	WE	-17 12 38	-41.6	+16 4.5	-0.27	14 35 36	89 12 45
12	TH	-17 29 16	-40.9	+15 58.1	-0.30	15 34 46	89 12 45
13	FR	-17 45 37	-40.1	+15 50.8	-0.34	16 33 58	89 12 46
14	SA	-18 1 39	-39.3	+15 42.7	-0.38	17 33 11	89 12 46
15	SU	-18 17 22	-38.5	+15 33.7	-0.41	18 32 24	89 12 46
16	MO	-18 32 47	-37.7	+15 23.8	-0.45	19 31 38	89 12 47
17	TU	-18 47 51	-36.8	+15 13.1	-0.48	20 30 52	89 12 47
18	WE	-19 2 36	-36.0	+15 1.5	-0.52	21 30 4	89 12 48
19	TH	-19 17 0	-35.2	+14 49.1	-0.55	22 29 16	89 12 48
20	FR	-19 31 4	-34.3	+14 35.8	-0.59	23 28 26	89 12 48
21	SA	-19 44 46	-33.4	+14 21.7	-0.62	24 27 37	89 12 48
22	SU	-19 58 7	-32.5	+14 6.8	-0.65	25 26 45	89 12 49
23	MO	-20 11 6	-31.5	+13 51.1	-0.69	26 25 55	89 12 49
24	TU	-20 23 42	-30.6	+13 34.6	-0.72	27 25 7	89 12 50
25	WE	-20 35 56	-29.6	+13 17.4	-0.75	28 24 21	89 12 50
26	TH	-20 47 47	-28.7	+12 59.3	-0.78	29 23 38	89 12 50
27	FR	-20 59 15	-27.7	+12 40.6	-0.81	30 22 56	89 12 51
28	SA	-21 10 19	-26.7	+12 21.2	-0.84	31 22 15	89 12 51
29	SU	-21 20 59	-25.7	+12 1.0	-0.87	32 21 35	89 12 51
30	MO	-21 31 14	-24.6	+11 40.2	-0.90	33 20 54	89 12 52

DECEMBER 1987

Date		THE SUN Declination At 0 hrs, UT ° ' "	Δ, 1 hr "	Equation of Time At 0 hrs, UT m s	Δ, 1 hr s	POLARIS GHA at 0 hr ° ' "	Declination ° ' "
1	TU	-21 41 5	-23.6	+11 18.7	-0.92	34 20 11	89 12 52
2	WE	-21 50 32	-22.5	+10 56.6	-0.95	35 19 28	89 12 52
3	TH	-21 59 33	-21.5	+10 33.8	-0.97	36 18 43	89 12 53
4	FR	-22 8 8	-20.5	+10 10.4	-1.00	37 17 58	89 12 53
5	SA	-22 16 19	-19.3	+ 9 46.5	-1.02	38 17 14	89 12 53
6	SU	-22 24 3	-18.3	+ 9 22.0	-1.05	39 16 30	89 12 54
7	MO	-22 31 21	-17.2	+ 8 56.9	-1.07	40 15 47	89 12 54
8	TU	-22 38 13	-16.0	+ 8 31.3	-1.09	41 15 6	89 12 54
9	WE	-22 44 38	-15.0	+ 8 5.2	-1.11	42 14 27	89 12 55
10	TH	-22 50 37	-13.8	+ 7 38.6	-1.13	43 13 49	89 12 55
11	FR	-22 56 9	-12.7	+ 7 11.6	-1.15	44 13 13	89 12 55
12	SA	-23 1 13	-11.5	+ 6 44.1	-1.16	45 12 37	89 12 56
13	SU	-23 5 50	-10.4	+ 6 16.2	-1.17	46 12 1	89 12 56
14	MO	-23 10 0	- 9.3	+ 5 48.0	-1.19	47 11 26	89 12 56
15	TU	-23 13 42	- 8.1	+ 5 19.5	-1.20	48 10 49	89 12 56
16	WE	-23 16 57	- 6.9	+ 4 50.6	-1.21	49 10 12	89 12 57
17	TH	-23 19 43	- 5.8	+ 4 21.5	-1.22	50 9 33	89 12 57
18	FR	-23 22 1	- 4.6	+ 3 52.1	-1.23	51 8 54	89 12 57
19	SA	-23 23 52	- 3.4	+ 3 22.5	-1.24	52 8 13	89 12 57
20	SU	-23 25 14	- 2.3	+ 2 52.8	-1.24	53 7 33	89 12 58
21	MO	-23 26 8	- 1.0	+ 2 23.0	-1.25	54 6 55	89 12 58
22	TU	-23 26 33	0.1	+ 1 53.0	-1.25	55 6 18	89 12 58
23	WE	-23 26 31	1.3	+ 1 23.1	-1.25	56 5 44	89 12 59
24	TH	-23 26 0	2.5	+ 0 53.1	-1.25	57 5 12	89 12 59
25	FR	-23 25 1	3.7	+ 0 23.2	-1.25	58 4 41	89 12 59
26	SA	-23 23 33	4.8	- 0 6.7	-1.24	59 4 11	89 12 59
27	SU	-23 21 37	6.0	- 0 36.4	-1.23	60 3 40	89 13 0
28	MO	-23 19 14	7.2	- 1 6.0	-1.23	61 3 7	89 13 0
29	TU	-23 16 22	8.3	- 1 35.5	-1.22	62 2 33	89 13 0
30	WE	-23 13 2	9.5	- 2 4.7	-1.20	63 1 58	89 13 0
31	TH	-23 9 14	10.7	- 2 33.6	-1.20	64 1 23	89 13 0

JANUARY 1988

Date		THE SUN Declination At 0 hrs, UT ° ' "	Δ, 1 hr "	Equation of Time At 0 hrs, UT m s	Δ, 1 hr s	POLARIS GHA at 0 hr ° ' "	Declination ° ' "
1	FR	-23 4 58	11.8	- 3 2.3	-1.19	65 0 47	89 13 1
2	SA	-23 0 15	13.0	- 3 30.8	-1.17	66 0 12	89 13 1
3	SU	-22 55 4	14.1	- 3 58.8	-1.16	66 59 38	89 13 1
4	MO	-22 49 26	15.2	- 4 26.6	-1.14	67 59 5	89 13 1
5	TU	-22 43 21	16.3	- 4 53.9	-1.12	68 58 34	89 13 1
6	WE	-22 36 49	17.5	- 5 20.8	-1.11	69 58 5	89 13 2
7	TH	-22 29 50	18.5	- 5 47.4	-1.08	70 57 36	89 13 2
8	FR	-22 22 25	19.7	- 6 13.4	-1.07	71 57 8	89 13 2
9	SA	-22 14 33	20.8	- 6 39.0	-1.05	72 56 41	89 13 2
10	SU	-22 6 14	21.8	- 7 4.1	-1.02	73 56 13	89 13 2
11	MO	-21 57 30	22.9	- 7 28.6	-1.00	74 55 45	89 13 3
12	TU	-21 48 20	24.0	- 7 52.6	-0.98	75 55 15	89 13 3
13	WE	-21 38 45	25.0	- 8 16.1	-0.95	76 54 44	89 13 3
14	TH	-21 28 44	26.1	- 8 38.9	-0.92	77 54 12	89 13 3
15	FR	-21 18 18	27.1	- 9 1.1	-0.90	78 53 39	89 13 3
16	SA	-21 7 28	28.1	- 9 22.7	-0.88	79 53 6	89 13 3
17	SU	-20 56 13	29.1	- 9 43.7	-0.84	80 52 33	89 13 3
18	MO	-20 44 35	30.1	-10 3.9	-0.82	81 52 2	89 13 3
19	TU	-20 32 32	31.1	-10 23.5	-0.78	82 51 33	89 13 3
20	WE	-20 20 6	32.0	-10 42.3	-0.75	83 51 7	89 13 4
21	TH	-20 7 18	33.0	-11 0.4	-0.72	84 50 41	89 13 4
22	FR	-19 54 6	33.9	-11 17.8	-0.69	85 50 16	89 13 4
23	SA	-19 40 33	34.8	-11 34.3	-0.66	86 49 51	89 13 4
24	SU	-19 26 37	35.7	-11 50.1	-0.63	87 49 24	89 13 4
25	MO	-19 12 20	36.6	-12 5.1	-0.59	88 48 55	89 13 4
26	TU	-18 57 42	37.5	-12 19.2	-0.56	89 48 25	89 13 4
27	WE	-18 42 43	38.3	-12 32.6	-0.52	90 47 54	89 13 4
28	TH	-18 27 24	39.1	-12 45.1	-0.49	91 47 22	89 13 4
29	FR	-18 11 45	40.0	-12 56.8	-0.45	92 46 51	89 13 4
30	SA	-17 55 46	40.8	-13 7.7	-0.42	93 46 21	89 13 4
31	SU	-17 39 28	41.5	-13 17.7	-0.38	94 45 51	89 13 4

FEBRUARY 1988

Date		THE SUN Declination At 0 hrs, UT ° ' "	Δ, 1 hr "	Equation of Time At 0 hrs, UT m s	Δ, 1 hr s	POLARIS GHA at 0 hr ° ' "	Declination ° ' "
1	MO	-17 22 52	42.3	-13 26.9	-0.35	95 45 23	89 13 4
2	TU	-17 5 57	43.0	-13 35.3	-0.32	96 44 56	89 13 4
3	WE	-16 48 44	43.8	-13 42.9	-0.28	97 44 30	89 13 4
4	TH	-16 31 13	44.5	-13 49.7	-0.25	98 44 5	89 13 4
5	FR	-16 13 25	45.2	-13 55.6	-0.22	99 43 40	89 13 4
6	SA	-15 55 20	45.9	-14 0.8	-0.18	100 43 14	89 13 4
7	SU	-15 36 59	46.5	-14 5.1	-0.15	101 42 47	89 13 4
8	MO	-15 18 22	47.2	-14 8.7	-0.12	102 42 20	89 13 4
9	TU	-14 59 29	47.8	-14 11.5	-0.08	103 41 51	89 13 4
10	WE	-14 40 21	48.5	-14 13.5	-0.05	104 41 20	89 13 4
11	TH	-14 20 58	49.0	-14 14.7	-0.02	105 40 49	89 13 4
12	FR	-14 1 21	49.7	-14 15.2	0.01	106 40 16	89 13 4
13	SA	-13 41 29	50.2	-14 14.9	0.04	107 39 44	89 13 3
14	SU	-13 21 24	50.8	-14 13.9	0.07	108 39 12	89 13 3
15	MO	-13 1 6	51.3	-14 12.2	0.10	109 38 42	89 13 3
16	TU	-12 40 35	51.8	-14 9.7	0.13	110 38 14	89 13 3
17	WE	-12 19 51	52.3	-14 6.6	0.16	111 37 47	89 13 3
18	TH	-11 58 56	52.8	-14 2.7	0.19	112 37 21	89 13 3
19	FR	-11 37 49	53.2	-13 58.1	0.22	113 36 55	89 13 3
20	SA	-11 16 32	53.7	-13 52.8	0.25	114 36 27	89 13 3
21	SU	-10 55 3	54.1	-13 46.8	0.27	115 35 58	89 13 3
22	MO	-10 33 25	54.5	-13 40.2	0.30	116 35 26	89 13 3
23	TU	-10 11 37	54.9	-13 32.9	0.33	117 34 52	89 13 3
24	WE	- 9 49 40	55.3	-13 25.0	0.36	118 34 18	89 13 2
25	TH	- 9 27 33	55.6	-13 16.4	0.38	119 33 44	89 13 2
26	FR	- 9 5 19	56.0	-13 7.2	0.41	120 33 10	89 13 2
27	SA	- 8 42 56	56.5	-12 57.4	0.43	121 32 37	89 13 2
28	SU	- 8 20 26	56.5	-12 47.1	0.45	122 32 5	89 13 2
29	MO	- 7 57 49	56.8	-12 36.2	0.48	123 31 34	89 13 2

MARCH 1988

Date		THE SUN Declination At 0 hrs, UT ° ' "	Δ, 1 hr "	Equation of Time At 0 hrs, UT m s	Δ, 1 hr s	POLARIS GHA at 0 hr ° ' "	Declination ° ' "
1	TU	- 7 35 5	57.1	-12 24.8	0.50	124 31 4	89 13 2
2	WE	- 7 12 14	57.4	-12 12.8	0.52	125 30 34	89 13 1
3	TH	- 6 49 17	57.6	-12 0.4	0.54	126 30 4	89 13 1
4	FR	- 6 26 14	57.8	-11 47.4	0.55	127 29 34	89 13 1
5	SA	- 6 3 7	58.0	-11 34.1	0.58	128 29 3	89 13 1
6	SU	- 5 39 54	58.2	-11 20.3	0.59	129 28 30	89 13 1
7	MO	- 5 16 36	58.4	-11 6.1	0.61	130 27 56	89 13 0
8	TU	- 4 53 14	58.6	-10 51.5	0.63	131 27 20	89 13 0
9	WE	- 4 29 48	58.7	-10 36.5	0.64	132 26 43	89 13 0
10	TH	- 4 6 19	58.8	-10 21.2	0.65	133 26 4	89 13 0
11	FR	- 3 42 47	59.0	-10 5.6	0.66	134 25 26	89 12 59
12	SA	- 3 19 11	59.1	- 9 49.7	0.67	135 24 47	89 12 59
13	SU	- 2 55 33	59.2	- 9 33.5	0.68	136 24 10	89 12 59
14	MO	- 2 31 53	59.2	- 9 17.1	0.70	137 23 34	89 12 59
15	TU	- 2 8 12	59.3	- 9 0.4	0.70	138 23 0	89 12 59
16	WE	- 1 44 29	59.3	- 8 43.5	0.71	139 22 26	89 12 58
17	TH	- 1 20 45	59.3	- 8 26.4	0.72	140 21 52	89 12 58
18	FR	- 0 57 1	59.3	- 8 9.1	0.73	141 21 18	89 12 58
19	SA	- 0 33 17	59.3	- 7 51.6	0.73	142 20 41	89 12 57
20	SU	- 0 9 33	59.3	- 7 34.0	0.74	143 20 1	89 12 57
21	MO	+ 0 14 11	59.2	- 7 16.3	0.75	144 19 20	89 12 57
22	TU	+ 0 37 53	59.2	- 6 58.4	0.75	145 18 37	89 12 57
23	WE	+ 1 1 34	59.1	- 6 40.4	0.75	146 17 53	89 12 56
24	TH	+ 1 25 12	59.0	- 6 22.3	0.75	147 17 10	89 12 56
25	FR	+ 1 48 49	58.9	- 6 4.0	0.76	148 16 28	89 12 56
26	SA	+ 2 12 23	58.8	- 5 45.6	0.76	149 15 46	89 12 56
27	SU	+ 2 35 54	58.7	- 5 27.8	0.76	150 15 5	89 12 55
28	MO	+ 2 59 22	58.6	- 5 9.6	0.76	151 14 25	89 12 55
29	TU	+ 3 22 46	58.3	- 4 51.4	0.76	152 13 46	89 12 55
30	WE	+ 3 46 6	58.3	- 4 33.2	0.75	153 13 6	89 12 55
31	TH	+ 4 9 22	58.0	- 4 15.1	0.75	154 12 26	89 12 54

APRIL 1988

Date		THE SUN Declination At 0 hrs, UT ° ' "	Δ, 1 hr "	Equation of Time At 0 hrs, UT m s	Δ, 1 hr s	POLARIS GHA at 0 hr ° '	Declination ° ' "
1	FR	+ 4 32 33	57.8	− 3 57.1	0.75	155 11 45	89 12 54
2	SA	+ 4 55 39	57.5	− 3 39.2	0.74	156 11 2	89 12 54
3	SU	+ 5 18 40	57.3	− 3 21.4	0.73	157 10 18	89 12 54
4	MO	+ 5 41 36	57.0	− 3 3.8	0.73	158 9 32	89 12 53
5	TU	+ 6 4 25	56.8	− 2 46.3	0.72	159 8 44	89 12 53
6	WE	+ 6 27 9	56.5	− 2 29.0	0.71	160 7 55	89 12 53
7	TH	+ 6 49 45	56.3	− 2 12.0	0.70	161 7 6	89 12 52
8	FR	+ 7 12 15	56.0	− 1 55.1	0.69	162 6 16	89 12 52
9	SA	+ 7 34 38	55.6	− 1 38.6	0.68	163 5 28	89 12 52
10	SU	+ 7 56 53	55.3	− 1 22.5	0.67	164 4 40	89 12 51
11	MO	+ 8 19 1	55.0	− 1 6.2	0.65	165 3 54	89 12 51
12	TU	+ 8 41 0	54.6	− 0 50.5	0.64	166 3 9	89 12 51
13	WE	+ 9 2 50	54.3	− 0 35.1	0.63	167 2 24	89 12 51
14	TH	+ 9 24 32	53.8	− 0 20.1	0.61	168 1 38	89 12 50
15	FR	+ 9 46 4	53.5	− 0 5.4	0.60	169 0 50	89 12 50
16	SA	+10 7 27	53.0	+ 0 8.9	0.58	170 0 0	89 12 50
17	SU	+10 28 39	52.6	+ 0 22.9	0.57	170 59 8	89 12 49
18	MO	+10 49 42	52.1	+ 0 36.5	0.55	171 58 13	89 12 49
19	TU	+11 10 33	51.7	+ 0 49.7	0.53	172 57 18	89 12 49
20	WE	+11 31 13	51.2	+ 1 2.5	0.52	173 56 23	89 12 48
21	TH	+11 51 42	50.7	+ 1 14.9	0.50	174 55 28	89 12 48
22	FR	+12 11 59	50.2	+ 1 26.8	0.48	175 54 34	89 12 48
23	SA	+12 32 4	49.7	+ 1 38.4	0.46	176 53 41	89 12 47
24	SU	+12 51 56	49.2	+ 1 49.5	0.44	177 52 48	89 12 47
25	MO	+13 11 36	48.6	+ 2 0.1	0.42	178 51 56	89 12 47
26	TU	+13 31 2	48.0	+ 2 10.2	0.40	179 51 5	89 12 47
27	WE	+13 50 15	47.5	+ 2 19.9	0.38	180 50 13	89 12 46
28	TH	+14 9 14	46.9	+ 2 29.1	0.36	181 49 20	89 12 46
29	FR	+14 27 59	46.3	+ 2 37.8	0.34	182 48 26	89 12 46
30	SA	+14 46 30	45.7	+ 2 46.0	0.32	183 47 30	89 12 45

MAY 1988

Date		THE SUN Declination At 0 hrs, UT ° ' "	Δ, 1 hr "	Equation of Time At 0 hrs, UT m s	Δ, 1 hr s	POLARIS GHA at 0 hr ° '	Declination ° ' "
1	SU	+15 4 46	45.0	+ 2 53.7	0.30	184 46 32	89 12 45
2	MO	+15 22 47	44.4	+ 3 0.8	0.28	185 45 33	89 12 45
3	TU	+15 40 33	43.8	+ 3 7.4	0.25	186 44 32	89 12 44
4	WE	+15 58 4	43.1	+ 3 13.4	0.23	187 43 30	89 12 44
5	TH	+16 15 18	42.5	+ 3 18.9	0.20	188 42 29	89 12 44
6	FR	+16 32 17	41.8	+ 3 23.8	0.18	189 41 28	89 12 44
7	SA	+16 48 59	41.1	+ 3 28.1	0.16	190 40 28	89 12 43
8	SU	+17 5 25	40.3	+ 3 31.9	0.13	191 39 30	89 12 43
9	MO	+17 21 33	39.6	+ 3 35.0	0.11	192 38 32	89 12 43
10	TU	+17 37 24	38.9	+ 3 37.6	0.08	193 37 35	89 12 43
11	WE	+17 52 58	38.1	+ 3 39.5	0.06	194 36 38	89 12 42
12	TH	+18 8 13	37.4	+ 3 40.9	0.03	195 35 39	89 12 42
13	FR	+18 23 11	36.8	+ 3 41.7	0.01	196 34 38	89 12 42
14	SA	+18 37 50	35.8	+ 3 41.9	−0.02	197 33 35	89 12 41
15	SU	+18 52 9	35.0	+ 3 41.5	−0.04	198 32 30	89 12 41
16	MO	+19 6 10	34.3	+ 3 40.6	−0.07	199 31 23	89 12 41
17	TU	+19 19 52	33.4	+ 3 39.0	−0.08	200 30 16	89 12 41
18	WE	+19 33 13	32.6	+ 3 37.0	−0.11	201 29 10	89 12 40
19	TH	+19 46 15	31.7	+ 3 34.4	−0.13	202 28 4	89 12 40
20	FR	+19 58 56	30.9	+ 3 31.2	−0.15	203 26 59	89 12 40
21	SA	+20 11 17	30.0	+ 3 27.5	−0.18	204 25 55	89 12 40
22	SU	+20 23 17	29.1	+ 3 23.3	−0.20	205 24 52	89 12 39
23	MO	+20 34 56	28.3	+ 3 18.6	−0.22	206 23 49	89 12 39
24	TU	+20 46 14	27.3	+ 3 13.4	−0.24	207 22 47	89 12 39
25	WE	+20 57 10	26.4	+ 3 7.7	−0.26	208 21 43	89 12 39
26	TH	+21 7 44	25.5	+ 3 1.5	−0.28	209 20 39	89 12 38
27	FR	+21 17 57	24.6	+ 2 54.9	−0.30	210 19 33	89 12 38
28	SA	+21 27 47	23.7	+ 2 47.8	−0.31	211 18 25	89 12 38
29	SU	+21 37 16	22.7	+ 2 40.3	−0.33	212 17 16	89 12 37
30	MO	+21 46 21	21.8	+ 2 32.3	−0.35	213 16 5	89 12 37
31	TU	+21 55 4	20.9	+ 2 23.9	−0.37	214 14 54	89 12 37

JUNE 1988

Date		THE SUN Declination At 0 hrs, UT ° ' "	Δ, 1 hr "	Equation of Time At 0 hrs, UT m s	Δ, 1 hr s	POLARIS GHA at 0 hr ° '	Declination ° ' "
1	WE	+22 3 25	19.9	+ 2 15.1	−0.38	215 13 42	89 12 37
2	TH	+22 11 22	18.9	+ 2 5.9	−0.40	216 12 30	89 12 37
3	FR	+22 18 56	18.0	+ 1 56.3	−0.41	217 11 20	89 12 37
4	SA	+22 26 7	17.0	+ 1 46.4	−0.43	218 10 12	89 12 37
5	SU	+22 32 54	16.0	+ 1 36.1	−0.45	219 9 5	89 12 36
6	MO	+22 39 18	15.0	+ 1 25.4	−0.46	220 7 58	89 12 36
7	TU	+22 45 17	14.0	+ 1 14.4	−0.47	221 6 52	89 12 36
8	WE	+22 50 53	13.0	+ 1 3.1	−0.48	222 5 44	89 12 36
9	TH	+22 56 5	12.0	+ 0 51.5	−0.50	223 4 34	89 12 36
10	FR	+23 0 53	11.0	+ 0 39.6	−0.50	224 3 23	89 12 35
11	SA	+23 5 16	10.0	+ 0 27.5	−0.51	225 2 9	89 12 35
12	SU	+23 9 15	8.9	+ 0 15.2	−0.52	226 0 55	89 12 35
13	MO	+23 12 49	7.9	+ 0 2.6	−0.52	226 59 39	89 12 35
14	TU	+23 15 59	6.9	− 0 10.0	−0.54	227 58 24	89 12 35
15	WE	+23 18 44	5.9	− 0 22.9	−0.54	228 57 10	89 12 35
16	TH	+23 21 5	4.8	− 0 35.8	−0.54	229 55 56	89 12 34
17	FR	+23 23 0	3.8	− 0 48.8	−0.55	230 54 44	89 12 34
18	SA	+23 24 31	2.8	− 1 1.9	−0.55	231 53 33	89 12 34
19	SU	+23 25 37	1.7	− 1 15.0	−0.55	232 52 23	89 12 34
20	MO	+23 26 18	0.7	− 1 28.1	−0.55	233 51 12	89 12 34
21	TU	+23 26 35	− 0.4	− 1 41.2	−0.54	234 50 2	89 12 34
22	WE	+23 26 26	− 1.4	− 1 54.3	−0.54	235 48 50	89 12 34
23	TH	+23 25 53	− 2.4	− 2 7.2	−0.54	236 47 38	89 12 34
24	FR	+23 24 55	− 3.5	− 2 20.1	−0.53	237 46 24	89 12 33
25	SA	+23 23 32	− 4.5	− 2 32.9	−0.53	238 45 8	89 12 33
26	SU	+23 21 45	− 5.5	− 2 45.5	−0.52	239 43 51	89 12 33
27	MO	+23 19 33	− 6.5	− 2 58.0	−0.51	240 42 33	89 12 33
28	TU	+23 16 57	− 7.5	− 3 10.3	−0.50	241 41 15	89 12 33
29	WE	+23 13 56	− 8.6	− 3 22.4	−0.49	242 39 57	89 12 33
30	TH	+23 10 30	− 9.5	− 3 34.2	−0.49	243 38 40	89 12 33

JULY 1988

Date		THE SUN Declination At 0 hrs, UT ° ' "	Δ, 1 hr "	Equation of Time At 0 hrs, UT m s	Δ, 1 hr s	POLARIS GHA at 0 hr ° ' "	Declination ° ' "
1	FR	+23 6 41	-10.6	- 3 45.9	-0.47	244 37 25	89 12 33
2	SA	+23 2 27	-11.6	- 3 57.2	-0.47	245 36 12	89 12 33
3	SU	+22 57 49	-12.6	- 4 8.4	-0.45	246 35 0	89 12 33
4	MO	+22 52 47	-13.6	- 4 19.2	-0.44	247 33 48	89 12 33
5	TU	+22 47 21	-14.5	- 4 29.7	-0.43	248 32 35	89 12 33
6	WE	+22 41 32	-15.6	- 4 40.0	-0.41	249 31 21	89 12 33
7	TH	+22 35 18	-16.5	- 4 49.8	-0.40	250 30 5	89 12 33
8	FR	+22 28 42	-17.5	- 4 59.4	-0.38	251 28 47	89 12 33
9	SA	+22 21 41	-18.5	- 5 8.5	-0.37	252 27 27	89 12 33
10	SU	+22 14 18	-19.5	- 5 17.3	-0.35	253 26 8	89 12 32
11	MO	+22 6 31	-20.4	- 5 25.6	-0.33	254 24 48	89 12 33
12	TU	+21 58 22	-21.3	- 5 33.5	-0.31	255 23 29	89 12 33
13	WE	+21 49 50	-22.3	- 5 41.0	-0.29	256 22 11	89 12 33
14	TH	+21 40 56	-23.2	- 5 48.0	-0.27	257 20 55	89 12 33
15	FR	+21 31 39	-24.1	- 5 54.5	-0.25	258 19 40	89 12 33
16	SA	+21 22 1	-25.0	- 6 0.4	-0.23	259 18 26	89 12 33
17	SU	+21 12 0	-25.9	- 6 5.9	-0.20	260 17 12	89 12 33
18	MO	+21 1 38	-26.8	- 6 10.8	-0.18	261 15 58	89 12 33
19	TU	+20 50 55	-27.7	- 6 15.2	-0.16	262 14 44	89 12 33
20	WE	+20 39 51	-28.5	- 6 19.0	-0.13	263 13 29	89 12 33
21	TH	+20 28 26	-29.4	- 6 22.2	-0.11	264 12 12	89 12 33
22	FR	+20 16 40	-30.3	- 6 24.9	-0.08	265 10 55	89 12 33
23	SA	+20 4 34	-31.1	- 6 26.9	-0.06	266 9 36	89 12 33
24	SU	+19 52 8	-31.9	- 6 28.4	-0.03	267 8 16	89 12 33
25	MO	+19 39 22	-32.7	- 6 29.2	-0.01	268 6 55	89 12 33
26	TU	+19 26 17	-33.5	- 6 29.5	0.02	269 5 35	89 12 33
27	WE	+19 12 52	-34.3	- 6 29.1	0.04	270 4 16	89 12 33
28	TH	+18 59 9	-35.1	- 6 28.1	0.06	271 2 58	89 12 33
29	FR	+18 45 7	-35.9	- 6 26.6	0.09	272 1 43	89 12 33
30	SA	+18 30 46	-36.6	- 6 24.4	0.12	273 0 29	89 12 33
31	SU	+18 16 8	-37.4	- 6 21.6	0.14	273 59 15	89 12 34

AUGUST 1988

Date		THE SUN Declination At 0 hrs, UT ° ' "	Δ, 1 hr "	Equation of Time At 0 hrs, UT m s	Δ, 1 hr s	POLARIS GHA at 0 hr ° ' "	Declination ° ' "
1	MO	+18 1 11	-38.1	- 6 18.2	0.17	274 58 2	89 12 34
2	TU	+17 45 51	-38.8	- 6 14.2	0.19	275 56 49	89 12 34
3	WE	+17 30 25	-39.5	- 6 9.6	0.21	276 55 31	89 12 34
4	TH	+17 14 36	-40.3	- 6 4.4	0.24	277 54 12	89 12 34
5	FR	+16 58 30	-40.9	- 5 58.7	0.26	278 52 53	89 12 34
6	SA	+16 42 8	-41.6	- 5 52.4	0.29	279 51 32	89 12 34
7	SU	+16 25 29	-42.3	- 5 45.5	0.31	280 50 12	89 12 34
8	MO	+16 8 34	-42.9	- 5 38.0	0.33	281 48 53	89 12 34
9	TU	+15 51 24	-43.6	- 5 30.0	0.36	282 47 35	89 12 34
10	WE	+15 33 58	-44.2	- 5 21.4	0.38	283 46 19	89 12 35
11	TH	+15 16 18	-44.8	- 5 12.2	0.40	284 45 4	89 12 35
12	FR	+14 58 22	-45.4	- 5 2.5	0.43	285 43 50	89 12 35
13	SA	+14 40 13	-46.0	- 4 52.2	0.45	286 42 36	89 12 35
14	SU	+14 21 49	-46.5	- 4 41.3	0.47	287 41 23	89 12 35
15	MO	+14 3 12	-47.1	- 4 29.9	0.50	288 40 10	89 12 35
16	TU	+13 44 21	-47.7	- 4 18.0	0.52	289 38 56	89 12 36
17	WE	+13 25 17	-48.2	- 4 5.5	0.54	290 37 41	89 12 36
18	TH	+13 6 1	-48.7	- 3 52.5	0.56	291 36 25	89 12 36
19	FR	+12 46 32	-49.2	- 3 39.0	0.58	292 35 8	89 12 36
20	SA	+12 26 51	-49.7	- 3 25.0	0.60	293 33 51	89 12 36
21	SU	+12 6 58	-50.2	- 3 10.5	0.63	294 32 32	89 12 36
22	MO	+11 46 54	-50.6	- 2 55.5	0.64	295 31 14	89 12 37
23	TU	+11 26 39	-51.0	- 2 40.1	0.66	296 29 55	89 12 37
24	WE	+11 6 14	-51.4	- 2 24.2	0.68	297 28 40	89 12 37
25	TH	+10 45 37	-51.8	- 2 7.8	0.70	298 27 26	89 12 37
26	FR	+10 24 51	-52.2	- 1 51.1	0.72	299 26 14	89 12 38
27	SA	+10 3 55	-52.8	- 1 33.9	0.73	300 25 3	89 12 38
28	SU	+ 9 42 49	-53.1	- 1 16.3	0.75	301 23 53	89 12 38
29	MO	+ 9 21 34	-53.5	- 0 58.4	0.76	302 22 41	89 12 38
30	TU	+ 9 0 10	-53.8	- 0 40.2	0.78	303 21 28	89 12 38
31	WE	+ 8 38 38	-54.2	- 0 21.6	0.78	304 20 14	89 12 39

SEPTEMBER 1988

Date		THE SUN Declination At 0 hrs, UT ° ' "	Δ, 1 hr "	Equation of Time At 0 hrs, UT m s	Δ, 1 hr s	POLARIS GHA at 0 hr ° ' "	Declination ° ' "
1	TH	+ 8 16 57	-54.6	- 0 2.8	0.80	305 18 57	89 12 39
2	FR	+ 7 55 7	-54.9	+ 0 16.3	0.81	306 17 40	89 12 39
3	SA	+ 7 33 10	-55.2	+ 0 35.8	0.82	307 16 23	89 12 39
4	SU	+ 7 11 6	-55.5	+ 0 55.4	0.83	308 15 7	89 12 40
5	MO	+ 6 48 54	-55.8	+ 1 15.3	0.84	309 13 53	89 12 40
6	TU	+ 6 26 36	-56.0	+ 1 35.4	0.85	310 12 40	89 12 40
7	WE	+ 6 4 11	-56.3	+ 1 55.7	0.85	311 11 29	89 12 40
8	TH	+ 5 41 40	-56.5	+ 2 16.1	0.86	312 10 18	89 12 41
9	FR	+ 5 19 3	-56.8	+ 2 36.6	0.87	313 9 9	89 12 41
10	SA	+ 4 56 21	-57.0	+ 2 57.6	0.87	314 8 1	89 12 41
11	SU	+ 4 33 34	-57.2	+ 3 18.5	0.88	315 6 52	89 12 42
12	MO	+ 4 10 42	-57.4	+ 3 39.5	0.88	316 5 43	89 12 42
13	TU	+ 3 47 45	-57.5	+ 4 0.6	0.88	317 4 33	89 12 42
14	WE	+ 3 24 45	-57.7	+ 4 21.8	0.89	318 3 22	89 12 42
15	TH	+ 3 1 41	-57.8	+ 4 43.1	0.89	319 2 10	89 12 43
16	FR	+ 2 38 33	-57.9	+ 5 4.4	0.89	320 0 58	89 12 43
17	SA	+ 2 15 23	-58.0	+ 5 25.8	0.89	320 59 45	89 12 43
18	SU	+ 1 52 10	-58.2	+ 5 47.2	0.89	321 58 32	89 12 44
19	MO	+ 1 28 54	-58.2	+ 6 8.5	0.89	322 57 20	89 12 44
20	TU	+ 1 5 37	-58.3	+ 6 29.9	0.88	323 56 9	89 12 44
21	WE	+ 0 42 18	-58.3	+ 6 51.1	0.88	324 55 0	89 12 45
22	TH	+ 0 18 58	-58.4	+ 7 12.4	0.88	325 53 53	89 12 45
23	FR	- 0 4 23	-58.4	+ 7 33.5	0.88	326 52 48	89 12 45
24	SA	- 0 27 45	-58.4	+ 7 54.5	0.87	327 51 43	89 12 46
25	SU	- 0 51 7	-58.4	+ 8 15.4	0.86	328 50 39	89 12 46
26	MO	- 1 14 29	-58.4	+ 8 36.1	0.86	329 49 33	89 12 46
27	TU	- 1 37 50	-58.4	+ 8 56.7	0.85	330 48 25	89 12 47
28	WE	- 2 1 11	-58.4	+ 9 17.0	0.84	331 47 16	89 12 47
29	TH	- 2 24 32	-58.3	+ 9 37.1	0.83	332 46 6	89 12 47
30	FR	- 2 47 51	-58.2	+ 9 56.9	0.82	333 44 55	89 12 47

OCTOBER 1988

Date		THE SUN Declination At 0 hrs, UT ° ' "	Δ, 1 hr "	Equation of Time At 0 hrs, UT m s	Δ, 1 hr s	POLARIS GHA at 0 hr ° ' "	Declination ° ' "
1	SA	− 3 11 8	−58.2	+10 16.5	0.80	334 43 45	89 12 48
2	SU	− 3 34 24	−58.0	+10 35.7	0.79	335 42 37	89 12 48
3	MO	− 3 57 37	−57.9	+10 54.6	0.78	336 41 31	89 12 49
4	TU	− 4 20 47	−57.8	+11 13.2	0.76	337 40 26	89 12 49
5	WE	− 4 43 55	−57.7	+11 31.4	0.74	338 39 23	89 12 49
6	TH	− 5 6 59	−57.5	+11 49.2	0.72	339 38 21	89 12 50
7	FR	− 5 30 0	−57.3	+12 6.6	0.71	340 37 20	89 12 50
8	SA	− 5 52 56	−57.2	+12 23.6	0.69	341 36 19	89 12 50
9	SU	− 6 15 48	−56.9	+12 40.2	0.67	342 35 18	89 12 51
10	MO	− 6 38 34	−56.8	+12 56.3	0.65	343 34 16	89 12 51
11	TU	− 7 1 16	−56.5	+13 11.9	0.63	344 33 13	89 12 51
12	WE	− 7 23 52	−56.2	+13 27.1	0.61	345 32 10	89 12 52
13	TH	− 7 46 21	−56.0	+13 41.8	0.59	346 31 6	89 12 52
14	FR	− 8 8 45	−55.7	+13 55.9	0.57	347 30 1	89 12 52
15	SA	− 8 31 1	−55.4	+14 9.5	0.55	348 28 56	89 12 53
16	SU	− 8 53 10	−55.0	+14 22.6	0.52	349 27 52	89 12 53
17	MO	− 9 15 11	−54.8	+14 35.0	0.50	350 26 50	89 12 53
18	TU	− 9 37 5	−54.3	+14 47.0	0.47	351 25 49	89 12 54
19	WE	− 9 58 49	−54.0	+14 58.3	0.45	352 24 50	89 12 54
20	TH	−10 20 25	−53.6	+15 9.0	0.42	353 23 53	89 12 55
21	FR	−10 41 52	−53.2	+15 19.1	0.40	354 22 57	89 12 55
22	SA	−11 3 9	−52.8	+15 28.6	0.36	355 22 2	89 12 56
23	SU	−11 24 16	−52.3	+15 37.3	0.34	356 21 6	89 12 56
24	MO	−11 45 12	−51.9	+15 45.5	0.31	357 20 8	89 12 56
25	TU	−12 5 58	−51.5	+15 52.9	0.28	358 19 9	89 12 57
26	WE	−12 26 33	−51.0	+15 59.6	0.25	359 18 8	89 12 57
27	TH	−12 46 57	−50.5	+16 5.5	0.22	0 17 7	89 12 58
28	FR	−13 7 8	−50.0	+16 10.7	0.19	1 16 6	89 12 58
29	SA	−13 27 8	−49.4	+16 15.2	0.15	2 15 7	89 12 58
30	SU	−13 46 54	−48.9	+16 18.9	0.12	3 14 10	89 12 58
31	MO	−14 6 28	−48.4	+16 21.7	0.09	4 13 14	89 12 59

NOVEMBER 1988

Date		THE SUN Declination At 0 hrs, UT ° ' "	Δ, 1 hr "	Equation of Time At 0 hrs, UT m s	Δ, 1 hr s	POLARIS GHA at 0 hr ° ' "	Declination ° ' "
1	TU	−14 25 49	−47.8	+16 23.8	−0.05	5 12 20	89 12 59
2	WE	−14 44 55	−47.2	+16 25.0	−0.02	6 11 28	89 13 0
3	TH	−15 3 47	−46.6	+16 25.7	−0.05	7 10 37	89 13 0
4	FR	−15 22 25	−46.0	+16 25.1	−0.05	8 9 46	89 13 0
5	SA	−15 40 48	−45.3	+16 23.8	−0.09	9 8 55	89 13 1
6	SU	−15 58 55	−44.6	+16 21.7	−0.12	10 8 3	89 13 1
7	MO	−16 16 46	−44.0	+16 18.8	−0.16	11 7 11	89 13 1
8	TU	−16 34 21	−43.3	+16 15.0	−0.19	12 6 18	89 13 2
9	WE	−16 51 39	−42.5	+16 10.4	−0.23	13 5 25	89 13 2
10	TH	−17 8 39	−41.8	+16 4.9	−0.27	14 4 30	89 13 3
11	FR	−17 25 23	−41.0	+15 58.5	−0.30	15 3 36	89 13 3
12	SA	−17 41 48	−40.3	+15 51.4	−0.34	16 2 42	89 13 4
13	SU	−17 57 54	−39.7	+15 43.3	−0.37	17 1 50	89 13 4
14	MO	−18 13 42	−38.7	+15 34.5	−0.40	18 0 59	89 13 4
15	TU	−18 29 11	−37.9	+15 24.8	−0.44	19 0 11	89 13 5
16	WE	−18 44 20	−37.0	+15 14.3	−0.47	19 59 24	89 13 5
17	TH	−18 59 9	−36.2	+15 2.9	−0.50	20 58 39	89 13 5
18	FR	−19 13 38	−35.3	+14 50.8	−0.54	21 57 54	89 13 6
19	SA	−19 27 46	−34.4	+14 37.8	−0.57	22 57 9	89 13 6
20	SU	−19 41 32	−33.6	+14 24.1	−0.61	23 56 23	89 13 6
21	MO	−19 54 58	−32.6	+14 9.5	−0.64	24 55 36	89 13 7
22	TU	−20 8 1	−31.7	+13 54.2	−0.67	25 54 47	89 13 7
23	WE	−20 20 42	−30.8	+13 38.1	−0.70	26 53 57	89 13 7
24	TH	−20 33 2	−29.8	+13 21.2	−0.74	27 53 7	89 13 8
25	FR	−20 44 57	−28.9	+13 3.5	−0.77	28 52 18	89 13 8
26	SA	−20 56 30	−27.9	+12 45.1	−0.80	29 51 31	89 13 8
27	SU	−21 7 40	−26.9	+12 25.9	−0.83	30 50 46	89 13 9
28	MO	−21 18 25	−25.9	+12 6.0	−0.86	31 50 2	89 13 9
29	TU	−21 28 47	−24.9	+11 45.3	−0.89	32 49 21	89 13 10
30	WE	−21 38 44	−23.9	+11 24.0	−0.92	33 48 40	89 13 10

DECEMBER 1988

Date		THE SUN Declination At 0 hrs, UT ° ' "	Δ, 1 hr "	Equation of Time At 0 hrs, UT m s	Δ, 1 hr s	POLARIS GHA at 0 hr ° ' "	Declination ° ' "
1	TH	−21 48 17	−22.8	+11 1.9	−0.94	34 48 0	89 13 10
2	FR	−21 57 25	−21.8	+10 39.3	−0.97	35 47 20	89 13 11
3	SA	−22 6 7	−20.7	+10 15.9	−1.00	36 46 40	89 13 11
4	SU	−22 14 24	−19.6	+ 9 52.0	−1.03	37 46 0	89 13 11
5	MO	−22 22 15	−18.5	+ 9 27.4	−1.05	38 45 18	89 13 12
6	TU	−22 29 39	−17.5	+ 9 2.3	−1.07	39 44 36	89 13 12
7	WE	−22 36 38	−16.3	+ 8 36.6	−1.09	40 43 53	89 13 12
8	TH	−22 43 10	−15.2	+ 8 10.5	−1.11	41 43 9	89 13 13
9	FR	−22 49 15	−14.1	+ 7 43.8	−1.13	42 42 26	89 13 13
10	SA	−22 54 53	−13.0	+ 7 16.7	−1.15	43 41 44	89 13 13
11	SU	−23 0 4	−11.8	+ 6 49.2	−1.16	44 41 4	89 13 13
12	MO	−23 4 47	−10.7	+ 6 21.4	−1.18	45 40 26	89 13 14
13	TU	−23 9 3	− 9.5	+ 5 53.1	−1.19	46 39 49	89 13 14
14	WE	−23 12 52	− 8.3	+ 5 24.6	−1.20	47 39 14	89 13 14
15	TH	−23 16 12	− 7.2	+ 4 55.8	−1.21	48 38 40	89 13 15
16	FR	−23 19 5	− 6.0	+ 4 26.8	−1.22	49 38 6	89 13 15
17	SA	−23 21 30	− 4.9	+ 3 57.5	−1.23	50 37 31	89 13 15
18	SU	−23 23 27	− 3.7	+ 3 28.1	−1.23	51 36 55	89 13 15
19	MO	−23 24 55	− 2.5	+ 2 58.6	−1.24	52 36 17	89 13 16
20	TU	−23 25 56	− 1.3	+ 2 28.9	−1.24	53 35 38	89 13 16
21	WE	−23 26 28	− 0.2	+ 1 59.2	−1.24	54 34 58	89 13 16
22	TH	−23 26 32	1.0	+ 1 29.5	−1.24	55 34 19	89 13 16
23	FR	−23 26 8	2.2	+ 0 59.7	−1.24	56 33 42	89 13 17
24	SA	−23 25 16	3.4	+ 0 29.9	−1.24	57 33 6	89 13 17
25	SU	−23 23 55	4.5	+ 0 0.2	−1.23	58 32 32	89 13 17
26	MO	−23 22 6	5.7	− 0 29.5	−1.23	59 32 0	89 13 17
27	TU	−23 19 50	6.9	− 0 59.0	−1.22	60 31 28	89 13 18
28	WE	−23 17 5	8.0	− 1 28.4	−1.22	61 30 58	89 13 18
29	TH	−23 13 52	9.2	− 1 57.7	−1.21	62 30 28	89 13 18
30	FR	−23 10 11	10.4	− 2 26.7	−1.20	63 29 57	89 13 18
31	SA	−23 6 2	11.5	− 2 55.6	−1.19	64 29 26	89 13 18

REFERENCES

1. American Congress on Surveying and Mapping. *Technical Standards for Property Surveys*. Falls Church: ACSM, 1968.

2. Brown, Curtis M. *Boundary Control and Legal Principles*. New York: John Wiley and Sons, Inc., 1969.

3. Brown, Curtis M. and Eldridge, Winfield H. *Evidence and Procedures for Boundary Location.* New York: John Wiley and Sons, Inc., 1962.

4. Buckner, R.B. *Astronomic Azimuth and State Plane Coordinate Control*. (Workshop notes for two-day seminar), Columbus, Ohio, 1980.

5. Buckner, R.B. *Preservation of Land Survey Evidence through Accurate Azimuth Determination*. (unpublished M.S. thesis) University of Illinois, 1969.

6. Buckner, R.B. "Preserving and Perpetuating Land Survey Evidence," *Surveying and Mapping*, Vol. XXXII, No. 4, 1972.

7. Buckner, R.B. "Reasons and Methods for Accurate Direction in Land Surveys," *Surveying and Mapping*, Vol. XXXV, No. 4, 1975.

8. Buckner, R.B. *Surveying Measurements and their Analysis*. Sacramento: Landmark Enterprises, 1983.

9. Buckner, R.B. *Understanding Surveying Measurement*. (Workshop notes for two day seminar) Columbus, Ohio, 1980.

10. Curtis, Kenneth S. *Meridian Determination by Solar and Polaris Observation*. Lafayette: Indiana Society of Professional Land Surveyors, 1975.

11. Kiely, Edmund R. *Surveying Instruments: Their History*. Columbus: CARBEN Surveying Reprints.

12. Mackie, J.B. *Astronomy for Surveyors*, Eighth Edition. London: Charles Griffin & Company, Ltd., 1978.

13. Mitchell, Hugh C. and Simmons, Lansing G. *The State Coordinate Systems*. Washington: U.S. Government Printing Office, Special Publication No. 235, 1974.

14. Mueller, I.I. *Spherical and Practical Astronomy as Applied to Geodesy*. New York: Frederick Ungar Publishing Co., 1969.

15. Mueller, I.I. *Introduction to Surveying*. New York: Frederick Ungar Publishing Co., 1979.

16. Nassau, Jason J. *Practical Astronomy*. New York: McGraw-Hill Book Company, Inc., 1948.

17. Poling, Arthur A., Jr. "Astronomic Azimuths for Local Control," <u>Surveying and Mapping</u>, Vol. XXVII, No. 3, 1967.

18. Senne, Joseph H. <u>Determination of North from Polaris Observations</u>. Rolla: Missouri Department of Natural Resources, Standards of Practice No. 3, 1977.

19. Vanderaa, Marion H. "The Solar Pointing Error," <u>Surveying and Mapping</u>, Vol. XXIV, No. 2, 1964.

20. U.S. Department of Commerce, Coast and Geodetic Survey. <u>Manual of Geodetic Astronomy</u>, Special Publication No. 237. Washington: U.S. Government Printing Office, 1952.

21. U.S. Department of Commerce, Coast and Geodetic Survey. <u>State Plane Coordinates by Automatic Data Processing</u>, Publication 62-4. Washington: U.S. Government Printing Office, 1968.

22. U.S. Department of Commerce, National Bureau of Standards. <u>NBS Time and Frequency Dissemination Services</u>. Boulder, Colorado: NBS, 1976.

23. U.S. Department of Commerce, NOAA. <u>Classification, Standards of Accuracy, and General Specifications of Geodetic Control Surveys</u>. Rockville, MD: U.S. Department of Commerce, 1974.

24. U.S. Department of the Interior. <u>Topographic Map Information and Symbols</u>. Reston, VA: Geological Survey, 1978.

25. U.S. Naval Observatory. <u>American Ephemeris and Nautical Almanac</u>. Washington: U.S Government Printing Office.

26. <u>World Almanac and Book of Facts</u>. New York: Newspaper Enterprise Association, Inc., 1971.

ADDRESSES

U.S. Government Printing Office, Superintendent of Documents, Washington, D.C. 20402

U.S. Department of Commerce, National Bureau of Standards, Washington, D.C. 20234 and Boulder, Colorado 80302

U.S. Department of the Interior, Bureau of Land Management, Federal Center Building 50, Denver, Colorado 80225

U.S. Department of the Interior, Geological Survey (see page 26)

National Geodetic Survey (see page 29)

INDEX

Adams, O.S. 108
altitude 14, 16, 78, 127-131, 146, 147
altitude method 12, 13, 126, 139, 140
anomalies, gravity 27
apparent solar day (see solar day)
apparent time 58, 59, 120
Aries 120
astrogeodetic conversions 183-216
astrolabe 1
atmospheric refraction (see refraction)
Autumnal equinox (see equinox)
azimuth (see north)
azimuth mark 111, 158
Big Dipper 17
bubble sensitivity 138, 145
Burt, William Austin 2
cardinal direction 118
Cassiopeia 18
celestial equator (see equator)
celestial pole 16, 18, 123
celestial sphere 14, 16, 17
central meridian 83, 84, 87, 108, 109
CHU radio station 22
civil time 58
compass, magnetic 2, 4
compass, solar 2
conical projection 83, 108
control data 113-115
convergence angle 83, 84, 162
convergence of meridians 5, 83-91
Copernicus, Nicolaus 1
culmination 17
cylindrical projection 83, 108, 109
declination (astronomic) 15, 16, 19, 56, 57, 60, 119, 120
declination (magnetic) 2, 4
deflection of the vertical 6, 29, 81, 82, 148
delta-alpha ($\Delta\alpha$) angle 84, 90, 162
diagonal eyepiece prism 134, 135
diameter of the sun (see semi-diameter)
dioptra 1
diurnal aberration 157
diurnal circle 17
Dixon, Jeremiah 2
DUT1 correction 23, 59, 151
ecliptic 119, 120
Ellicott, Andrew 2
ellipsoid 81
elongation 17
ephemeris 12, 20, 55-60, 229-251
equation of time 19, 58-60, 122

equator 14, 118, 120
equatorial system 119, 120
equinoctial colure 119, 121
equinoxes 17, 57, 119, 120
Erathosthenes 1
errors and error sources
    bubble centering 32, 33, 36, 134, 137, 138, 144, 145, 154, 156
    instrument centering 142, 144, 154, 159
    Laplace correction 148
    latitude 25, 26, 146, 148-150, 155, 158
    longitude 25, 26, 146, 149, 150, 155, 158
    pointing 143, 154, 158
    propagation 154-157
    random 142, 163
    reading 142, 143, 154, 158
    round-off 61, 159, 229
    sun's diameter 152, 153
    systematic 142, 163
    target centering 142, 144, 154, 159
    time 150, 151, 155, 157
eyepiece prism 134, 135
field notes 41, 69, 73, 75
geoid 81
gravity anomalies 27
great circle 118
Greenwich Apparent Time 59, 60-62, 122
Greenwich Hour Angle 60-63, 121, 122
Greenwich Mean Time 22
grid distance 110, 111
grid factor 111, 112
horizon 14
horizon system 119
horizontal angle 67
hour angle 15, 16, 64, 65, 121
hour angle method 12, 13, 124, 126
hour angle system 119
hour circle 120
infinite focus 43, 53
inverted image 52
isogram 82
Lambert Conformal 83, 85, 87, 108, 109
land data systems 117
Laplace correction 6, 12, 29, 81, 82, 84, 90, 183-216
latitude 25, 118
limb pointings 41
local attraction 4
Local Apparent Time 59
Local Hour Angle 15, 61, 64, 121
lodestone 2

INDEX

longitude 25, 118
magnetic compass 2, 4
mapping angle 83
Mansfield, Jared 2
Mason, Charles 2
mean solar day (see solar day)
mean time 58, 59, 120
meridian 118
meridian angle (see hour angle)
meridian of observer 14
meridian of star 14
nadir 14
National Almanac Office 229
National Bureau of Standards 22, 23
National Geodetic Survey 6, 82, 111
National Map Accuracy 146
north
    arbitrary, assumed 4
    astronomic 6, 14, 81
    geodetic 6, 81, 83, 89
    grid 5, 83-85, 89, 90, 111, 114, 161
    magnetic 4
    record 5
    true 4, 6
obliquity of the ecliptic 119, 120
observer's meridian 14
parallax (sun) 55, 128, 129
plane azimuth (see north, grid)
plumb line 14, 81
polar distance 15, 16
polar variation 157
preserving evidence 117
pressure, barometric 127
prime meridian 118
prime vertical 17
projection surface 108
radius of earth 110, 116
refraction, atmospheric 55, 127, 129
rheostat 48
right ascension 55, 119, 120, 122
Rittenhouse, Benjamin, David 2
Roelofs solar prism (see solar prism)
scale factor 108-110, 112
sea level factor 110, 112
second-order azimuth 8, 84, 107, 117, 160
second term correction 84, 85, 117
semi-diameter of sun 55, 129-132, 152
sextant 1
sidereal day 19, 63, 120-122
sidereal hour angle 119, 121
significant figures 61

skew normal 157
solar compass 2
solar day 19, 56, 58, 63, 120, 121
solar eyepiece filter 3, 37, 40
solar prism 133, 135
solstices 19, 57, 119
specifications 158-160
sphere 118
spherical triangle 15, 16, 123
spherical trigonometry 123
state plane coordinates 5, 83, 107-117
stop watches 38, 41, 42
striding level 134, 136-138
summer solstice (see solstices)
sunrise, sunset 34
targets, night 134, 136
theta angle (θ) 85, 88, 162
third-order azimuth 8, 30, 84, 111, 158, 159
time
    apparent 58, 59, 120
    civil 58
    daylight savings 122
    Greenwich 59-62, 122
    local 58, 59
    Local Apparent 59
    mean 58, 59, 120
    sidereal 19, 63, 120-122
    solar 19, 56, 58, 63, 120, 121
    standard 122
    universal 22, 59, 68
    zones 22, 122, 123, 141
timing 50, 51
timeKube 3, 24, 38
transit (upper, lower) 61, 118
Transverse Mercator 84, 86, 108, 109
traversing 112
Universal Time 22, 59, 68
Ursa Major 18
Ursa Minor 18
USGS maps 20, 25, 78, 84, 110, 146
U.S. Naval Observatory 229
Vernal equinox (see equinox)
vertical circle 118
weather information 127
well-defined target 36, 143
winter solstice (see solstice)
WWV radio station 22, 23
year 121
zenith 14, 20
zenith angle (distance) 78, 128-131
zone (SPC) 83, 84, 86

Also by R. B. Buckner

SURVEYING MEASUREMENTS
and their Analysis

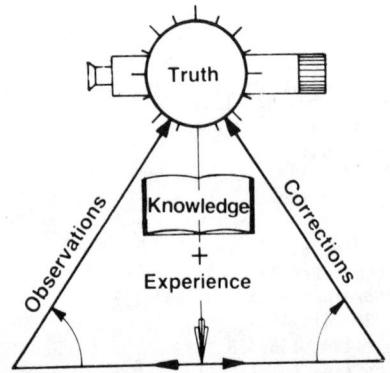

by R. B. Buckner
Order #480

The author's logo explains in graphical form how the "true value" in any measurement is always equal to the reading plus the corrections to that reading, and that experience and knowledge gained from practice and a study of the theory of measurement lead the surveyor to the corrections to be applied. The concept of "true value" is one of many theories explained in this book. The more complex theory of estimating random errors is another important theme and is presented in a meticulous way designed to instill in the reader a deep appreciation for measurement as both an art and science. Surveyors, surveying students, engineers, mathematics teachers, and others associated with measurement will find a style and depth of presentation on surveying measurements heretofore unavailable. For the surveyor, an appreciation of the analysis phase of measurements will be gained. For the layman and allied professional, a reading of a few basic chapters will lend much understanding as to why surveyors rarely agree on values for measurements of the same quanitity. For mathematics teachers, an appreciation of how to teach calculations employing measurements (rather than counts) will be gained so that the general public will hopefully be better enlightened in the future when working with numbers.

WHAT WILL THIS BOOK DO FOR ME?

- ■ *Explain the nature of surveying measurement*
- ■ *Develop expertise in measurements*

WHO WILL BENEFIT?

- ■ *Surveyors, survey students, civil engineers, mathematics teachers and others interested in the mystery of measurement.*

UNIQUE FEATURES OF THE BOOK . . . . .

- ■ *Explains the art and science of measurements and bridges the gap between elementary books and adjustment books.*
- ■ *Four chapters are especially suited for people who seek an appreciation but not a complete expertise in measurement.*

For more information on this and several hundred other books, request a free catalog from
LANDMARK ENTERPRISES,
10324 Newton Way, Rancho Cordova CA 95670